Pirone's Tree Maintenance
Seventh Edition

P. P. Pirone examining the trunk of a very old London planetree at the New York Botanical Garden. The tree has a diameter of 4 feet, 2 inches at the $4^1/2$-foot level, a height of 90 feet, and a branch spread of 60 feet.

Pirone's Tree Maintenance

Seventh Edition

John R. Hartman
Thomas P. Pirone
Mary Ann Sall

OXFORD
UNIVERSITY PRESS
2000

OXFORD
UNIVERSITY PRESS

Oxford New York

Athens Auckland Bangkok Bogotá Buenos Aires Calcutta
Cape Town Chennai Dar es Salaam Delhi Florence Hong Kong Istanbul
Karachi Kuala Lumpur Madrid Melbourne Mexico City Mumbai
Nairobi Paris São Paulo Singapore Taipei Tokyo Toronto Warsaw

and associated companies in

Berlin Ibadan

Copyright ©2000 by Oxford University Press, Inc.

Published by Oxford University Press, Inc.
198 Madison Avenue, New York, New York 10016

Oxford is a registered trademark of Oxford University Press

3 2530 60542 7200

Library of Congress Cataloging-in-Publication Data
Hartman, John Richard, 1943–
Pirone's tree maintenance / John R. Hartman, Thomas P. Pirone,
Mary Ann Sall. — 7th ed.
        p. cm.
Rev. ed. of: Maintenance of shade and ornamental trees /
Pascal Pompey Pirone. 2d ed. 1948.
Includes bibliographical references.
ISBN 0-19-511991-6
1. Ornamental trees—Diseases and pests. 2. Trees, care of.
I. Pirone, T. P. II. Sall, Mary Ann. III. Pirone, Pascal Pompey, 1907–
Maintenance of shade and ornamental trees. IV. Title.
SB761.H34 2000
635.977—dc21
99-21143

635.977
HART

1 3 5 7 9 8 6 4 2
Printed in the United States of America
on acid free paper

Contents

Preface

Since the first edition of this book, then entitled *Maintenance of Shade and Ornamental Trees*, in 1941, P. P. Pirone has been either the sole author, or, for the sixth edition, a co-author. He has now reached the age of 92, and, while still in good physical and mental health, he feels that he has not kept up enough with recent developments in tree care to have meaningful input into this revision. Hence it was decided that the title be changed to *Pirone's Tree Maintenance* to acknowledge his authorship of the original book and his many years of input, while at the same time crediting the senior authorship to John Hartman, who was primarily responsible for this revision.

In keeping with previous editions, we have attempted to follow a course between the technical and the popular in order to serve the needs of as broad a clientele and readership as possible. While we have retained the general structure of the sixth edition, there are several topics in this edition that have received greater attention than in previous versions of the book.

In recent years there has developed an increased awareness of the responsibilities of arborists and tree owners to maintain trees that are not dangerous to people and property. This edition provides expanded coverage of the topic of hazardous trees; this information can be found in Part II, Chapter 9, "Diagnosing Tree Problems." This edition also develops more fully the topic of pollarding in Part I, Chapter 7, "Pruning." Most readers are well aware of the reasons for not topping landscape trees. At the same time, increased travel opportunities exposes many of them to pollarding, which is widely practiced in Europe. We try to explain the difference between pollarding and topping and why one works while the other does not.

The use of pesticides in Part II is treated differently in this edition than in past editions. Recognizing that availability and legality of specific pesticides for specific trees and pests changes fairly rapidly, we have placed most of our suggestions on use of pesticides in Chapters 11, 12, and 13. Rather than our listing specific chemicals to use for each problem, readers where appropriate will refer back to the designated chapter to obtain general information on which chemicals might best be used for a particular pest or disease. We still feel that the use of resistant cultivars, pest exclusion, cultural practices, and biological controls are preferable to application of chemical pesticides in the landscape. Chapter 13, "Coping with Tree Pests and Diseases," has been extensively revised to reflect these views.

Each year, new outbreaks of pests, diseases, and tree problems appear or old problems become more or less important. Thus, as in previous editions, Part III, Abnormalities of Specific Trees, has been revised throughout. In this section, it is difficult to choose which pests and diseases are likely to be important to most readers. This section will continue to undergo revision as trees and their problems change over time.

We wish to express our gratitude to the professional arborists and scientists whose publications are listed in the various bibliographies in this book. Without their contributions a book like this could not have been written. We wish to single out particularly the *Journal of Arboriculture* for its excellent articles on all phases of tree care.

Finally, we wish to thank Jannine Baker and Sherrill Pirone for proofreading and editing assistance, Jennifer Walls for typing, and the following persons and organizations for the use of illustrations:

Bartlett Tree Expert Co., Stamford, Conn.: Figs. 12-7, 12-18, III-5, and III-51.

Dominick Basile, Lehman College, New York City: Frontispiece.

John Bean Div., FMC, Tipton, Ind.: Fig. 13-4.

J. C. Carter, Illinois Natural History Survey, Urbana, Ill.: Figs. III-46 and III-94.

R. A. Cool, Lansing, Mich.: Fig. 5-5.

Thomas Corell, Cooperative Extension Association of Suffolk County, N.Y.: Fig. III-103.

Robert d'Ambrosio, Ambrose Laboratories, Ltd., Eastchester, N.Y.: Figs. 10-5, 10-11, 11-4, 11-5, 13-3, and III-38.

R. W. Doherty, Sleepy Hollow Restorations, Tarrytown, N.Y.: Fig. 8-8.

T. H. Everett, New York Botanical Garden, N.Y.: Figs, 1-1, 1-6, 4-2, 5-2, and III-113.

E. F. Guba, Massachusetts Agricultural Experiment Station, Amherst, Mass.: Fig. III-49.

J. W. Hendrix, University of Kentucky, Lexington, Ky.: Fig. 2-5.

George Hepting, Asheville, N.C.: Fig. 12-20.

Ulla Jarlfors, University of Kentucky, Lexington, Ky.: Fig. 12-10.

Cheryl Kaiser, Lexington, Ky.: Figs. 10-6, 12-3, III-7, III-48, III-60, III-90, III-91, and III-106.

Jannine Baker, University of Kentucky, Lexington, Ky.: jacket, Figs. 5-15, 6-1, 6-2, 7-6 left, 7-7, 10-18, 10-28, 13-2, and 13-3.

Robert McNiel, University of Kentucky, Lexington, Ky.: Figs. 5-4 and 12-6.

Nassau County Cooperative Extension Association, New York: Fig. III-119.

D. A. Potter, University of Kentucky, Lexington, Ky.: Figs. 11-3, 11-6, 11-7, 11-10, 13-1, III-1, III-33, III-44, III-57, III-61, III-67, III-79, III-107, III-117, and III-125.

William A. Rae, Frost and Higgins Co., Burlington, Mass.: Fig. 5-6.

Bob Ray, Bob Ray Co., Louisville, Ky.: Figs. 10-1 and 10-2.

Edward Scanlon Associates, Olmstead Falls, Ohio: Figs. 1-7 and 4-3.

John C. Schread, Connecticut Agricultural Experiment Station, New Haven, Conn.: Figs. III-10, III-29, III-30, III-70, III-114, III-133, III-134, III-136, and III-137.

Robert Southerland, Salem, N.Y.: Fig. 1-4.

John Strang, University of Kentucky, Lexington, Ky.: Fig. 10-20.

Richard Stuckey, C.A.S.T., Ames, Iowa: Fig. 3-4.

Alfred G. Wheeler, Jr., Bureau of Plant Industry, Pennsylvania Department of Agriculture, Harrisburg, Penn.: Fig. III-68 and III-69.

Lexington, Kentucky   T. P. P.

July 1999

I

General Maintenance Practices

1

The Value
of Trees

Almost everyone likes trees for one reason or another. They provide cool shade during hot summer days, and whether in the yard or the park, they offer a serene setting to relieve the tensions of modern life. They add to the beauty and value of property. The splendor of a springtime floral display and the pageantry of autumn coloration make trees a delight to the eyes (Fig. 1-1). Even in winter, tree crown silhouettes and textured bark details provide enjoyment (Fig. 1-2). Trees evoke sentiment perhaps because Grandpa planted them when he and Grandma moved into the neighborhood many years before or because some historic event took place beneath their boughs. No matter what the reason—aesthetic, financial, or sentimental—a tree is a sound investment. A house surrounded by large trees is worth more money than a house without them. A house with no trees near it looks hot in summer, appears unbalanced, and suggests, however unjustly, a lack of interest on the part of those who live in it. Established residential areas are usually more inviting and more restful than new housing developments for one important reason: they are well supplied with large trees.

Trees have other beneficial effects:

First, they help supply the oxygen we need to breathe. Each year an acre of trees can produce enough oxygen to keep eighteen people alive. They also help refresh our air supply by using up some of the carbon dioxide that we exhale and that factories and engines emit. Prehistoric tree forms that made up the immense forests of the Carboniferous period (whose fossil remains we now mine as coal) are thought to have been important as air purifiers. The huge *Equisetum* and arboreal ferns, now mostly extinct, fixed large amounts of carbon dioxide and released oxygen for millions of years. The resulting shift of gas concentrations and the production of cleaner air influenced the development of present-day plant and animal forms.

Second, trees sequester carbon dioxide. Indeed, the current worldwide loss of trees has the potential to increase atmospheric carbon dioxide concentrations to levels that could create for the earth an undesirable "greenhouse effect" that could lead to global warming. Some cities are increasing their tree planting and preservation efforts as part of a local "greenhouse gas" containment policy,

Fig. 1-1. A splendid springtime display of flowering dogwood.

which reduces carbon dioxide by sequestering it in municipal trees. Urban forests in the United States are estimated to hold approximately 900 million metric tons of carbon.

Third, trees reduce noise pollution by acting as barriers to sound. Each 100-foot width of trees can absorb about 6 to 8 decibels of sound intensity.

Fourth, trees trap particulate air pollutants and absorb others, some of which are used as nutrients for growth. Air quality improvements have been attributed to urban trees, and in large cities such as Chicago, pollutants removed by trees are thought to amount to thousands of tons annually.

Fifth, trees alter the microclimate of the site where they grow. Trees may be effectively used to keep a house warmer in winter and cooler in summer. Evergreen trees planted on the north side of a house will shield it from cold winds. Deciduous trees planted to the south of a house allow much of the winter sun's rays to reach the house, thus providing some warmth. In summer, the same trees will shade the house and make the rooms on that side cooler. When air temperature is 84° F (29° C), street surface temperature may be as high as 108° F (42° C), but on a street lined with trees, the surface temperature is just 88° F (31°C) because heat rays are reflected off the surface of leaves, thus making it more com-

Fig. 1-2. Showy, whitish bark of Melaleuca.

fortable for pedestrians and travelers in automobiles. Trees also reduce glare and reflection from the sun and from artificial lights.

Sixth, trees have value in improving the aesthetics of our surroundings. Trees can enhance or soften the architecture; trees can frame or provide background to pleasing views or they can block objectionable views. Trees can provide privacy, control traffic patterns, and control soil erosion. Pleasing patterns of tree placement (Fig. 1-3), decorative tree walls, and shapes of individual trees all improve our aesthetic outlook.

Seventh, trees enhance outdoor urban spaces, making them more attractive to people, which in turn helps to build stronger communities. In at least one study, this greater sense of community was associated with reduced violence in the home. Thus, with other factors such as building design and demographics being equal, people living in housing with trees in the open spaces suffered less domestic violence than people living in similar housing without trees. So trees have social value.

Finally, the character of a city is changed by an abundance or a dearth of trees. Cities that spend liberal amounts of money to maintain their old trees and to plant new ones are generally considered nicer places in which to live. The urban forest, although not a natural ecosystem, is nevertheless a biological community of species highly dependent on the physical environment, much of it created by

Fig. 1-3. This allée in France is lined with planetrees.

people (Fig. 1-4). Trees, environment, and every form of life within an ecosystem are interrelated and interdependent. Implicit in the concept of an urban forest ecosystem, then, is the idea that people have an effect on trees and trees an effect on people. In addition, trees planted along streets and parkways of towns and cities represent a considerable financial investment. New York City alone has several million trees.

Value of Individual Trees

Trees on private property contribute to the value of the real estate, increasing values 5 to 20 percent. On public or commercial property, trees have a value of their own apart from the real estate value.

When a tree is suddenly or unexpectedly damaged or destroyed, tree owners in some circumstances can seek reimbursement from insurance companies. To obtain reimbursement, it is necessary to appraise the value of a tree. Appraisal may also be needed to determine the damages one must pay for injuring or killing someone else's tree. Homeowners may want to know how much to deduct from their income tax for a tree damaged or destroyed by hurricane, ice storm, lightning, or fire.

To aid in this process, the council of Tree and Landscape Appraisers produced a book entitled *Guide for Plant Appraisal*. Experienced and knowledgeable individuals representing five organizations in the landscape plant industry collaborated in the development of the *Guide* and its revisions, the most recent of which was published in 1992. The five organizations contributing expertise to the *Guide* are American Nursery and Landscape Association, American Society of Consulting Arborists, Associated Landscape Contractors of America, International Society of Arboriculture, and National Arborists Association.

Tree Evaluation Based on the Guide. The *Guide* first makes a case for the concept that trees have value. Trees to be evaluated are then subdivided into two groups: those that can be replaced because of their relatively small size, and those that are too large to replace. For large trees, several methods for estimating value are described. Detailed evaluation procedures for each method are then presented. The following information generally describes the procedure used. Those who are professionally involved in tree appraisal will want to obtain and study the *Guide* for details and to participate in training courses and practice sessions offered by one of the sponsoring organizations.

Establishing Replacement Value. If transplantable trees are available, lost trees of similar size can be replaced. The tree evaluator needs to compute the value based on local costs and labor rates. Such costs should include: (1) cost of the transplant plus shipping; (2) cost of removal of the damaged or dead tree; (3) planting costs, including digging, backfilling, and soil amendments; (4) cost of postplanting maintenance. Problems related to trying to obtain a tree of the same species and size must be overcome. The size of easily transplantable trees is usually expressed as the trunk diameter 6 inches above ground, or 12 inches above ground if the tree is over 4 inches in diameter.

Fig. 1-4. Trees along Independence Mall in Philadelphia.

Once the basic replacement cost is established, the evaluator can proceed using the *Guide* to scale down the tree value by taking into account the condition and location of the original tree as described for large trees in the next section. This appraisal value would be used to determine compensation.

If a tree is actually replaced, there is unfortunately no way to compensate for the loss in future growth that the transplanted tree will experience. For example, a healthy 4-inch red oak is lost and replaced by a healthy transplant of the same species in the same location. During the next several years, while adjusting to transplanting, the new tree will not grow as fast as the old tree would have had it not been damaged. Thus, the owner of the damaged tree will have suffered a loss of future growth, and the optimum mature size reached by the replacement will have been delayed compared to the original tree. This loss may not be accounted for in most tree evaluation schemes because it is difficult to quantify.

Determining Values of Large Trees

A tree of a size too large to transplant should have a value greater than that of a smaller tree that is transplanted as a replacement. A fairly complex formula is used to compute the value of large trees. Using the formula, the four factors that determine tree value are: (1) diameter of the trunk; (2) species; (3) condition; and (4) location. Collecting the information needed to make use of the formula requires knowledge, precision, keen observation, and varying amounts of subjective judgment. Training and experience are also essential and can be obtained at tree evaluation workshops offered to arborists by professionals in the industry.

Tree Size. Tree size is expressed as the number of square inches calculated for the area of the trunk cross section at 4.5 feet above ground. A price based on cost per unit of trunk area can be determined by calculating the cost per square inch of a transplantable tree. For example, if the cost of a 3-inch diameter tree (cross sectional area of 7 square inches) from the nursery is $300, then the cost per square inch is about $42.50. The size-based value of a large, 16-inch diameter tree (trunk cross section area at 4.5 feet above ground is 201 square inches) can be determined by multiplying $42.50 times 201, which computes to $8,545. When trees reach a certain size, additional trunk diameter does not add appreciably to the value, so adjustments need to be made in these calculations. Also, trees with multiple trunks or low branching require special consideration. These issues are addressed in the *Guide*. The tree value one obtains will need to be adjusted for species, condition, and location to make the final appraisal.

Tree Species. It is generally agreed that not all types of trees of the same size have the same value. In fact, most arborists working in a specific geographic area tend to agree on which species should be considered high- or low-value trees. Trees that are long-lived, hardy, durable, strong, adaptable, clean, beautiful, and pest-resistant are considered high-value trees. Tree species and cultivars can be assigned percentage values based on whether they have desirable traits. These values are generally given in increments of 5 or 10 percent and in some regions can be obtained from charts that often give a percentage range for each tree type.

Good examples of species differences that could account for different rating percentages are easy to imagine. For example, in most circumstances, Chinese elm

Fig. 1-5. The American elm, one of our most beautiful trees, has been virtually eliminated in many areas by the Dutch elm disease.

could deserve a 90 percent rating whereas Siberian elm would not be worth more than a 30 or 40 percent rating because the latter is structurally weak and subject to storm damage. On the other hand, a single species, American elm, may be worth 90 percent where Dutch elm disease is not likely to be a problem but worth only 10 percent where the disease threatens (Fig. 1-5). Honey locust, with its formidable trunk thorns, would deserve a 20 percent rating, but its thornless cultivars have a higher value, perhaps 60 percent. Flowering crabapple cultivars such as 'Hopa' or 'Almey', which are extremely susceptible to scab disease in humid temperate regions, might deserve a rating of only 30 percent, whereas more tolerant or resistant cultivars such as 'Mary Potter' or 'Prairiefire' may rate 70 to 90 percent. These percentages are used to modify the size-related values previously determined.

To remove some of the subjectivity inherent in making these kinds of quality ratings, published lists of trees with their ratings have been developed. Several state and regional chapters of the International Society of Arboriculture (ISA) have developed guides for tree and plant rating in their regions. The guide reflects the collective wisdom of that particular chapter and is most useful in that part of the country.

Tree Condition. Determining the condition of a tree requires an analysis based on careful examination of the tree. Many aspects of this examination are utilized when diagnosing tree problems (Chapter 9).

Fig. 1-6. Norway maple 'Cavalier' has a compact round top and is suitable for large open areas.

First, an overview of the situation needs to be made to see whether the tree is typical for that species in that area. The crown should be examined for dead wood, foliage density, and structural weakness. The foliage should be scrutinized for abnormal color, size, and shape, which indicate problems, as well as for symptoms and signs of diseases and insects.

Yearly twig growth and bud size and condition need to be considered as well as twig abnormalities relating to attack by disease organisms and insects. Trunk problems including cavities, cracks, injuries, and presence of diseases or insect pests need to be noted. Condition of roots and soil-related growing conditions that might affect tree condition should also be considered. In any case, the evaluator needs to systematically determine whether a tree is in good condition and, if not, the nature and seriousness of the problem.

A percentage can be assigned (100 = perfect condition) and used in calculating the overall value. The *Guide* provides a detailed checklist of factors to consider in determining plant condition.

Tree Location. The tree location value refers to what it is that the tree does and the impression it makes in the place where it is growing. The kind, value, and use of the property where the tree is growing should be determined. A tree growing in the woods or along the highway would have less value than a tree in a residential neighborhood. Inherent attractive features of the tree such as flowers and foliage are also important. How the tree affects the environment of the persons interacting with it also is important (Fig. 1-6). This can involve everything from providing shade where needed to providing architectural features such as screening out undesirable views and accenting buildings. Again, a percentage is assigned, in this case subjectively, because what one might call a beautiful character or attribute of a tree in a certain location may not appear that way at all to another individual. The *Guide* provides a detailed checklist of what to look for in establishing a percentage relative to location.

Calculating the Value. The value of the tree is calculated by multiplying the size-based dollar amount by the percentage allowed for species, and that amount in turn by the percentage allowed for condition. This amount is then multiplied by the percentage allowed for location. For example, to determine loss, a 25-inch diameter bur oak may have a base value of about $23,000. (In this example, the base value is calculated by determining the value per square inch of a 4-inch transplantable bur oak from the nursery which costs $600. The cross-sectional area is about 12.5 square inches, or a value of $48 per square inch. At $48/in², a 25-inch tree would have a value of about $23,600, so that if the tree is replaced, the loss is $23,000.) Thus, a $23,000 bur oak (80 percent species value) in fair condition (50 percent) and in a good location (75 percent) would have a value of $23,600 x 0.8 x 0.5 x 0.75 = $6,900.

Additional skill and judgment are required for tree evaluation when the following circumstances occur:

- Trees have negative value because they are entirely inappropriate or a hazard;
- Trees are located in historic or very important locations;
- Rare or unusual specimen trees are being considered;
- Multi-trunked trees are being evaluated (Fig. 1-7);
- Trees are no longer at the site but are being evaluated after the damage has been done;
- It will not be known for several years whether a tree has suffered permanent injury;
- Trees are declining; and
- Potential exists for biased evaluation. Since three of the four evaluation criteria are based largely on subjective judgment, it is important that evaluators honestly arrive at the same value no matter what their professional orientation or which side of a dispute they favor. Unfortunately, in practice, different evaluators using the same criteria often do not arrive at similar values.

Fig. 1-7. Multitrunked trees are a pleasant addition to the landscape.

In order to make this appraisal method work, especially for large trees, the evaluator needs to know the cost of the largest normally available transplantable tree of any given species so the cost per unit trunk area can be computed. In addition, one needs to have species ratings developed and agreed upon by tree professionals in the region, and one needs to have training and experience in assessing tree condition and the relative merits of various tree locations. Perhaps most important, the value of a tree must not exceed the value of the property upon which it once grew. Indeed, if all plants on a property of a certain value are lost, their appraisal should typically not exceed 15 percent, a figure sometimes used to represent the enhancement of property values by landscaping.

Other Methods for Determining the Value of Trees. The *Guide* also describes guidelines for determining monetary loss when trees are damaged or destroyed. A cost of repair method is used when it is determined that repairing the damage might be best under the circumstances. A cost of cure method is applied when treatments are needed to return the property to a reasonable level of its original condition. A method for evaluating palms is also given in the *Guide*. Trees do have value, and methods that are standardized and easily proved are desirable. Obviously, the consulting arborist who is appraising values of lost trees must use good judgment to determine a fair and reasonable value for trees being examined.

How Tree Values Are Used

Tree evaluation that is conducted with the hope of collecting damages or tax relief following a casualty loss is often unrewarded, for not all circumstances dictate that there be such compensation. The professional who is involved in tree evaluation must be knowledgeable about insurance and tax laws and know what is and is not allowed. The results of court cases involving tree losses suggest that the involvement of professionals who are qualified witnesses is important.

Tree Maintenance, Arboriculture, and Urban Forestry

The cultivation and management of trees has been documented by humankind for thousands of years. Many of the earliest efforts at tree maintenance were centered on the production of economically important products such as wood and fruit. The science and art of individual tree care was probably applied first to fruit trees. Early farmers and foresters also managed trees for livestock shade, sprouts, timber, and firewood.

During the last several hundred years in Europe, and later in America, trees with amenity value were planted in cities and towns and on estates. Provoked in this century by the loss of trees due to catastrophic diseases such as Dutch elm disease, interest in amenity trees expanded and the science of woody plant care became known as arboriculture. In many countries, the word arboriculture connotes the science of growing fruit trees, so perhaps tree maintenance is still connected to its origins in this way. In recent decades, the term urban forestry has been used to describe the maintenance of urban trees as part of a total ecosystem rather than on an individual tree basis. Indeed, urban forestry often includes the management of wildlife habitats, open spaces, parks, greenbelts and other spaces that support vegetation as well as individual streetside and landscape trees.

Arboriculture, urban forestry, and tree maintenance all deal with the cultivation and management of amenity trees. In the mind and imagination of the public, the term arboriculture is probably not as well understood as urban forestry. Nevertheless, in this book, for the most part, the field of study of those who do tree maintenance is called arboriculture, and its practitioners are called arborists, but this field of study does not exclude urban forestry and the activities of urban foresters. Tree maintenance activities also lie within the realm of community forestry, environmental horticulture, landscape ecology, landscape horticulture, landscape maintenance, landscape management, and urban horticulture.

Arboricultural Practitioners

Tree owners often are faced with the problem of finding someone competent to care for their trees. Knowing that their tree has value, and knowing that it is possible for an incompetent individual to harm their tree, the selection of the right person or company becomes an important decision. Consumers need to be wary and take the time to examine an arborist's previous work and ask for recommendations before any tree work is agreed upon. In almost any urban area, persons can be seen working with trees. Some, just new in the business or operating part-time or during periods of unemployment, may be equipped only with a pick-up truck and a chain saw. Others, having a more substantial investment, may utilize bucket trucks, climbers with ropes and saddles, and chippers and chipper trucks. Still others work for multistate corporations and are often involved in utility line clearance. They can frequently be recognized from one area to another by the similarly painted and equipped trucks and machinery. Many of these tree workers belong to firms that hold membership in organizations of arborists and tree-related professionals within state or national areas. Although size and equipment do not provide a good means of determining whether a tree-care firm is competent, sometimes membership in at least one professional organization is a clue to a firm's competence. Perhaps more important, some firms employ or are managed by certified arborists. Arborist certification should assist consumers in sorting out those who are competent to work on their valuable trees and those who are not. The primary activities of members of the following organizations include many of the major tree-maintenance operations. The brief description of each organization does not describe all of its activities.

American Society of Consulting Arborists (ASCA)

This organization—5130 West 101st Circle, Westminster, CO 80030-2314—consists of professional arborists who evaluate shade trees, diagnose tree problems, and consult on all matters relating to planting, maintenance, and preservation of trees. This well-trained and experienced group has about 150 members, primarily in North America. Some members own tree-care companies, thus offering services related to their consulting recommendations. ASCA members also conduct training sessions related to shade tree evaluation.

International Society of Arboriculture (ISA)

The ISA—P. O. Box 3129, Champaign, IL 61826—primarily devotes itself, through its scientific journal, annual conference, and research funding, to educational programs on all phases of tree maintenance. The various interest groups within ISA reflect the organization's diversity. These groups include the Utility Arborist Association, Society of Commercial Arboriculture, Municipal Arborists and Urban Foresters Society, Arboricultural Research and Education Academy, and Student Society of Arboriculture. In addition, ISA is represented geographically by chapters consisting of individual or groups of states, provinces, or countries. An ISA chapter or international conference is likely to be attended by individuals in all phases of tree care, from the tree climber who has participated in a tree-climbing competition to the scientist who has delivered a research paper. ISA consists of more than 10,000 members belonging to chapters on four continents, but most of them are from North America.

The ISA has developed a certified arborist program to establish meaningful arboricultural standards, encourage continuing arboricultural education, and identify competent arborists. Certified arborists are required to pass written examinations and attend approved educational sessions to obtain and maintain their certification. The International Society of Arboriculture has developed a study guide for this certification. A certified arborist has at least demonstrated some competence and might be considered a good choice for tree work.

The ISA has produced several publications on various phases of tree maintenance including a plant health care manual, a book on arborists' equipment, and other educational materials, including children's literature on trees and instructional videotapes. Through its research trust, the ISA sponsors scientific research on trees, the results of which are published in their journal, the *Journal of Arboriculture.*

National Arborists Association (NAA)

The NAA—P.O. Box 1094, Amherst, NH 03031—consists of commercial arborists interested in improving the way professionals care for trees. The NAA series of slide, cassette, and video programs provide valuable instruction for tree workers. These educational programs and the NAA Book of Standards, which are used by the tree care industry, define proper tree care techniques for pruning; cabling, bracing, and guying; fertilizing; installing lightning protection; and applying pesticides. NAA membership is open to all commercial arborists, and the organization has been a strong advocate for them.

Other Organizations

The following groups also have an interest in preserving and maintaining trees:
- American Nursery and Landscape Association, 1250 I Street NW, Suite 500, Washington, DC 20005. This is the main voice of the nursery industry, having interests in education and research on growing and marketing of trees and other plant materials.
- American Forestry Association, 1319 Eighteenth Street NW, Washington, DC 20036. It is the oldest citizen's organization advocating trees, forests, and conservation. This group publishes a national register of big trees. Their Global Releaf program exemplifies their continuing interest in urban forestry and the role citizens can play in tree maintenance.
- Arboricultural Association, Amphield House, Romsey, Hants SO51 9PA, England, is a British organization interested in the maintenance of amenity trees. This group has developed standards for arboricultural practice and prints a scientific publication, *Arboricultural Journal.*
- Associated Landscape Contractors of America, 12200 Sunrise Valley Drive, Suite 150, Reston, VA 22091. Landscape installation is an important activity of their membership.
- Council of Tree and Landscape Appraisers, 1250 I Street NW, Suite 504, Washington, DC 20005. The council, representing several groups, published the shade tree evaluation guide.
- Society of Municipal Arborists, 7747 Old Dayton Road, Dayton, OH 45427. Municipal arborists and others belong to this group, which promotes street tree management.
- National Arbor Day Foundation, Arbor Lodge 100, Nebraska City, NE 68410. This foundation sponsors the Tree City USA program.
- Society of American Foresters, 5400 Grosvenor Lane, Bethesda, MD 20814. This society for professional foresters has members with interests in urban forestry.

• Trees for Life, 3006 W. St. Louis, Wichita, KS 67203-5129 provides assistance for planting and care of food bearing trees in developing countries.

Arboricultural Periodicals

Arborists and others interested in trees can learn more about trees and keep up with some of the latest developments by subscribing to, or obtaining from their municipal or college library, periodicals dealing with tree care. These publications have information ranging from popular or practical articles on tree maintenance to scientific studies of trees. The following are suggested:

Arbor Age
Arboricultural Journal, The International Journal of Urban Forestry
Arborists News
American Forests
American Nurseryman
Grounds Maintenance
Journal of Arboriculture
Journal of Forestry
Weeds, Trees, and Turf

Selected Bibliography

Chapin, R. R., and P. C. Kozel. 1975. Shade tree evaluation studies at the Ohio Agricultural Research and Development Center. OARDC Research Bulletin 1074. Wooster, Ohio. 46 pp.

Council of Tree and Landscape Appraisers. 1992. Guide for plant appraisal, 8th ed. International Society of Arboriculture, Champaign, Ill. 103 pp.

Grey, G. W., and F. J. Demke. 1978. Urban forestry. Wiley, New York. 299 pp.

Robinette, G. O. 1972. Plants/people/and environmental quality. U.S. Department of the Interior, National Park Service, Washington, D.C. 139 pp.

2

The Structure of the Tree and Function of Its Parts

The successful establishment and maintenance of trees are greatly aided by an understanding of the anatomy and functions of a normal, healthy tree. Without such knowledge it is difficult to diagnose accurately the many abnormalities that commonly occur in trees or to appreciate various practices undertaken to maintain or restore their vigor.

Normality in any living organism is so vague and relative that it almost defies definition. Perhaps a scientifically accurate definition might be derived by applying to the tree something of the psychologist's concept of the normal human being. By this rule, the normal tree would fall within certain limits of health and vigor that include the majority of trees in a specified range.

For our purposes, however, we cannot go far astray if we are guided in our evaluation of a healthy tree by the desirable external appearances with which we are all familiar. Since a tree is most admired when it has a symmetrical branch system and abundant, attractive foliage, we include these two characteristics among our standards of perfection. Healthy roots must also be included, for without them proper branch and foliage growth cannot continue.

To gain a working knowledge of the proper care of trees, we must know something of the structure of each of the different parts that constitute a normal shade tree, the functions these parts perform in the growth process, and the relation between the parts.

Parts of a Tree

The roots, stems, and leaves are vegetative structures that perform specific functions to maintain the life, health, and vigor of the tree. The flowers, fruit, and seeds are reproductive structures and have evolved to ensure the continuation of the tree species.

The structure and functions of angiosperms (which produce true flowers) and gymnosperms (which bear cones) are generally similar, although there are specific differences in cells, tissue, and organs between the two groups. The angiosperms typically have broad, flat leaves and are deciduous whereas the gymnosperms typically bear needles and are evergreen.

Fig. 2-1. Lateral roots of a hackberry, exposed by a high-pressure stream of water. These relatively shallow roots are growing into a layer of topsoil that had been spread over subsoil just before planting.

Roots

There is no organ more vital to a healthy tree than its roots, but because the root system is largely unseen, its importance seldom is appreciated by laymen. Many believe that root distribution somehow mirrors the branching pattern of the aerial portion of the tree or that roots extend out only to the edge of the tree crown, as far as the "dripline." In fact, for most trees, roots extend laterally far beyond the tree crown but do not extend to depths equivalent to the tree height.

In tree seedlings, there are two general types of genetically determined rooting patterns: *taproots* and *branched roots*. Taproot systems are characterized by a fast-growing primary root that penetrates deeply. Hickory, walnut, sweetgum, pines, and many oaks have taproot systems. Branched, surface root systems are characterized by slower growing shallow primary roots with more rapidly growing lateral branches. Elm, spruce, and maple are branched, surface-rooted species.

Although the root systems of seedlings exhibit the genetically inherited pattern, roots of older trees are greatly influenced by soil and site characteristics, so the inherited pattern is often obscured. The two most important environmental factors that determine root distribution are oxygen and water. In general, tree roots will extend to greater distances both vertically and laterally in well aerated soils. Observations on fruit trees have revealed lateral spread of roots equivalent to three times tree height in sandy soil, two times in loam, and one and one-half times in clay. Likewise in well-aerated soils, roots may extend down 30 feet; however, under most conditions most of the roots will be found in the upper 18 inches of soil, and virtually all will occur in the upper 3 feet of soil. Site characteristics such as hard pan, clay layer, or perched water table restrict root development. Root development also is impeded by surface coverings such as concrete or asphalt.

Under favorable circumstances the root system of some trees is enormous. It may comprise one-third to one-half of the entire volume of the tree. The total length of all roots of a large spreading oak runs into hundreds of miles. Roots branch and rebranch. One researcher estimated that a mature red oak had at least 500 million root tips!

Structure of Root Systems. The taproot is generally cut off when the tree is transplanted, or, if it survives, it becomes less dominant as the tree gets older. Taproots that may have dominated the root system of seedlings are barely visible in most circumstances when trees reach maturity.

Lateral roots in young trees originate from the taproot, the trunk, or the first branched surface roots. They are attached to the trunk at the soil line and become the major roots that form the root flare at the base of the tree. In mature trees, the major share of the root system is made up of lateral roots that extend from the trunk in all directions at a depth of a foot or so (Fig. 2-1), depending on species and growing environment. The fine *absorptive roots* that arise from the lateral roots frequently grow upward into the top few inches of the soil. Lateral roots produce and support smaller diameter *sinker roots* that, depending on soil environment, grow straight down several feet, which may aid the tree in anchorage or water absorption.

Oblique roots in young trees originate close to the trunk from the first roots or from the laterals near the trunk. They grow downward at a steep angle and may penetrate several feet into the soil. They bear few fine roots or lateral roots and may function to anchor the young tree and to take up water. As the tree matures, these roots comprise less and less of the total root system.

At the point of attachment to the trunk, the lateral roots are relatively few and

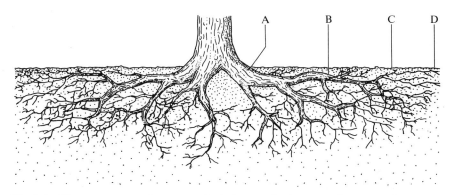

Fig. 2-2. The root system of most trees includes A, a few large woody roots; B, long rope-like lateral roots; C, fine absorptive nonwoody roots; and D, a zone of root hairs near the root tips.

large. These woody roots taper rapidly into long ropelike strands from which further lateral roots branch. The lateral roots divide and redivide, becoming progressively smaller in diameter until they become fine, absorptive, nonwoody roots at some distance from the trunk (Fig. 2-2). Nonwoody roots may convert to permanent woody roots in response to injury to the main root.

The tip of a nonwoody root is covered by a group of cells called the *root cap*. It functions to protect the growing tip from injury, and cells near the outside of the cap have mucilaginous walls that help ease the progress of the tip through the soil. A short distance back from the tip, the rootlet is covered by numerous, very fine outgrowths of thin-walled cells, called root hairs (Fig. 2-3). More than 10,000 root hairs per square inch have been found on certain roots. They are especially abundant on young trees, but may be rare on mature trees in nature. The absorptive, nonwoody roots are composed of tissues derived by cell division within a special area called the *apical meristem*. The apical meristem is located just behind the root cap.

The tissues of the root are organized more or less concentrically (Fig. 2-4). At the center of the cylinder are *xylem* cells, specialized for conducting water and minerals, and *phloem* cells, specialized for conducting food materials. This core is

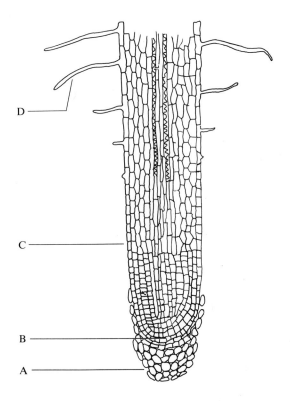

Fig. 2-3. Longitudinal section of a root tip showing A, root cap; B, apical meristem where cell division occurs; C, zone of cell elongation; and D, root hair.

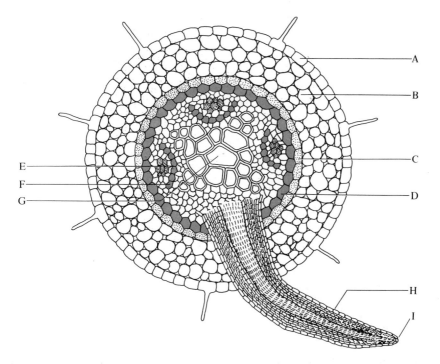

Fig. 2-4. Cross section of a young root showing A, epidermis; B, cortex; C, endodermis; D, pericycle; E, phloem; F, cambium; G, xylem; H, lateral root; and I, root cap. The lateral root arises from internal meristematic tissues, represented here by the cambium.

surrounded by the *pericycle, endodermis, cortex,* and *epidermis* on the outside. Lateral roots arise from the pericycle. The endodermis is a single row of cells with a bandlike thickened suberized area, known as the *casparian strip*. The strip is generally thought to decrease the ability of water to enter the pericycle. Thus the endodermis may exercise control over water and dissolved substances entering and leaving the root. The cortex is composed chiefly of storage cells with large intercellular spaces. Finally, the epidermis serves as covering over all. Young epidermis has root hairs and older epidermis may develop a cuticle and become lignified as well.

In cross section woody roots resemble woody stems. Annual rings are less pronounced in roots than in stems, partly because seasonal changes in environment are buffered by the soil. Long-lived woody roots develop a distinct core of heartwood just as stems do. Root tissues are adapted for storage and have less need for strong supporting tissues.

Root Growth. The root tip is pushed through the soil by the process of cell division, which takes place in the apical meristem and by elongation of the newly formed cells.

The root system of trees includes long-lived large, woody roots and short-lived small, nonwoody roots. The large, main, woody roots are essentially perennial, and unless killed by unfavorable environmental conditions, insects, or disease, they will endure as long as the tree. The longevity of the small absorptive roots varies with species, environmental conditions, and tree health, but generally they live only a few weeks to a year or sometimes several years.

Some roots undergo secondary growth, which causes an increase in root diameter and may start during the first or second year of root life and continue thereafter. These roots become the perennial woody component of the root system. Roots near the soil surface are the first to begin growth each spring; those at greater depths begin later. More wood is produced near the soil surface, so older roots are often oval in cross section.

Function of the Root System. The primary functions of roots are anchorage, absorption, conduction, storage, and synthesis of hormones.

Woody roots ramify through the soil, thus firmly anchoring trees. Diseased or pruned roots fail to anchor trees, just as water-soaked soil does.

Trees derive most of their water and mineral salts from soil. Water primarily enters through root hairs, specialized roots (*mycorrhizae*), and epidermal cells of the absorptive roots, although some enters through older roots as well. The root hairs and mycorrhizae act to increase the surface area of the roots and come in close contact with the film of water that surrounds soil particles. Water enters root cells through diffusion while the uptake of mineral salts is a relatively selective process dependent on metabolic activity.

Roots conduct water, mineral salts, and sometimes stored foods to stems and leaves. Conversely, they also conduct foods from the leaves and stems to all parts of the system underground.

When sugars move into roots more rapidly than they can be used by growing cells, they may be converted to starch and stored for a time. During the dormant season, large quantities of starch are stored in the woody roots of trees. In the spring when active growth resumes, this food reserve is mobilized and conducted back into the stems and leaves.

Specialized Roots. The roots of many vascular plants harbor filamentous fungi, forming symbiotic associations called mycorrhizae (Fig. 2-5). Mycorrhizae fall into two categories: *ectotrophic* forms in which the fungus exists both inside and outside of the root and *endotrophic* forms in which the fungus lives entirely within the host cells. Forms of mycorrhizae intermediate between the two are also found. Ectotrophic mycorrhizae are commonly found associated with the absorptive roots of many tree species. The fungus penetrates the cell wall of the cortex and covers the outside of the root with a network of fungal strands. Ectotrophic forms are found on many important trees, including alder, beech, birch, chestnut, oak, pecan, poplar, willow, cedar, Douglas-fir, fir, larch, pine, and spruce. The endotrophic form is associated with finely divided roots, and the fungus lives intracellularly in cortical cells. The endotrophic form is less prevalent and is commonly found on tuliptree, maple, and sweetgum. Some fungi can form mycorrhizal associations with tree roots of many species, while others may be very specific to only one species.

The mycorrhizae are important in tree physiology. They act to increase the absorption surface area for minerals, particularly phosphorus. In most sites, mycorrhizae appear to be essential for tree survival. The fungus benefits by getting

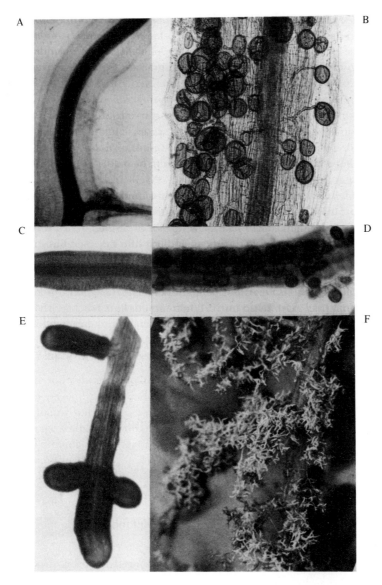

Fig. 2-5. Mycorrhizae: A, uncolonized black cherry root; B, black cherry root infected with endomycorrhizae; C, uncolonized sweetgum root; D, sweetgum root infected with endomycorrhizae; E, ectomycorrhizae formed on Norway spruce (note presence of a fungus mantle around the root); and F, ectomycorrhizae on pin oak (note presence of abundant short roots). (From Maronek, D. M., J. W. Hendrix, and J. Kiernan. 1981. Mycorrhizal fungi and their importance in horticultural crop production. Horticulture Reviews 3: 172–213.)

carbohydrates from the tree. The tree benefits by the increased mineral absorption and perhaps by some degree of root disease protection.

There are commercial products such as MycorTree available for adding mycorrhizae to tree root systems. Containing a mixture of mycorrhizal fungal spores, these products are intended to benefit newly transplanted trees as well as to improve the health of established trees. In poor soils, mycorrhizal treatment may provide benefits, but most good soils already contain naturally occurring populations of mycorrhizae. There is not a large body of research on the value of mycorrhizal treatments for landscape trees. Pines planted on strip mine spoils with very poor growing conditions have benefitted from application of mycorrhizal fungi to their roots.

Natural grafts between roots are common. The graft involves a vascular connection that is the result of a union of cambium, phloem, and xylem of previously unconnected roots. For a graft to occur the roots must be compatible, usually of the same species (although interspecies grafts occur occasionally), and there must be growth pressure. Often a community of trees is functionally united by numerous within-species grafts. Water, minerals, metabolites, and sometimes disease organisms pass freely through the graft unions.

Root nodules form on many leguminous woody plants and many other trees, notably alder and ceanothus, as well. Associated with these nodules are microorganisms that have the ability to fix nitrogen. In the case of the legumes, the organisms are bacteria of the genus *Rhizobium*. For nonlegumes, the organisms are most commonly actinomycetes. The association between actinomycetes and tree roots may be called actinorrhizae. Usually the microorganisms invade the roots through root hairs or small wounds and infect the cortical parenchyma cells.

Stems

The stem—the woody framework of the tree—includes the tree trunk, or bole, which begins just above the root flare, and the branches. It is constituted to endure the strains placed upon it by the varying conditions of its environment. For a seemingly rigid structure, the stem is amazingly flexible and will often bend to a surprising degree before breaking.

Structure of the Stem. The largest part of the stem or trunk is composed of a wood cylinder made up of tissues called *xylem* (Fig. 2-6). The xylem is composed of long tubelike cells through which water containing dissolved mineral salts is conducted from the roots to the leaves; *parenchyma* cells, which are involved in storage of food materials and lateral transport of materials; and *wood fibers*, which lend tensile strength to the tree. The tubelike cells are of two types: thick-walled *tracheids* and thinner walled *vessel elements*. Both types of cell are found in angiosperms, whereas in gymnosperms only tracheids are present. Functional vessels and tracheids have no living contents. Indeed in the functional area of the xylem only about 10 percent of the cells are alive. The parenchyma cells in sapwood are living and form a network of connected living cells via the rays from the sapwood into the vascular cambium, phloem, and cork cambium.

In most mature trees, a dark inner zone called *heartwood* can be seen. The heartwood is composed entirely of dead cells and as such is physiologically inactive, although it may be chemically reactive when invaded by decay fungi. It serves

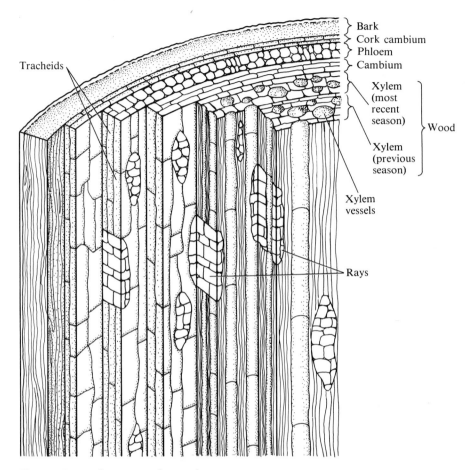

Fig. 2-6. Internal anatomy of a woody stem.

as a mechanical support for the tree. However, many trees are able to survive for years after the heartwood is completely rotted out. Heartwood is often more resistant to decay because it contains less moisture and lacks nitrogen and carbohydrate reserves needed for decay fungi to proceed. It must also be said that not all discolored wood is heartwood—some may be sapwood reacting to wounds and branch death.

The outer portion of the wood cylinder is called the *sapwood*. The sapwood is usually lighter in color and in density than the heartwood. The functional water-conducting and physiologically active cells are located in the sapwood. As the cells on the inner edge of the sapwood mature and die or become plugged they are converted into heartwood. The boundary between the sapwood and the heartwood is often irregular, extending several annual rings closer to the outside at some points than others. The inner part of the sapwood is continuously transformed into heartwood. Thus the heartwood constitutes a progressively expanding inner core. The sapwood, however, in mature trees tends to remain about the same thickness.

The *pith* occupies the center portion of the stem. The pith has no specific function in mature trees and is frequently so crowded by growth pressure that in older stems it is nearly lost.

Immediately outside of the xylem is a narrow band of cells known as the *vascular cambium*. During the growing season, the cells of the cambium are stimulated to divide, giving rise to the xylem tissue on the interior and *phloem* tissue on the exterior. Thus new xylem cells are added to the outside of old xylem cells and new phloem cells are added to the inside of old phloem cells. Although the cambial cells, by repeated divisions, are differentiating into xylem and phloem, some daughter cells remain as cambium, so a cambium layer is always between the xylem and phloem. As the stem increases in circumference, the cambium keeps step with this increase by radial divisions, thus increasing the number of cells and enlarging the circumference of the cambium layer. Generally, the thicker this layer of cambium, the more vigorous the tree.

All tissues lying outside of the cambium are considered to constitute the *bark*. The bark has a more complex tissue structure than the wood and includes the phloem, cortex, and periderm.

The phloem constitutes a fairly narrow band composed of a variety of cells, most prominently *sieve-tube members,* or *sieve cells, companion cells,* fibers, and parenchyma. The sieve tubes or sieve cells serve as passageways through which food materials are conducted. Most of the normal cell contents are missing from mature sieve-tube members. With the exception of a few trees such as palms and basswood, sieve-tube members live for only one to three years. In most species, nonfunctional phloem cells are crushed by growth pressures.

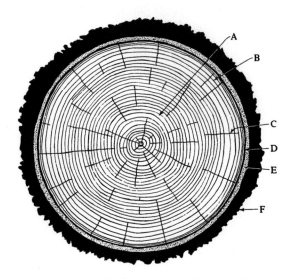

Fig. 2-7. Cross section of a tree trunk, showing A, the heartwood, composed of dead cells; B, the sapwood, which has many living xylem cells that conduct water and minerals upward; C, medullary rays, which transfer water and food radially; D, the cambium, from which are formed xylem cells on the inside and phloem tissue on the outside; E, the phloem cells, which conduct elaborated food; and F, the outer bark.

The *cortex* is a complex tissue located between the phloem and outer tissues. Its function in young stems is to store food and support the stem. This tissue is not found in mature stems.

The *phellogen*, or cork cambium, produces outer bark tissues. It is located outside of the cortex and can be found in mature trees. The phellogen functions in much the same way as the cambium, except that it produces cork externally and *phelloderm* internally. Some species produce no phelloderm. The phellogen and the tissues it produces are collectively called the *periderm*.

The cork cells die soon after formation and form a waterproof protective tissue. Their walls consist largely of *suberin*, a waxy substance. The continued increase in circumference of the tree causes the cork to rupture and fall off. Eventually the loss of dead bark about equals the formation of new bark. Many species can be identified by their unique bark appearance.

Inspection of the cross section of any tree trunk (Fig. 2-7) reveals numerous concentric lines, the *annual rings*, which are plainly visible because of the alternating large and small cells. These rings represent the annual increase in the diameter of the tree. The large cells are formed by rapid growth in the spring; the smaller, by slower growth in the summer. As the season progresses, the new cells become smaller and smaller, until growth stops altogether in the fall. This compact region of small cells, known as *summer wood*, makes the annual ring. It is in sharp contrast to the large cells, or *spring wood*, formed by rapid growth during the spring.

If a tree is leaning, the new xylem formed tends to straighten up the tree, so the annual rings may not be the same distance apart all the way around the stem. The wood formed in response to a lean is called *reaction wood*, and the size of the annual ring in the reaction zone is larger than on the opposite side of the stem. In conifers, the reaction wood is called *compression wood*, and it is formed on the underside of the leaning trunk. The greater elongation of xylem cells in compression wood tends to force the trunk upright. In broadleaved plants the reaction wood is called *tension wood* and is found on the top side of the leaning trunk (Fig. 2-8). The internal contraction of tension wood tends to pull the trunk upright. Reaction wood also exists in branches to pull or push them to their original angle of attachment.

Across the annual rings many radial lines are visible, running from the outer edge of the wood cylinder toward the center. These lines, called *medullary rays*, are groups of cells extending through the many cylinders of wood; they provide for radial transfer of water and food.

Stem Growth. The stem increases both in length (primary growth) and diameter (secondary growth) by the functioning of actively dividing cells known as *meristematic tissues* and by the enlarging of newly formed cells before secondary wall formation. The meristematic tissues make up only a small portion of the total mass of the tree. They are located in buds and stems. In the stems, the cambium and phellogen are meristematic tissues. Meristematic tissues are the main source of the hormone auxin, which plays an essential role in regulating tree growth.

The growth of the shoot begins with division of cells of the apical meristem, which is located in the terminal bud. Subsequently the cells, which do not continue dividing, will elongate, differentiate into specialized cell types, and mature.

Fig. 2-8. Tension wood (top of figure) on one side of this honeysuckle trunk developed because the trunk was leaning.

The increase in shoot length results almost entirely from the elongation process. The mature cells cease to enlarge. Thus the apical meristem is constantly carried upward and outward by the growth of cells it produces.

Trees produce several types of buds, which are categorized by position (terminal or lateral), contents (vegetative, floral, or mixed), or activity (dormant or active). Within the terminal bud, cells divide to produce leaf primordia and lateral or axillary buds. Each lateral bud is potentially the terminal bud of a new branch shoot or a flower. Under some circumstances, especially after injury, buds may be produced in locations other than at a leaf axis. They are called adventitious buds and originate from a variety of tissues including wound callus, cambium, or mature nonmeristematic tissues.

The lateral buds may continue growth without interruption and develop into leafy branches or flowers. More frequently they become dormant after developing to a length of about a quarter of an inch. The buds may remain dormant or resume active growth after a time. The dormancy or activity of lateral buds is governed by hormones produced by the tree.

Most woody plants grow intermittently, with one or more flushes of growth within a distinct growing season, rather than continuously. Typically, rapid periods of stem elongation are followed by periods of little or no elongation even when favorable environmental conditions exist. In the tropics, some tree species are found to grow continuously; in the temperate zones, however, all trees grow intermittently.

Under exceptional conditions trees normally having one growth flush per year may resume growth within the same season. Young trees, particularly oak and maple, may have a second growth flush if they are heavily fertilized. Older trees may resume growth if excessive rainfall follows a period of drought.

Water sprouts arise from dormant buds on the main trunk or branches of trees. They may be stimulated to grow by sudden exposure to light, as when an adjacent tree or shading branches are removed. Water sprouts are usually vigorous and tend to grow vertically.

Branches are often shed by trees. This occurs by two mechanisms: abscission by a physiological process similar to leaf abscission and natural pruning by death of branches, during which no abscission zone is formed. Abscission of twigs is a response to adverse environmental conditions and to aging and subsequent loss of branch vigor. It occurs in a well-delineated abscission zone and is preceded by periderm formation. Mature white oak, poplar, willow, maple, walnut, and ash commonly undergo abscission. Most shed twigs in the autumn, but maples shed largely in the spring and summer.

Natural pruning generally begins with branches of the lower crown. Death usually is caused by low light intensity. The dead branch is sealed off from the living tree stem by gums or plugs called tyloses in angiosperms or resins in gymnosperms. The tree has no mechanism to cause the dead branch to fall off, but saprophytic fungi and insects attack the dead branch and in most cases it will eventually fall off.

In the spring, the resumption of stem cambial activity follows renewed activity in the buds and the development of leaves. Activity begins at the top of the tree and proceeds downward. During the early stages of cambial activity, the cells enlarge because of water uptake and their walls become thinner. The bark may be easily peeled at that time.

The onset of cambial activity appears to be closely related to the beginning of shoot growth although shoot growth in most tree species stops before diameter growth. Even in trees that have only one flush of shoot growth in the spring, cambial activity continues until late summer or early fall. Cambial activity apparently ceases first in the upper crown, then in the stem, and finally in the roots.

Stem Functions. The functions of the stem, or trunk, of a tree are structural support; conduction of water, food, hormones, and nutrients; and storage of food materials. One of the chief services rendered by the much-branched stem or trunk is support of the leaves to provide them maximum benefit from the sun's energy. It is essential to the growth of the tree that as many leaves as possible be exposed to the sun. This exposure is accomplished by means of multiple branching of the tree trunk into limbs, branches, and finally twigs. Structural support may fail if the stem is decayed or severely injured.

The trunk and its subdivisions are the connecting links between the roots, where most water and mineral nutrients are absorbed, and the leaves, where food is manufactured. It is the function of these connecting links to translocate the water and mineral elements in water solution from the roots to the leaves and to distribute elaborated food materials to every growing part of the tree, including twigs and root tips.

The large size of trees indicates that they have highly developed translocation

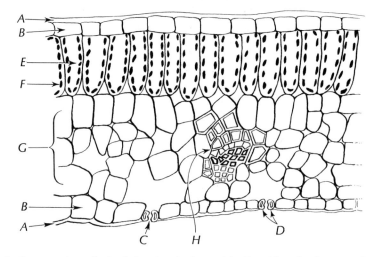

Fig. 2-9. Cross section of a leaf, showing A, the cuticle; B, epidermis; C, stoma; D, guard cells; E, palisade cells; F, chloroplasts; G, the spongy mesophyll; and H, the vascular tissues (xylem and phloem).

systems. Water movement is along the path of least resistance in wood. Most trees show spiral grain to some extent; thus water usually ascends in a helical pattern. The angle of the spiral grain changes from year to year so that over a few years' time a single root may be connected to a number of different branches.

Sugars manufactured in the leaves are translocated in the phloem. The direction of movement changes during the season. In the spring, sugars are initially transported into growing leaves. At a certain stage of maturity when leaves produce more sugars than are needed for growth, the leaves begin to export sugars and the direction of flow in the supporting twig is reversed. Leaves near the stem tip may export sugars both up to the expanding terminal bud and down to the roots.

In all deciduous trees, food materials are gradually conducted from the leaves to twigs. These foods are partially assimilated for immediate growth. The surplus is stored in twigs, trunk, and roots. Late in the season, when the leaves become less active in the manufacture of food, part of this surplus is reabsorbed for assimilation by the growing parts. The initial twig and root growth and, frequently, flower production, which takes place in the spring before leaves are expanded and active, are entirely dependent upon food stored in twigs, stems, and roots.

Leaves

Leaves function as solar radiation receptors and as sites of sugar manufacture. A very large surface area is necessary to perform these functions. In forests it has been estimated that the surface area of leaves is four to six times greater than that of the ground.

Structure of Leaves. The typical leaf of a broad-leaved tree consists of a blade and a stalk called the *petiole*. In cross section (Fig. 2-9), we see the surface cells, or *epidermis*, the outer walls of which are coated with a waxy covering, or *cuticle.* The cuticle is a noncellular layer consisting of wax and cutin and is attached to the epidermis by a layer of pectin. This covering is perforated by small openings called *stomata* (singular *stoma*). Each stoma is surrounded by two kidney-shaped movable cells, the *guard cells.* Through these pores moisture evaporates from the leaf and gases are exchanged between the atmosphere and the interior of the leaf. Moisture loss from the leaf is also controlled by the leaf waxes.

Beneath the epidermis is one layer or more of long, cylindrical cells, the palisade cells. These contain many minute *chloroplasts,* which contain the green substance known as chlorophyll. Numerous large, loosely arranged cells, the *spongy mesophyll,* lie between the palisade cells and the lower epidermis. The high proportion of intercellular space facilitates the diffusion of gases between the stomata and the palisade cells.

Embedded in the mesophyll is the veinal or vascular tissue. The veins of angiosperm leaves generally form an interconnecting branching system. A vein typically consists of a strand of xylem and a strand of phloem surrounded by a layer of cells called the *bundle sheath.* The phloem is typically on the lower side and the xylem on the upper. The arrangement of veins of leaf blades varies in detail from species to species.

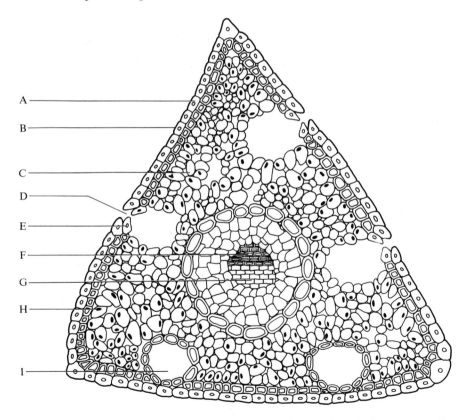

Fig. 2-10. Cross section of a needle, showing A, epidermis; B, cuticle; C, chlorenchyma; D, stoma; E, guard cell; F, xylem; G, phloem; H, chloroplast; and I, resin duct.

The leaves of most gymnosperms are either needlelike or scalelike; however, a few, for example, ginkgo, have broad, flat leaves. The leaves are generally evergreen, except in larch, bald cypress, and a few others. Generally the leaves are firm, with small, thick-walled epidermal cells and a thick cuticle (Fig. 2-10). The stomata are usually sunken and are typically found in longitudinal bands.

Leaf Growth. Leaf growth is initiated by the meristematic tissue of the shoot near the apex. In many species, cell division is completed in the fall while the leaves are still enclosed in the bud. The growth of leaves in the spring may be entirely the result of cell enlargement.

The seasonal pattern of leaf growth varies widely among species. Most commonly, new leaves appear in one or more distinct flushes of growth. Some species produce new leaves more or less continually throughout the growing season. Individual angiosperm leaves tend to expand rapidly, taking from 2 to 40 days (average 14 days) to fully expand. Leaves of evergreen species grow out more slowly than those of deciduous trees. Individual leaves of gymnosperms may continue to expand from early spring through late summer. Indeed, in some species leaves may continue to increase in length for more than a year.

Leaf life has an inherently limited duration. Leaves of deciduous species last only one growing season, whereas those of evergreen species persist at least until the leaves of the following season are produced and may last several years. The leaves of needle evergreens generally last 2 to 3 years, although some live as long as 10 to 15 years.

Leaves of most woody perennial plants are shed as a result of changes in an abscission zone at their base, often before the leaf dies. The abscission zone is the weakest part of the petiole. It contains a high proportion of parenchyma cells and few thick-walled cells. Before leaf fall, changes occur in the zone that result in the formation of a cork layer over the leaf scar and the plugging of exposed vessels. Thus the abscission zone functions to bring about leaf fall and to protect the shoot from invasion of pathogenic organisms through the leaf scar.

Leaf fall is under the control of plant hormones but is thought to be triggered by adverse environmental conditions, such as drought, temperature changes, or shortened days. These conditions stimulate leaf senescence, which always precedes natural abscission.

Function of Leaves. The primary function of the leaf is *photosynthesis* (photo = light; synthesis = manufacture), the manufacture of sugars through the use of energy provided by sunlight. Photosynthesis is the source of energy not only for trees and other plants, but indirectly for virtually all other living things through their consumption of plants. Oxygen is an important product of photosynthesis, as it is essential to all living organisms, primarily for its role in *respiration.*

The manufacture of sugars is accomplished in the chloroplasts, through a highly complicated process in which carbon dioxide, which enters the leaf through the stomata, is combined with water, which is taken up through the root system. The sugars are then used by the tree as a source of energy, through the process of respiration, for further growth and development. Respiration takes place in living cells of the stem, roots, and reproductive structures as well as in the leaves.

Sugars also provide the starting material for the synthesis of other essential plant constituents such as starch, cellulose, lignin, proteins, fats, and oils. Formation of these compounds involves highly complicated processes that may also take place in living cells of organs other than leaves.

The process of *transpiration* occurs as the result of the evaporation of water from leaf surfaces, primarily through the stomata. The rate of transpiration is influenced by a number of factors including humidity, temperature, and wind, but the most important factor is the condition of the stomata. All other factors being equal, water loss is least when stomata are completely closed. Closed stomata, however, do not allow the entry of carbon dioxide essential for photosynthesis; hence, for the tree to be healthy, stomata cannot remain closed. Transpiration can thus be regarded as a necessary consequence of the need for photosynthesis.

Water lost through transpiration is replaced by water taken up from the soil solution by the roots and conducted to the leaves through the xylem. The processes by which this occurs, particularly in tall trees, has been the subject of some controversy, but capillary action by which water is drawn upward is generally thought to play a role.

Reproductive Organs

Angiosperms. Typically a flower is composed of four whorls of modified leaves: sepals, petals, stamens, and carpel(s), all attached to the receptacle, the modified stem end that supports these structures (Fig. 2-11). The stem terminates in the flower and does not continue to grow. The stamens are the male organs of the flower and produce the pollen. The carpels are the female organs and bear the ovaries.

Many trees have single-sex flowers that contain either stamens or carpels. In some cases these single-sex flowers are borne on the same tree, as in birch and alder. In other cases, such as holly, persimmon, poplar, and willow, they are borne on separate staminate and pistillate trees. Buckeye produces complete flowers, staminate flowers, and pistillate flowers on the same tree. Ash produces complete flowers plus either pistillate or staminate flowers on the same tree. Most fruit trees produce only complete flowers.

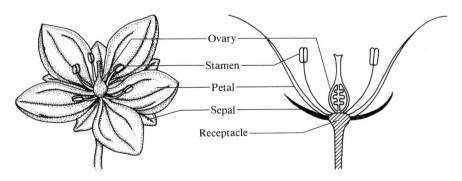

Ovary
Stamen
Petal
Sepal
Receptacle

Fig. 2-11. Structure of a typical angiosperm flower.

Trees do not flower until they are mature, typically after 5 to 20 years. The flower buds of many trees are initiated during the previous season, so abundant bloom depends upon the tree's health during the previous year.

The development of fruit usually requires fertilization of the egg, contained in the carpel, by the sperm, borne by the pollen. However, in a few species, fruit may mature without fertilization and/or seed development. These *parthenocarpic* fruit occur in some species or cultivars of fig, apple, peach, cherry, plum, citrus, maple, elm, ash, birch, and tuliptree. The period from flowering until fruit ripening varies greatly among species and cultivars and, in some cases, because of environmental conditions and cultural practices. For many species, this interval is about 15 weeks.

Gymnosperms. The flowers of gymnosperms consist of pollen cones and seed cones, which in most species are borne on the same tree, although one exception is juniper. The pollen cones are usually from $^{1}/_{4}$ inch to 4 inches long. They are brightly colored, often yellow, purple, or red. Seed cones are much larger, ranging from $^{1}/_{2}$ inch to more than a foot long. Cones are usually initiated during the season before pollination occurs.

Cross-pollination in gymnosperms is common because female cones are generally concentrated in the upper branches and male cones in the lower ones. Wind-transported pollen is shed over a 2- or 3-day period for most trees. Pollination occurs in most species in early spring, soon after the seed cones emerge from the dormant buds.

The cones begin to enlarge after pollination and the scales close. In the case of Douglas-fir, development from pollination to seed maturation is completed in about 4 months. For most pines, the process takes 2 years.

Selected Bibliography

Cronquist, A. 1971. Introductory botany. 2nd ed. Harper & Row, New York. 885 pp.

Gartner, B. L. (ed.). 1995. Plant stems: physiology and functional morphology. Academic Press, San Diego. 440 pp.

Kozlowski, T. T. 1971. Growth and development of trees. II. Academic Press, New York. 514 pp.

Kozlowski, T. T., and S. G. Pallardy. 1997. Physiology of woody plants. Academic Press, San Diego. 411 pp.

Romberger, J. A., Z. Hejnowicz, and J. F. Hill. 1993. Plant structure, function and development. Springer-Verlag, Berlin. 524 pp.

Shigo, A. L. 1994. Tree anatomy. Shigo and Trees, Associates, Durham, N.H. 104 pp.

Thomas, H., and J. L. Stoddart. 1980. Leaf senescence. Ann. Rev. Plant Physiol. 31:83–111.

Weier, T. E., C. R. Stocking, and M. G. Barbour. 1974. Botany: An introduction to plant biology. Wiley, New York. 693 pp.

Zimmermann, M. H., and C. L. Brown. 1971. Tree structure and function. Springer-Verlag, Berlin. 336 pp.

3

The Soil and Its Relation to Trees

No book on the maintenance of shade and ornamental trees would be complete without some discussion of that important natural body, the soil. Usually considered by the layman to be an inert mass, soil is actually a dynamic entity in which physical, chemical, and biological changes are constantly occurring. A detailed treatment of these changes is beyond the scope of this book, but a brief presentation of the more important ones, particularly as they relate to trees, is essential.

Soil serves as a storehouse for mineral nutrients, a habitat for microorganisms, and a reservoir of water for tree growth. It also provides anchorage for the tree.

Soil Components

The soil is composed of three distinct parts or phases: gaseous, the soil air; liquid, the soil water; and solid, the soil minerals and organic matter. These parts are interrelated.

The Soil Air. The pore space in soils amounts to 30 to 50 percent of the volume. The quantity of air contained in the soil is governed primarily by the extent to which this pore space is filled with moisture. Ordinarily, air comprises about 20 percent of the volume of a well-tilled soil.

Oxygen, one of the most important components of soil air, is essential for respiration in roots and supports the activity of numerous beneficial microorganisms. Nitrogen, another component of the soil air, is used as a raw material by nitrogen-fixing bacteria to manufacture protein materials. These later decompose to yield nitrates, which nourish the tree.

All tree roots and soil microorganisms respire; that is, they continuously consume oxygen and emit carbon dioxide. In poorly aerated soil, where gaseous exchange between the soil air and the atmosphere is retarded, the oxygen supply decreases, whereas the quantity of carbon dioxide increases. This unfavorable oxygen–carbon dioxide balance may retard or, in extreme cases, completely check root growth, since roots do not thrive in such areas. This, in fact, explains why the more active roots of city trees are often confined to soil areas between the curb and the sidewalk and to nearby lawn areas. Because the concentration of carbon dioxide must be very high, over 20 percent, to cause direct injury to tree roots, lack of oxygen probably is the more common cause for restricted growth or death of roots.

Although oxygen is absolutely necessary for a healthy root system, an insufficiency of this element for relatively short periods will not result in serious damage. Flooding of soil in winter when roots are dormant is not as serious as it is during summer, when the roots are making their most active growth. Roots of apple trees can grow slowly with as little as 3 percent oxygen, but they require at least 10 percent for good growth.

Capillary and gravitational movements of water and the daily changes in temperature and barometric pressure are responsible for the aeration of soils. The oxygen supply is often renewed too rapidly in sandy soils and too slowly in heavy clay soils. In either case, the situation is improved by adding organic matter or by mixing the two classes of soil. Aeration of soils is also increased by structural changes caused by the use of lime and certain fertilizer salts. The channels remaining after the roots of plants decay and those caused by the burrowing of earthworms also aid in ventilating the soil.

Some trees can thrive under conditions of relatively poor soil aeration. Willows, for example, will grow along streams and ponds where the oxygen supply is relatively low. Some scientists believe that in willows oxygen produced by photosynthesis in the leaves is transferred internally to the root cells. The roots of one species of willow, *Salix nigra*, were found to be capable of growing and absorbing nutrients in almost complete absence of oxygen. Cypress grows naturally in swamps where the soil is flooded for long periods. The roots of this tree have pyramid-shaped protuberances, called "knees," which extend as high as 10 feet into the air. The knees are composed of light, soft, spongy wood and bark, which permit air to reach the roots during the weeks and even months when the swamps are covered with water.

On the other hand, plants like yew and cherry thrive only in soils that are well drained and well aerated. Such trees are killed or severely damaged if the soil is waterlogged for only a few days during the growing season.

Some trees can survive prolonged periods of soil flooding; others cannot. New York state nurseryman Philip M. White observed the response of trees whose roots were flooded in from 4 to 15 inches of standing water for 10 days as the aftermath of hurricane Agnes in western New York state in 1972. Listed among the survivors were:

Acer rubrum	Red maple
Cornus mas	Cornelian cherry
Fraxinus americana	White ash
Gleditsia triacanthos inermis	Thornless honey locust
Juglans nigra	Black walnut
Juniperus chinensis 'Pfitzeriana'	Pfitzer juniper
Juniperus virginiana	Red cedar
Malus dolgo	Dolgo crabapple
Morus alba	Mulberry
Platanus occidentalis	Sycamore
Populus deltoides	Cottonwood

Salix alba and *S. discolor*	White willow and pussy willow
Tilia cordata	European littleleaf linden

Other trees known to survive prolonged flooding include:

Acer negundo, A. saccharinum	Box elder, silver maple
Aesculus glabra	Ohio buckeye
Alnus glutinosa	European alder
Aralia spinosa	Devil's walking stick
Betula nigra	River birch
Celtis occidentalis, C. laevigata	Hackberry, sugarberry
Fraxinus pennsylvanica, F. velutina	Green ash, velvet ash
Magnolia virginiana	Sweetbay magnolia
Melaleuca quinquenervia	Melaleuca
Metasequoia glyptostroboides	Dawn redwood
Nyssa sylvatica	Blackgum
Quercus bicolor, Q. lyrata, Q. macro-carpa, Q. nigra, Q. palustris	Swamp white oak, overcup oak, bur oak, water oak, pin oak
Sequoia sempervirens	Coast redwood
Taxodium ascendens, T. distichum	Pond cypress, bald cypress
Ulmus alata, U. americana	Winged elm, American elm

Trees that did not survive the prolonged flooding were:

Acer saccharum	Sugar maple
Acer platanoides	Norway maple
Betula papyrifera	White birch
Betula populifolia	Gray birch
Cercis canadensis	Redbud
Cladrastis lutea	Yellowwood
Cornus florida	White flowering dogwood
Cornus florida 'Rubra'	Red flowering dogwood
Cornus florida 'Cloud 9'	'Cherokee Chief' dogwood
Crataegus phaenopyrum	Washington hawthorn
Crataegus lavallei	Lavalle hawthorn
Magnolia soulangiana	Saucer magnolia
Malus sp. 'Hopa', 'Lodi', 'McIntosh', 'Radiant'	Apple and crabapple
Picea abies	Norway spruce
Picea pungens	Colorado spruce
Picea pungens 'Glauca'	Colorado blue spruce
Prunus persica	Flowering peach
Prunus serotina	Black cherry
Prunus subhirtella 'Pendula'	Weeping cherry
Quercus borealis	Red oak
Robinia pseudoacacia	Black locust
Sorbus aucuparia	European mountain-ash

Taxus cuspidata	Upright yew
Taxus cuspidata 'Expansa'	Spreading yew
Taxus media 'Hicksi'	Hick's yew
Thuja occidentalis	American arborvitae
Tsuga canadensis	Canadian hemlock

Inadequate aeration of the soil may affect the roots in an indirect manner. It may change the kind of organisms in the soil flora and make the roots more susceptible to attack by parasitic fungi. For example, the fungus *Phytophthora cinnamomi* may become so abundant in a poorly drained soil that it will cause severe damage to tree roots. Flooded roots often exude materials that attract the swimming spores of the *Phytophthora* fungus. Moreover, certain organisms thrive under conditions of little or no oxygen. They produce compounds such as hydrogen sulfide and nitrites, which may harm the roots.

Aside from direct injury to the roots, poor aeration also prevents the roots from absorbing water and minerals. A deficiency of oxygen and an excess of carbon dioxide reduce the permeability of the roots to water and prevent the roots from absorbing minerals from the soil solution.

Poor aeration of the soil results in development of roots near the soil surface. Such a shallow root system makes a tree more susceptible to toppling by winds when the soil is wet and to droughts during the summer.

The single most critical factor in the success or failure of tree planting is soil drainage and aeration. Trees are killed more frequently by poor soil drainage than anything else. We can avoid this to some extent by selecting species more tolerant of poor soil drainage. However, no tree grows best in a totally saturated soil.

Soil Water. Trees depend on the water contained in the pore space of soil in which they are growing. The roots absorb the water and pass it on to the leaves, where it functions both as a basic component of the tree and in the absorption of carbon dioxide from the air. Water is an essential part of the protoplasm, or living matter, of every cell. It also functions in the soil as a solvent of the necessary mineral nutrients and as a medium by which they enter the plant. Further, water transports other raw materials and elaborated food from one part of the tree to another.

Water may exist in any soil in the *hygroscopic, capillary,* or *free water* state. Hygroscopic water occurs in the form of a very thin film over the soil particles where the moisture in the soil is in equilibrium with that in the air. Capillary water exists as a thicker film, which is attracted to the particle by its own surface tension. This tension is stronger than the pull of gravity. When two soil particles are in contact, and one has a thinner film of capillary water than the other, the drier surface has a greater attraction for the water. As a result, the water moves from the wetter to the drier surface until both particles are surrounded by films of equal thickness. When the film of capillary water reaches the maximum thickness that can be maintained by surface tension, any additional water responds to the pull of gravity and flows downward into the soil to become free water. Free water occupies the spaces between the moist soil particles, just as the air does when only hygroscopic and capillary water are present. When free water occupies

Fig. 3-1. These trees died because the high water table excluded air from their roots.

all the pore spaces in the soil or any horizon of the soil, its upper surface is known as the *water table*. Because of lack of air, roots cannot survive long in such a water-saturated zone (Fig. 3-1).

When soil water comes in contact with the complex organic compounds that comprise the cell walls of the root hairs, it is absorbed by *imbibition*. From this point the water now passes on, by a process known as *osmosis*, through the plasma membrane lining the cell walls and becomes part of the cell sap. The concentration of soluble substances in the cell sap is usually greater than in the soil solution. The movement of water into the cell tends to equalize the concentrations on both sides of the cell wall. The pull exerted by the water lost from the leaves during transpiration further aids the movement of water from cell to cell and the influx of water from the soil into the root hairs.

When the soil surrounding the root hairs becomes very dry, as during a drought, the concentration of the solution in the soil may be greater than that within the cell. Then the reverse process takes place. Water moves out of the root hairs and plant cells, causing their collapse, and the continuous water column from roots to leaves cannot be maintained. Consequently, the plant wilts. Wilting also occurs if the quantity of water lost from the leaves is greater than that gained by the roots. Such reverses are often evident in shade trees during the summer.

Proper and timely application of water is critical to the survival and growth of newly transplanted trees, which have a limited root system. This topic is discussed in Chapter 5. Established trees growing in climates that typically receive summer rains rarely need irrigation except during periods of extended drought unless they are growing at a site that limits penetration of water into the soil. Tree owners wishing to optimize tree growth should see to it that 1 inch of water per week is applied to the root zone either by rainfall or by sprinkler. Trees growing in regions with hot, dry summers are more likely to experience drought stress. Since the stresses caused by prolonged lack of water may ultimately cause a tree to decline and die, installation of a permanent irrigation system is needed in such locations.

Symptoms of lack of water include wilting of leaves and scorching, in which areas of the leaf between the veins or along the margins turn light or dark brown. See Chapter 10 for a further discussion of this topic. Prompt application of water to such trees is necessary to alleviate the drought stress and prevent long-term damage to the tree.

Trees vary with respect to their moisture requirements. Red maple, pin oak, sweetgum, and most species of willow prefer soils with a high moisture content, whereas red oak, chestnut oak, bur oak, ash, and poplar grow best in soils containing smaller amounts of water. Some trees found in Mediterranean climates, such as Brazilian and California pepper trees, California bay laurel, coast live oak, and carob, are drought tolerant. Spruces can tolerate wetter soils than most other species of conifers.

A change in the water table often seriously affects a tree's health. Although not evident at the surface, a rise in the water table may cause drowning of the deeper roots. This happened to London planetrees planted in Battery Park at the southern tip of Manhattan Island and to Hyssop crabapple trees planted in New York

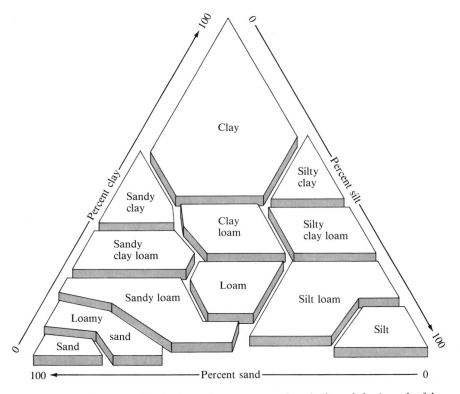

Fig. 3-2. The soil texture triangle shows the percentage of sand, silt, and clay in each of the textural classes. (From Stefferud, A. [ed.]. Soil: The yearbook of agriculture 1957, p. 32. U.S.D.A. House Document No. 30. Washington, D. C.)

City's Stuyvesant Town housing development many years ago. A sudden lowering of the water table, caused by drainage of a wetland, could provoke drought in trees with shallow root systems nearby.

Soil Minerals. The solid body of the soil consists of varying proportions of coarse grains (sand), medium-sized grains (silt), and fine particles (clay), all of which are mineral particles originating from the disintegration of rocks. The relative proportions of three components determine the soil type (Fig. 3-2). In addition, the soil contains varying amounts of organic matter, humus substances formed by the microbial decomposition of plant and animal remains.

The smaller-sized particles of the clay portion of the soil are known collectively as *colloidal clay.* These particles aid greatly in the retention of water and mineral elements needed by plants in the soil. The working qualities of a soil depend largely on its content of colloidal clay and the condition in which this exists.

Chemically, colloidal clay consists mainly of iron and aluminum silicates and oxides. The silicates are the seat of retentive or "exchange" processes in soils. Surrounding each tiny fragment of the exchange mineral is a layer of absorbed water containing both *cations* and *anions.* Cations are positively charged (+) and anions are negatively charged (–) particles that are produced when soluble soil nutrients dissolve in water. The cations consist chiefly of hydrogen (H), calcium (Ca), magnesium (Mg), potassium (K), and sodium (Na); the anions are hydroxyl (OH), phosphate (PO_4), nitrate (NO_3), sulfate (SO_4), and chloride (Cl) ions. Most of these ions are plant nutrients.

As a root hair enters the zone of absorbed water, an exchange of ions takes place between this water layer and the root hairs. As a result, the mineral nutrients of the soil solution move onto the surface of the roots, and the carbonic acid of the plant is released into the soil. The uptake of mineral nutrients by plant cells is relatively selective, and the process depends on metabolic activity.

Soil Organic Matter. The second solid component of soil, organic matter, is usually concentrated in the upper layers. It is essential for the maintenance of soil microorganisms and it makes the soil porous and friable. This increases its water-holding capacity and improves its aeration. In its decay, organic matter yields up mineral nutrients and nitrogen for the use of the currently growing plants. In addition organic matter has a high "exchange capacity" acting much like the exchange minerals in colloidal clay.

Soil Organisms. In addition to the familiar macroorganisms, such as earthworms and insects, the soil is inhabited by a wide variety of microorganisms, numbering in the billions per cubic inch of soil. These microorganisms, which include protozoa, bacteria, actinomycetes, fungi, and nematodes, are all involved in the decomposition of organic matter, making nutrients available for uptake by roots. Some bacteria and actinomycetes can fix nitrogen, either alone or in symbiosis with roots of certain plants. Serious diseases of trees and other plants are caused by soil-inhabiting fungi, bacteria, and nematodes. These are discussed in Chapter 12.

Certain fungi are able to form symbiotic relationships with roots; the structure and function of these mycorrhizae are discussed in Chapter 2, under Specialized Roots. The failure of certain tree species, notably beech, to grow well when planted in particular soils has been attributed to the absence of mycorrhizal fungi in these soils. Mycorrhizal fungi can be introduced into such soils by incorporating soil from natural habitats in which these tree species thrive, or by incorporating commercial products containing mycorrhizae.

Elements Essential for Tree Growth

For growth and development trees require the major elements carbon, oxygen, hydrogen, nitrogen, potassium, phosphorus, sulfur, calcium, magnesium, and the trace elements iron, boron, manganese, copper, zinc, chlorine, and molybdenum. Plants use elements in four primary ways: to form structural units, such as carbon in cellulose and nitrogen in protein; to incorporate into molecules important in metabolism, like magnesium in chlorophyll; to act as catalysts in enzymatic reactions, as several of the trace elements do; and to help maintain osmotic balance, as does potassium in stomatal guard cells. Each essential element functions uniquely and cannot be replaced by another.

As already discussed in Chapter 2, carbon and oxygen are provided by the carbon dioxide in the atmosphere, which enters the leaves through the stomata.

Hydrogen is obtained from water absorbed through roots and carried to the leaves by the water-conducting system.

Nitrogen is also a component of air but higher plants cannot use the form in which it occurs in the atmosphere. In nature it is taken from the air and assimilated by certain types of soil bacteria and bluegreen algae. Trees in turn obtain nitrogen from decaying organic material in the soil.

All other elements required by trees are taken from the soil. They are absorbed by the roots only in solution, in the manner already described in Chapter 2. From the standpoint of their relative value as raw materials for plant use, the elements in the soil may be divided into three classes: nonessential; essential and abundant; and essential and critical.

Nonessential elements include silicon, aluminum, sodium, and some of the rarer elements. These elements are believed to have little or no role in the nutrition of plants.

The second group, essential and abundant, includes calcium, iron, magnesium, and sulfur. They are usually present in ample amounts in soil, although occasionally soil may be deficient in one or more of these elements, or these elements may be unavailable.

Elements that are essential and critical include nitrogen, phosphorus, and potassium, and in some sites magnesium, manganese, and boron. These elements are sometimes present in soil in insufficient quantities and may be the limiting factor in plant growth. This is especially true of nitrogen. Commercial fertilizers have as their main constituents the first three elements of this important group.

Influence of Soil on Nutrient Availability

Soil Type. The physical and chemical properties of soil influence the amount of mineral nutrients held by the soil and to some extent the availability of nutrients to plants. Essential plant nutrients, present as inorganic ions, are absorbed onto the surface of soil particles and organic matter or are free in the soil solution. The soil-absorbed ions are not readily leached but are available for plants to absorb. The capacity of a soil to absorb cations (positively charged ions) is called cation exchange capacity, and this is a good measure of soil fertility. Soils containing a large percentage of clay have a high cation exchange capacity because clay, consisting of large numbers of small particles, has a large surface area. Soil organic material, humus, also has a high nutrient exchange capacity (five times more than clay). Sandy loam soils therefore must be fertilized more frequently than clay loam soils or soils with a high percentage of organic matter.

Soil Reaction (pH). Soils are acid, neutral, or alkaline in reaction. The degree of acidity or alkalinity is expressed in terms of pH values, very much as temperature is expressed in degrees Celsius or Fahrenheit. Temperature scales center on the freezing point of water, whereas the acidity-alkalinity scale is divided into fourteen major units centering on a neutral point (pH 7). Values greater than pH 7 are alkaline, and those less than pH 7 are acid. The intensity of the alkalinity increases tenfold with each unit increase in pH. For example, the alkalinity at pH 8 is ten times as great as that at pH 7. Similarly, the acidity increases ten times for each unit decrease in the pH scale. Thus the acidity at pH 6 is ten times as great

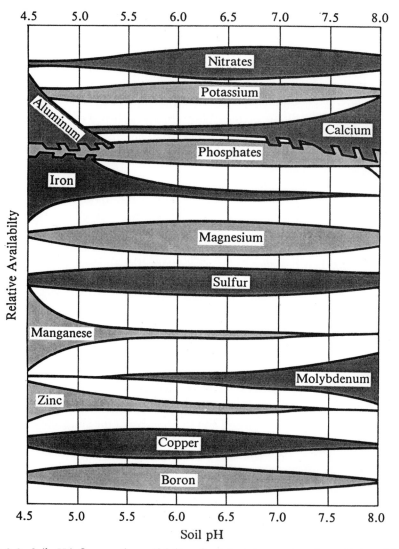

Fig. 3-3. Soil pH influences the availability of nutrients to growing plants. (Adapted from Publication AGR-19, Liming acid soils, University of Kentucky Cooperative Extension Service, University of Kentucky, Lexington, KY 40546.)

as that at pH 7, and at pH 5 it is ten times as great as that at pH 6 and one hundred times as great as that at pH 7.

The pH scale refers not to the total amount of acid or alkaline constituents in the soil, but to those portions of the acid and alkali that are in a free or dissociated state. A pH reading gives an indication of the relative number of free *acid* (H+) and *alkaline* (OH–) ions in the soil, and not of the combined or undissociated molecules.

Many tree soils are acid in reaction because a large portion of the alkaline exchange cations (those of calcium, magnesium, potassium, and sodium) have been either absorbed by the trees or lost through leaching, leaving acid exchange minerals behind. In addition, the continued use of acid-forming fertilizers and sprays (especially those containing sulfur) tends to increase acidity. In cities, certain gases such as ammonia and sulfur dioxide, which are carried down by rainwater, further tend to increase the acidity of the soil. The pH values of most eastern tree soils lie between 5 and 6.5. They are seldom higher than 7, unless their content of limestone is naturally high or large quantities of lime have been added to them. In soils of midwestern landscapes and in parts of the west, soil pH levels may be 7 or higher. Where soil pH levels exceed 7.5 or 8, the choice of trees to grow may be very limited.

Although soil acidity is not the only factor governing the availability of nutrients, it nevertheless plays an important part (Fig. 3-3). For example, an increase in acidity beyond certain limits means a decrease in the supply of certain plant nutrients, the hydrogen having taken the place of the cations in the exchange complex. The exchanged cations may have been removed by the trees or may have been carried away by the drainage water. At the other extreme, as the alkalinity increases, the roots are less able to absorb iron, manganese, zinc, copper, and boron, because these minerals are insoluble under such conditions. A good case in point is that of the availability of iron salts, which are absorbed by the plant at a pH of 6.7 or below. If a tree is growing in an alkaline soil (above pH 7), there is a strong possibility that it will suffer from iron deficiency and that the leaves will become chlorotic (see Chapter 10).

In general, most plants grow best on soils having a slightly acid reaction. A pH value between 5.5 and 7 seems most favorable for nearly all trees. Growth is best when the soils in which trees are planted and grown are maintained within the pH range to which they are adapted. Relatively few trees seem to prefer a distinctly acid soil. Among those that do are silverbell (*Halesia*), sorreltree (*Oxydendrum*), yellowwood (*Cladrastis*), and Franklin-tree (*Franklinia*).

Adjusting Soil pH. Soil acidity or alkalinity can be changed by the use of chemicals applied to the surface or incorporated into the soil. To increase soil pH (sweeten acid soil), lime is used either as ground limestone (calcium carbonate or calcium magnesium carbonate), quicklime (calcium and magnesium oxides), or slaked lime (calcium and magnesium hydroxides). Limestone is the most convenient form to use, but the other forms are faster acting. The amount of lime needed to produce the desired change in soil pH depends on soil texture, organic matter, and the material being used. Soils with fine texture or with high levels of organic matter have a greater buffering capacity and require more lime to achieve the same adjustment as coarser soils.

Soil test results often give two values: the water pH and the buffer pH. The buffer pH takes into account the buffering capacity of soil and thus gives a better estimate of how much lime is needed. Table 3-1 gives the amount of lime needed for the indicated pH change based on water pH values. These values are for a silt-loam soil and would be somewhat different for soil with more or less clay. Table 3-2 gives information based on buffer pH values, which is applicable to any soil type.

Table 3-1. Amount of agricultural limestone needed to raise the pH of a silt-loam soil based on water pH measurement.

Water pH value	Required amount of agricultural limestone needed to raise the soil pH to 6.4 (lb/1000 ft²)
Above 6.4	0
6.4–5.8	0–100
5.8–5.2	100–200
Below 5.2	200

Table 3-2. Amount of agricultural limestone needed to raise soil pH, based on buffer pH measurement.

Buffer pH value	Required amount of agricultural limestone needed to raise the soil pH to 6.4 (lb/1000 ft²)
6.7	70
6.6	100
6.5	115
6.4	140
6.3	160
6.2	190
6.1	210
6.0	230
5.9	245
5.8	255
5.7	280
5.6	315
5.5	325

Source: Publication ID-72, Principles of home landscape fertilization, University of Kentucky Cooperative Extension Service, University of Kentucky, Lexington, KY 40546.

Table 3-3. Suggested application of powdered sulfur to reduce the pH of an 8-inch layer of soil.

Original pH of Soil (Based on Water pH Value)	Pints of Sulfur Needed per 100 ft² to Reach pH of									
	4.5		5.0		5.5		6.0		6.5	
	Sand	Loam	Sand	Loam	Sand	Loam	Sand	Loam	Sand	Loam
5.0	$2/3$	2	-	-	-	-	-	-	-	-
5.5	$1^1/3$	4	$2/3$	2	-	-	-	-	-	-
6.0	2	$5^1/2$	$1^1/3$	4	$2/3$	2	-	-	-	-
6.5	$2^1/2$	8	2	$5^1/2$	$1^1/3$	4	$2/3$	2	-	-
7.0	3	10	$2^1/2$	8	2	$5^1/2$	$1^1/3$	4	$2/3$	2

Source: University of Kentucky Cooperative Extension Publication ID-72.

Soil pH may be decreased (reduced in alkalinity) by the addition of sulfur, though the effects may not be as long lasting as that of lime for increasing soil pH. Agricultural sulfur is the most convenient form to use, but finer ground dusting or wettable sulfurs are faster acting. Aluminum sulfate is often recommended, but it may have a toxic effect on plants if used in large amounts. About 7 pounds of aluminum sulfate are required to achieve the same effect as 1 pound of sulfur. To avoid toxic effects on vegetation, sulfur should not be used at temperatures above 80° F. The amount of sulfur needed to make desired pH changes is given in Table 3-3.

Lime and sulfur both need to be applied over a broad area to be effective; application should not be confined to the area near the trunk or planting hole. If applications are made before planting, the material should be worked into the soil down to a depth of 8 inches. After planting, the materials should be spread evenly over the soil surface and may be worked into the top 2 to 4 inches, but care must be taken to avoid damaging the tree's roots. If the soil needs pH adjustment and material is incorporated only into the backfill soil when planting a tree, then nutritional deficiency symptoms could appear as soon as the new roots spread into the untreated soil. Lime and sulfur do not effect permanent pH changes. Retreatments must be made from time to time.

County agricultural agents, agricultural experiment station chemists, and field representatives of agricultural chemical companies offer soil-testing services and supply information as to the amount of lime, sulfur, or aluminum sulfate that must be added to produce the desired change in pH on a specific soil. Alternatively, inexpensive soil-testing kits can be purchased for individual use.

Table 3-4 lists the most favorable pH range of many shade and ornamental trees.

Table 3-4. Soil reaction adaptations of specific trees.

Common Name	Botanical Name	Most Favorable pH Range
Acacia	*Acacia*	6.5 - 7.5
Alder	*Alnus,* named sp.	6.5 - 7.5
Almond, flowering	*Prunus glandulosa*	6.5 - 7.5
Apple	*Malus* sp.	6.5 - 7.5
Arborvitae	*Thuja* sp.	6.5 - 7.5
Ash	*Fraxinus* sp.	6.0 - 7.5
Bald cypress, common	*Taxodium distichum*	6.5 - 7.5
Beech	*Fagus,* named sp. and var	6.5 - 7.5
Birch, sweet	*Betula lenta*	4.0 - 5.0
Boxwood, common	*Buxus sempervirens*	6.5 - 7.5
Buckeye, Ohio	*Aesculus glabra*	6.5 - 7.5
Buckeye, red	*Aesculus pavia*	6.0 - 6.5
Catalpa	*Catalpa,* named sp.	6.5 - 7.5
Cherry	*Prunus* sp.	6.5 - 7.5
Cherry, choke	*Prunus virginiana*	6.5 - 7.5
Chestnut, American	*Castanea dentata*	4.0 - 6.0
Chinquapin	*Castanea pumila*	4.0 - 6.5
Cypress	*Chamaecyparis,* named sp. and var	4.0 - 7.5.

Table 3-4. Soil reaction adaptations of specific trees (continued).

Common Name	Botanical Name	Most Favorable pH Range
Dogwood, flowering	*Cornus florida*	5.0 - 6.5
Douglas-fir	*Pseudotsuga menziesii*	6.0 - 6.5
Elm	*Ulmus* sp.	6.5 - 7.5
Empress-tree	*Paulownia tomentosa*	6.5 - 7.5
Eucalyptus	*Eucalyptus* sp.	6.5 - 7.5
Fir	*Abies,* named sp.	4.0 - 6.5
Franklin-tree	*Franklinia alatamaha*	4.0 - 6.0
Fringe-Tree	*Chionanthus virginica*	4.0 - 6.0
Ginkgo	*Ginkgo biloba*	6.0 - 6.5
Hackberry	*Celtis* sp.	6.5 - 7.5
Hawthorn	*Crataegus* sp.	6.0 - 7.5
Hemlock, Canada	*Tsuga canadensis*	4.0 - 6.5
Hemlock, Carolina	*Tsuga caroliniana*	4.0 - 5.0
Hickory, shagbark	*Carya ovata*	6.0 - 6.5
Holly, American	*Ilex opaca*	4.0 - 6.0
Hop-Hornbeam, American	*Ostrya virginiana*	6.0 - 6.5
Hornbeam	*Carpinus,* named sp.	6.5 - 7.5
Horsechestnut	*Aesculus hippocastanum*	6.0 - 7.0
Juniper	*Juniperus,* many sp.	6.5 - 7.5
Juniper, common	*Juniperus communis* and var.	6.0 - 6.5
Juniper, mountain	*Juniperus communis saxatilis*	4.0 - 5.0
Kentucky coffee tree	*Gymnocladus dioica*	6.5 - 7.5
Larch	*Larix,* named sp.	6.5 - 7.5
Linden (basswood)	*Tilia* sp.	6.5 - 7.5
Loblolly bay	*Gordonia lasianthus*	4.0 - 6.0
Locust, black	*Robinia pseudoacacia*	5.0 - 7.5
Locust, honey	*Gleditsia,* named sp.	6.5 - 7.5
Magnolia	*Magnolia* sp.	4.0 - 7.0
Maple	*Acer,* many sp.	6.5 - 7.5
Maple, mountain	*Acer spicatum*	4.0 - 5.0
Maple, red	*Acer rubrum*	4.5 - 7.5
Maple, striped	*Acer pennsylvanicum*	4.0 - 5.0
Mock-orange	*Philadelphus* sp.	6.0 - 7.5
Mountain-ash, American	*Sorbus americana*	5.0 - 7.0
Mountain-ash, European	*Sorbus aucuparia*	6.5 - 7.5
Mountain laurel	*Kalmia latifolia*	4.0 - 6.5
Mulberry	*Morus,* named sp. and var.	6.5 - 7.5
Oak, black	*Quercus velutina*	6.0 - 6.5
Oak, blackjack	*Quercus marilandica*	4.0 - 5.0
Oak, chestnut	*Quercus prinus*	6.0 - 6.5
Oak, English	*Quercus robur* and var.	6.5 - 7.5
Oak, European turkey	*Quercus cerris*	6.5 - 7.5
Oak, overcup	*Quercus macrocarpa*	4.0 - 5.0
Oak, pin	*Quercus palustris*	5.5 - 6.5
Oak, post	*Quercus stellata*	4.0 - 6.5
Oak, red	*Quercus borealis maxima*	4.5 - 6.0

Table 3-4. Soil reaction adaptations of specific trees (continued).

Common Name	Botanical Name	Most favorable pH Range
Oak, sand blackjack	Quercus catesbaei	4.0 - 5.0
Oak, scarlet	Quercus coccinea	6.0 - 6,5
Oak, shingle	Quercus imbricaria	4.0 - 5.0
Oak, shrub	Quercus ilicifolia	5.0 - 6.0
Oak, southern red	Quercus falcata	4.0 - 5.0
Oak, swamp white	Quercus bicolor	6.5 - 7.5
Oak, white	Quercus alba	6.5 - 7.5
Oak, willow	Quercus phellos	4.0 - 6.5
Oak, yellow	Quercus muhlenbergii	4.0 - 5.0
Peach	Prunus persica	6.5 - 7.5
Pear	Pyrus sp.	6.5 - 7.5
Pine	Pinus, many sp.	4.0 - 6.5
Planetree	Platanus, named sp.	6.5 - 7.5
Plum, American	Prunus americana	6.5 - 7.5
Plum, beach	Prunus maritima	6.5 - 7.5
Plum, common	Prunus domestica	6.5 - 7.5
Plum, purple-leaf	Prunus cerasifera pissardi	6.5 - 7.5
Poplar	Populus, named sp.	6.5 - 7.5
Quince, flowering	Chaenomeles japonica	6.5 - 7.5
Redbud	Cercis sp.	6.5 - 7.5
Red cedar	Juniperus virginiana and var.	6.0 - 6.5
Savin	Juniperus sabina and var.	6.0 - 6.5
Serviceberry	Amelanchier sp.	6.0 - 6.5
Silverbell	Halesia tetraptera	4.0 - 6.0
Sorrel-tree	Oxydendrum arboreum	4.0 - 6.0
Spruce	Picea sp.	4.0 - 6.5
Spruce, Colorado	Picea pungens and var.	6.0 - 6.5
Stewartia	Stewartia malacodendron	4.0 - 6.0
Styrax	Styrax americana	4.0 - 6.0
Sweetgum	Liquidambar styraciflua	6.0 - 6.5
Tree-of-heaven	Ailanthus altissima	6.5 - 7.5
Tuliptree	Liriodendron tulipifera	6.0 - 6.5
Tupelo	Nyssa sylvatica	5.0 - 6.0
Walnut	Juglans, many sp.	6.5 - 7.5
White cedar	Chamaecyparis thyoides	4.0 - 5.5
Willow	Salix, many sp.	6.5 - 7.5
Willow, creeping	Salix repens	4.0 - 5.0
Yew	Taxus sp.	6.0 - 6.5

Soil Temperature. Well-established trees, as well as those recently transplanted, are adversely affected by extremes in soil temperature. Low soil temperature retards or even prevents water and mineral absorption by the roots. If the air temperature is high enough to allow loss of water from the leaves and twigs while the soil temperature is low, serious injury may result. The loss of more moisture than can be replaced through the roots often results in drying and

Fig. 3-4. Poor soil anchorage due to rain-saturated soils contributed to the demise of this wind-thrown tree. Root rot disease also contributes to poor anchorage by tree roots.

browning of both narrow- and broad-leaved evergreens (see Winter Drying of Evergreens, Chapter 10).

Because soil temperature normally drops slowly with the approach of fall and winter, heat can be conserved for a long period by early fall mulching. This practice prevents deep freezing of the soil and reduces the amount of winter drying of small trees. The mulching materials should be removed early the following spring; otherwise they will screen out the warm rays of the sun.

The Soil as an Anchor

Embedding themselves in the soil, the roots are able to hold the tree erect. Trees tend to develop anchorage where it is most needed. The tendency of isolated trees is to develop anchorage rather equally all around, with perhaps a slightly stronger development on the side toward the prevailing strong winds. The better protected a tree is from wind, the less it depends on secure anchorage and the fewer roots it seems to provide for such anchorage.

The factors that are important to a tree from the standpoint of anchorage are the depth and the texture of the soil, the relative abundance of such mechanical allies as stones, and the roots of other trees and shrubs. Sufficient room for the roots to spread horizontally must also be available.

Trees growing in wet soils are more easily blown over during windstorms than are trees growing in well-aerated soils (Fig. 3-4). One reason is that the lubricating action of the wet soil particles lowers the trees' resistance. Another is the absence of deep anchor roots, which are unable to develop because of lack of air in the deeper layers. In wet soils most of the tree roots are confined to the upper 12 inches; in well-aerated soils the roots may penetrate much deeper.

Urban Soils

Soils in urban areas are usually very different from rural soils. Disturbances caused by human activities may result in a high degree of variability (Fig. 3-5). Thus a tree planted at one site may be in a very different soil environment than one planted only a few yards away due, for example, to the presence of construction debris at one site and not another. Modification of the soil structure can lead to compaction and restricted aeration and water drainage. Materials unfavorable or even toxic to trees and to soil microorganisms may be present in such soils. These and other negative features associated with urban soils should be considered when selecting sites for tree planting. An awareness that such factors can cause poor growth or death of trees will also aid in the diagnosis of tree problems.

On a practical level, arborists and urban foresters should be aware that the local USDA soils maps may not correctly describe the soils in a specific planting site. Soil pH levels can easily be measured on-site to be sure that soil pH levels are not extreme in either direction and are suitable for the trees to be planted.

Many soils being developed for landscape use are very compacted. Relative soil compaction can be determined using a 2-foot-long soil probe, or soil coring tube. With experience, soils can be probed to determine where impervious layers or hardpans exist. Soil compaction is dependent on the bulk density (g/cm^3) and the texture of the soil. Both can be determined by a soil testing laboratory. The soil must drain well. A hole dug in the planting site and filled with water should empty in a few hours—otherwise impervious layers will need to be broken up by tillage tools such as a chisel plow that is ripped through the soil, or tiles or drainage tubes will need to be installed. In some circumstances, soil soluble salts and laboratory soil nutrient analysis will be needed.

Soil Improvement Prior to Planting

Probably the best way to improve the physical condition of the soil before trees are planted is to incorporate large amounts of organic matter (Fig. 3-6). Such material not only improves the structure of the soil but also increases its water-holding and ion-exchange capacity. Disappointing or unprofitable results may ensue, however, if the organic matter is added to infertile, very wet, or extremely acid soils. Such adverse conditions must be rectified before much benefit is derived from the organic matter. Wet soils, for instance, may be improved by the insertion of drain tiles to draw off the excess water, thus increasing the aeration. Shallow soils similarly will respond better to organic matter incorporations if any underlying hardpan or impervious subsoil is first broken up by the use of special digging tools or by blasting with dynamite. Incorporating ground limestone into the subsoil also markedly improves the physical condition of the soil, especially where the subsoil is very acid.

Fig. 3-5. Urban soils often include construction debris, which is unsuitable for tree root growth and development.

Of the many sources of organic matter, those most commonly used for soil improvement are well-rotted stable manure and two distinct types of peat, peat moss and sedge peat.

Peat moss, also known as sphagnum peat or peat, is very acid in reaction (pH 3.0–4.5) and is used principally for acid-loving plants, such as rhododendrons and azaleas, hemlocks, pines, and spruces. It is fibrous in nature, light brown in color, low in nitrogen content (around 1 percent on a dry basis), and has a high water-absorbing capacity. It should never be applied in the form in which it is purchased, but must be thoroughly soaked beforehand, sometimes up to several days. When large quantities of this type of peat are used, some nitrogen in the form of nitrate salts should be applied the following summer to supply both the trees and the bacteria that decompose the peat. This practice will compensate for

Fig. 3-6. Effect of peat moss on root development of dogwood. Left: Tree grown in soil without the addition of peat moss. Notice the few scraggly roots. Right: Tree grown in soil into which peat moss has been added. Notice the heavy mass of fibrous roots. Organic soil amendments may enhance air movement into and root penetration through the soil.

the leaching due to the frequent watering necessary for transplanted trees, and for the consumption by bacteria.

Sedge peat, also known as reed peat, muck, or humus, may be substituted for peat moss. Large deposits of this type of peat are available for exploitation. As offered for sale, sedge peat is dark brown to black and relatively high in nitrogen (2.0–3.5 percent). It is less acid in reaction (pH 4.5–6.8) and has a lower water-absorbing capacity than peat moss. It can be safely used around most trees.

Composted yard wastes, especially those containing high levels of chopped woody plant wastes such as prunings, are increasingly popular as soil amend-

Fig. 3-7. Sidewalk trees on a downtown street in Lexington, Kentucky. The tree in the foreground is much too large for the tree pit provided. It is likely that the tree roots have broken out from the pit and are exploiting suitable soil nearby. Growth of the adjacent tree appears to be limited by soil volume.

ments. Depending on the mix of woody and green (e.g., grass clippings) materi-al that goes into the compost, these wastes can be low or moderately rich in nitro-gen, and possibly neutral to high in pH. Composted yard wastes are popular because they reduce municipal landfill costs and their supply is consistent, though somewhat variable.

Soil Volume

Tree plantings in many urban areas are limited by soil volume. Tree planting pits set in paved landscapes, above-ground containers, and narrow treelawns between sidewalk and curb must contain an adequate volume of well-drained, good qual-ity soil for the tree to perform well (Fig. 3-7). A survey of tree planting pits was made in several eastern U.S. cities, and the researcher concluded that where tree planting sites are surrounded by pavement, a soil volume of at least 100 cubic feet was necessary to sustain long-term tree growth. That means that if tree roots in a pit can exploit soil to a depth of 2 feet, the area of the pit should be 5 by 10 feet. Where the standard tree pit opening is about 4 feet by 4 feet, considerable root-ing space is needed under the pavement.

There are many designs for tree planting pits, and trees can grow well in those that are well designed and well maintained. Far too often, trees are planted in a poorly designed planting pit site only to suffer poor growth for several years and then removal. A well-designed tree pit, tree-growing container, or in some cases a treelawn surrounded by pavement should provide for the following:

- An adequate soil volume, whether exposed to the surface or hidden under the surrounding pavement. Soil should not be more than 3 feet deep.
- A uniform loamy soil mix.
- Good drainage so that the soil does not become saturated following rains.
- A system to provide air to the rooting soil. Perforated ventilation tubes are use-ful in tree planting pits, particularly if they allow "flow through" air movement in contrast to a vertical perforated tube closed at the bottom.
- Irrigation, either done manually on a routine maintenance schedule or auto-matically, if needed.
- Pavement sloping away from the planting pit so that salts and other wastes do not accumulate in the pit.

An alternative to individual tree planting pits or containers is the construction of tree planting beds where clusters of trees and other plants can be grown (Fig. 3-8). For example, if a design calls for a dozen trees in a plaza, why not designate two larger planting beds with room for 6 trees each rather than 12 individual tree pits. This provides physically more space for tree roots than the smaller tree pit. In any case, the existence of trees in tree pits or containers will require much more maintenance (e.g., regular watering, pruning, protection from traffic, etc.) than trees grown in an open setting such as a park.

Engineered, Load-bearing Soils

In order to get better performance from trees in planting pits, tree roots must be able to exploit additional space—ideally, the space under pavements. With coarse,

Fig. 3-8. A group of urban trees growing in a bed providing a large soil volume for root growth.

angular gravel, one can create a compacted pavement bed suitable for a sidewalk or roadway, yet with large pore space (40 percent by volume) suitable for tree root penetration. By mixing some soil with the coarse gravel, a medium suitable for dual use—tree root exploitation and paving support—results. If one uses clay loam soil, at the rate of 15–20 percent by weight of the coarse gravel (0.75–1.5-inch size, with no dust or fine gravel mixed in), plus hydrogel (30 grams/kg of stone) and water to help the soil stick to the stone, one can create a structural soil suitable for both tree roots and pavement support. This works because the angular stones, while resting one upon the other, provide all the structural support needed while allowing plenty of pore space. As long as that space is not entirely filled with soil, the base will not become severely compacted and tree roots will penetrate the medium. During installation, this mixture resembles muddy gravel.

The muddy gravel mix described above, developed at Cornell University, was shown in experiments to have potential benefits to trees. The research confirmed that the specially engineered base allowed roots to grow outside the planting pit and penetrate deeply throughout the base beneath sidewalk slabs. These trees performed as well as trees not growing in the pits. In contrast, trees growing in pits surrounded by standard sidewalk-bearing compacted bases did not perform as well, and their roots tended to grow between the sidewalk slab and the compacted base. This growth pattern often leads to future problems with sidewalk lifting.

In Amsterdam, The Netherlands, a similar approach is being used to incorporate tree rooting space into the area surrounding tree planting pits. Amsterdam tree soil is an engineered soil of medium coarse sand with 5 percent medium organic matter and 4 percent clay. It can be compacted slightly—enough to support brick or concrete pavers suitable for pedestrian or bicycle traffic—yet with air- and water-holding and draining characteristics suitable for tree root growth. In over 20 years of use, pavers are not being lifted by tree roots where this soil mix is used under them.

Thus, it is possible to integrate pavements and trees into the same design, rather than just digging tree pits and surrounding them with standard tightly compacted pavement base in urban areas. This approach may also be useful for expanding the rooting space of trees in typical suburban tree lawns.

Soil Improvement for Established Trees

Trees in heavily used landscapes such as parks, playgrounds, and even home yards frequently suffer from the effects of soil compaction. Compacted soils are poorly aerated and are difficult to water and fertilize. Trees in compacted soils gradually show symptoms of poor vigor, dieback, and decline.

Fig. 3-9. A power auger can be used for vertical mulching.

Fig. 3-10. Grow gun, used to loosen compacted soils.

Many arborists use a technique called vertical mulching to alleviate the effects of soil compaction and restore tree vigor. Vertical mulching involves drilling holes in the soil throughout the tree root zone and backfilling with porous soil amendments. Holes, 18 inches deep, are made in the soil of the tree root zone using a power drill having a 2- to 4-inch auger (Fig. 3-9). These holes are spaced 2 to 3 feet apart in a grid pattern where feeder roots grow both inside and outside the extent of the tree canopy. The holes are filled with sand, perlite, or other porous material. Organic amendments such as peat moss or wood chips may be used, but they eventually break down and can become compressed. Fertilizer can also be added to the holes, if it is needed.

Vertical mulching should provide more air to the soil's lower depths and should permit greater penetration of water and mineral nutrients into the soil. Arborists using the technique report improvement in tree health, but little scientific research documents the benefits of this procedure. Some root injury is likely to occur when an auger is used throughout the root zone. However, the benefits of improved air, water, and mineral nutrient movement should also be realized.

Compacted soils may also be loosened by aerification of the soil. With a machine such as the Grow Gun (Fig. 3-10) in which a water-assisted probe is inserted into the soil, a sudden burst of pressurized air plus soil amendment is blown into the cracks and crevices created by the blast. The premise is that these newly formed cracks and crevices will provide unimpeded channels for air, water, and mineral element movement and root development, thus improving the health of the tree. Again, although the research is limited, some preliminary results suggest this approach is useful.

Radial trenching is another approach to providing a better environment for root growth. In this procedure, a series of narrow trenches is dug from the trunk out to at least the dripline in all directions. The trench is refilled with amended soil, which provides all the benefits of an improved rooting medium. This approach is still relatively new but may show some benefit.

Profile, a plant growth regulator product containing paclobutrazole, is labeled for use in stimulation of tree roots and has been shown experimentally to increase root density when injected into the soil. When applied to transplanted trees, leaf area and twig growth increased the first season. In a long-term study, Profile soil injection is being tested for use in improving the rooting status of declining mature trees. Similarly, application of mycorrhizae to transplants and established trees in the landscape is used to stimulate development of better functioning roots (See Chapter 2).

Selected Bibliography

Anonymous. 1957. Soil: The 1957 yearbook of agriculture. U.S. Superintendent of Documents, Washington, D.C. 784 pp.

Craul, P. J. 1985. A description of urban soils and their desired characteristics. J. Arboric. 11: 330–39.

Hole, F. D., and J. B. Campbell. 1985. Soil landscape analysis. Rowman and Allanheld, Totowa, N.J. 196 pp.

Kozlowski, T. T. 1985. Soil aeration, flooding and tree growth. J. Arboric. 11:85–88.

Thompson, L. M., and F. R. Troeh. 1978. Soils and soil fertility. McGraw-Hill, New York. 516 pp.

Tisdale, S. L, W. L. Nelson, and J. D. Beaton. 1985. Soil fertility and fertilizers. Macmillan, New York. 754 pp.

Urban, J. 1989. New techniques in urban tree planting. J. Arboric. 15:281–84.

4

Factors Important for Tree Selection

Tree species vary greatly in their ability to grow in different environments. The selection of an appropriate tree for a specific site requires consideration of the characteristics of the site, the adaptation of tree species to these characteristics, the function of the tree in the landscape, and its availability.

Characteristics such as soil type, drainage, exposure, elevation, climate, air pollution, and surrounding vegetation are important considerations in determining which tree species can be successfully grown at a specific site. Although horticultural selections may vary, some tree species are better adapted to growing, for example, under wet conditions, in very acid or alkaline soils, in the presence of air pollutants or salt spray, or in shade. Some species are extremely sensitive to frost, whereas others require chilling to break dormancy. Many species are hardy in a broad range of climatic zones. However, growers should be aware that biotypes of many species have, through the process of natural selection, arisen in specific localities.

The presence of endemic insect or disease problems may limit the success of some species in some areas. Chestnut blight, Dutch elm disease, western spruce budworm, and bronze birch borer have drastically reduced the planting of the affected species over large regions.

Plant ecologists refer to the ability of a plant to grow within a certain set of conditions as its fitness. In natural plant communities, the species present are those that are most fit for the circumstances. For example, in some parts of the Eastern North American forest, oak and hickory predominate in shallow, relatively dry soils of ridge crests or southwest slopes; beech, linden, and maple thrive in deeper and more moist soils present in valleys and northeast slopes; and in moist alluvial soils, sweetgum, bald cypress, pin oak, pecan, and swamp chestnut oak may be present.

One of the most important considerations in tree selection is cold-hardiness. A system of designating cold-hardiness zones was developed by the U.S. Department of Agriculture (Fig. 4-1). These zones are defined by average minimum (winter) annual temperatures and trees grow in particular zones because of their inherent hardiness to winter cold. In North America, zones 1 (cold, below –50° F) through

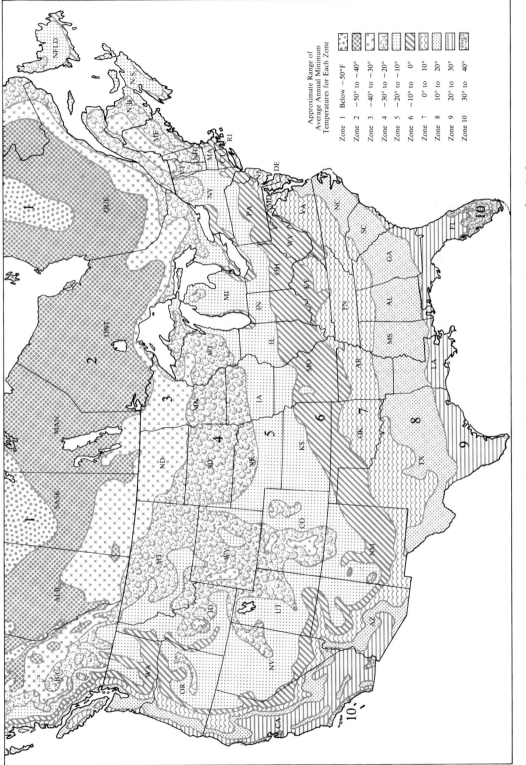

Fig. 4-1. Zones of plant hardiness. (From U.S. National Arboretum. Plant Hardiness zone map. U.S Department of Agriculture. Miscellaneous Publication No. 814.)

Approximate Range of
Average Annual Minimum
Temperatures for Each Zone

Zone 1   Below −50°F
Zone 2   −50° to −40°
Zone 3   −40° to −30°
Zone 4   −30° to −20°
Zone 5   −20° to −10°
Zone 6   −10° to 0°
Zone 7   0° to 10°
Zone 8   10° to 20°
Zone 9   20° to 30°
Zone 10 30° to 40°

10 (mild, 30–40° F) are defined roughly from north to south (with coastal regions milder, and higher elevations colder) in 10-degree increments. In the Eastern North American deciduous forest, flowering dogwood (*Cornus florida*) is not hardy in zone 4, where minimum temperatures typically range from –30 to –20° F, but thrives in zone 6 (–10 to 0° F) and in warmer locations. This system is not perfect, because for some species an inability to tolerate sudden changes in temperature such as a cold period following a warm period can cause more winter injury than the same temperature following a cool period. Furthermore, the cold-hardiness zone concept deals with average minimum temperatures and not with the extremes that really test a plant's hardiness. Topographical features may make an area warmer or colder than the general climate zone. Landscape designs involving aboveground planters and berms may expose plants to harsher than natural conditions. Nevertheless, these hardiness zones are helpful in designating the kinds of trees that will grow in a geographic region and where exotic trees introduced from other continents will survive. Lists of adapted trees for these zones are available from nurseries and books on tree selection.

Similar hardiness zone maps showing maximum summer temperatures as well as typical summer warmth and winter chilling data are being developed. Thus, a nursery might be able to offer information on cold or hot extremes as well as minimal summer warmth or winter chilling needed by a tree to thrive. Those who plant trees will be able to select species best adapted to their local climatology.

By observing the selection of species that are already established and thriving in a given area, one can get a rough idea of which trees are likely to be successful. Care must be taken to make observations at sites that are similar in as many aspects as possible to the intended planting location. Species native to an area may usually be relied on to grow well, require less maintenance, and have few insect and disease problems. However, many well-adapted exotics also are available (Fig. 4-2). Advice from cooperative extension agents, local arborists, or nurseries can help prevent costly mistakes.

Special situations existing at a site may also influence tree selection. If the tree is to be located in confined planting pits or near sidewalks or underground utility conduits, root distribution must be considered. Willow, poplar, and silver maple are said to have particularly invasive roots where leaky sewer pipes are concerned. Trees planted near parking areas should not produce dripping exudates or messy fruit, and trees requiring a lot of maintenance should not be used where maintenance is not affordable. Plantings under utility lines should be selected to minimize interference with these lines; low- or slow-growing species are preferred. Advanced planning is necessary to select trees that will be compatible not only with the environment of the site but also with its functions.

Consideration of the function of the tree in the landscape will further narrow the list of candidate trees. Functions include providing shade, wind break, noise barrier, visual screen, or fruit. The function may be strictly aesthetic: trees may be selected for their desirable growth habit such as crown shape, size, and density of branches or foliage or for their color of flower, foliage, or bark (Fig. 4-3).

Because the selection of suitable trees depends on so many factors and because

Fig. 4-2. The cork tree, *Phellodendron lavallei,* from eastern Asia is an excellent tree for streetside or lawn planting. Female cork trees produce fruit that litters the street, so only male trees are recommended.

such a variety of trees is available (some suitable only for specific regions), we do not attempt to discuss or recommend specific species. The bibliography at the end of this chapter includes a wide range of approaches to tree selection.

There are, however, some trees that have characteristics that make them less desirable, especially for street trees. These are listed below, with the reasons for so classifying them:

Boxelder (*Acer negundo*): a weedy tree attractive to insects.

Silver maple (*Acer saccharinum*): soft and brittle wood highly susceptible to storm damage.

Yellow buckeye (*Aesculus octandra*): loses foliage early, becoming unsightly.

Tree-of-heaven (*Ailanthus altissima*): brittle wood; male flowers have offensive odor.

Silk-tree, or Mimosa (*Albizia julibrissin*): short-lived, brittle wood; subject to insects and diseases.

Southern catalpa (*Catalpa bignonioides*): flowers and seed pods produce unsightly litter; subject to caterpillar attack.

Western catalpa (*Catalpa speciosa*): seed pods produce unsightly litter; short-lived weak wood.

Persimmon (*Diospyros virginiana*): fruits produce unsightly litter.

Russian olive (*Eleagnus angustifolia*): susceptible to canker diseases.

Honey locust (*Gleditsia triacanthos*): objectionable thorns and seed pods (thornless, seedless clones good).

Black walnut (*Juglans nigra*): nut husks stain sidewalks and shoes; roots secrete substance toxic to rhododendrons, tomatoes, and other species.

Osage-orange (*Maclura pomifera*): large useless fruits that interfere with lawn mowing.

Fruiting mulberries (*Morus* sp.): fruits produce unsightly litter that attracts insects.

Empress-tree (*Paulownia tomentosa*): unsightly seed pods; coarse leaves.

Most species of poplar (*Populus* sp.): weak wood; susceptible to diseases.

Black cherry (*Prunus serotina*): fruits produce unsightly litter; subject to tent caterpillars.

Black locust (*Robinia pseudoacacia*): subject to borers.

Willow (*Salix* sp.): Some are messy, weak-wooded trees.

Siberian elm (*Ulmus pumila*): weak-wooded; susceptible to leaf beetles.

In cold, dry climates, some of the trees listed here, such as Siberian elm, *Prunus* and *Populus* species, are desired because they at least survive the harsh environment. Others, such as tree-of-heaven, silver maple, poplars, and mulberry, are well adapted to air pollutants, drought, low humidity, heat, wind, and other rigors of the inner city and industrial areas. They will thrive with little to no maintenance and might be considered for use in vacant lots, industrial areas, and various bits of "leftover" land. Professor Russell A. Beatty of the Department of Landscape Architecture at the University of California, Berkeley, says their lack of aesthetic appeal is balanced by their ability to "bring a measure of amenity to otherwise ugly and unhealthy landscapes."

Problems Peculiar to Streetside Trees

Since trees are more difficult to grow in the unnatural environment of cities, successful culture of streetside trees involves many problems not usually encountered in forest and country plantings.

It is particularly difficult for trees in large cities to do their best. City conditions are so unfavorable for most trees that it is a wonder they grow at all. Smoke and other air pollutants, mechanical injuries, a disrupted water table, highly compacted soil, lack of organic matter, limited root space, reflected heat from buildings, visitations by dogs, lack of water, use of salt on sidewalks to melt ice—these are but a few of the external forces that make the lot of a city tree so difficult. Add to these invasions by insects and infection of the roots by soil-inhabiting parasitic fungi, and one is almost ready to forgo planting trees in cities. Despite these hazards and handicaps, however, millions of trees do grow in large metropolitan areas, including Brooklyn.

In cities, the runoff of rainfall is nearly 100 percent. Hence the small open soil area between the street curb and the sidewalk, where most city trees are planted, has little chance to absorb rainwater.

Another important point to bear in mind is that soil that may have had adequate aeration at the time a tree was planted may become compacted and may deteriorate until the aeration is entirely inadequate. Trees growing along city

Fig. 4-3. A superior cone-shaped form of ornamental pear (*Pyrus calleryana*) is the cultivar 'Chanticleer'.

streets offer excellent proof of this. When the tree is first planted in the area between the street and the sidewalk, soil conditions in the tree lawn may be satisfactory. As the roots extend into the areas actually covered by the street and sidewalk, conditions become less favorable. In addition, trampling of the open soil area by pedestrians compacts the soil and makes it as nearly impervious to air and water as concrete. Street trees are more likely to decline from lack of soil air, water, and fertility than are lawn trees, because the soil mass in which their roots can develop is more restricted than it is for lawn trees. As discussed in Chapter 3, this problem can be largely overcome by planting city trees in beds (Fig. 4-4).

Three factors largely govern success in planting trees along public rights-of-way: selection of the proper species, good growing conditions, and reasonable aftercare and protection.

Considerations in Choosing Trees for City Streets

Tall-growing trees such as the London planetree, the ginkgo, and the honey locust are known to tolerate city conditions better than trees like the sugar maple. The explanation is often that the foliage of the first three is more tolerant of air pollutants. The authors are of the opinion that some deciduous trees are more tolerant because their roots can survive under the extremely poor soil conditions prevalent in most cities; that is, they are better able to tolerate excessively dry soils, poor soil aeration, high concentrations of soluble salts, and other adverse below-ground factors.

Tall- Versus Low-Growing Trees. When P. P. Pirone wrote his first book on trees over 50 years ago, the consensus among shade-tree commissions, park department officials, arboricultural firms, and other agencies empowered to regulate the selection, planting, and care of trees on public property was that tall-growing trees such as elms, oaks, planetrees, and maples were best suited for planting along city streets (Fig. 4-5).

Times have changed and so have ideas on the proper kinds of trees to plant. Many officials now realize that it is best to select trees that will give the least trou-

Fig. 4-4. Sidewalk trees in a bed on a downtown street in Lexington, Kentucky.

Fig. 4-5. Sixty-year-old pin oaks (*Quercus palustris*) line a street in Lexington, Kentucky. Note the wide tree-lawns.

ble and require the lowest maintenance costs in future years. While many of the tall-growing or standard shade and ornamental trees have long been favorites on wide city streets, they are rapidly losing their popularity because of high maintenance costs. Such trees are more expensive to spray and prune, and during severe wind, ice, electrical, and snow storms or hurricanes they cause great damage to power and telephone wires. Moreover, tall-growing trees have a greater tendency to crumble curbs or push them out of line, crack and raise sidewalks (Fig. 4-6), and clog sewers.

Tall-growing trees should be planted where they have plenty of room to grow and where they will not interfere with public and private utility services and solar collection devices. The modern city planner realizes that trees should be fitted into the available space.

Tall-growing trees such as elms, pin oaks, and planetrees were popular in the past for several reasons. They are easy to reproduce either from seeds or from cuttings, grow rapidly, transplant readily, and require relatively little aftercare. Certain low-growing trees, on the other hand, may require more skill to grow and maintain. The initial cost of such trees of streetside planting size may be greater than that of the standard tall-growing types because of the high cost of their production, as well as supply and demand factors. Overall maintenance and, when necessary, removal costs are much lower, however, than those for tall-growing trees. Low-growing trees can be used under electric utility lines, resulting in decreased right-of-way maintenance costs (Fig. 4-7).

Proponents of the tall-growing trees claim that low-growing trees have several serious drawbacks: their relative scarcity as compared with tall-growing kinds;

Fig. 4-6. Tree roots damaging a sidewalk.

Fig. 4-7. Many flowering crabapple cultivars can be grown under utility wires where they would not interfere with the wires or require excessive pruning.

their low-hanging branches, which interfere with vehicular and pedestrian traf-fic; and their inability to provide as much shade as tall-growing kinds, especially for cars parked along city streets. As more nurseries grow the low-growing forms, the first drawback is being overcome. The second problem can be overcome if low-growing trees are selected and pruned properly with a central leader or, if grafted, the scion is placed on the rootstock 6 to 7 feet above the ground level. The drawback of low-hanging branches might also be overcome where communities require the planting of trees inside the property line instead of between the street and sidewalk. The third drawback is not solved easily, but where more shade is desired, close planting of low-growing trees is suggested.

The ideal small-type tree should grow rapidly; it should reach a useful size soon after being transplanted to its permanent site and should then abruptly slow down. It should be reasonably free from insect pests and diseases and should have pleasing growth habits.

Persons entrusted with the selection and care of streetside trees should adopt programs for proper selection and maintenance. Some of the more desirable low-growing trees need more frequent spraying than the older tall-growing kinds; if available, disease- and insect-resistant types should be selected to avoid this problem.

Size and Spacing of Trees. The best size for street planting varies from $1^1/4$ to $1^1/2$ inches in caliper at the base. Larger trees are needed in areas where vandal-ism is possible, but they are more expensive and become established more slowly.

Where large-growing trees are used along city streets, a spacing of 40 to 60 feet is a typical distance that allows for fairly open growth. This might require more pruning of lower branches. Use of closer spacings of around 25 feet is more expensive but could encourage a more desirable upright growth habit, resulting in less need for pruning and more rapid shade development. Narrow streets are best planted in ginkgo, European linden, or some of the upright cultivars of maple, oak, elm, and hornbeam.

Location. Street trees are customarily placed between the walk and the street curb. Planting a tall-growing tree in this area is bound, in time, to conflict with overhead wires and with the sidewalks and curbs. This area can be properly plant-ed with the kind of tree that will not lead to such difficulties. The decision on where to locate the tree depends on site factors.

Many subdivisions are being developed with underground electrical utilities so that overhead wire interference is not a problem. Potential sidewalk and curb damage by tree roots can be avoided if they are properly constructed. Proper con-struction of the concrete slab and the support base under the slab (discussed in Chapter 3) may allow for root penetration without sidewalk lifting. Mechanical root barriers, which force the roots to grow deeper into the soil before they grow in the direction of the walk, have been tried. These root deflectors can be installed at planting or fit into an existing tree lawn following root pruning, if needed. The barriers have vertically oriented ribs or embedded herbicides that deflect tree roots downward. Most research done with root barriers shows that the roots just dive under and grow back up close to the surface on the other side.

Where tall-growing trees are used, planting street trees in the area between the

Fig. 4-8. Sometimes the ideal location for street trees is not along the curb but back in the front lawn. Trees here are less subject to vehicular injuries and have more favorable growing conditions. However, pedestrians are more vulnerable to motor vehicles when trees are placed away from the street.

walk and the residence is more satisfactory (Fig. 4-8). This, of course, is feasible only where the house is set back some distance from the walk. The advantages of such an arrangement are many. It removes trees from the area where they are continually subject to injury by street traffic, the trees have more favorable soil conditions, they are less likely to tangle with overhead wires, the crowns do not form over the street area, and the street looks wider. Moreover, the tree will be damaged less severely if the street is widened at a later date. A disadvantage of locating trees in the residential yard, especially if it is on private property, is that the jurisdiction over such trees is taken away from the Shade-Tree Commissioner or Department of Parks. In Massachusetts, municipally managed trees can be planted as far as 20 feet within private property.

However, a good case can be made that it is best to plant trees between the sidewalk and curb if only because a row of tree trunks provides a barrier between pedestrians and vehicular traffic. Even if the barrier is only psychological, strollers, joggers, and parents of children playing on the walk have an increased sense of security from busy street traffic when trees are planted in the "treelawn" space between sidewalk and street.

Selecting Good Quality Trees to Plant

It is important for those responsible for selecting trees for city streets, parks, and private residences to purchase trees from a reliable source. Reliable nurseries will be sure that the tree species and cultivars they sell are accurately named and labeled; are derived from locally or regionally adapted seed source; are grafted properly, carefully matching scion and rootstock to avoid future problems; and are selected for genetically high vigor. Reliable nurseries do not sell trees with girdling roots or root deformities and soil, if it is included, much different from the soil in the new well-prepared site. Periodic root pruning of trees in the nursery adds expense but should result in a denser root system and better survival. Insect, disease, and weed control are essential, and the fertilization programs used to encourage rapid early growth should have been reduced in the years before digging. Reliable nurseries dig the trees with root balls and use harvesting and handling techniques that meet industry standards, as discussed in the next chapter. In addition, trees from a reliable nursery are structurally sound, having been pruned for good branch spacing and attachment. For good survival in the landscape, the tree buyer must pay attention to nursery production details as well as to transplanting techniques.

Selected Bibliography

Anonymous. 1973. Trees for polluted air. U.S.D.A. Forest Service, Misc. Publ. 1230, Washington, D.C. 12 pp.

Anonymous. 1974. Trees for New Jersey streets. New Jersey Federation of Shade Tree Commissions, New Brunswick, N.J. 42 pp.

Anonymous. 1977. Trees for New York City. New York Department of City Planning, New York Botanical Garden, New York. 27 pp.

Baker, R. L., J. E. Kissida, Jr., W. Gould, and D. G. Pitt. 1979. Trees in the landscape. University of Maryland Bull. 183. 16 pp.

Daniels, R. 1975. Street trees. Pennsylvania State University. University Park. 47 pp.

Dirr, M. A. 1990. Manual of woody landscape plants. Stipes Publishing, Champaign, Ill. 1007 pp.

Ferguson, B. (ed.). 1982. All about trees. Ortho Books, San Francisco. 112 pp.

Gilman, E. F. 1997. Trees for urban and suburban landscapes. Delmar Publishers, Albany, N.Y. 662 pp.

Grey, G. W., and F. J. Deneke. 1986. Urban forestry. 2nd ed. Wiley, New York. 299 pp.

Harris, R. W. 1992. Arboriculture: Integrated management of landscape trees, shrubs and vines. 2nd ed. Prentice-Hall, Englewood Cliffs, N.J. 674 pp.

Hudak, J. 1980. Trees for every purpose. McGraw-Hill, New York. 229 pp.

Koller, G. L., and M. A. Dirr. 1979. Arnoldia 39 (3). 237 pp. (entire issue).

Krussman, G. 1984. Manual of cultivated broad-leaved trees and shrubs. Timber Press, Beaverton, Ore. (3 vols.).

Littlefield, L. 1975. Woody plants for landscape planting in Maine. University of Maine Bull. 506. 58 pp.

May, C. 1972. Shade trees for the home. U.S.D.A. Agriculture Handbook 425, Washington, D.C. 48 pp.

Nelson, Eileen (ed). 1980. Source book for shade tree management. Cornell University Cooperative Extension Service, Ithaca, NY. 50 pp.

Perry, B. 1981. Trees and shrubs for dry California landscapes. Land Design Publications, San Dimas, Calif. 184 pp.

Rehder, A. 1940. Manual of cultivated trees and shrubs hardy in North America, exclusive of subtropical and warmer temperate regions. Macmillan, New York. 996 pp.

Reisch, K. W., P. C. Kozel, and G. A. Weinstein. 1975. Woody ornamentals for the midwest. Kendall/Hunt Publications, Dubuque, Iowa. 293 pp.

Whitcomb, C. E. 1976. Know it and grow it: A guide to the identification and use of landscape plants in the southern states. Whitcomb, Tulsa, Okla. 500 pp.

Wray P. H., and C. W. Mige. 1985. Species adapted for street-tree environments in Iowa. J. Arboric. 11:249–52.

5

Transplanting Trees

Transplanting is the moving of a tree from one location to another in such a manner that it will continue to grow. Success, however, is not determined by whether the tree merely survives but rather by its ability to resume growth and development with the least interruption. Transplanting includes moving both field-grown and container-grown trees.

Importance of Good Transplanting

Although the time it takes to transplant a tree is only a fraction of its anticipated life span, the future health, beauty, and utility of the tree can be greatly influenced by the methods used in digging, transporting, and planting. No matter how carefully transplanting is accomplished, root disturbance is inevitable. Many roots are killed in the process of digging and moving field-grown trees. Transplant shock—the retardation of growth and development occurring after transplanting—is caused largely by the physiological stress placed on the tree by the destruction of a large portion of its root system and consequent reduction in uptake of water and minerals. For normal growth and development to resume, new roots must be generated to replace those damaged during transplanting. For container-grown trees, new roots must grow out into the surrounding soil if the tree is to become successfully established. Thus careful handling will minimize physiological stress, whereas faulty transplanting can cause considerable retardation in growth or even death of the tree.

Transplantability

Some trees can be moved more successfully than others. The transplantability of an individual tree is governed by a number of factors, including inherent characteristics of its species or clone, its size and health, and its former habitat. As a rule, trees with compact, fibrous root systems are more successfully moved than those with large taproots or scraggly root systems. Deciduous trees are more easily transplanted than evergreens and small trees are easier to move than large trees.

Experience indicates that some species require more care during and after moving to ensure survival. Table 5-1 lists a number of commonly grown species

Table 5-1. Relative transplantability of some common trees.

High	Medium	Low
Alder	American holly	American hornbeam
American elm	Apple, Crabapple	Beech
American linden	Buckeye	Birch
Callery peach	Bur oak	Dogwood
Callery pear	Black oak	Hawthorn
Catalpa	Bald cypress	Hickory
Common pear	Cherry	Japanese pagoda-tree
Cork tree	Chestnut oak	Magnolia
Eastern red cedar	English oak	Pawpaw
Empress-tree	Ginkgo	Pecan
Fringe-tree	Golden-chain	Pin cherry
Fir	Golden larch	Sassafras
Green ash	Hemlock	Scarlet oak
Hackberry	Katsura-tree	Sourgum, Tupelo
Honey locust	Kentucky coffee tree	Swamp oak
Hop-hornbeam	Larch	Sweetgum
Jack pine	Redbud	Tuliptree
Locust	Red maple	Walnut
Mountain ash	Red oak	White oak
Osage-orange	Shingle oak	Willow oak
Pin oak	Shumard oak	
Planetree	Spruce	
Poplar	Willow	
Red pine	Yellow chestnut oak	
Silver maple	Yellowwood	
Sugar maple		
Sumac		
Tree-of-heaven		
Yew		

This table is an adaptation from published lists, reflecting the likelihood of successful transplanting of various tree species.

according to their success in being transplanted. Those in the high category require normal care and are most likely to survive a move in suboptimal conditions. Those in the medium category require normal care and are less likely to survive suboptimal conditions. Those in the low category require great care and should not be moved in suboptimal conditions. Individual experience may differ because of different circumstances surrounding the transplant, so the list should be taken only as a guide.

Large trees can be moved less successfully than small trees for obvious logistical reasons and some not so obvious reasons. Modern equipment enables the lifting and transporting of large trees, but such trees often suffer from prolonged periods of transplant shock. Although under standard nursery practices the root ball of transplanted trees is proportional to the crown size, larger trees lose a much greater mass of roots. Thus trees over 4 inches in diameter often do not grow or grow at a very slow rate for several years after transplanting, while smaller trees may surpass them in size in a few years. For example, if trees 2 inches and 6 inches in diameter are transplanted at the same time using standard nursery practices, and horizontal spread of roots resumes at an average 18 inches per year, then the smaller tree would recover all of the root system lost in transplanting in $2^1/_2$ years, whereas it would take the larger tree 7 years. It is generally good practice to transplant the smallest acceptable tree.

Healthy trees with good reserves of stored carbohydrates are better candidates for transplanting than are spindly trees. Stout, well-formed twig growth is indicative of good levels of stored carbohydrates. The quick resumption of root growth, which is essential to overcome transplant shock, requires reserve energy. Also, healthy trees are less likely to harbor pests and diseases, which may be introduced into previously uncontaminated areas on unhealthy stock.

The site where the candidate tree has grown influences its root development and hence its transplantability. Root systems develop over a greater area and to greater depths in sandy, well-drained soils. Thus more roots will be damaged when transplanting from a sandy site. Generally, greater success is expected when plants are moved from heavy to light soils than the reverse. Nursery-grown trees usually have better root systems for transplanting than trees grown without cultivation.

Container-grown trees usually suffer less transplant shock than field-grown trees. However, container-grown trees with girdling roots or with disproportionately large tops are poor candidates for transplanting.

Season for Transplanting

By using sufficient care and properly prepared plants, most trees can be transplanted in any season. However, the time of year influences the likelihood of success. Most arborists agree that deciduous trees can be moved in the fall, in the winter before the soil freezes, or in the spring before growth begins. Fall planting has a number of advantages. In the fall the cells of most woody plants are lignified and less subject to stress due to lack of moisture. Fall-planted trees will survive as long as the trees are dormant and soil temperature is favorable (usually above 40° F) for at least a month after transplanting. Research results suggest that for many species, fall-planted trees have the advantage over spring-planted trees in number of new roots, stem diameter, and height. However, a dry fall followed by a cold winter may result in the death of a high percentage of fall-transplanted trees.

Some deciduous trees survive best if they are planted in the spring. Trees with fleshy root systems, such as dogwood, magnolia, willow oak, tuliptree, and yellowwood, are more successfully moved in the spring. Thin-barked trees such as birch should be transplanted in the spring or should have protective trunk wraps if they are moved in the fall. Other deciduous trees that are best moved in the spring are beech, white fringe-tree, ginkgo, goldenrain-tree, hickory, American hornbeam, red and white oaks, pecan, persimmon, prunus, sassafras, sourwood, sweetgum, tupelo, walnut, and willow.

With care, conifers can be moved at any season of the year. Best success, however, follows their transplanting during August and September when the soil is warm (60–70° F at 6–12 in. soil depth). Some nurserymen feel they should never

Table 5-2. Recommended minimum diameter and depth of soil balls for deciduous trees of various sizes

Tree Trunk Diameter in Inches*	Diameter of Ball (in.)	Depth of Ball (in.)**
¹/₂–³/₄	14	11
³/₄–1	16	12
1–1¹/₄	18	13.5
1¹/₄–1¹/₂	20	14
1¹/₂–1³/₄	22	15
1³/₄–2	24	16
2–2¹/₂	28	19
2¹/₂–3	32	20
3–3¹/₂	38	23
3¹/₂–4	42	26
4–4¹/₂	48	29
4¹/₂–5	54	31
5–5¹/₂	57	32
5¹/₂–6	60	33
over 6	diameter in inches times 10	proportional

* Diameter of a single-stemmed shade tree measured at 6 inches above ground for trees under 4 inches; 1 foot above ground for larger trees.
**Depth is measured from the depth of the uppermost main roots coming off the buttress roots, if present. Scrape away excess topsoil to expose buttress roots before digging.

be moved after October 1. Exceptions include cypress, firs, hemlock, larches and Austrian pine, which are best moved in the spring after soil temperatures have begun to rise.

Broad-leaved evergreens are most successfully transplanted in the spring. They should be moved in the fall only if they can be protected through the winter.

Circumstances may dictate a need for midsummer transplanting. With intensive effort and care, almost any tree can be moved in summer. When a tree is dug in midsummer, the soil in the ball of earth should be near its maximum water-holding capacity. This can be assured by trenching and watering the tree the day before the actual digging. After digging, the tree should be lifted at once and taken to a sheltered area. Make the ball 2 to 3 inches larger than normal (see Table 5-2). To reduce transpiration, trim foliage by pruning weak, injured, or broken branches or use an antitranspirant. Plant the tree as soon as possible and water well. If air temperature is high or humidity low, top sprinkle several times each day for two weeks.

Transplanting can proceed through the winter in mild climates as long as the soil does not become too muddy for operation of the equipment. Where the soil freezes, digging may proceed, but the frozen ball must be protected to prevent root damage. The soil will lose moisture even when frozen and roots will be killed by desiccation. Japanese cherry, tuliptree, magnolia, and willow oak should not be transplanted in frozen balls. Most oaks, elm, linden, sugar maple, and beech requiring 6- to 14-inch balls can be successfully moved in frozen soil.

Although balled and burlapped and container-grown trees can be transplanted in any suitable season, bare-root trees are best planted in the spring. As buds swell in the spring, trees begin developing roots. If trees have already leafed out, the stimulus to root development is lost. Thus trees should be planted before the buds break.

Preparing Trees for Transplanting

Nursery-Grown Trees. Properly grown nursery trees are root-pruned every few years, not only to confine the greater part of the root system within a small area, but also to encourage the roots to branch and develop a greater abundance of small roots closer to the trunk. Trees that have been root-pruned can be transplanted more efficiently and with a greater chance of success.

Established or Wilding Trees. Sometimes established trees are moved within or removed from a landscape to be used elsewhere. In addition, many trees growing naturally in fields or woods without previous cultivation are transplanted to new sites to provide shade or ornament. Small trees need little special preparation, but large trees should be root-pruned to stimulate the formation of a compact root system. Best results are obtained when such trees are root-pruned, as shown in Figure 5-1, for two successive years before they are moved. In the spring or fall a circular area around the tree at some distance from the trunk (approximately 5 inches for each inch of trunk diameter) is marked off. Three ditches, covering half the circumference, are dug with a spade, and all the exposed roots are severed. A year later, the remaining area is trenched. The soil excavated each time should be mixed with manure, leaf mold, or a commercial fertilizer before it is replaced in the open trenches. The tree is ready for moving one year after the second series of trenches is dug and refilled. Trenching and root-pruning in this manner encourage greater fine root development within the trench area. Also, the shock to the tree is spread over a 3-year period.

Antitranspirants. Loss of water from evergreen trees or deciduous trees in leaf can be partly overcome by spraying with an antitranspirant just before the tree is moved. Trees sprayed with these materials must still be watered in their new location, but they may become established more quickly than untreated trees. Many professional arborists and nurserymen use antitranspirants whenever they move trees in the summer or whenever they move large trees in leaf.

Antitranspirants form a film of wax, plastic, or resin on the leaves, blocking stomatal pores and coating leaf cells with a waterproof film. This protection from water loss may last for a few weeks. These materials have not been found to be effective under all circumstances. If used improperly, antitranspirants can be toxic to leaves, particularly evergreens under some conditions, or they may block normal transpiration and movement into the leaf of carbon dioxide needed for photosynthesis. If they are to be used, they should first be applied on a trial basis to part of the canopy to determine whether toxicity occurs. In that way, extensive damage to a valuable plant can be avoided.

Fig. 5-1. Method of root-pruning an established or wilding tree before it is moved. The soil is dug from sectors A, exposed roots are pruned, and the soil is replaced after it has been amended with organic material. One year later the same procedure is followed in sectors B. The tree should be moved a year after the pruning at sectors B. Upper Left: Sketch shows the extent of roots before pruning. Upper right: After root-pruning.

Digging Trees

Proper digging includes the conservation of as much of the root system as possible, particularly of the smaller and finer roots. Any appreciable reduction in such roots, which are the roots most active in absorbing water and nutrients, results in a considerable retardation of the tree's growth and reduces the speed of reestablishment.

Trees may be moved with bare roots or with a soil ball. The former method leads to greater transplant shock and slower reestablishment; however, handling bare-root trees is often simpler and less costly. Bare-root planting is mainly used for deciduous trees up to 2 inches in diameter planted in the fall or spring. Larger deciduous trees, evergreens, and trees moved in the winter or summer should be dug with a soil ball adhering to their roots.

Before any digging begins it is advisable to tie the tree's branches with heavy twine or quarter-inch rope. This will reduce the chances of damage or breakage of the branches.

Bare Root. Transplanting trees with bare roots is usually limited to small (less than 2-in. diameter) deciduous trees. A trench is dug around the tree to a depth just below the greater part of the root system. With most small trees this depth is 15 to 18 inches. The trench should be dug just beyond the last root-pruning zone for well-grown nursery trees or at a distance from the trunk of 12 inches for each inch of trunk diameter. The roots are then combed free of soil by the use of a spading fork.

As the roots are exposed, they should be covered with wet burlap to prevent drying and mechanical injury. The soil close to the trunk should be left intact to hold the tree erect until the lateral roots have been uncovered. Then the tree may be tipped over gradually and the soil removed from beneath the trunk. Taproots should be cut at a depth of 15 to 18 inches.

The tree should be gripped by the trunk at the root crown and lifted from the soil. After the tree has been lifted, care must be taken to prevent root death. Roots should be covered with wet burlap, damp straw, or loose soil. Research has shown that lateral roots are easily damaged under conditions of high temperature and low humidity, even if the drying period is only 30 minutes. However, damage can be reduced if dried roots are soaked for several hours before planting. Trees should be stored in a sheltered place out of direct sun and wind.

An alternative method, pulling the tree from moist soil using a pin and clevis inserted through the trunk, has been tested for use with small trees. This method works only in light soil and is very injurious to the trunk. A high degree of top and root damage is to be expected with this method.

Ball and Burlap. Whenever possible, deciduous trees over 2 inches in diameter, evergreens, and small trees in leaf should be moved with soil adhering to the roots. The diameter and depth of the soil ball depend largely on the type of soil, the root habit, the type of tree, and the size of the tree. Although small trees with shallow roots would seem to require a flat ball, a flat "pancake" ball is not stable and could break easily, so a mostly spherical ball is used. Larger trees are often moved with a ball that is broader than it is deep, but the balls are made deep enough to not break easily. The main objective governing the size of the earth ball is the protection of the greatest number of roots in the smallest soil mass.

The soil ball for deciduous trees is usually 9 to 12 inches in diameter for each inch of trunk diameter. Table 5-2 gives the general guidelines. As a rule, evergreens require a ball of smaller diameter.

Since most soils weigh about 110 pounds per cubic foot, no more soil than is absolutely necessary should be allowed to remain around the ball. A rapid method for computing the weight of a soil ball is to square the diameter of the ball in feet, multiply this figure by the depth in feet, then take two-thirds of this total and multiply by 110. The resulting figure will give the approximate weight of the soil in pounds.

Digging should not be attempted unless the soil is moist. If the soil is dry, it should be watered thoroughly at least 2 days before digging. Unless this is done the soil ball will break and the roots will be completely separated from the soil. Such a tree will have much less chance of surviving in its new location.

The first step in digging of the soil ball is the cutting of a circle around the tree 6 to 8 inches farther away from the trunk than the intended soil ball. Trench down to approximately three-fourths of the final ball depth on the outside of the initial circle. This will allow space for shaping the ball. It is best to work with the back of the spade toward the tree to avoid loosening the soil ball. Large roots should be cut cleanly with hand shears. Moist burlap should be used to cover exposed roots during the digging operation.

Even when mechanical trenchers are used, final shaping of the soil ball is done by hand. Beginning at the top of the soil ball, excess soil that may have accumulated near the trunk during cultivation in the nursery is carefully shaved off to reveal the buttress roots. If the tree is small (less than 2-in. diameter), a sharp spade can be used to undercut the soil ball and sever any remaining roots. Undercutting larger trees is accomplished by running a small steel cable beneath the burlap at the base of the ball. Both ends of the cable are attached to a hook on another cable running to a winch. The winch will pull the cable to sever the roots and detach the soil ball from the ground. Keep the ends of the cable low in the hole to get a flat cut. After cutting, secure loose burlap to the underside of the ball with nail pins.

The soil ball may then be lifted from the hole and wrapped with burlap. There is a great temptation to grab the top of a tree with a small soil ball and yank it from the hole. Many soil balls are jarred loose and root systems damaged by this practice. Small balls should either be levered out of the hole using a spade or lifted out using a burlap sling. After lifting, the ball should be placed in the middle of a burlap square and opposite corners should be tied across the top of the ball. Loose sections should be pinned snugly using sixpenny or eightpenny balling nails to make a neat package. If the soil is loose or sandy, the ball should be reinforced with rope or twine.

For larger trees, burlap strips are pinned around the sides of the soil ball before the tree is lifted. The strips should overlap the top of the ball by several inches and be sufficiently wide to cover the bottom of the ball after it is undercut. Again, the burlap should be pleated and pinned to make a neat package. Rope lacing further reinforces the ball (Fig. 5-2). When the soil is loose or the ball is large, wire baskets should be used to cover the burlap before the rope is laced on or sometimes as a substitute for the rope (Fig. 5-3).

Fig. 5-3. Balled and burlapped deciduous tree. Wire basket is held by lacing twine. Paper trunk wrap is for protection during shipment and removed at transplanting.

Fig. 5-2. Balled and burlapped evergreen tree ready for transplanting.

Large soil balls will require mechanical assistance for lifting. A chain sling attached to a front-end loader or crane works best under most circumstances. If ropes or chains are secured to the base of the tree, pad it well. Some arborists lift the tree using one or two heavy steel pins inserted through the base of the trunk. This method is very risky and often results in trunk and ball damage.

After lifting, the trees should be protected against excessive drying. Move the trees to a sheltered area, place them close together, sprinkle, and cover the soil with wet mulch or plastic. If deciduous trees are dug in full leaf, they should be held 24 to 48 hours before replanting to acclimate them to root loss and thus reduce transplant shock.

Instead of burlap, wooden boxes may be built in the hole surrounding the soil ball. This method is used where soil is very sandy or when trees are to be held for an extended time before planting (Fig. 5-4).

Frozen Soil Ball. Where the soil freezes to depths of 1 foot or more, trees can be moved with frozen root balls. Such balls are less likely to be damaged in handling and thus require less wrapping. The tree and the planting site should be mulched before the soil freezes. The tree is dug in the conventional manner, but it should not be undercut until just before the move. The ball can easily crack, so care must be taken during handling. The planting hole should be dug only shortly before the move to prevent the inside of the hole from freezing. After moving, the tree should be watered only enough to settle the soil around the ball, and mulch should be applied to prevent further freezing. Move trees only when the temperature exceeds 23° F.

Fig. 5-4. In sandy soils, a wooden box may be used to contain roots and soil during transplanting.

Mechanical Tree Digging. Mechanical spades are available that are designed to dig and lift trees with a ball of soil surrounding their roots. In some cases the tree spade may also be used to transport the tree, eliminating the necessity of burlap covering. However, when several trees are to be dug before transporting, the tree spade is used to set the trees into burlap squares for conventional wrapping. Trees larger than the machine can adequately handle should not be moved this way. Workers must be specially trained in the maintenance and operation of the equipment to ensure its safe and efficient use.

The tree spade encircles the tree, and four pointed blades are hydraulically forced into the soil (Fig. 5-5). The blades form the soil ball container when they meet below the tree. The soil ball is longer than on conventionally dug trees, although the diameter should be the same for both (Table 5-2). Most of the roots are cut off cleanly. Occasionally a taproot is caught between the blade tips at the bottom, or a large lateral root is wedged between two blades. The soil ball can be badly damaged and roots mangled when this happens. When roots are shredded and split rather than cut cleanly, they will die back several inches from the cut end. This death intensifies transplant shock and delays the establishment of the tree. Thus it is vital that the blades be kept sharp.

Fig. 5-5. A mechanical tree spade digging a maple tree in a nursery. In the planting site, a ball of soil would have already been removed by the tree spade to provide a same-sized planting hole.

When properly done, mechanical tree digging can be more successful than bare-root and comparable to ball and burlap transplanting. Tests conducted by the Forestry Division of the city of Lansing, Michigan, revealed that the mortality rate for streetside trees planted bare root was 28 percent, whereas only 1 percent of the trees planted with a tree spade died.

In-Field Container-Grown Trees. Containers set in the ground allow trees to be lifted from their growing site with much less effort than conventional digging methods. The trees are grown in wire baskets lined with filter fabric and filled with field soil. Although some roots penetrate the fabric bag, they are restricted in size. Most of the roots are confined to the container. Trees are mechanically lifted using a device that grips both the trunk and the wire basket. Trees up to 10 feet tall have been moved by this method. Trees can also be grown in rigid plastic containers that are nested inside a slightly larger container already set in the soil. Lifting these "pot-in-pot" trees for transplanting is a relatively easy matter.

Transporting Trees

While small trees are transported very easily, large trees usually require special types of transportation equipment. When trees are to be transported relatively short distances, a sled with wooden runners can be used. For longer distances, a number of well-designed and efficient tree-moving machines are available. These are used mainly by commercial arborists, park superintendents, and other people who move large trees. The machines are mounted either on trucks or tractor trailers (Fig. 5-6).

Fig. 5-6. Loading a large multistemmed Japanese maple, 22 feet high and 24 feet wide, with a $9^{1}/_{2}$-foot root ball, at a nursery.

The limit to the size of tree that can be moved is governed by the amount of road clearance over which the tree is to be transported and the weight the roads and bridges will stand.

Considerable injury may occur during the moving of large trees unless the following precautions are taken:

1. Pad unprotected areas with burlap, canvas, or some other material to avoid bruising and slipping the bark of the trunk and branches.
2. Carefully tie in all loose ends with a soft rope to avoid twig and branch breakage.
3. Keep the earth ball moist or cover bare roots with wet burlap or moist sphagnum moss.
4. Avoid excessive drying of the tops, especially of evergreens, while the trees are in transit by covering them with canvas or wet burlap or spraying them with an antitranspirant.
5. Protect the root system and top from drying out if trees are to be stored for some time before planting at the new site.

Planting Trees

Preparation of the Hole. The care taken in the preparation of the new site and the hole is a major factor ensuring the ultimate success in transplanting. The site was discussed in Chapter 3. Three factors should be considered in preparing the hole: size and depth, soil quality, and drainage.

For bare-root trees, the hole must be sufficiently deep and wide to accommodate the full root system without bending or cramping the roots. For balled and burlapped or container-grown trees, the hole should be a foot or several feet wider, especially at the top of the hole, than the soil ball to allow backfill to be worked in around the ball and to provide aerated soil for the new roots. In well-drained soils the bottom of the hole should be just deep enough to place the top of the ball (base of the root flares) at the soil surface. Digging deeper would put the tree at risk from settling of the backfill.

Trees planted in heavy or poorly drained soils will not thrive unless some precautions are taken prior to planting. The drainage capacity of a site can be determined by drilling a hole 2 feet deep and filling it with water. If the hole does not drain appreciably within 30 minutes, steps must be taken to improve drainage.

Improving soil drainage is usually beneficial; however, efforts to improve drainage may be expensive and may not always ensure success. Agricultural drainage tiles or perforated flexible plastic tubing may be placed in the bottom of the planting hole (Fig. 5-7) to drain away excess water. If clay fill-soil is encountered at the bottom of the planting hole, drainage holes could be drilled through the clay layer and filled with pea gravel. It is tempting to place a layer of gravel in the bottom of the hole in attempt to improve drainage. This could worsen the problem because the backfill soil and soil in the root ball are of a finer texture than the gravel. Excess water will actually be retained by the root ball and backfill soil and not drain into the gravel layer in such cases. If the drainage problem is not too severe or is localized, the tree could be planted above grade within a soil mound.

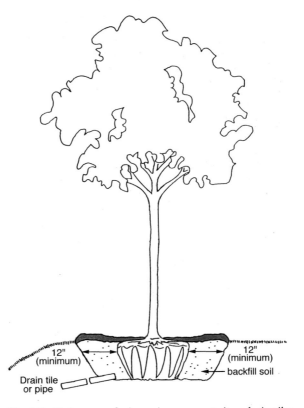

Fig. 5-7. To improve drainage in some wet sites, drain tiles are placed in the planting hole, The base of the tree trunk with root flare should be just at or above ground level.

Where it is not physically or economically possible to correct drainage, the best alternative is to plant trees that will tolerate saturated soils, such as bald cypress, sweetgum, willow, or tupelo. Avoid planting evergreens, especially yew and broad-leaved types.

When the mechanical tree spade is used to dig the hole before transplanting a tree with the spade, the sides of the hole and the face of the soil ball may become glazed. Soil glazing may retard root penetration and development into surrounding soil. However, the glazed surface in the hole can be roughened by use of hand tools to avoid the problem.

Setting the Tree. Before placing the tree in the hole, determine how the tree was oriented in the nursery. If it is known, place the transplant so that the south side is still facing south to preserve its original sun, wind, and light relationships. The depth of planting must be close to the original so that the soil-stained ring at the base of the trunk is level with the surrounding soil surface. Be sure that the soil-stained ring is not the result of excess soil thrown against the trunk during nursery operations. On larger trees, dig into the top of the root ball for evidence of flaring buttress roots and plant at the level of exposed buttress roots. Experienced arborists have found that if a tree is set a bit higher in its new location than in its former site, the chances of continued good growth are increased because the tree may settle some after watering. Trees set too deeply will not thrive. Absorptive roots require oxygen to function and thus will die if placed at a soil depth where oxygen levels are too low. Flowering dogwood, American beech, oaks, and narrow-leaved evergreens are especially sensitive to planting that is too deep.

Because the roots of trees grow downward at a slight angle from the horizontal, it is wise to make a cone-shaped mound of soil at the bottom of the hole when planting bare-root trees (Fig. 5-8). Before planting, diseased, kinked, and broken roots should be pruned. The root crown can then be set on this mound and the roots spread over and down the sides, thus assuring close contact with the soil along their entire length.

Balled trees need to be set into the hole at the proper depth (Fig. 5-9). Then the rope holding the burlap should be removed and the burlap loosened. The burlap need not be removed from the bottom of the ball; indeed removal may damage the ball and the roots. It should be rolled back from the top, slit along the sides, and pulled back away from the sides of the root ball. Plastic wrapping material must be completely removed. Wire baskets, essential to lowering the tree into the hole, are no longer needed, and wire strands should be cut with a bolt cutter as far down the sides of the ball as possible.

A

B

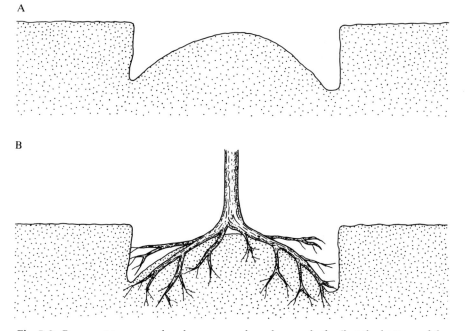

Fig. 5-8. Bare-root trees are placed on a cone-shaped mound of soil at the bottom of the planting hole. This allows closer contact of the roots with soil and thus reduces air pockets.

Conventional container-grown trees should be tapped out of their containers carefully. Root systems of trees grown in containers have been forced to conform to the shape of the container (Fig. 5-10 A). This encircling outer layer of roots should be cut in several places (Fig. 5-10 B), or removed to encourage new roots to grow out into the surrounding soil. If not removed, such roots may eventually girdle the tree.

The wire basket and filter cloth sack must be removed from in-field container-grown trees prior to planting. Prune any upper roots that show circular growth habits. No other root pruning is necessary.

Any wires that hold labels should be loosened or removed so that they do not girdle the trunk or branches as they increase in girth.

Hormone Treatments. In some tests, hormones applied as sprays, root dips, and soaks have been shown to stimulate regeneration from medium and large roots and to result in significantly greater new leaf growth. Results were most favorable when fall-dug trees were treated in the spring. Attempts to stimulate roots with exogenously applied hormones have had mixed results, but the root

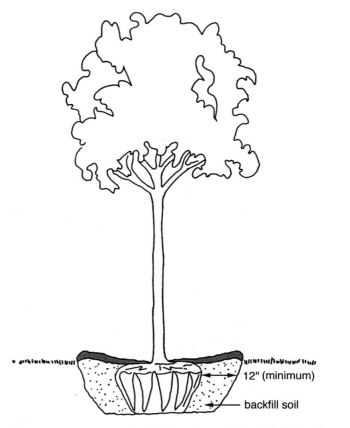

12" (minimum)

backfill soil

Fig. 5-9. In well-drained soils, the planting hole for balled and burlapped trees should be just deep enough to place the top of the soil ball at the soil surface.

Fig. 5-10. A, encircling roots of a container-grown tree; B, encircling roots severed before planting.

spray method may be more practical than others attempted. Further testing is needed to determine if these approaches are cost-effective.

Backfilling. Traditionally it was recommended that the soil removed from the hole be improved by amending with organic or coarse mineral materials such as peat, pine bark, sand, and vermiculite. However, most research shows that back-filling with unamended soil results in better tree growth. Where the backfilled soil is improved and the landscape soil is poorly drained, the roots may have too much water in wet periods and too little in dry periods. It is important, however, to improve aeration of unamended soil before using it as backfill. Break up clods and compacted soil to provide small pores without large air pockets.

Once the tree has been set in the hole at the proper depth and orientation, the soil should be added gradually. For bare-root trees, work the first lot in firmly at the base of the roots. Then add more soil and work it under and around the lower roots. The tree may be gently raised and lowered during the filling process to eliminate air pockets and bring the roots in close contact with the soil. When the roots are just covered, tamp the entire area firmly with the feet. Heavy soils should not be tamped too firmly around the roots, since this will prevent oxygen and water from reaching the roots. Add water, let the soil settle, and finish filling the hole with loose soil.

For trees transplanted with a soil ball, work the soil beneath the ball to leave no air pockets. Add soil until the hole is about half filled, tamp firmly, then water to settle the soil and finish filling with loose soil.

Fig. 5-11. Heavy-duty tree protection in Paris, France.

Postplanting Operations

Supporting. Newly planted trees may need artificial support to prevent excessive swaying in the wind, to promote upright growth, or to guard against mechanical damage. However, staking is not necessary for many trees and can have undesirable effects. Staked trees develop a smaller root system and a decreased trunk taper. They are often injured by ties that may rub against or girdle the trunk or limbs. Their tops offer more wind resistance than tops that are free to bend. The decision to stake a tree should not be automatic but rather should be made after consideration of the strength of the trunk, wind conditions, and traffic patterns. Proper staking techniques can lessen or prevent problems.

Consider support for newly planted trees having a large canopy in relation to the root ball, a wet planting site, or a windy planting site. Even then, larger trees may only need to be supported for a few months while roots become established. Trees grown without support and with good spacing in the nursery should not require support after transplanting. In some urban environments, semipermanent support and trunk protection may be needed to protect trees from vehicles and pedestrians (Fig. 5-11).

Trees with strong trunks may need stakes only to prevent mower damage and to provide anchorage until the roots grow into the surrounding soil. Two or three stakes 3 to 4 feet long are driven 1 foot into the ground (Fig. 5-12). The stakes should be placed about a foot from the trunk. Ties consisting of fabric bands should be fastened to allow some movement of the trunk.

Trees with weak trunks may need stakes for support. For trees up to 20 feet in height, one or two strong stakes or poles, 6 to 8 feet in length, can be driven 2 feet or so into the ground and 6 to 12 inches from the trunk. A wide fabric band, attached to the stake, is looped around the tree to support the tree (Fig. 5-13). Although it is common to see a short length of garden hose with a wire run through it placed around the trunk to attach the tree to the stake, the hose does not protect the bark well, and too often this device is left on until the tree is girdled.

A number of devices are commercially available to attach trees to supporting stakes. Some automatically adjust as the tree grows, eliminating the necessity of frequent changes of ties.

Guy wires or cables are often used to support larger trees. They should slope from about halfway up the trunk to the ground at an angle of about 45 degrees.

Fig. 5-12. Trees with strong trunks need stakes only to protect them from mower damage and to provide anchorage. The stakes should be fastened to allow some trunk movement.

The upper ends of wires are attached to fabric bands that encircle the trunk with a loop. The lower ends of the wires are anchored to stakes driven deeply into the soil. From one to four wires or cables are usually placed around a large tree. Guy wires or cables are rarely placed around trees set along city streets or in public parks. Unless protected or readily visible, especially at night, they are a constant source of danger to passers-by.

Pruning. Newly transplanted trees should not be pruned except to remove diseased or broken branches. It was once thought that the tops of transplanted trees should be pruned to compensate for the loss of roots and to maintain a balance between the two parts. However, recent studies indicate that this practice may not be beneficial in preventing transplant shock and may be detrimental by reducing the capacity for photosynthesis and for hormone production. High levels of auxin produced in the foliage stimulate cytokinin hormones needed for more rapid root formation. Apparently, the added moisture stress resulting from leaving the plant top intact is offset by more rapid development of the supporting root system from the additional carbohydrate and hormone produced. It is still necessary to conserve most of the roots during digging and to provide adequate water and aftercare. Additional pruning beyond broken, weak, diseased, or interfering branches should be delayed until after the first growing season. Pruning of the young tree for optimal branch spacing and tree balance can wait until after the transplant is established.

Little or no pruning of balled evergreens is needed except to remove broken or severely injured branches. If the leading shoot is broken during moving, one of two practices may be followed: (1) one of the next nearest lateral branches may

Fig. 5-13. A fabric band attached to the stake can be tied around the tree trunk for support.

be bent upward and held in place by tying to a small piece of wood; or (2) all the nearest laterals but one may be removed, so that the lone lateral will tend to grow upright and will soon replace the broken leader.

Watering. The soil around the tree must be watered thoroughly after the tree is set in place. Additional water must be applied from time to time until the tree is firmly established. It is most important that water be applied directly to the root ball area the first year, because that is where most of the tree roots are. The frequency of watering depends on the type of soil, the size of the tree, and the amount of rainfall. However, in general the soil should be wetted to a depth of 15 inches in well-drained soils at approximately weekly intervals, or less often for heavier soils. Deciduous trees moved during late fall or winter require very little water, at least until new leaves emerge the following spring.

The water must soak into the ground slowly to allow it to moisten down to the roots. The depth of water penetration can best be determined with a soil auger or by digging with a narrow spade. The tree benefits little from frequent light waterings that moisten the soil to a depth of only a few inches. Water may be effectively applied with a bucket with holes in the bottom set close to the trunk. In this way, water is supplied gradually. Refill the bucket two or three times to provide the tree roots a complete soaking, and then wait a week before watering again. Plastic irrigation bags are specially made to slowly release water on the root ball.

One common error of inexperienced gardeners is to overwater the soil around the newly transplanted tree. This drives out essential oxygen from the soil, and it also favors the development of root-decaying fungi. After the initial watering the soil around newly planted trees should not be heavily watered in spring until root growth starts and the leaves begin to emerge.

The soil ball around a large newly transplanted tree can dry out in a surprisingly short time. Summer showers cannot be depended upon to supply sufficient moisture to keep the soil ball moist. Hence an occasional good soaking with the garden hose or bucket is necessary. To get good penetration into the ball itself, it may be necessary to punch holes into the ball with a crowbar or to bore holes of varying depths with a soil auger before applying the water.

Frequent syringing of the leaves of newly planted evergreens on cloudy days helps to cut down water loss by the leaves and in some cases washes off spider mites and soot.

Surfactants or wetting agents may enable water to penetrate soil more quickly and more uniformly. Use of such materials may be of particular benefit in hard-to-wet soils.

Gel-forming polyacrylamide polymers (hydrogels) added to the planting soil are sometimes said to reduce the need for watering. Research data for this treatment are inconsistent, thus they are not recommended.

Mulching. Newly planted trees will perform better with mulching (Fig. 5-14). Mulches placed on the soil over the rooting zone provide several benefits. Mulch suppresses growth of turfgrass and weeds that compete with the tree for soil moisture and nutrients; elimination of vegetation near the tree thwarts potential injury to the trunk from lawn mowers and string trimmers. Mulches help conserve soil moisture and reduce soil erosion while lessening soil temperature

Fig. 5-14. Mulch placed around the base of a newly planted ash.

extremes. Some mulches may also increase soil fertility, improve soil structure, and reduce soil compaction. Although increased rooting may occur in the mulch layer, mulches are also thought to improve depth of rooting. Mulches not only benefit newly planted trees but also established trees.

The most often used mulches are organic materials such as shredded hardwood bark, wood chips, pine bark, cypress bark, organic compost, conifer needles and leaves. Other types of mulching materials include sand, gravel, rubber mats, plastic sheeting, and landscape fabrics. Fabric and plastic mulches are usually covered over with one of the other mulch materials. The choice of mulching materials depends on the desired appearance of the landscape and on what is locally available. Mulches should generally be applied 2–4 inches thick in as wide a circle as possible over the root system, but not directly against the trunk. Use a thinner organic mulch layer when underlying soil is poorly drained or when mulch materials are fine textured, and avoid placing mulch closer than 2–3 inches from the trunks of small trees and closer than 6 inches from the trunks of large trees.

Organic mulches are sometimes overused, being piled against the trunk and piled too high. Mulches applied against the trunk can lead to death of buried trunk phloem tissues due to lack of oxygen or excess heat of decomposition,

Fig. 5-15. Base of a young maple covered with wire screen wrapping to protect the bark from the sun.

increased risks of pathogenic microbes causing decay of moist trunk bark, and increased risk of chewing damage by rodents finding protection in the mulch. Roots may be damaged from excess mulch due to insufficient oxygen where soils beneath the mulch are poorly drained. In addition, decomposing mulch may tie up nitrogen and change the soil pH. Some sources of hardwood bark may increase soil pH, and pine needles may decrease soil pH. Depending on the source, mulches can be modified with the addition of nitrogen, lime, or sulfur to avoid some of the soil-related problems. Where pH changes are a concern, monitor the soil with periodic soil tests. Where nitrogen deficiency is a concern, observe plant growth to determine whether or not to add nitrogen to the mulch.

Fig. 5-16. Lawn mower guard, made from plastic drainage pipe. Tree guards must be removed before the tree trunk diameter grows to the size of the guard.

Wrapping. In most tree transplanting circumstances, tree trunk wrappings are not needed, yet they are often done. The trunk and larger branches of a transplanted tree may be wrapped with paper wrap fastened with twine, fine mesh wire screen, or some other material for the purpose of preventing sunscald and frost cracks (Fig. 5-15). In northern regions, trunk wraps are used on newly transplanted trees to prevent injury to the cambium on the southwest side due to the heat of the sun. A protective covering is especially desirable on trees collected from shaded woods, since the bark of such trees is thought to be more sensitive to sunscald. If wraps are used, they should be securely fastened and left on through one year and then removed. In most cases, they are not needed and can lead to problems.

Wrapping the trunk of a newly transplanted tree does not decrease potential borer insect damage, and in some cases has increased damage compared to trees not wrapped. Keeping the trunk wet beneath the wrapping could foster the development of fungus cankers. P. P. Pirone found that pin oaks are especially sensitive to such cankers.

To protect young dogwood, flowering crabapple, and other susceptible trees from rabbit and mouse feeding, wrap some hardware cloth or metal screening around the base of the tree in late fall.

Trees planted in lawns may need lawn mower guards to prevent basal injuries (Fig. 5-16).

Tree shelters. Tree shelters are tan or light green translucent tubes placed around tree seedlings planted in reforestation projects, or sometimes around the base of the trunk of newly transplanted trees (Figs. 5-17, 5-18). Widely used in cool, moist regions of Europe, tree shelters provide not only physical protection for seedlings from destructive rodents, but also a greenhouse-like effect inside them. Increased temperature, humidity and carbon dioxide often result in increased height growth, but reduced stem caliper. Tree shelters may be 3–5 feet high, thus providing benefits to seedlings for many years. If they are removed too soon, as they are apt to be in a landscape situation, trees may need additional staking. Tree shelters are probably best used to help tree seedlings become established for reforestation.

Fertilizing. As a rule, fertilizer that is applied at transplanting will not be used by the tree in the first season's growth, so high levels of nitrogen should not be applied until new roots are formed. Indeed, research has shown that high levels of added nitrogen have a negative effect on root growth, so may prolong transplant shock. One study found no benefit from nitrogen fertilization during the first three seasons of growth. Depending on soil conditions, applications may begin at the end of the first growing season. If fertilizer is to be used at transplanting, low levels are suggested, especially if the native backfill soils are low in nitrogen. Where plantings are inconvenient or expensive to fertilize regularly or are in sandy soils receiving heavy rainfall, slow-release fertilizers may be used. They should be placed in the hole and covered by 4 to 6 inches of soil before the tree is placed in position.

Fig. 5-17. The seedling tree inside this 5-foot tree shelter will be protected and have enhanced height growth for several years.

Fig. 5-18. Tree shelter used to protect the base of a newly transplanted tree near Runnymeade, England.

Selected Bibliography

Anonymous. 1997. American standard for nursery stock. ANSI Z60.1-1996, American National Standards Institute, New York. 57 pp.

Berdel, R., C. Whitcomb, and B. L. Appleton. 1983. Planting techniques for tree spade dug trees. J. Arboric. 9:282–84.

Gowin, F. R. 1983. Girdling by roots and ropes. J. Environ. Hortic. 1:50–52.

Shoup, S., R. Reavis, and C. E. Whitcomb. 1981. Effects of pruning and fertilizers on establishment of bareroot deciduous trees. J. Arboric. 7:155–57.

Watson, G. 1985. Tree size affects root regeneration and top growth after transplanting. J. Arboric. 11:37–40.

Watson, G. W., and E. B. Himelick. 1983. Root regeneration of shade trees following transplanting. J. Environ. Hortic. 1:52–54.

———. 1997. Principles and practice of planting trees and shrubs. International Society of Arboriculture, Champaign, Ill. 200 pp.

Whitcomb, C. E. 1979. Factors affecting the establishment of urban trees. J. Arboric. 5:217–19.

Whitcomb, C. E., R. Reiger, and M. Hanks. 1981. Growing trees in wire baskets. J. Arboric. 7:158–60.

6

Fertilizers and Their Use

Woodland trees thrive despite the absence of artificial fertilization, but many of our shade and ornamental trees do not grow in such a favorable environment. Under forest conditions, decayed leaves and dead plants replace mineral elements (referred to here as nutrient elements, or nutrients) taken up by living plants. However, in urban settings, leaves and lawn clippings are usually raked up and removed. The soils near buildings are often disturbed by construction activities, causing the loss of topsoil, compaction, and the presence of clay subsoil within the rooting zone. Turfgrass competes with trees for nutrients and water, and pavement frequently covers all or part of root systems. In woodlands, natural selection ensures that native trees can survive under local fertility conditions; however, people often attempt to grow species that are not adapted to the soil composition of their landscape.

Benefits of Fertilizer Application

The effects of undernourishment may be expressed as failure of a plant to grow as rapidly as expected or as a gradual weakening due to a shortage of an essential element. This drastic loss of vigor can weaken the tree so that it is less able to withstand biological and environmental pressures that can result in tree decline. In some cases, the deficiency of a specific nutrient will cause a reduction in the beauty and utility of trees and may cause symptoms that can be confused with infectious diseases. These special cases are discussed more fully in Chapter 10.

Although fertilizer application (or in some cases, a decision not to apply) does not constitute the whole of soil improvement, it is an important part of any successful tree maintenance program. The objectives of tree fertilization are fourfold:

1. To increase the size of small trees as rapidly as possible;
2. To maintain the healthy appearance and vigor of mature trees in poor soils;
3. To rescue weakened, nutrient-deprived trees; and
4. To cure specific nutrient deficiencies.

Species response is variable; however, nearly all trees respond to nitrogen fertilizer applications by making significant growth increases, especially when they are young. Fertilizer research trials generate mixed results because they are conducted for varying time periods with different species of different ages and in soils with different fertility levels. In one long-term test, fertilizer did not enhance the growth of crabapples and sugar maples. In another test, however, young trees receiving regular nitrogen fertilizer applications were taller, had greater canopy spread with larger and darker green leaves, and had greater root area than unfertilized trees. For example, treated 8-year-old tuliptrees produced four times as much shade as untreated ones. An increase in growth rate of 170 percent was observed in swamp oak, 61 percent for red maple, 30 percent for red oak, and 15 percent for linden. Generally, trees that have the slowest growth rates when untreated had the greatest relative response when fertilized.

But, are faster-growing trees healthier? With fertilizer stimulating top growth more than root growth, water demand is increased and unirrigated trees suffer drought stress sooner. Compared to trees growing moderately well, abundantly fertilized trees do not necessarily undertake more photosynthesis. Less carbohydrate manufacture coupled with increased drought stress could make trees even more vulnerable to pests. Trees slightly underfertilized, but not obviously nutrient deficient, will develop more extensive root systems to maintain their health.

Mature trees deficient in mineral elements may benefit from nutrient addition. Leaf size and color improve so the tree can gain more energy from photosynthesis, which should enhance the speed of wound closing and defenses against attack by wound-invading decay organisms. These responses require energy, and healthy trees are able to respond more efficiently than those that are chronically malnourished.

Occasionally, weakened or diseased trees can be rescued by timely fertilizer application. Although long-neglected trees can be an unpleasant sight in the landscape, careful pruning and fertilizing may be all that is needed to improve tree appearance. Before a declining large, sound, well-placed tree is removed from a landscape, the cause of decline needs to be determined, and if lack of fertility is involved, rescue may be attempted. There are two opposing viewpoints regarding the fertilization of declining trees. On the one hand, it is thought that nitrogen fertilizers primarily promote shoot growth at the expense of root growth and therefore may do more harm than good. On the other hand, without additional foliage to produce food and energy reserves, how can the tree be expected to grow additional roots? Fertilizing trees under stress apparently makes them more susceptible to borers and bark beetles, so fertilizing stressed trees could require greater pesticide applications.

Applications of high levels of nitrogen fertilizer have been used with success to "rescue" Verticillium wilt–infected maples and goldenrain-trees. However, fertilizer application is not a cure for trees affected by fungal leaf spot, heart or root rot, or bark or cambial diseases. There is a danger in overstimulation of tree growth with excess fertilization. Succulent shoot growth is more susceptible to fire blight and powdery mildews. The growth rate of certain insect pests such as eastern tent caterpillar may also be enhanced.

There are well-documented examples of declining trees' being deficient in a specific mineral element. Iron deficiency chlorosis affects many shade tree species, especially under high soil pH situations. Similarly, tip dieback and small crinkled leaves are associated with zinc deficiency, and chlorosis and decline have been associated with manganese deficiency. Applications of specific nutrient compounds may alleviate such conditions, although in some instances deficiencies may result from an improper balance of nutrients or from improper soil pH rather than lack of the specific nutrient in the soil.

Determining the Need for Fertilizer

The mineral element status is important to the overall health of trees. In the manual *Plant Health Care for Woody Ornamentals*, published by the International Society of Arboriculture, the issue of tree fertilization is addressed. The manual makes the case for examining carefully the need for fertilizers before using them on trees. Moderately nutrient-stressed trees may make better use of water, develop larger root systems, and contain more carbohydrates and defensive chemicals. Thus, in a low-maintenance landscape with soils of average fertility, greater stress tolerance and insect and disease resistance might be enhanced by not fertilizing.

Fertilizers are sometimes used to excess, causing unnecessary expense, occasional damage to sensitive plants, and potential water pollution. Before deciding whether to fertilize, arborists should determine what is to be accomplished by fertilization. In the landscape, a young sapling, a mature healthy tree, a weakened tree, and a chlorotic tree would each have different needs.

Look for Subnormal Growth. The most direct way to determine if fertilizers are needed is to observe the growth rate and leaf color of the trees. Young shade trees with twig growth of 9 to 12 inches per year and mature trees with growth of 6 to 8 inches per year will probably not be helped by additional nutrients. Leaf color should be typical of the species; watch for unusual yellowing, especially between the veins, which may indicate a specific nutritional deficiency. Note also whether new or old foliage is affected by the problem.

Use a Soil Test. Soil tests are also helpful in determining if fertilizer application is warranted. A soil test can be obtained for a nominal fee from most state land-grant university county extension offices or from various private laboratories. A typical agricultural soil test will show soil pH values as well as levels of potassium and phosphorus. Additional tests of soil samples can determine calcium, magnesium, and other element levels. Nitrogen amounts are normally not determined because soil nitrogen levels change relatively rapidly. The pH value is probably the most useful information obtained from soil tests.

Since different laboratories use different tests, which yield different nutrient values, interpretations of results, if not provided by the testing laboratory, may require the assistance of a knowledgeable expert. Precise soil sampling is important to accurate soil testing. Determine the extent of the root system of the tree(s) in question (usually the tree height or twice the distance of the branch spread) and take a small amount of soil from each of eight to ten representative areas in the root zone of the tree; combine and mix the samples and use a pint of the mixture for testing. Soil sampling tubes or augers can be used to obtain soil for testing. Collect samples at the depth at which the roots are growing. Although the roots are confined to the top 1 to 6 inches of wet or compacted soil, in most soils tree roots will be in the top 1 to 12 inches or deeper in some cases.

Analyze Leaf Tissues. Nutritional problems may also be diagnosed through

Table 6-1. Amounts of mineral elements typically found in leaves of normal and deficient woody plants.*

| | Proportion of Dry Leaf Tissue | | | |
| | Normal | | Deficient | |
Mineral	Percentage	Parts per Million	Percentage	Parts per Million
Nitrogen	2.0–4.0		1.5	
Potassium	0.75–2.5		0.3–0.6	
Calcium	0.7–2.5		0.2–0.5	
Magnesium	0.2–0.6		0.05–0.2	
Phosphorus	0.12–0.5		0.08–0.1	
Sulfur	0.2–0.5		0.12–0.14	
Chlorine	0.01	100		
Iron		50–400		33
Manganese		20–800		7–18
Zinc		15–100		<15
Boron		15–100		8
Copper		5–20		<5
Molybdenum		0.1–1.5		<0.1

*Values were derived from a variety of reference sources and are composite figures. Mineral levels for specific healthy and deficient landscape trees may or may not fall within these ranges.

analysis of tree foliage. Usually, recently matured leaves are taken from various parts of the tree with a suspected problem and from similar healthy trees nearby. These samples are then analyzed by an appropriate laboratory for mineral content.

Each laboratory has specific sampling procedures, but the following suggestions generally apply:

1. Use the mailers and information supplied by the laboratory.
2. Although collecting newly matured leaves of current season's growth is often suggested, some laboratories prefer mature leaves from the middle or base of the current season's growth.
3. Take samples from comparable healthy trees nearby.
4. Wash dust and debris from freshly collected leaves.
5. Do not collect diseased, injured, or dead leaves. Try to collect leaves as soon as possible after symptoms become visible.
6. Supply the laboratory with all the information requested and send partially air-dried specimens to the laboratory.
7. Collect soil samples for each tree at the same time leaves are collected for analysis.
8. Get professional help in interpreting the results of tissue and soil tests.

Typical levels of mineral nutrients found in healthy and deficient leaves of a range of woody plants are listed in Table 6-1. The range for healthy leaves can be considered a sufficiency range and values below and above the range may be deficient or in excess, depending on the species being examined. For trees growing in good soils at the correct pH, most foliar nutrients will be sufficient if the soil test shows them to be sufficient. Again, comparison of analyses of affected and healthy trees can be of much more value than comparison of an analysis to published average values. In addition, normal mineral element values for most tree species are not known.

It is important to remember that fertilizers will not solve problems caused by inadequate sunlight or water, air pollution, plant diseases, or insect attack.

Specific Nutrient Elements

Nitrogen is the element most frequently in short supply for optimum tree growth, although the other nutrients may be deficient under special circumstances. These are summarized in Table 6-2. The deficiency or sufficiency of nutrients in the soil is affected by the mobility of the nutrients and is strongly influenced by soil factors such as texture and pH. More nutrients become mobile as the soil becomes coarser and the pH decreases. Mobility also determines how nutrients should be applied for maximum effectiveness. Nutrients that move readily can be applied to the soil surface and watered in. Those that bind tightly to soil particles need to be incorporated into the soil or sprayed onto foliage. The following brief discussion describes the availability of the essential elements and how a deficiency of the element affects trees.

Nitrogen. This element is nearly always in short supply for maximum tree growth. Lack of nitrogen results in small yellow foliage and generally reduced growth. Nitrogen is readily lost by leaching, volatilization, or denitrification. Other plants, notably turf, compete with trees for available soil nitrogen.

Table 6-2. Soil conditions under which nutrient element deficiencies are likely to occur.

Soil	Unavailable Nutrient
High pH (alkaline)	Boron, calcium, copper, iron, manganese, phosphorus, zinc
Low pH (acid)	Boron, calcium, molybdenum, phosphorus, potassium
Coarse sandy	Boron, calcium, copper, magnesium, manganese, nitrogen, potassium, zinc
Leached (high rainfall)	Calcium, molybdenum, nitrogen
Wet (poorly drained)	Iron, manganese, nitrogen
Parent mineral material lacking	Phosphorus, calcium, boron
Serpentine (high in magnesium silicate)	Calcium, molybdenum
Organic	Copper, zinc, manganese
Low organic matter	Potassium
Sodic	Calcium
Calcareous	Copper, iron, manganese, zinc, magnesium, molybdenum

Phosphorus. Adequate levels (30 or more pounds per acre [lb/A] soil test) of phosphorus for tree growth are present in nearly all soils. However, when the soil pH is out of the range 5 to 7, phosphorus becomes increasingly unavailable to plants. Phosphorus, which is relatively immobile in soil, is involved in many essential plant processes.

Potassium. Most soils contain enough potassium (165+ lb/A soil test) for tree growth. Deficiencies of potassium occur where soils are acid, sandy, low in organic matter, and low in cation exchange capacity. Potassium is involved in the regulation of a number of essential plant processes. Its deficiency usually produces no distinctive symptoms other than a reduction in growth. Potassium is relatively immobile in soil.

Calcium. Levels of calcium adequate to support tree growth are found in most soils. Deficiencies occur in very acid soils, sandy soils in areas with high rainfall, serpentine soils, and alkaline or sodic soils. Calcium is important both in the cell structure of the tree and in the soil, where it affects pH and availability of other nutrients.

Magnesium. Most soils have adequate levels of magnesium to support tree growth. Magnesium may be leached out of sandy, acid soils; in calcareous soils it is tied up in insoluble forms that are not available for root uptake. Magnesium is a constituent of chlorophyll, and its deficiency results in yellowing of leaves. In most cases the most efficient method to increase the availability and retention of magnesium is to adjust soil pH and to increase the cation exchange capacity by adding organic matter.

Sulfur. Sulfur dioxide from air pollution provides most plants with adequate levels of sulfur. Sulfur is an essential constituent of enzymes and its deficiency results in a general retardation of physiological processes, often manifested as chlorosis.

Micronutrients. Most soils contain sufficient amounts of available trace elements to promote tree growth. Deficiencies, however, do occur. They are most commonly observed in trees growing in unusual soil circumstances, especially on soils outside of a pH range 6.0 to 7.0 (see Table 6-2). Under these conditions many elements are fixed into insoluble compounds and are thus unavailable for root uptake. Iron deficiency is probably the most common and conspicuous micronutrient deficiency in trees, and it occurs primarily in soils of high pH.

Less commonly, deficiencies occur in sandy soils where heavy rainfall depletes mobile ions by leaching. Boron and molybdenum are particularly prone to loss in this way. In some soils the presence of other elements may interfere in the uptake of a micronutrient. Finally, some soils may intrinsically lack particular trace elements. It is important to differentiate between soils actually deficient in an element and those where the element is present but in a form unusable by higher plants. Generally, where soil pH is causing a nutrient to be unavailable, a longer-lasting solution to the problem is to adjust the soil pH by liming or adding sulfur. However, some soils are difficult to treat and pH adjustment is not always successful. Direct application of the nutrient to the soil or to tree foliage or trunk may be needed to avoid problems associated with inappropriate soil pH. It is extremely important to accurately identify the specific micronutrient deficiency

Table 6-3. Average analysis of organic materials.

Organic Concentrate	Nitrogen (%N)	Phosphorus (%P₂O₅)	Potassium (%K₂O)
Dried blood	13.0	1.5	
Fish meal	10.4	5.9	
Activated sewage sludge	6.5	3.4	0.3
Tankage	7.0	8.6	1.5
Cottonseed meal	6.5	3.0	1.5
Linseed meal	6.0	1.0	1.0
Bat guano	13.0	5.0	2.0

Source: Western fertilizer association. 1980. Western fertilizer handbook, 6th ed. Interstate Printers and Publishers, Inc., Danville, Ill. p. 132.

before applying corrective elements, because applying the wrong material may make the problem worse.

Fertilizer Formulations

Organic Concentrates. Activated sewage sludge, tankage, blood, fish and seed meals, and bat guano are used as specialty fertilizers. In general these materials are richer in nitrogen than in phosphorous or potassium and contain a higher percentage of nitrogen than most manures, crop residues, or composts (see Table 6-3). With the exception of urea, organic nitrogen must be decomposed into forms usable by higher plants. The activity of soil microorganisms, which are decomposition agents, is dependent upon favorable soil temperature and moisture. Thus nutrients become available in flushes whenever soil microbe growing conditions are near optimum. Often these conditions also favor leaching and loss of nutrients. Research has shown conclusively that organic nutrients are not superior to inorganic nutrients for encouraging plant growth. The concentrations of nutrients, soluble salts, and trace elements in organic fertilizers vary tremendously, making optimum application rates difficult to calculate. Some organic concentrates such as sewage sludge contain heavy metals, which may accumulate in the soil to levels toxic to trees.

Bulky Organic Materials. Organic amendments most commonly used by home gardeners and horticulturists are animal manures, digested sawdust, leaf mold, peat, and compost. These materials increase the quality of mineral soils by adding organic matter, which increases aeration, water penetration, and cation exchange potential. They are, however, low-grade fertilizers that contain less than 1 to 2 percent nitrogen, phosphorus, or potassium and hence are of little benefit to soils with adequate organic material. Wood residues are low in plant nutrients, and if incorporated into the soil, they can increase the need for nitrogen and phosphorus, which are needed for their decomposition.

Commercial Dry Fertilizers. The form of fertilizer most familiar to homeowners and arborists is dry inorganic fertilizer. These commercial products are

Table 6-4. Composition of common fertilizer sources.

Compound	Formula	Nitrogen (%N)	Phosphorus (%P$_2$O$_5$)	Potassium (%K$_2$O)
Ammonium nitrate	NH$_4$NO$_3$	35.0		
Ammonium sulfate	(NH$_4$)$_2$SO$_4$	21.2		
Sodium nitrate	NaNO$_3$	16.5		
Potassium nitrate	KNO$_3$	13.8		46.6
Urea	H$_2$NCONH$_2$	46.6		
Natural organic matter	---	4.0		
Monoammonium phosphate	NH$_4$H$_2$ PO$_4$	12.2	61.7	
Diammonium phosphate	(NH$_4$)$_2$HPO$_4$	21.2	53.8	
Superphosphate	Ca(H$_2$ PO$_4$)$_2$ + CaSO$_4$		20.2	
Triple superphosphate	Ca(H$_2$ PO$_4$)$_2$		48.0	
Monopotassium phosphate	KH$_2$ PO$_4$		52.2	34.6
Potassium chloride	KCl			60.0
Potassium sulfate	K$_2$SO$_4$			54.0

Source: Adapted from L. F. Rader, Jr., L. M. White, and C. W. Whittaker. 1943. Soil Science 55:201–18.

either made by chemical reactions or taken from mineral deposits. They may be complete fertilizers, which provide all three primary plant nutrients—nitrogen, phosphorus, and potassium—with or without a selection of trace elements. The percentage by weight of each is indicated by the formula on the label such as 10-10-10 or 10-8-6. The first number denotes the percentage of nitrogen; the second, phosphorus (P$_2$O$_5$ or equivalent); and the third, potassium (K$_2$O or equivalent). Alternatively, they may be simple fertilizers providing only one essential element (Table 6-4). There is a broad range of essential element ratios available. Since there is also a range of percentage by weight of these elements, cost comparisons between formulations is sometimes difficult.

Commercial fertilizers may be formulated as homogeneous products or bulk blends. In homogeneous products each granule or pellet has the same analysis. Bulk blends are mixtures of two or more fertilizer materials.

Commercial inorganic fertilizers are formulated to provide plant nutrients in an immediately available form. Water to put the nutrients into solution and soil temperatures favorable for root metabolism are required for root uptake. Thus inorganic fertilizers applied when soil temperatures are too low for root metabolism may be lost by leaching before the plant is able to use them.

Liquid Formulations. Liquid formulations are designed to be applied to the soil in much the same way as commercial dry formulations, to be sprayed on the foliage, or to be injected in the trunk. When applied to the soil they behave in a manner similar to the dry materials.

Liquid formulations applied to the foliage or injected or implanted in the wood effectively provide nutrients to trees growing in soils where these nutrients would be unavailable because of soil factors and are not as subject to leaching as soil-applied materials. They are, however, too costly and inconvenient for general use.

Foliar sprays allow nutrient elements to be absorbed by the leaves. Response to nutrients is evident within a few days. Usually the response does not last more than one year and reapplications are needed. The effectiveness of foliar feeding is governed by environmental factors, plant age and species, and the formulation of the material. Results vary depending on circumstances.

Response to water-soluble implants or injection of liquid formulations into the trunk may be evident within a few weeks. For large trees that have inaccessible root systems and that are difficult to spray with liquid fertilizers, trunk treatments provide a good alternative. Response to trunk feeding usually continues for several years. There must be a strong need for fertilization in such cases, because repeated injections can cause serious wounds.

Controlled-release Formulations. To combat the problem of leaching losses, fertilizers have been formulated that are able to release nutrients slowly over a long period of time. Since only a small percentage of the nutrients will be lost at any one time, more nutrients are available for plant uptake and application frequency can thus be reduced. The controlled-release formulations include three main types: (1) low-solubility coating over soluble fertilizer; (2) fertilizer of low solubility; and (3) fertilizers of low solubility that require microbial action to release nutrients.

All forms of controlled-release fertilizers provide plant nutrients at a fairly uniform level over a longer period of time from a single application than do most other formulations. Timing of application is less critical because such fertilizers are less vulnerable to leaching. The controlled-release products are expensive, and their use may not be justified for ordinary landscape tree care. However, in areas of low fertility and high rainfall or where economics or inaccessibility dictates an infrequent fertilizer schedule, these materials have the advantage over other forms of fertilizer.

Treatment Area

The area to be fertilized is governed by the location of the absorptive roots. It was once thought that most tree roots occurred within the dripline, the line defined by the outer edge of the foliage. However, it is now known that tree roots commonly exist within a circle with a radius two times the crown radius, which will encompass an area well beyond the dripline. Species vary greatly in root extension. Roots of narrow or columnar trees extend well beyond twice the crown radius. Measurements of sugar maple indicate that the root area is $1^1/4$ times larger than the crown area and for tuliptree it is $2^1/2$ times larger, whereas for pin oak root area is about the same as crown area. With most trees, absorptive roots lie mainly in the outer two-thirds of the circle. Trees growing in sandy well-drained soils have more extensive root systems than those in finer-textured soils. Roots of large urban trees commonly occupy an entire lawn or backyard of urban lots and beyond if physical barriers are absent.

Fig. 6-1. To fertilize the trees in this typical front yard, the entire lawn area should be treated.

Fig. 6-2. Lawn care company applying fertilizer.

Fertilizers should be applied throughout the estimated root zone. For example, if a tree has a trunk-to-dripline branch spread of 15 feet, then the roots probably extend outward from the trunk 30 feet. The circular area around the tree to be fertilized would be approximately 2800 square feet (Table 6-5). For most homeowners having large trees the entire lawn will be treated (Fig. 6-1). However, do not apply fertilizers within a foot of tree trunks. The high concentration of mineral salts may damage the root collar and trunk base.

Be aware that soil surface fertilizer applications will affect turf grass. Amounts desired for trees in some circumstances can be harmful to the grass. Similarly, fertilizer applied to turf grass (Fig. 6-2) may affect tree performance. Turf growing in dense shade either should not be treated or should be treated only sparingly. If high levels of nitrogen are applied in dense shade, the grass is forced to metabolize that nitrogen, thus requiring more light energy than is available, so that grass may die. It is also best to treat the lawn beyond the root zone area to avoid a green island appearance.

Soil Application Methods

The method of application that is most efficient under a given set of conditions depends largely upon the objectives of fertilization, the soil conditions, and economics. Although many methods may work equally well, it is foolish to waste money on complicated techniques that give no better results than simple ones.

Soil Surface—Broadcast. Application of dry fertilizer to soil surface is referred to as broadcast. This method is appropriate for nutrients that move readily in the soil, notably nitrogen and most chelated micronutrients. Fertilizers may be applied using a cyclone seeder or lawn spreader. To ensure uniform coverage, it is best to divide the amount to be applied in half and apply the first half in even swaths across the area. Then apply the second half at right angles to the first. After application, the area should be watered thoroughly to move the nutrients into the root zone. A second watering on the next day will further aid distribution of the material, especially in heavy turf.

Soil Surface—Drench. Liquid fertilizers may be incorporated in irrigation water and applied to the soil surface. This method is appropriate for nutrients that are mobile in the soil. A proportioner permits soluble fertilizer to flow accurately into irrigation lines. Fertilizer is applied over the same area as described for broadcast applications.

Liquid formulations have no special advantages over dry formulations when applied to the soil surface. Both require water to carry them to the rooting zone. Liquid formulations do not penetrate any deeper in the rooting zone than do properly watered-in dry fertilizers. They are generally more expensive than the dry formulations and when used with irrigation water there is the risk that the nutrients will be flushed to depths below the root zone.

Soil Incorporation—Drill Holes. One method to incorporate either dry or liquid formulations into soil is to place them in drill holes. Holes spaced about 2 feet apart and about 18 inches deep are made in concentric rings around the tree throughout the rooting area (approximately 250 holes per 1000 square feet) (Fig. 6-3). They may be made with a punch bar, manually operated soil auger, or power-operated auger. The augers are preferred over the punch bar because the punch bar compacts the soil at the bottom and sides of the hole, which impedes nutrient dispersal. The holes should have 2-inch diameters. Avoid making holes within 2 feet of the trunk for trees 12 inches in diameter and within 3 feet for trees above 18 inches in diameter. This avoids damaging the root collar with high concentrations of fertilizer and severing main transport roots. The material to be applied is divided between the holes and poured into the holes through a funnel. Either liquid or dry formulations may be used. Automatic devices for drilling holes and inserting fertilizer have been developed. After application, water thoroughly. Soil plugs may be replaced before watering but if they can be left out, better aeration will result.

Table 6-5. Estimated fertilizer application area around trees of selected branch spread.

Total branch spread (ft)	10	20	30	40	50	60
Distance (ft) from the trunk to the extended branch tips	5	10	15	20	25	30
Probable extent of root zone (distance [ft] from the trunk)	10	20	30	40	50	60
Landscape area to be fertilized around the tree (X 1000 ft^2) *	0.3	1.2	2.8	5.0	7.9	11.3

*Fertilizer rates are normally given as pounds of nitrogen or other mineral element per 1000 ft^2.

Fig. 6-3. Drill holes for soil incorporation of fertilizer should be placed concentrically at 2-foot intervals. They should not be placed within 2 feet of the tree trunk and should extend out to a distance twice the radius of the tree crown. The holes should be about 18 inches deep.

Studies have shown that in most cases this method of applying nitrogen provides no more tree growth stimulation than broadcast treatments. However, where the tree is growing in thick sod or compacted soil, the added aeration may be as beneficial to tree growth as the fertilizer.

A similar fertilization method involves the use of tree spikes. These are made from concentrated solid fertilizer and are driven into the ground. These spikes are more expensive than other forms of dry fertilizer. At recommended rates, they produce an uneven distribution of the immobile nutrients, e.g., phosphorus and potassium. They do have the same advantage as drill holes of enabling one to incorporate nitrogen fertilizer into the soil, and they may require less labor to apply. Although it is tempting to think that placing fertilizer deep in the soil will provide nutrients for the tree and not for the grass, in fact, many grasses have roots that go as deep or deeper into the soil than tree roots, so the grass will get some of the fertilizer too.

Soil Incorporation—Injection. Air or water pressure may be used to inject fertilizer into the soil (Fig. 6-4). The air injection method forces dry fertilizer along with sand into the soil to a depth of 15 to 18 inches. After treatment the area should be well watered. The distribution of the injection sites should be the same as with the drill holes.

Liquid formulations or any soluble dry fertilizer may be injected into the soil with water. Injection sites should be 2 to 3 feet apart, about 160 sites and 200 gallons per 1000 square feet. The material should be injected to a depth of 15 to 18 inches. Application rates should not exceed $1^1/_2$ gallons per hole. Liquid should be forced into the soil under moderately high pressure of 150 to 200 pounds per square inch, but not higher because this will cause air pockets to form around the root system.

Injection into soil is an efficient method of incorporating fertilizer into the soil. The equipment may be used for other purposes, and the method is less tedious than drilling holes and funneling dry fertilizer. However, water-soluble forms of phosphorus and potassium are more expensive than the less soluble

Fig. 6-4. Tree-feeding needle attached to a pressure sprayer may be used to apply fertilizer.

compounds. Also, fertilizer solutions are notoriously corrosive to equipment. On balance, the benefit of ease of application probably outweighs the disadvantages.

Again, for optimum tree growth, the injection methods have generally not been shown to be superior to surface broadcast of nitrogen fertilizer. Soil injection does provide improvement of aeration and water penetration in compacted soils.

Soil Incorporation—Radial Trenching. The radial trench method was developed in an effort to place nutrients nearer to the root system and also to provide a less compacted soil for roots to expoit. A series of trenches $1/2$ to 1 foot wide are dug in a pattern radiating out from the trunk to well beyond the outer spread of the branches. The trenches are begun far enough away from the trunk to minimize damage to the major buttress roots. This area is then filled with rich composted soil to which fertilizers have been added. The advantage to the use of radial trenching is that some relief from soil compaction is achieved. However, there are some disadvantages also: (1) the soil is enriched in only a small part of the total area covered by the roots; (2) some roots are injured when the trench is dug; and (3) large amounts of sod must be lifted and replaced where trees are growing in lawns.

Soil-Applied Fertilizer Rates

Fertilizer rates are based on the number of pounds or ounces needed per 1000 square feet of root area. Whether wet or dry, watered in from the surface or incorporated, the amount of fertilizer to use can be calculated from the estimated root area (Table 6-5). For large trees, assume the area to be the entire part of the yard where the tree is located.

The amount of any fertilizer to apply largely depends on the objective of fertilization, the soil type, formulation of the fertilizer, and tree species. The following section describes appropriate rates under most circumstances, and conditions under which greater or lesser amounts should be given. However, these rates are general guidelines only; advice from local arborists or cooperative extension personnel should be sought to determine appropriate rates for local conditions. If established trees are growing well and show no mineral element deficiencies, then no additional fertilizer is needed.

Nitrogen. There are numerous forms of nitrogen fertilizer, but to be usable by higher plants, all forms must be converted to nitrate (NO_3), ammonium (NH_4), or urea.

Urea is an organic form of nitrogen that may be absorbed directly by roots and is soluble in water. Two different synthetic organic nitrogen sources containing urea are urea formaldehyde (UF) and isobutylidene diurea (IBDU). These are mixtures of bound and free urea, making some urea available immediately while the rest is made available over a period of time. After urea is watered into the soil, it is converted by microorganisms first into ammonium and then into nitrates. Although ammonium can easily be leached, urea cannot, so some of the nitrogen applied as urea will remain in the soil even if heavy rain follows application.

Various formulas for computing the ratio of nitrogen fertilizer needed in relation to trunk diameter have been devised. However, in landscapes where trees of different sizes are to be treated at the same time, it is simpler to apply fertilizer per unit area over the entire range of tree root extension. Thus trees with larger root systems will receive larger amounts of nitrogen.

Table 6-6. Consequences of applying different annual levels of nitrogen to a typical urban landscape.

Annual Nitrogen Application (lb/1000ft²)	Effect on Turf Grass	Effect on Woody Plants
0	Quality and growth minimum. Weeds may become dominant. Tall fescue more tolerant of low fertility.	Mature healthy plantings will continue moderate growth. Immature woody plants will not make rapid growth.
2	Good quality and growth. Optimum for most turf grass sites that cannot be irrigated.	Mature healthy plantings will make good growth. Little effect on attempts to "rescue" declining plants. May stimulate some growth of young plants.
4	Lush, high-maintenance turf grass. May be detrimental if timing and culture are mismanaged or heavy shade exists.	May push unneeded growth for mature plants. When nitrogen fertilizer is needed to reverse a decline this rate may help. Young plants can more rapidly attain size at this rate.
6	Problematic for bluegrass and fescue lawns. Thatch accumulates in bluegrass. Excessive growth is succulent and susceptible to disease. Root system tends to be shallow, thereby increasing drought susceptibility. Excessive top growth is at the expense of good root development. Without irrigation during dry periods, bluegrass lawn will die. Unless applied in three equal doses at three separate intervals (minimum of 6 weeks apart), or unless applied during cold weather, lawn will be burned with excess fertilizer.	Will cause shoots to lengthen considerably and the succulent growth may be more susceptible to disease. Difficult to know how the woody plant actually uses the available nitrogen in root vs. shoot growth. When extension growth is important (e.g., small landscape tree needing maximum extension growth to provide shade for house), this amount of fertilizer may be warranted. Regular irrigation is a must when following this program.

Source: Publication ID-72, Principles of home landscape fertilization, University of Kentucky Cooperative Extension Service, University of Kentucky, Lexington, KY 40546.

For most soils, the typical rate of shade tree fertilization adequate to maintain tree health is 1 to 2 pounds of nitrogen per 1000 square feet (87–174 lb/A) per year. Experimentally, trees have been shown to respond to rates from 1 to 12 pounds, but response is not proportionately greater when more than 6 pounds of nitrogen is applied. In some circumstances, rates over 6 pounds can be injurious. Soils with high levels of organic matter need less nitrogen, whereas sandy soils, especially in areas with high rainfall, require greater amounts. Where turf is present, the total yearly amount of fertilizer should be calculated to include the amount that is applied for the turf grass. If the lawn is already being fertilized, it is unlikely that the tree will need much more fertilizer. If trees are being fertilized at a high rate to correct a specific problem or to promote rapid growth, the amount of fertilizer used should be divided into two or more applications to avoid turfgrass burn. Evergreens require lower levels of nitrogen than deciduous trees. Overfertilization leads to open growth with widely spaced branches, and thus may ruin the typical crown shape of conifers. Overfertilization of deciduous trees may increase foliage density to the point where interior foliage may be weakened by heavy shade.

The number of pounds of fertilizer required to get a specific rate of nitrogen can be calculated easily. For example, urea is 45 percent nitrogen; to get 2 pounds of nitrogen one would need to apply 4.4 pounds of urea (2 divided by 0.45 = 4.4). Ammonium sulfate is 33.5 percent nitrogen and sodium nitrate is 16 percent nitrogen. If one is using a complete N-P-K fertilizer, the percentage of nitrogen is given by the first number of the analysis.

Table 6-6 describes the consequences of different levels of annual fertilization on turfgrass and woody plants in a typical landscape. It is important to determine the fertilization objectives before deciding how much nitrogen to use.

Phosphorus and Potassium. These elements are rarely in short supply, and overuse of fertilizers containing phosphorus or potassium may lead to toxicity symptoms on plants and to water pollution. Therefore, it is important to use these fertilizers only when needed and in the right amounts. Soil tests showing phosphorus levels below 30 pounds per acre or potassium levels below 165 pounds per acre indicate these elements are in short supply. Application of phosphorus is also warranted when foliar analysis indicates that the current season's foliage is below 0.1 percent phosphorus. Foliar analysis will not give reliable indications for potassium because it is very mobile in the plant. Phosphorus and potassium are largely immobile in soils; hence these fertilizers should be incorporated into the soil rather than applied to the surface.

There are various forms of potassium fertilizer. Potassium chloride (KCl) may be used where salinity is not a problem, but chloride may be toxic if used in arid regions. Potassium nitrate (KNO_3) may cause nitrogen excess problems if it is used to supply large amounts of potassium. Potassium sulfate (K_2SO_4) is the preferred form for most uses. Wood ashes are a traditional source of potassium, but they can raise the pH of alkaline soils to undesirable levels.

Table 6-7 gives the amount of fertilizer of various percentages of phosphorus and potassium needed to correct deficiencies. These rates are for middle-range soil textures. Sandy loam soils require about 50 percent less and clay loam soils need about 50 percent more.

Other Essential Elements. Materials containing the other essential elements should be applied only if foliar analysis and/or symptoms reveal a deficiency. If needed, soil-applied nutrients should be applied as follows:

Calcium: Apply according to soil test. If deficiency exists, up to 150 pounds of gypsum ($CaSO_4$) or 30 pounds of sulfur may be needed per 1000 square feet on alkaline soils. Apply lime (Tables 3-1 and 3-2) if soil is below pH 6.0. Calcium is relatively immobile in soil and should be incorporated into the soil to be effective. Treatment should be made prior to planting and the material should be incorporated to a depth of 18 inches. If treatment is made after planting, then materials should be worked into the top 4 to 6 inches of soil.

Magnesium: Apply according to soil test. If deficiency exists, up to 50 pounds of Epsom salts ($MgSO_4 \cdot 7H_2O$) per 1000 square feet may be needed. Apply dolomitic limestone (Tables 3-1 and 3-2) to acid soils.

Sulfur: Present in many fertilizers, additional sulfur is probably not needed, but it can be applied as gypsum at 15 pounds per 1000 square feet.

Iron, Manganese, Zinc: Chelated forms may be applied to soils at 0.5 to 4 pounds per 1000 square feet. Consult product labels for exact rates. Sulfates of iron or manganese can also be used at rates ranging from 4 to 30 pounds per 1000 square feet, depending on need.

Boron: Borax may be used at the rate of 1 pound per 1000 square feet.

Copper: Copper sulfate, applied at 3 pounds per 1000 square feet, is normally sufficient.

Molybdenum: Sodium or ammonium molybdate may be applied at 4 to 40 pounds per 1000 square feet.

Arborists should be sure of the identity of the specific problem before applying nutrients. In addition, whenever possible, adjustment of soil pH is preferable to and usually more effective over time than application of specific mineral elements.

Foliage or Dormant Twig Spray

Application to aerial parts of the tree is an efficient way to provide some micronutrients. Soil treatments will not supply materials to roots where soil conditions render the nutrients immobile.

Table 6-7. Rates of phosphorus and potassium fertilizer needed to correct deficient soils.

Soil Test Results (lb/A)	Fertilizer Needed (lb/1000ft^2)		
Phosphorus	15%P	20%P	46%P
0–9	52	34	15
10–19	34	23	10
20–29	17	11	5
30+	0	0	0
Potassium	15%K	60%K	
0–99	52	14	
100–149	34	11	
150–199	17	10	
200+	0	0	

Source: After E. M. Smith and C. H. Gilliam. 1980. How and when to fertilize field grown nursery stock. Am. Nurseryman 151(2):8, 68–69, 72–73.

Dilute solutions are sprayed on the foliage until they just begin to drip off. Coverage needs to be uniform to correct deficiencies of elements that are not readily transported between plant parts. Nitrogen, potassium, phosphorus, chlorine, and sulfur are very mobile within the plant. Zinc, copper, manganese, iron, and molybdenum are only partially mobile. Calcium and magnesium are largely immobile. The presence of a surfactant in the mixture will increase the effectiveness of most sprays. The amount of surfactant used is important because too much will reduce absorption and may cause burning of leaves. Consult the label for recommended amounts.

Materials penetrate the leaf cuticle and epidermal walls by a diffusive process which is influenced by temperature and the concentration of the spray. They are absorbed onto the surface of the plasma membranes and pass through the membranes to enter the cytoplasm via active pathways. Thus environmental factors that influence the rate of plant metabolism will also influence the efficiency of absorption. In general, high temperatures (above 85° F) and low humidity reduce absorption, and young leaves absorb more readily than old leaves.

The chemical form of nutrients may also influence absorption. Chelated forms of micronutrients are generally absorbed well. However, different chemical forms of phosphorus give dramatically different test results. In descending order of efficiency of absorption the forms of phosphorus are H_3PO_4, K_2HPO_4, NaH_2PO_4, KH_2PO_4, $Ca(H_2PO_4)_2$. Sprays of 2600 parts of nitrogen per million parts of water in the form of urea, calcium nitrate, and ammonium sulfate all give the same results, which implies that the nitrogen source does not influence absorption. The addition of urea to spray solutions increases the effectiveness of foliar sprays containing phosphorus, manganese, sulfur, magnesium, and iron.

Chelated forms of zinc and zinc sulfate may be applied to dormant twigs to correct zinc deficiency. Because zinc is only partially mobile in plants, coverage must be very thorough. Dilute formulations are sprayed until the material just begins to drip off.

Spray treatments are effective for temporarily correcting nutrient deficiencies. Improvement in foliar color is usually evident within a few days. Thus sprays can be used to assist in diagnosing mineral deficiency problems. Spray treatments must be repeated annually. They are best used for correcting micronutrient deficiencies where soil conditions preclude efficient soil treatments.

For nutrients such as nitrogen, which are required in larger amounts foliar applications are inconvenient, messy, and expensive. Uniform coverage of mature trees is nearly impossible. Fertilizer applications to foliage can cause leaf or fruit burn under high temperature; if the solution is too concentrated, they leave a whitish residue on foliage. Although many chelated forms can be combined with routine pesticide sprays, urea is incompatible with some pesticides. For major nutrients this method of application should be considered as a temporary measure that supplements soil treatment. Rates for solutions of selected mineral nutrients applied as sprays to tree foliage follow:

Magnesium: Use 3.3 ounces (19 teaspoons) of Epsom salts ($MgSO_4 \cdot 7H_2O$) per gallon of water. Magnesium sulfate is incompatible with some pesticides.

Fig. 6-5. Trunk implant used for correcting mineral deficiency.

Boron: Use 1 to 2 teaspoons of boric acid per gallon of water.
Copper: Use 3 to 6 teaspoons (1/2 to 1 oz) of copper sulfate ($CuSO_4 \cdot 5H_2O$) per gallon of water.
Iron: Use 4 teaspoons of ferric sulfate ($FeSO_4 \cdot H_2O$) per gallon of water.
Manganese: Use 2 to 8 teaspoons of manganous sulfate ($MnSO_4 \cdot 2H_2O$) per gallon of water.
Zinc: Use 4 to 6 teaspoons of zinc sulfate ($ZnSO_4 \cdot 7H_2O$) per gallon of water.
Molybdenum: Use 1/10 to 1/2 teaspoon of sodium molybdate ($NaMoO_4 \cdot H_2O$) or ammonium molybdate [$(NH_4)_2MoO_4 \cdot 2H_2O$] per gallon of water.

Trunk Implants and Injections

Injection or implants in the stem or trunk can provide a rapid response and are especially useful where other methods fail or where roots are inaccessible. The process creates tree wounds and should not be considered as a first alternative. These methods are appropriate only for application of micronutrients.

Implants are made by placing gelatin capsules containing small amounts of soluble nutrients into holes drilled into trunks or branches (Fig. 6-5). The holes are made in a spiral pattern around the trunk. Their number depends on the size of the tree (see Table 6-8), and their diameter depends on the size of capsule used, in most cases $^3/_8$ to $^1/_2$ inch. The holes should be drilled deep enough to seat the entire capsule plus a plug entirely in the wood. Do not leave an air pocket around the capsule. Some capsules are manufactured to seal the hole, so do not require a plug. However, in most cases a good-fitting plug placed with its surface flush with the cambium layer will promote wound covering and exclude most rot organisms. The holes are usually covered by tree callus within 6 months.

Material may also be implanted in trees without gelatin capsules. Ferric citrate (an iron chelate) has been used to correct iron chlorosis in this manner. A

Table 6-8. Number of trunk implants to use for various tree sizes.

Trunk Diameter (in.)	Capsule Size	Distance between Holes (in.)	Number of Capsules per Tree
1-4	No. 2	2	2-6
4-12	No. 000	3	4-12
>12	$1/8$ oz.	4	> 9

Source: After D. Neely. 1976. Iron deficiency chlorosis of shade trees. J. Arboric. 2:128–30.

Table 6-9. Quantities of material and numbers of holes needed for low-pressure injections of iron sulfate.

Tree Diameter (in.)	Number of Holes per Tree	Amount of $FeSO_4$ (oz.)	Number of Refills
1–4	4	0.5	1
5–10	4–5	2.0	1
10–15	5–6	4.5	1-2
15–20	6–7	5.5	2-3
20–25	7–8	6.5	3-4
25–30	8–9	6.5	4-5
30–35	9–10	7.5	5-6

Source: J. W. Fischbach and B. Webster. 1982. New method of injecting iron into pin oaks. J. Arboric. 8:240.

$7/16$-inch spiral wood bit is used to drill holes into the wood to a depth of $1^1/2$ inches. A plastic syringe with the tip removed is used to inject $1^1/2$ grams of powdered ferric citrate into the hole. The hole is filled with a tight-fitting cork. The placement and number of holes is the same as with gelatin implants.

Soluble forms of micronutrients may be injected in solution into the trunk either by gravity feed or under pressure. Holes are drilled in the trunk root flare region using a standard high-speed drill bit. Sawdust must be cleaned from the holes before they are fitted with lag screw injection heads. It usually takes two or three turns of the screw in the wood to seal the hole and hold the screw in place.

With the gravity feed system the holes should be deep enough to seat the screw securely; small trees may be impossible to inject. The soluble material is placed in an elevated reservoir filled with about $1/2$ gallon of water. It is allowed to trickle into the hole. After the reservoir is nearly empty, plain water should be added to flush the residue into the tree. It takes about 24 hours to treat a tree in this manner. The number of holes needed depends on the size of the tree; holes should be 3 inches apart for trees less than 4 inches in diameter and 8 to 10 inches apart for larger trees (see Table 6-9).

It takes much less time to inject the solution into the tree when the solution is under pressure. Using pressures of 100 to 200 pounds per square inch, 1 quart can be injected in from $^1/_2$ to 5 minutes, depending on the tree species and time of year. Holes are drilled $^3/_5$ to $1^1/_2$ inches deep; if deeper, the injection time is increased. Injection holes should be placed directly under scaffold branches at or just above the root flare. For trees over 16 inches in diameter, holes should be made every 6 inches around the trunk. For most trees, 1 to 2 quarts of solution can be injected. After material is injected, leave the hole exposed; apply no sealants. Many species have been successfully treated in this manner; however, butternut, shagbark hickory, black cherry, white ash, maples, firs, and pines may not accept injection solutions readily.

Even smaller volumes of liquid having greater nutrient concentrations can be injected into trees under pressure using a plastic capsule and feeder tube system (Fig. 6-6). In this case, small-diameter holes are drilled into the root flare area, a thin plastic tube is inserted into each hole, and a pressurized plastic capsule (Mauget system) containing the nutrient solution is impaled on the tube. The punctured plastic capsule delivers a small volume of solution into the tree.

Injection rates are higher in trees after new leaves expand than during dormancy or early in the growth period. Presumably, negative xylem pressure potential associated with transpiration is responsible for this advantage. Likewise, injection rates are faster from midmorning to early evening, and with mild wind. Root injection produces more uniform response in the top of the tree but is often inconvenient.

Fig. 6-6. Mauget capsule for injection of trees with mineral elements.

The injection and implant methods require holes to be drilled in the trunk. These holes, aside from being unsightly, may be ports of entry for rot organisms. Even when rot organisms do not invade, limited areas of the wood are often discolored and nonfunctional. The injured tissue associated with all wounds is limited by the tree to the wood present at the time of injection (see Chapter 7). Thus dieback, staining, or infection is usually confined to a small column. However, retreatment may be required every 3 to 5 years, so considerable damage could be caused by injection over the life of a tree. When injecting or implanting, (1) keep the holes as small as possible; (2) drill as low on the tree as possible; (3) avoid vertical alignment of holes from previous treatments; (4) remove external fixtures as soon after treatment as possible; and (5) implant or inject only those substances that have been proven to be effective and noninjurious to the tree. Damage to the tree will be minimized if these precautions are followed.

Implants and injections have been used successfully to treat deficiencies of micronutrients, most notably iron and zinc. Response to treatment is usually evident within 3 weeks, and retreatment is required only every 3 to 5 years. Some researchers have found that injections produce more consistent response than implants. Injection of dilute zinc sulfate, for example, was judged superior in treating zinc deficiency over implants of chelated zinc cartridges. The cartridges, when used according to manufacturers' directions, delivered too little zinc to overcome symptoms, However, implants require less equipment and can be successfully made by less highly skilled workers. Labels and instructions of implant and injection products should be studied and understood before applications are made. Again, implants and injections should only be considered as a last resort if other methods of fertilizer application have failed.

Recommendations for Newly Transplanted Trees

Historically, if trees were fertilized at all they were treated with a complete fertilizer added to the backfill when transplanted. We now know that most soils have adequate potassium and phosphorus, and that the nitrogen needed for growth the first year is already in the transplant. Researchers have found that for most normal soils newly transplanted trees do not respond to fertilizer until several years after moving. Moderate levels of nitrogen fertilizer applied at transplanting time may provide future benefits. High levels of nitrogen at planting may have a negative effect on root growth, thus prolonging transplant shock.

Once established, trees respond significantly to nitrogen except in very fertile soils. In one study, fertilization had little effect on tree growth until the third year after transplanting. Then the trees responded with vigorous growth to additions of nitrogen but not to phosphorus or potassium. In most soils, nitrogen is the only element that is chronically in short supply.

Trees newly planted in sandy soil where rainfall is high may require additions of nitrogen before establishment. Slow-release formulations added to the planting hole and then covered with several inches of soil will provide a continuous supply of nutrients for up to 18 months. Slow-release forms should also be used where inaccessibility or expense necessitates an infrequent application schedule.

Timing of Tree Fertilization

Applications of fertilizer should be timed so that nutrients are available for periods of rapid growth. Some trees, for instance, spruce and fir, have a single flush of growth in the spring. Many more—yew, holly, and most shade trees, for example—have two or more periods of growth, the most important in the spring and another in midsummer. A few, such as juniper, grow continuously during the growing season.

In general, to stimulate spring growth of trees, fertilizer applications to the soil should be made in the late summer or early fall. For later growth flushes, fertilizers may be applied after the spring flush. Fertilizer should not be applied in early spring; applications made in spring are sometimes lost through leaching by spring rains. Fertilizers may also be applied to the soil after the top of the tree is fully dormant in the late fall, but such applications will not stimulate growth the following spring. However, this is the best time to fertilize turfgrasses. Fall fertilization is said to promote the growth of turfgrass root systems whereas spring fertilization promotes growth of turfgrass foliage. Tree roots will continue to absorb nutrients until soil temperatures fall below 40° F.

Foliar sprays should be made after leaves begin to expand, whereas trunk implants and injections are most successful when done after leaves are fully expanded.

The frequency of application depends on the kind of tree, the objective of fertilization, and other factors. It may vary from small, multiple treatments each year to one made every 3 or 4 years. Most trees will respond to annual applications of nitrogen. Trees being treated for micronutrient deficiency using foliar sprays need annual applications. Where trunk injections or implants are used, retreatment should be made when symptoms of deficiency reappear. For phosphorus and potassium, soil tests are needed to accurately time the frequency with accuracy. It is possible to kill trees with too much fertilizer and also to waste fertilizer, so it is better to wait for symptoms of deficiency for nutrients to appear than to apply fertilizers indiscriminately.

Selected Bibliography

Hamilton, D. F., M. E. C. Graca, and S. D. Verkade. 1981. Critical effects of fertility on root and shoot growth of selected landscape plants. J. Arboric. 7:281–90.

Himelick, E. B., and K. J. Himelick. 1980. Systemic treatment for chlorotic trees. J. Arboric. 6:192–96.

Lloyd, J. 1997. Plant health care for woody ornamentals. International Society of Arboriculture. Champaign, Ill. 223 pp.

Maynard, D. N., and O. A. Lorenz. 1979. Controlled-release fertilizers for horticultural crops. Hortic. Rev. 1:79–140.

Neely, D. 1976. Iron deficiency chlorosis of shade trees. J. Arboric. 2:128–30.

———. 1980. Trunk and soil chlorosis treatments of pin oak. J. Arboric. 6:298–99.

———, E. B. Himelick, and W. R. Crowley, Jr. 1970. Fertilization of established trees: A report of field studies. Illinois Nat. Hist. Surv. Bull. 30:235–66.

Reil, W. O. 1979. Pressure-injecting chemicals into trees. Calif. Agric. 33(6):16–19.

Swietlik, D., and M. Faust. 1984. Foliar nutrition of fruit crops. Hortic. Rev. 6:287–355.

Van de Werken, H. 1984. Fertilization practices as they influence growth rate of young shade trees. J. Environ. Hortic. 2:64–69.

7

Pruning

The pruning of trees is probably the most noticeable and important of all tree maintenance practices. Thoughtful pruning produces a structurally sound tree that can better withstand adverse environmental conditions. In addition, properly pruned trees require less cabling, bracing, and sometimes require less managing of pests to maintain good health. Appropriate pruning will add to the aesthetics and prolong the useful life of trees.

Need for Pruning

Trees are pruned principally to provide training when they are young, to preserve their health and stability, to improve appearance, and to prevent damage to human life and property.

Pruning for Training. Pruning young trees in the nursery and landscape has consequences that will last for the life of the tree. Early pruning often makes the difference between whether a tree will live out its maturity relatively problem free, or will be a constant source of trouble for its owner. Pruning a tree while it is young to improve its structure and therefore its usefulness and longevity is far less expensive and more effective than waiting until the tree is mature.

Pruning for Health and Stability. Broken, dead, or diseased branches are pruned to prevent pathogenic organisms from penetrating into adjacent parts of the tree and to reduce the amount of inoculum available for spread to other trees. Indeed, pruning is an important control measure for fire blight, black knot, several twig blights, some cankers, and dwarf mistletoe. It can be of some benefit in controlling Dutch elm disease and crown gall.

Live branches are removed to permit penetration of sunlight and circulation of air through the canopy. An open canopy makes a less favorable site for powdery mildew. It also favors formation of flower buds and allows pesticide sprays to cover leaves more evenly.

Proper pruning of the tree crown can reduce wind resistance and help prevent breakage. Branches that form an acute angle of attachment are removed because they are especially prone to breakage. When branches having narrow crotch angles get larger, stress on the crotch increases. Also, acutely angled branches often have bark embedded in the branch attachment, which causes a weak joint. These weak attachments are prone to splitting and breaking out during ice storms or high winds. Thus pruning can increase the structural stability of trees.

Pruning for Appearance. For centuries, trees in formal gardens have been carefully pruned to produce desired—often fantastic—shapes. In most landscapes, however, pruning is used to maintain or restore the characteristic crown form. Occasionally it is desirable to keep normally large-growing shade or ornamental trees within the restricted bounds of small properties. Judicious pruning of new growth during summer will produce the desired dwarfing effect with no harm to the tree. The Japanese art of bonsai is the extreme example of this practice. Pruning influences the balance between vegetative growth and flower bud production. In some species, this influence can be used to invigorate a declining tree or to produce fewer but larger flowers and fruit.

Pruning for Safety. Dead, split, and broken branches are a constant hazard to human life and property. Danger from falling limbs is always greatest in trees along city streets and in public parks. Low-hanging live branches must be removed to a height of 10 to 12 feet when they interfere with pedestrian and vehicular traffic. Branches that obscure clear vision of warning signs or of traffic must also be removed.

Trees are pruned to prevent them from interfering with energized lines. Branches touching lines can interrupt service, and wind-thrown limbs can knock down electrical and telephone lines.

Response of Trees to Pruning

Effect on Tree Shape. The growth habit of a woody plant is largely genetically determined. The characteristic shape is influenced by the location of leaf and flower buds, the pattern of budbreak along the branches, and differential elongation of branches in various positions on the stem. These factors produce three basic tree forms: columnar, excurrent, and decurrent (Fig. 7-1).

In the technical sense, the term columnar shape is used to describe trees whose terminal growth is concentrated in a single growing point, for example, single-trunk palm trees. Columnar is also used to describe growth shown by deciduous trees whose branches are parallel and erect, with tree shape tapering to a point, for example, Lombardy poplar.

Most conifers and some deciduous trees have an excurrent form. In these trees the terminal bud exerts weak apical dominance over the lateral buds. Thus various numbers of lateral buds will grow in the same season as the shoots on which they are formed. The terminal shoot, however, is usually able to elongate more than the lateral shoots, so a conical shape develops.

The decurrent form is produced by most shade trees. Terminal buds exert strong apical dominance so few or no lateral buds develop in the year that the shoot on which they grow is produced. However, in the following year the lateral buds are released and may outgrow the terminal shoot. Thus the trees become round-headed with many shoots elongating to a similar degree.

Pruning interferes with the hormonal dominance system of trees and thus has a significant influence on tree shape even beyond the simple removal of wood.

Fig. 7-1. Crown shapes: A, columnar (palm); B, excurrent (fir); C, decurrent (zelkova).

Although the one-time removal of the terminal bud cluster on an excurrent tree will not change it to a decurrent form, repeated removals will alter its growth habit. The shoots from buds nearest to the point of removal will elongate to a greater degree than shoots farther down, giving the tree a bushy appearance. Similarly, the removal of the terminal bud cluster on a shoot with strong apical dominance will allow one or more buds just below the cut to grow. However, buds farther down the shoot will still be repressed. Consequently, the new branches will be concentrated right below the cut.

The removal of a lateral branch with an acute angle of attachment from a young leader branch will often force an accessory bud to grow from the same node. The resulting branch is usually attached at a wider angle to the leader than was the original branch.

When a lateral is removed at its point of origin, or the length of a branch is reduced by cutting just above a lateral large enough to assume the terminal role, new growth will be distributed over the entire remaining branch. The bushy regrowth that follows removal of terminal buds will be avoided and a more natural crown shape will be retained.

Effect on Growth. Lush regrowth on young trees following pruning is a familiar sight. Since the same root system is providing water and minerals to fewer branches, the remaining shoots are able to grow vigorously. This gives rise to the notion that pruning is invigorating. However, when a substantial portion of the top is removed from young trees, less food is produced by the trees for root and future top growth. If the tree's carbohydrate reserves are already depleted, heavy pruning could jeopardize tree health. Severe pruning, especially of young trees, while invigorating to individual branches may be dwarfing to the entire tree.

Moderate pruning of mature trees is generally invigorating. Trees that produce fruit on one-year-old wood show marked shoot growth increase when pruned during dormancy. The pruning removes potential fruit, allowing more of the tree's energy to be channeled into foliage and shoot production. Crown thinning allows more light to penetrate to interior and lower leaves. Experiments have shown that the photosynthesis rate within the crown of thinned apple trees is four times higher than that of unthinned trees. The increased photosynthesis rate of these leaves makes up for removed foliage and the remaining branches are able to respond vigorously following pruning.

Effect on Flowers and Fruit. Pruning influences the number, size, and quality of flowers and fruit. Dormant pruning of trees such as apples, which produce fruit on one-year-old wood, reduces the number of flowers and fruit, but those remaining grow larger. Thinning of these trees designed to increase light penetration to the inner crown will increase flower bud formation and hence increase the crop in the following year.

Dormant pruning of trees such as goldenrain-tree, which produce flowers on current season's growth, increases the number and size of flowers and fruit on the remaining branches.

Wound Reactions. Pruning produces wounds; thus the tree becomes vulnerable to attack by wood-rotting fungi. However, trees are not passive and are able to marshal both chemical and physical defenses. The work of Dr. Alex Shigo and his colleagues at the U.S. Forest Service has led to a better understanding of these processes. According to Dr. Shigo, chemical boundaries are formed in the tissues present at the time of injury. Food reserves in sapwood are converted to resinous materials and phenols, which discolor the wood and cause it to become more decay resistant. This localized resistance keeps the decay from spreading internally. Vessels are also blocked, which slows vertical spread. The cambium responds by laying down a thin layer of unique cells called a barrier zone. This zone separates the wood formed before wounding from the wood formed after wounding. The zone acts as a barrier to protect the new wood and restricts the development of decay organisms to the tissue formed before wounding. Callus tissue forms at the margin of the injury and may eventually close the wound, but trees do not heal in the conventional sense of replacement of damaged tissue. The pruning wound will remain embedded in the tree as new tissue grows over it.

The success of the response in preventing decay depends on the vigor of the tree and the virulence of the invading organisms. The genetic constitution of individual trees may enable them to marshal their defenses more efficiently. In many cases, the defenses are breached by the invaders and new barrier zones are formed.

Types of Pruning Schemes

The two basic pruning schemes are defined in terms of the tree's reaction: thinning out and heading back.

Thinning Cuts. There are situations where reducing the tree crown is desirable and necessary. The best way to accomplish this is to use a thinning cut, in which entire lateral branches are removed to their point of origin. Alternatively, the length of a branch is reduced by cutting just above a lateral that is large enough to assume the terminal role, a practice also known as drop crotching. Thinning cuts can be used to reduce the tree's height and spread or its density

Fig. 7-2. Before (left) and after (right) thinning cuts.

Fig. 7-3. Tree after topping. Notice the large branch stubs remaining.

while retaining its natural shape (Fig. 7-2). Only selected portions of the tree's canopy are removed, reducing the possibility of sunscald damage. Pruning cuts are made close to the trunk, leaving only the collar of the removed branch instead of stubs. The pruning cuts are less conspicuous than those left from heading cuts and they "close" more rapidly and completely. Regrowth tends to be scattered naturally along the length of the remaining branches. Use of thinning cuts requires greater skill and time than heading cuts, but in most situations it is well worth it.

Heading Cuts. This pruning scheme may involve pinching, tip pruning, shearing, pollarding, dehorning, stubbing, and topping. Pinching and tip pruning are used for training young trees. Cuts are made so that buds or branches below the cut are encouraged to grow in a desired direction. Shearing, or heading back, of young shoot tips of needle evergreens such as hemlocks creates bushy regrowth and a denser canopy. Pollarding starts out as tip pruning. These are examples of desirable heading cuts.

Topping, stubbing, and dehorning are examples of drastic removal or cutting back of large branches in mature trees (Fig. 7-3). Major branches are cut so that stubs several feet long are left without any important side branches attached to them. Topping is not desirable and usually harms the tree.

Many homeowners have their trees topped often by so-called professionals when their trees have reached heights which they consider unsafe. They fear a strong wind might blow over these large trees. This fear is largely unjustified. The extensive root system of a healthy tree, if left relatively undisturbed, provides adequate support for the tree. An old healthy tree with a good root system is actually less likely to blow over than a smaller tree with its smaller, less developed root system.

Some homeowners believe that the stimulation of new growth associated with topping is actually beneficial to the tree. Although the tree appears rejuvenated with new foliage and branches, this only masks the real damage topping does to the tree.

Trees may also be topped to remove potentially hazardous dead and diseased branches which may break off during ice storms or windstorms. Unfortunately, topping removes healthy as well as unhealthy limbs. The hazardous limbs are best removed by selective pruning instead of topping.

Large, mature trees are often topped to prevent interference with overhead utility wires. They are also topped when they block views, interfere with buildings or other trees, or shade solar collectors or other areas (e.g., lawns and gardens) where sunlight is wanted. In some of these situations, removing large limbs may be necessary; however, alternatives such as proper early training, selective thinning out of branches and limbs, or whole tree removal should be considered and adopted where feasible.

Removing much of the tree canopy upsets the crown-to-root ratio and seriously affects the tree's food supply. A 20-year-old tree has developed 20 years' worth of leaf surface area. This leaf surface is needed to manufacture sufficient food to feed and support 20 years' worth of branches, trunks, and roots. Topping not only cuts off a major portion of the tree's food-making potential, it also severely depletes the tree's stored reserves. It is an open invitation for the tree's slow starvation.

Removing the tree's normal canopy suddenly exposes the newly injured cambium to the heat of the sun, drying it out and killing these regenerating tissues back further.

Topping removes all the existing buds that would ordinarily produce normal sturdy branches.

Large branch stubs left from topping seldom close or callus. Nutrients are no longer transported to the large stubs and that part of the tree becomes unable to seal off the injury. This leaves the stubs vulnerable to insect invasion and fungal decay. Once decay has begun in a branch stub, it may spread into the main trunk, ultimately killing the tree. Fruiting bodies of decay fungi are often visible on the bark of decaying trees.

Topping stimulates the regrowth of dense upright branches just below the pruning cut (Fig. 7-4). These new shoots, referred to as water sprouts, are not as structurally sound as are the naturally occurring branches. The water sprouts often consist of succulent growth that is more susceptible to diseases (such as fire blight of rosaceous trees) and herbivorous insects (such as aphids and caterpillars).

Since water sprout regrowth is generally rapid and vigorous, a topped tree often will grow back to its original height faster and denser than a tree that has been properly pruned or thinned. This makes topping, at best, only a temporary solution to oversized trees. Some tree species (e.g., sugar maple, oak, and beech) do not readily produce water sprouts. Without the resulting foliage, a bare trunk

Fig. 7-4. Prolific sprouting follows topping. These sprouts will be weakly attached to the topped limbs.

results and the tree quickly dies. Deteriorating branch stubs, along with weak sucker growth, make topped trees highly vulnerable to wind and ice damage.

From an aesthetic aspect, topping disfigures the tree. Unsightly branch stubs, conspicuous pruning cuts, and a broomlike branch growth replace its natural beauty and form. The practice of topping may have started out of ignorance when people, not wanting a large tree, attempted to reduce crown height by emulating pollarding practices, common in Europe. However, pollarding and topping are not the same thing. In addition, some nonprofessional tree trimmers saw topping as an easy way to prune trees without needing to possess much skill or knowledge. Since a good selection of low-growing landscape trees is available now, there is no need for constant pruning to maintain low growth. Topped trees are unnatural substitutes for shade trees meant to offer several lifetimes of beauty and enjoyment.

Specific Pruning Situations

Newly Transplanted Trees. For years the idea prevailed that trees should be severely pruned at time of transplanting to compensate for root loss and to reestablish a favorable shoot-to-root ratio. This practice may be detrimental to rapid establishment following transplanting. In many cases the additional moisture stress from leaving the top intact is offset by the more rapid development of the supporting root system from the additional carbohydrates produced. At the time of transplanting, pruning should be limited to removal of dead, broken, diseased, or interfering branches. If the central leader is cut or broken, the largest vertical branch must be trained to become the new leader, and it may need support to maintain its new position. If the newly transplanted tree has double or multiple leaders, the best one must be selected and the others removed, or possibly selected as scaffold branches. It is best to leave small shoots along the trunk for later removal because they protect the trunk from sunburn and aid in the development of proper trunk taper.

Young Trees. After a tree becomes established, usually 1 to 3 years after transplanting, it may be pruned to achieve a desired shape. First, dead, diseased, misshapen, or crisscrossing interfering branches are removed. Temporary branches left along the trunk should be headed back but not removed until a sturdy trunk is established. After the tree is well established, the temporary branches may be removed. Trees that have a strong central trunk and a conical shape usually require no additional pruning, whereas those with a round top benefit from early training to establish well-spaced scaffold branches.

The scaffold branches are the main branches that make up the framework of the tree. For trees that will become large, branches should be selected that are at least 18 inches apart on the trunk; for small trees the distance is 6–12 inches.

Fig. 7-5. Undesirable branches interfere with the normal development of major branches. Note that branches are crowded, one has included bark, and two branches are at the same node.

Scaffold branches should also be distributed radially in all directions. In no case should the selected scaffold branch be more than one half the diameter of the trunk. Similarly, smaller branches developing from scaffold limbs should not be more than one half the limb diameter. The establishment of a well-developed framework requires several years of selective pruning. When the tree is first planted, more potential scaffold branches may be present than will be ultimately desired. Some will dominate, making final selection easier.

During the first few years of training, remove branches with acute angles of attachment, branches that are likely to interfere with or compete for the same growing space as potential scaffolds (Fig. 7-5), and branches growing from the same node as potential scaffolds. Branches that are a threat to compete with the main leader and become codominant must be pruned back, if not removed. Such branches can be reduced in size and influence by reducing their length or by removing branches attached to them so they grow more slowly in the future. This technique of slowing the growth of potential competing branches to favor the central leader is sometimes called subordination. Branches below the base of the permanent crown should be removed, though they may be left temporarily for a year or two to enhance trunk taper. Eventually select five to seven well-formed scaffold branches with even radial and vertical distribution.

Mature Trees. If the early training is properly done, trees will not need to be pruned for many years. However, they should be inspected annually to identify and remove hazards. Mature trees may benefit from one or more kinds of pruning operations. *Crown cleaning* involves periodical removal of broken, dead, diseased, poorly placed, or weak limbs. Weak, declining limbs often develop from self-shading by other foliage on the tree. *Crown thinning* helps to increase light penetration for better flower and fruit production, allow better coverage of spray material, or reduce water needs. When crown thinning occurs, selected branches all along a limb should be removed, and not just those at the base of the limb. This balances the weight of branches and foliage all along the limb. *Crown raising* may be needed when changes in land use require greater clearance under the tree. Limbs would be removed at the trunk or branches from the parent limb. *Crown reduction* is done to reduce the height of the tree. To maintain a natural crown appearance and to avoid bushy regrowth, mature trees should be thinned out rather than headed back. The lateral branch to which the main branch is cut must be at least one third of the diameter of the branch being removed. When pruning trees, no more than one fourth of the foliage should be removed in any given year. Drastic pruning weakens trees.

Conifers. Needled evergreens usually do not require pruning to maintain their characteristic crown shape. Top breakage, often caused by ice storms, may result in an undesirable forked leader. It is best to prune competing laterals and maintain a single central stem.

Some evergreens are pruned to produce a more bushy or compact plant. Pruning the growth at the ends of young shoots forces the plant to make new growth along the branches.

Not all evergreens can be pruned in the same manner at the same time. The method depends on the species and the type of growth.

Pines and spruces make only one flush of growth during the year. They can be

Fig. 7-6. Proper pruning cut is made just outside the branch collar (A-B). Left: Conifer. Right: Hardwood.

trimmed at any time, but best results are obtained when the new growth is soft. In pruning spruces, especially the blue spruce, the long ends should be removed at the point where side shoots are formed. This will check the terminal growth for a year or two and will enable the branches to fill out. Most pines are best pruned before the new needles start to emerge, that is, when the new growth still gives the appearance of candles. No damage is done to the needles, and the tree will retain its original healthy appearance.

Most of the other evergreens, including arborvitae, junipers, yew, cypress, and false cypress, continue to develop during the entire growing season. These may be trimmed until midsummer, although early summer pruning is preferable, since new growth will then cover the cuts before the end of the growing season.

Winter pruning is sometimes preferable when an evergreen has become straggly as a result of long neglect. The long branches are cut to encourage new bud development early in the next growing season. When pruning is postponed until summer, new buds will not develop until the following spring, and in the interim the plant will appear unattractive.

Evergreens with a distinctive shape should be pruned to retain the desirable effect. Only the long shoots should be removed. Formal evergreen hedges are best pruned with power or manual hedge shears. Hemlocks may well be pruned to increase their branching and fullness. If they are pruned before new growth starts, by early fall the effects of pruning will be well hidden by the new growth.

It is sometimes suggested that foliage be cut wet to prevent the cut ends from turning pink; however, this practice increases the risk of spreading foliar, twig, and branch diseases and should be avoided.

How to Make the Pruning Cut

Laterals. Research concerning the wound responses of trees has shown that most of the damage caused by pruning can be avoided by cutting along those lines that the tree forms for natural shedding. When a branch dies, the trunk tissue responds by walling off the dead branch. Thus proper pruning should only remove branch tissue, leaving the trunk intact.

How does one tell where the branch ends and the trunk begins? Fortunately the tree provides the necessary clues. Every branch has internal tissues that separate it from the trunk. As this tissue forms, the bark is forced upward to form a raised ridge on the trunk. This ridge is referred to as the *branch–bark ridge*. The raised or swollen area surrounding the base of the branch is made up of trunk tissue and is referred to as the *branch collar* or *shoulder ring*. When a branch dies, the collar continues to grow.

Pruning by making use of these clues is sometimes referred to as "natural target pruning" (Fig. 7-6). The pruning cut should be made at an angle from a point just beyond the branch–bark ridge to the junction of the lower part of the branch and the branch collar. Occasionally the branch collar is difficult to identify. Then the cut is made at an angle opposite to that of the branch–bark ridge. The exact angle of the pruning cut will vary with tree species. The resulting wound should be approximately circular in cross section; in later years, the callus tissue closing

Fig. 7-7. Handheld shears, positioned to make a pruning cut.

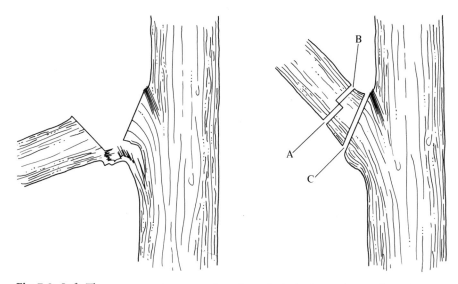

Fig. 7-8. Left: The wrong way to prune a large branch. A single cut close to the main stem may result in tearing of the bark. Right: The correct three-cut system for pruning a large branch: a preliminary undercut is made at A; second cut is made at B, to sever the main part of the branch. The remaining stub is removed by cutting at C.

the wound will resemble a doughnut. Injury to the branch–bark ridge or the branch collar should be avoided. Do not make flush cuts. Do not leave stubs. Either extreme increases the risk of internal decay for the tree and may delay callus closing of the wound.

When cutting small branches with handheld shears, place the cutting blade beneath or on the side of the branch with the slicing blade positioned nearest the main branch (Fig. 7-7). The cut is made either upward or diagonally in one stroke with minimal twisting. The cut should not be made downward because the trunk is more likely to be damaged in this manner.

To avoid damage to the tree, large branches are removed using a three-cut system (Fig. 7-8). This method removes most of the weight of the limb before the critical final cut is made and does so in a manner to prevent the bark from tearing. The first cut is made on the branch underside out several inches from the branch collar. It should be made about one-fourth of the way through the branch. The second cut is made from above and is made slightly farther out on the limb. The weight of the branch will cause the branch to fall and break off at the point where the bottom cut was made. If the limb is large and likely to cause injury or damage, it should be secured with ropes before cutting and then lowered to the ground. The final cut is made as described previously.

Main Stems. It is occasionally necessary to remove a terminal branch (Fig. 7-9). Damage to the tree can be lessened by cutting the main stem off just above a lateral and at an angle parallel to the branch–bark ridge of the lateral. When cutting a young tree, leave a short stub to prevent the new terminal from

YOUNG TREE

Proper cut

1/4″

Improper cuts

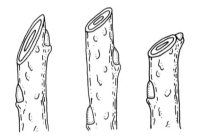

MATURE TREE

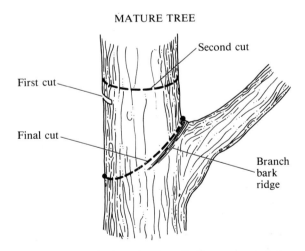

Second cut

First cut

Final cut

Branch bark ridge

Fig. 7-9. Removal of terminal branch: for a young tree, leave a short stub; for a mature tree, use three-cut system and make final cut at an angle parallel to the branch–bark ridge of the highest major lateral branch, leaving no stem stub.

splitting out. For larger trees, make an initial topping cut several inches above the final critical cut to remove the weight of the terminal branch. Be sure to select a lateral branch that is not less than a third of the size (diameter) of the main stem that has been removed. Small branches and twigs are not sufficient to assume the terminal role for a larger tree.

Where two main stems occur, it is sometimes necessary to remove one of them, especially if they form a narrow-angled crotch. In this situation, a branch collar may not exist, and finding the branch–bark ridge can be difficult. The bark from each stem may be concealed within the crotch, so it is hard to know just where the fork begins. It is best to prune as would be done for removing a main stem as described previously. Even though a proper pruning cut is made in this situation, decay is almost inevitable, as it is for topping cuts. It is best to recognize narrow crotch angles when the tree is young and remove one of the branches then.

The young terminal growth (candle) of a pine tree may be removed by snapping off the tender tissue with the fingers.

The International Society of Arboriculture has published tree-pruning guidelines, cited at the end of this chapter, which describe the principles and practice of tree pruning.

Wound Dressings

Arborists and gardeners have been advised over the years to use various wound dressings ranging from commercial tree paints, orange shellac, asphalt paints, creosote paints, and house paint, to grafting wax. Current research indicates that these materials are not effective in protecting trees from wood-rotting organisms. Indeed they may prolong the period of susceptibility. When these coatings crack, as they frequently do, moisture may enter the area between the wood and the coating. Keeping the wounded area wet fosters the growth of decay organisms. Thus wound dressings should be considered only as a cosmetic treatment and should be used sparingly in a thin coat, if at all. Although wound dressings are now thought to be useless, if not harmful, development of a beneficial wound dressing is possible. A paste-like wound dressing product, Lac Balsam, shows promise in some tests.

An important exception to the prohibition against treatments is the use of tree wound dressings on wounds of oak wilt–susceptible trees that have been pruned in late spring. Apparently, treated trees are less likely to get oak wilt because they are less attractive to the insect vectors. This may be the only beneficial use of tree wound dressings.

Plant hormones have been found to be variable in their effect on callus production and to be ineffective in preventing decay.

When to Prune

Although circumstances beyond the control of arborists often dictate the timing of tree maintenance procedures, there are optimum times to prune to achieve desired effects. Jobs such as removal of safety hazards, dead branches, or diseased branches, however, can be done anytime.

Generally, trees are best pruned in winter, when they are dormant. Large cuts may be made in the late winter or early spring before bud break, when hydrosta-

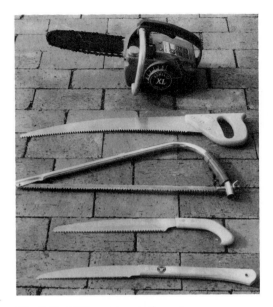

Fig. 7-10. Different types of pruning saws: gasoline-powered chain saw, hand saws.

tic pressure is greater than atmospheric pressure, i.e., sap pressure is positive. A wound made at that time will cause the tree to "bleed." Although some believe the bleeding to be unsightly, it may have a positive benefit in reducing the probability of colonization of the pruning cut by decay organisms. In the spring the rate of callus production is more rapid, further reducing the chance of invasion. Midsummer is also an acceptable pruning time.

Spring pruning after budbreak is undesirable, because during this time the bark is at its tenderest and damage to the bark is most likely. In addition, the food reserves of the tree are being directed toward new growth, leaving less energy available for wound repair. Pruning in the late summer is also undesirable because it removes foliar carbohydrates that would be sent to the tree before the leaves fall. This reduces food storage necessary for growth in the following spring and slows root production. In addition, a number of important decay fungi have been observed to produce their largest numbers of spores during the fall; this is another factor favoring the winter for major structural reshaping.

Pruning to produce a desired shape or to correct a misshapen crown is often most readily accomplished during the growing season. Also, dead or diseased branches are more conspicuous then.

Equipment

Effective and useful equipment is essential for good pruning (Fig. 7-10). Equipment must be maintained in first-class condition both for the safety of the operator and for the good of the tree. The International Society of Arboriculture has published an interesting and readable guide, cited at the end of this chapter, on the subject of arborist equipment.

Fig. 7-11. Chain saw operator in bucket truck.

It is important to select the right-sized equipment to fit the job. One-hand shears of the scissors (curved blade) type may be used to cut branches up to $1/2$ inch across. The scissors-type is preferable to the anvil-type because it can cut closer to the trunk and does not crush the wood tissue. Long-handled loppers (usually curved blade) provide added leverage and enable one to cut limbs up to $1^1/4$ inches across. Most pole pruners can also be used to cut $1^1/4$-inch limbs but allow the operator to work at some distance from the cut. Larger branches require a saw. One should not force shears to cut limbs larger than they were designed to cut. The twisting of the blades that accompanies the outsized cut will damage the nearby bark tissues, produce an imprecise cut, and may damage the shears.

Both hand and power saws are available in a variety of lengths and widths. For efficiency of use, precision of cut, and safety of the operator it is important that the saw be neither too large nor too small for the task.

Three types of hand saws are commonly used in tree work: pruning, pole, and crosscut. For small cuts, pruning saws with teeth that cut with the pulling motion are used. This type of saw is particularly suitable for overhead cutting.

A pole saw may be used for small cuts made at some distance from the operator. This saw also cuts with the pulling motion. Although pole saws allow cuts to be made from the ground, such cuts require more time and energy to make. Thus there is an efficiency tradeoff to be made, and this should be considered in planning equipment use.

Crosscut saws, either bow or hand, are appropriate for larger cuts. Because they are most commonly used to cut green wood, a fleam tooth pattern, which cuts equally well in both directions, is preferred. To prevent the saw from binding, the set should be wider than the thickness of the blade. A finer set is used for soft woods, a wider set for hard woods.

A wide variety of powered pruning tools is available. The most common type is the chain saw powered by a direct-drive two-stroke gasoline engine. Hydraulic and pneumatic power systems, which allow for pole handle mounting of the cutting bar, are also used. Chain saws range in size from lightweight, short bar types suitable for use by climbers in the tree to powerful saws having long chain bars suitable for felling large trees. In addition, hydraulic boom-mounted circular saws extending from a large vehicle are used for trimming trees along rural rights of way.

Improvements in power tools are constantly being made. The best information concerning type and size of equipment is found in manufacturers' instructions.

Arborists must be equipped to ascend the tree where pruning is to be done. Sturdy stepladders and extension ladders are good for pruning low branches. Bucket trucks (Fig. 7-11) or ropes and saddles (Fig. 7-12) are needed to attain greater heights. Climbing spurs, useful for scaling utility poles, should not be used in trees. The injuries produced by spurs provide openings for tree decay organisms.

Tree climbers should know how to use their climbing equipment, including how to properly tie knots and how to lower heavy limbs, equipment, and themselves from the tree. Climbing equipment must be in good condition and well designed. One of the authors once found himself dangling 30 feet off the ground in an old climbing saddle that had slipped up around his chest following a fall from a limb. This frightening experience could have been prevented if the saddle in use had been designed with leg straps attached to the butt strap. Needless to say, poor or damaged equipment is not acceptable for tree work.

Safety in Pruning

People engaged in tree pruning have an obligation to prevent injury or damage to themselves, their crew, passersby, other trees, and surrounding property. In 1968

Fig. 7-12. Professional tree worker properly equipped to prune landscape trees.

Fig. 7-13. Arborist secured in tree with climbing line and lanyard while the limb to be removed is supported by two separate ropes. This climber is wearing spurs (for tree removal only).

Fig. 7-14. Rope brake, a lowering device providing control of friction for lowering heavy limbs.

the American National Standards Committee on Safety in Tree Trimming Operations, Z-133 was organized in response to the efforts of a mother whose son had died while trimming trees. The committee produced a publication that details safety standards to be maintained while engaged in pruning, trimming, repairing, maintaining, and removing trees, and cutting brush. It is available from the International Society of Arboriculture or from the American National Standards Institute, 1430 Broadway, New York, NY 10018. Commercial operators should follow these standards. Homeowners pruning their own trees should also be aware of safe procedures.

Eye, head, and hearing protection with safety goggles, hard hat, and ear plugs constitute the most important basic safety equipment needed during all pruning operations. When one is working in the aerial part of the tree, secure footing and a safety harness or line are essential to prevent falls (Fig. 7-13). One should know the type of wood in the tree, since greater precautions are necessary in pruning weak-wooded trees, such as poplars, silver maple, willow, and tuliptree, than in pruning elm, oak, hickory, and planetree, which have strong, flexible wood. Bark peeling and fungal growths are signs of dead and dying branches; limbs with these symptoms should never be depended upon for support. Climbing danger is greatest when branches are wet and when temperatures are low. Large limbs should be roped and lowered to the ground and not left to fall, which could cause injury and damage. Unskilled amateurs should never climb; inexperienced professionals should not climb when carrying power tools unless they are properly trained in climbing and use of power tools. Any power tool heavier than 15 pounds should have a separate support line.

All overhead lines and cables should be considered to be energized and should never be touched directly or indirectly. Amateurs should never work closer than 10 feet from any overhead line. Workers who prune trees for line clearance must be specially trained. Other professionals should notify the utility company before any work is done within 10 feet of a power line. After storms, damaged lines may energize metal conductors, such as wire fences, at some distance. Special care is needed under these circumstances, and workers must be alerted.

Removing Prunings from the Tree and Tree Removal

Arborists are often faced with removal of large limbs weighing several hundred pounds, or even removal of entire trees in situations where just dropping them could cause injury to people or damage to trees and property nearby. Small branches, especially those pruned with handheld human-powered equipment, are simply allowed to drop. When a tree needs to be dismantled or when large limbs are removed, considerable skill and judgment are needed to safely lower the pieces of wood to the ground. Arborists must be familiar with ropes, knots, lowering devices (Fig. 7-14), and rigging techniques for securing the cut limbs, and the use of notched facecuts and backcuts to prevent accidental splitting and kicking back of cut ends. If done properly, large trees and limbs can be removed safely.

Removal of tree prunings from the property is facilitated by cutting limbs into firewood-sized logs and running smaller prunings through a chipping machine

Fig. 7-15. Brush chipper and dump truck.

(Fig. 7-15). When trees are removed, stumps and larger surface roots are normally ground up using a stump grinding machine.

Pruning for Artificial Size and Shape

In contrast to pruning to enhance the tree's natural form, pruning may be used to create trees with unusual and artistic sizes and shapes. These methods may include pollarding, architectural pruning, espalier, pleaching, and topiary. These ancient pruning concepts are thought to be Mediterranean in origin, are often derived from pruning concepts used in fruit production, and are labor intensive. The senior author observed firsthand the use and misuse of some of these techniques in France. In France, the word "arboriculteur" is ordinarily the term used to describe a grower of tree fruits such as apples, and not, as we would expect, the term for one who cares for amenity trees. The terms arboriculture, arborist, and arboriculteur come from the same Latin root.

Pollarding. Pollarding is a pruning system in which, starting at a young age, the tree is headed back every year to approximately the same place. The ends of the framework branches become knob-shaped and most of the regrowth branches arise from these knobs. The height and spread of the tree is kept uniform year after year. Pollarded trees may make attractive formal plantings.

In some circumstances, landscaping objectives might require the creation of stout tree columns along a street or in a plaza to complement the nearby buildings or to provide shade for pedestrians but not to have tall trees blocking the view of a building. Naturally low-growing trees such as flowering crabapples or dogwoods do not have straight, stout trunks, so pollarding, sometimes called "short pruning," or "head pruning" of otherwise tall-growing trees is used. Although this method is not to everyone's taste, one can appreciate the idea of having the columns of pollarded trees complement the columns of a classical building or those inside a grand cathedral. Pollarding is also done to show off the particular form of the tree framework as a vase, a parasol, or a pyramid, particularly in winter. There could also be practical reasons to pollard trees. Properly done, the trees are healthy, safe, and provide shade. Because their height is limited, damage from branches that break and fall is likely to be limited.

Tree species used for pollarding include London plane, horsechestnut, linden, black locust, and fruitless mulberry. These trees sprout easily behind pruning cuts and are easy to maintain for pollarding. For the most common pollard designs, training is begun in the nursery, where the slender tip of the trunk is cut off at a predetermined height to encourage lateral branches that will form the framework of the pollarded tree. Through annual pruning in the nursery and after establishment in the landscape, trees are carefully trained for many years to obtain the desired framework of main branches, usually several dozen (Fig. 7-16). Occasionally, the pollard design framework consists of headed branches arranged vertically, or even a trunk with a single pollard head. When the trees have developed the desired form and have just surpassed the desired height and width, the ends of these framework branches, still small in diameter, are cut off. In

Fig. 7-16. Pollarded linden trees along a roadside in France.

Fig. 7-17. Pollard heads of linden trees with one year's sprouting.

Fig. 7-18. Pollarded linden trees with summer foliage.

subsequent years, multiple branches arise from buds and latent buds just behind these cut ends, and they are pruned back to the framework every year or two. Thus, regular maintenance pruning removes all the branches back to their branch collars, leaving only the framework limbs. The ends of these limbs enlarge and form knobs, or pollard heads, sometimes called "cat's heads" (Fig. 7-17). Although branches may be removed from the cat's heads annually, arborists are careful to leave the branch collars and to never cut into the cat's heads. In summer, new branches, often several feet long, mask the cat's heads, give new shape to the tree, and provide shade (Fig. 7-18).

When first observed, pollarding looks like tree topping, which, as we know, is detrimental to the health of landscape trees. Pollarding is not tree topping. The important distinction is the size of branches removed—pruning of cat's heads is done annually or every other year so only small branches are removed, while tree topping decapitates large limbs. Does pollarding harm trees the way topping does? Apparently not, according to research done by scientists at the University of Paris. In naturally growing trees, carbohydrate reserves are distributed from the roots to the top of the tree, with concentrations highest in the large roots and trunk base and progressively diminishing as one moves to the limbs, large branches, small branches, and finally the twigs. Even within the limbs and branches, carbohydrates are concentrated towards the bases, as they are in the trunk. In a pollarded tree, the large roots, trunk base, and especially the cat's heads are rich in carbohydrate reserves, while the young branches emanating from them have few reserves. When one tops a tree, one loses significant carbohydrate reserves in the detached limbs, but when one pollards a tree, one removes few carbohydrate reserves. Furthermore, high carbohydrate reserves in the cat's heads of pollarded trees facilitate rapid closing of the relatively small wounds made by regular maintenance pruning, whereas the large wounds created by topping have relatively lower carbohydrate reserves available for wound closing and decay prevention. Thus, pollarding is much less harmful to the health of trees than is topping.

When the management of pollarded trees is changed due to a change in tree maintenance philosophy or to a lack of funds for annual maintenance, the trees are greatly harmed. Small branches left unpruned for several years soon become large branches that are poorly attached to the cat's heads (Fig. 7-19). Furthermore, carbohydrate reserves, once concentrated in the cat's heads, migrate to the enlarging branches. If a pollarded tree is allowed to "go natural" for many years, it eventually becomes hazardous because of poor branch attachment. If the trees are pruned belatedly in an attempt to return to pollarding, these trees will lose carbohydrate reserves in the removed limbs and the large wounds will close poorly because of depleted reserves in the cat's heads. Similarly, one should not attempt to pollard a large, naturally growing tree because of the wood decay risks associated with topping. Thus, one must not alternate between natural pruning and short pruning (pollarding) for reasons of tree health. Tree maintenance really involves the management of the tree's energy reserves. A decision to pollard is a major maintenance decision and must be made for the life of the tree and while the tree is still very young. In the gardens of the Château of Villandry, France, every winter a crew of four arborists prunes all of the 1,260 linden trees by hand,

Fig. 7-19. Tree, once pollarded, left to grow naturally, Angers, France.

using raised platforms to access the cat's heads. The city of Nantes, France, uses ladders and small chain saws when trees are dormant to pollard miles of plane-trees along city streets and avenues. This annual commitment of funds for tree maintenance, while not uncommon in France or other European countries, would be extremely rare in the United States.

Important principles to remember when considering the use of pollarding for tree maintenance: 1) The commitment to pollard a tree must be made for the life of the tree. 2) The branching framework must be established when the tree is young. 3) The ultimate tree size must be determined when the tree is young. 4) Pruning of wood more than two years old must be prohibited; cuts should not be made into the cat's heads or remaining branch collars. 5) Tree pollarding is labor intensive. 6) One cannot shift from natural pruning to pollarding, or from pollarding to natural pruning without severe damage to the tree. 7) Depending on circumstances, pollarding may be cost effective or wasteful of limited tree maintenance resources.

Architectural Pruning. In special circumstances, groups (usually rows) of trees may be pruned to form a curtain, an arch, a screen, or other architectural shape. These rather formal structures are meant to enhance the appearance of formal gardens, nearby buildings, plazas or avenues. In the garden, a tree screen or curtain may frame a landscape perspective or may serve as a backdrop for ornaments such as fountains or statues. Trees such as elm, horsechestnut, plane-tree, linden, hornbeam, and beech are often used for tree screens. Unlike in pollarding, tree crowns are trained to fill a space with many branches. When the tree crown reaches the desired size and geometric shape, it is pruned regularly, one to three times per season, to a multitude of small branches. Although architectural pruning was traditionally done by hand, modern arboriculturists are using laser-guided cutting machines to maintain perfectly squared and straight tree screens. Each time the screen is pruned, the twigs are pruned perhaps an inch beyond the last cut to encourage even more branching. Thus, the branching density increases and architectural shape enlarges slightly year by year. Every five years or so, it is necessary during the dormant season to prune the tree back to its original size and density so the process can begin anew. Like pollarding, architectural pruning is very labor intensive.

Topiary. The fantastic shapes of trees formed into geometric, animal, or other forms are created by a pruning and training system referred to as topiary (Fig. 7-20). The best trees for this use are small-leaved evergreens with readily

Fig. 7-20. Topiary pruning, Château Villandry, France.

growing latent buds. As with architectural pruning, the trees must be trimmed at least once or several times each season to be kept in shape. Often, forms are fitted over the tree to guide the pruners in making consistent, uniform shapes.

Espalier. This training technique results in a tree taking on the shape of its support, be it a wall, a trellis, or even a pole. Once the framework is defined, maintenance pruning keeps the tree the same size. The tree framework can vary, depending on the desired effect. The following are examples of different espalier shapes. Vertical branches may be trained to ascend from laterals attached to a short trunk, making the tree resemble a candelabra. A single unbranched trunk made to grow horizontally, sometimes called a cordon, is sometimes used in formal gardens to form a low barrier and also to produce flowers and fruit. Branches emanating from a short trunk may be trained to radiate out fanlike in all directions along the same vertical plane. A common espalier framework consists of well-spaced horizontal branches attached to a single upright trunk (Fig. 7-21). Branches may be trained in an informal pattern along a wall or trellis. Priority is often given to the conservation and promotion of flowering and fruiting spurs within the framework. These techniques are used primarily for fruit trees and vines.

Pleach. A pergola, bower, or arbor can be formed by pleaching, the weaving of branches of nearby trees together to form a covered space. Similarly, a kind of fence can be constructed by weaving closely planted lines of supple trees together (Fig. 7-22).

Fig. 7-21. Fruit tree espalier at Wisley Gardens, England.

Fig. 7-22. Willows woven together to form a lattice at Chaumont sur Loire, France.

Pruning for Disease Control

Very often it is necessary to prune trees that are partially affected by diseases such as fire blight of apple, pear, and mountain-ash; sycamore anthracnose; canker of blue spruce; tip blights of conifers; wilt of maples and elms; and twig blight of chestnut oak. In such cases, the infected twigs or branches should be removed well below the point of visible infection.

Extreme caution must be exercised to avoid transmission of the causal organism from a diseased to a healthy tree by means of the pruning tools. Indeed, Dutch elm disease can be spread by contaminated sawdust on pruning saws. After use on a tree suspected of being diseased, the cutting edges of all tools should be thoroughly disinfected by application of Lysol or 70 percent denatured ethyl alcohol. Furthermore, pruning when the leaves are wet should be avoided, since parasitic organisms are easily spread under such circumstances.

For further discussion of the role of pruning in disease control, see Chapter 13.

Pruning for Line Clearance

Electric utilities must maintain adequate clearance of overhead circuits to prevent interruptions to their customers' service. Tree branches can cause interruptions

Fig. 7-23. Trees can be pruned to avoid interference with power lines.

by contacting power lines. Sometimes inexpert pruners drastically prune trees by topping or heading back large limbs during right of way clearance, causing public outrage and increased pruning costs for the utilities. Trees do interfere with utility lines, and maintenance of these lines to prevent power outages is costly to the consumer. There are ways to minimize tree interference with utility lines.

Trees are either top, side, overhang (under), or through trimmed to allow them to grow 2 to 4 years before they again become hazardous to the line (Fig. 7-23). Careful use of thinning cuts to provide directional pruning will direct growth away from wires while retaining a more natural crown appearance. However, in some cases the tree crown may take on a V or L shape, which in the direction of the utility line may appear unsightly. Nevertheless, this kind of pruning, using proper limb removal methods is preferable to the alternative, which is topping.

Thinning cuts are much preferred to heading cuts or topping because this method does not promote unsightly bushy regrowth and it leaves fewer open wounds that do not close easily. The regrowth after topping usually interferes with the electrical lines sooner and will require even more cutting on revisits. One utility company discovered that switching from topping to thinning cuts reduced their pruning expenses by 50 percent. Others have reduced pruning costs by offering to remove customers' trees and replace them with desirable low-growing trees that would not become a line clearance problem.

Use of Chemical Growth Retardants

Because of the high cost of pruning trees manually, less expensive ways are being investigated. One is the use of growth regulators. They are primarily being used by utility companies to increase the length of time between trimming visits. Trees are treated just after pruning to retard sprouting. The growth regulators are particularly useful where topping is implemented.

Growth-retarding chemicals are being applied by overhead foliar sprays, trunk injection, basal spray–bark banding, and painting of cut surfaces. Some of the methods have resulted in reduced costs for utility companies.

Scheduling Pruning for Municipalities

Many cities have assumed an exclusive duty to maintain street trees. Pruning operations in these localities are more than aesthetically desirable; they are critical to ensure public safety and avoid municipal liability. Courts have ruled that trees must be maintained in safe condition, and any damages arising from street trees are attributable to the city.

Cities can schedule pruning operations by the following methods:

1. Request of residents.
2. Crisis—only when an emergency exists.
3. Task pruning—performing certain functions such as clearance of right of way obstructions.
4. Species pruning—scheduling common service needs by tree type.
5. Programmed maintenance—scheduling service needs by localities.

Most cities use a combination of these schedules. It has been found that using only request and crisis scheduling is inefficient and may increase the municipality's liability for damages occurring due to unidentified hazards. Modesto, California, found that the programmed maintenance schedule allows the entire city to be covered in a 7-year cycle. This schedule utilizes personnel in the most efficient manner and reduces the number of calls for crisis situations. An economic evaluation of pruning cycles using computer optimization techniques found that a 4- to 5-year cycle maximizes the return from pruning while minimizing the cost for most cities.

Selected Bibliography

Anonymous. 1994. American national standard for tree care operations—pruning, trimming, repairing, maintaining and removing trees, and cutting brush—safety requirements. American National Standards Institute, ANSI Z133.1-1994, New York. 22 pp.

Anonymous. 1995. American national standard for tree care operations—tree, shrub, and other woody plant maintenance—standard practices. American National Standards Institute, ANSI A300-1995, New York. 9 pp.

Blair, D. F. 1995. Arborist equipment: A guide to the tools and equipment of tree maintenance and removal. International Society of Arboriculture, Champaign, Ill. 291 pp.

Bory, G., G. Herbert, N. Macle, and D. Clair-Maczulajtys. 1995. Physiological consequences of architectural pruning on trees. In: The tree in its various states. Proceedings of the Second European Congress in Arboriculture. Versailles, Sept. 1995, pp. 111–19.

Bowles, H. 1985. Growth retardant use by utility companies. J. Arboric. 11:59–60.

Bridgeman, P. H. 1976. Tree surgery. David and Charles, North Pomfret, Vt. 144 pp.

Britton, J. C., D. W. Ceplecha, J. Goodfellow, R. W. Harris, T. A. Johnson, J. NcNeary, B. L. Morris, and K. A. Ottman. 1995. Tree pruning guidelines. International Society of Arboriculture, Champaign, Ill. 14 pp.

Domir, S. C., and B. R. Roberts. 1983. Tree growth retardation by injection of chemicals. J. Arboric. 9:217–24.

Giles, F. A., and W. B. Siefert. 1971. Pruning evergreens and deciduous trees and shrubs. University of Illinois Coop. Ext. Serv. Circular 1033. 57 pp.

Gilman, E. F. 1997. Trees for urban and suburban landscapes: An illustrated guide for pruning. Delmar Publishers, Albany, N.Y. 178 pp.

Leben, C. 1985. Sap pressure may affect decay in wounded trees. Am. Nurseryman 161(7):59–63.

Shigo, A. L. 1984. Compartmentalization: A conceptual framework for understanding how trees grow and defend themselves. Ann. Rev. Phytopathol. 22:189–214.

———. 1986. A new tree biology. Shigo and Trees, Associates, Durham, N.H. 619 pages.

———. 1989. Tree pruning: A worldwide photo guide. Shigo and Trees, Associates, Durham, N.H. 188 pp.

———. 1991. Modern arboriculture. Shigo and Trees, Associates, Durham, N.H. 424 pp.

Smithyman, S. J. 1985. The tree worker's manual. Ohio Agric. Educ. Curric. Materials Serv., Columbus, Ohio. 145 pp.

Steffek, E. F. 1982. The pruning manual, 2nd ed. Van Nostrand Reinhold, New York. 152 pp.

Stefulesco, C. 1995. Traditional pruning in France. In: The tree in its various states. Proceedings of the Second European Congress in Arboriculture. Versailles, Sept. 1995, pp. 108–10.

8

Tree Preservation and Repair

Mature trees, in addition to providing shade, beauty, and a relaxing atmosphere, enhance the value of property. Therefore, it is important to preserve trees when their roots, trunk, or limbs are threatened by the actions of nature or man. Physical damage to tree structure is usually obvious: broken limbs, damaged bark, exposed roots. However, lives of trees are more often threatened by damage that cannot be seen on casual inspection: internal decay, boring or mining insects, root suffocation. Some forms of damage are preventable; some are not. The preservation of mature trees depends on preventing damage where possible and ameliorating the impact of damage once it has occurred.

Is the Tree Worth Preserving?

When considering a course of action, whether to prevent or treat injury, the first question should be: Is the tree worth saving? The answer depends on the value of the tree (see Chapter 1) and the cost of preserving it.

Preventing and Treating Structural Damage

Damage to the structure of the tree may result from external forces such as windstorms or ice storms, from vehicle collision, or from the internal action of boring insects and decay fungi. Trunk, limbs, or branches may crack, split, or break.

In many cases structural damage caused by windstorms or ice storms can be prevented by careful pruning, such as the early removal of limbs forming narrow crotches. However, in some cases artificial support such as that provided by braces and cables may be required to preserve structurally weak or injured trees. Some trees, such as oak, beech, and locust, rarely need artificial support because they consist of tough wood fibers and usually have a single major stem from which branches arise at approximately right angles. Despite the absence of tough wood fibers, even pine, spruce, and fir seldom need support because of their single stem and type of branching. Other trees, such as elm, maple, and poplar, which have inherent structural weakness resulting from the manner in which their branches arise, often require considerable mechanical support.

Fig. 8-1. Left: Narrow branch crotch with embedded bark. The union is inherently weak. Right: Preferred branch attachment is strong.

When Is Artificial Support Justified?

Artificial supports prevent such injuries as crotch splitting and branch breakage, both of which ruin the tree form and eventually lead to wood decay. They may also prolong the life of many trees after decay or extensive cavities have already developed. Such supports may strengthen an otherwise weakened trunk. However, if artificial support is being considered, it signifies an implied acknowledgment that a potential hazard exists. Arborists need to be confident that the installation they are attempting will improve the safety of the tree and need to consider tree removal as possibly a better option.

The proper installation of cables or braces is often helpful in prolonging the life of trees having tight crotches, split crotches, deep trunk cavities, or inherent susceptibility to breakage. Changes in the tree's surroundings may also create a need for artificial supports due to exposure to winds that had not previously affected it.

Narrow Crotches. Narrow crotches commonly occur in certain trees in the course of normal development (Fig. 8-1). In the narrow angle between two sharply forking major branches, the bark and cambium are hindered from developing normally or may be killed by the pinching action of the branches as they increase in size. This results in a weak union of the type that readily splits when subjected to stress. Bracing or cabling can prevent splitting.

Split Crotches. After some splitting, artificial supports can prevent further damage and loss of one or both branches.

Deep Trunk Cavities. Cavities are formed by extensive decay of internal wood of trunk and limbs. Because the decayed wood no longer provides structural support for the tree, artificial bracing may be needed.

Susceptibility to Breakage. Because of inherent properties such as heavy foliage production and brittleness of the wood, some species of trees are more susceptible than others to branch and limb breakage by wind and ice storms.

Trees that are highly susceptible to breakage by winds are 'Bradford' pear, chestnut oak, honey locust, horsechestnut, mimosa (silk-tree), poplar, maple (red, silver, and sugar), sassafras, Siberian elm, tuliptree, willow, and yellowwood. Although the following are more resistant to wind damage, they too can lose branches and be toppled by winds of very high velocity: American beech, American elm, ginkgo, hackberry, littleleaf linden, London planetree, Norway maple, red oak, sweetgum, and white oak. Cabling large limbs may prevent disfigurement of susceptible species.

Land Use Changes. Wind patterns may be changed by construction or removal of buildings or the removal of nearby trees. These changes may subject a mature tree to greater wind stresses than previously present. Without judicious pruning and artificial supports, trees may suffer breakage. Also, artificial supports may be required to preserve trees after the loss of some large roots during wall or curb construction or grade changes.

Types of Artificial Support

Tree supports are of two types: rigid and flexible bracing. Rigid bracing involves the use of bolts and threaded rods for supporting weak or split crotches, long cracks in the trunk or branches, and cavities. Flexible bracing involves the installation of wire cables high in the tree to reduce the load on weak crotches and long arching limbs without hindering the normal branch sway.

Rigid Bracing. Lengths of steel rods with lag threads or long lag bolts are used for rigid bracing in sound wood. The holding power of such rods is similar to that of ordinary wood screws used in fastening two boards together. The lag rod is set in a hole drilled $1/16$ inch smaller than the rod. The wood surrounding the hardware will discolor, but if it is sound to begin with, it will seldom decay to any great extent.

Coarse-threaded rods or bolts with nuts and washers on each end may be used instead of rods with lag threads. They are sometimes used when decay is present or where considerable "squeeze" must be applied to close a gap. The bolts are inserted into holes drilled $1/16$ inch larger than the bolt. The washers should be seated just on the wood but not sunk into it. Some arborists seat the washers directly on the bark surface because placing washers there may create less injury than countersunk washers seated on the wood. Long-term research will be needed to finally resolve this question. A wood chisel or wood spade bit the size of the washer should be used to carve out a smooth-edged hole in the bark to seat the washer. Washers should be round and not diamond-shaped. Where decay is present, the bolts may break through the boundary zone (see Chapter 7) and cause the decay to spread. The spreading decay will loosen the bolt so its holding power is reduced to that of the washers seated on both sides of the limb. In most cases, removal of extensively decayed limbs should be considered before bracing. However, braces may extend the time that even considerably decayed trees remain safe and attractive.

Braces should be installed during the dormant season. In the spring, the bark

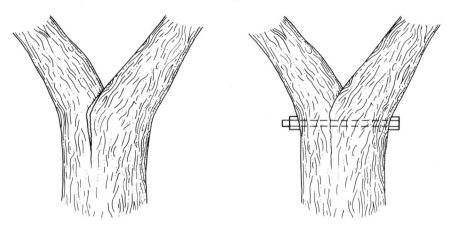

Fig. 8-2. Rods should be placed at or just below the split crotch. If two rods are used, they should be placed parallel to each other at the same level.

is easily injured and may split. Holes made during the fall are subjected to high levels of spores of decay fungi and thus a stand a good chance of becoming infected. Holes made directly above major roots often result in less chance of injury to the tree than holes made between major roots. Holes made closer to the base of the tree result in less injury than those made higher up on the trunk. Trees that are good compartmentalizers of decay (Chapter 7) and trees that are free of stress sustain less injury from bracing than slower growing trees. Wound dressings are not necessary and may promote infection.

Crotch bracing. The installation of artificial support may be justified on sound trees with weak crotches, although in many cases this results in crotch splitting elsewhere in the tree. To provide additional support, wire cables should be installed higher in the tree whenever the branches extend more than 20 feet above the crotch. In the past it was recommended that one or more parallel rods be inserted above the crotch. However, authorities now believe that rods should be placed at or just below the crotch (Fig. 8-2). If trees are bolted above the crotch, gusty winds may cause further splitting of the crotch.

Split limbs or trunks. Screw rods may be used to hold split limbs together or to support the sides of long splits in the trunk. They are usually placed about a foot apart and are staggered so as not to be in the same direct line of sap flow. Again, a fastened split may simply split elsewhere.

Rubbing limbs. Limbs that rub against each other can be braced together or kept separated by proper bracing. A single rod screwed through the center of each limb at the point of contact will hold the limbs rigidly together and thus eliminate rubbing. If it is desirable to keep such limbs separated, a length of pipe equal to the desired space between the two limbs can be placed over the rod before the branches are fastened together. If the limbs are to be separated and free movement between them is desired, the installation of a U-bolt in the upper limb, which rests on a wood block attached to the lower limb, may prove satisfactory for a short time. Eventually, the wood block will need replacement because of wear. If aesthetics and function permit, removal of one of the two rubbing limbs is advised.

Flexible Bracing. Galvanized steel strands (usually 7) twisted into cables are used to brace limbs together and to strengthen weak crotches. These generally are placed high in the tree to provide maximum support from the use of materials of minimum size and cost. Galvanized steel wire comes in common and extra-high-strength types, with the higher-strength wire being more brittle. The common wire cable is easier to use and to splice, so it is often preferred over the extra-high-strength wire.

Flexible bracing systems commonly employed are of several types. In the simple direct system, a single cable is installed to support two limbs arising from a single crotch (Fig. 8-3A). In the box system, cables are attached to all large limbs of a tree in a rotary manner. This arrangement provides lateral support and

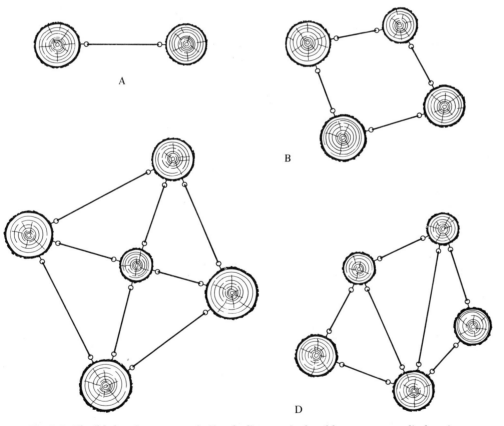

Fig. 8-3. Flexible bracing systems. A, Simple direct: a single cable supports two limbs arising from a single crotch. B, Box: cables attached to all large limbs in a rotary manner. C, Hub and spoke: cables radiate from a limb or metal ring in the center of the main branches. D, Triangular: cables placed in triangular form.

maximum crown movement, but little support to weak crotches (Fig. 8-3B). The hub and spoke system, in which cables radiate from a limb or metal ring in the center of the main branches (Fig. 8-3C), provides little direct or lateral support. In the triangular system, cables are placed as shown in Figure 8-3D to provide direct support to weak or split crotches and lateral support to the branches. Triangular bracing is probably the most efficient of the various types described.

The wire cables used for flexible bracing are placed at a point approximately two-thirds the distance from the crotch to the top of the tree. They should not rub against each other.

Lag hooks, hook bolts, bent eyebolts, and other materials are most often used to anchor the cables. Such anchors should be well separated (4–6 in.) when more than one is used on a single branch and should be placed in a straight line with the cable and as nearly at right angles to the branches as possible. Cables, once installed should be without slack and at the same time should have very little tension on them. They should prevent branches from spreading farther apart when they are subjected to heavy loads or strong winds.

Lag-hook installation. Cadmium-plated steel hooks ($1/2$ to $5/8$ in. diameter and 6-in. thread) are recommended as anchors for most branches. Rods should be inserted in branches at least 4 inches in diameter and to a depth of two-thirds of the diameter of the limb. The exact size selected is governed by the amount of stress and the size of the branches. Holes about $1/16$ inch smaller than the hooks are bored at the proper anchorage points, and the hooks are screwed in just far enough to allow a cable to be slipped over the open ends. After the lag hooks are in place, the limbs are pulled together sufficiently so there will be no slack in the cable after it is hooked on, but they are not pulled tight. Lag hooks are available that screw in opposite directions so that further tightening at each end of the cable can provide the right tension on the cable without unraveling it. A thimble is inserted over the hook and cable, and the cable is then passed through it and spliced to form a permanent union between the two anchors. All spliced areas should be coated with a metal preservative paint.

Eye bolt installation. Where wood decay is present or considerable stress occurs, eye bolts or hook bolts are generally used as anchor units. The hole for these is drilled completely through the wood, the bark is countersunk to a depth just below the cambium on the wood surface, and the bolt, washer, and nut are put in place. The cable ends are then inserted through the eyes and properly spliced.

Cabling systems should be checked every 10 years and either moved or removed.

Guying. Guy cables are used when other trees or branches are not available for cable attachment. Cabling to the ground may be necessary when other trees are removed and the retained tree is subjected to high levels of wind stress. The cables are attached to three separate points on the tree, usually on the main stem at the base of the crown. Compression springs added to the cable system will allow the tree to sway and reduce the danger of breakage at the cable attachment site. The cables are then attached to a soil anchor. The type of anchor required depends on the soil type and the level of anticipated strain. Simple screw-type anchors are appropriate for loam or clay soils. Cast iron cone anchors are useful in lighter soils. The common impact anchor driven into the ground by means of a sledge hammer may be used for low-strain situations, but this is ineffective where the guy cable is greater than 45 degrees to the plane of the ground.

Treating Fresh Wounds

After a windstorm that has left broken limbs, a lightning strike that has split open the trunk, or a vehicle collision that has ripped off a large section of bark, it is natural to want to do something for the injured tree. Good intentions very frequently lead to increased damage.

Broken Limbs. Broken limbs should be removed using the same guidelines given in Chapter 7 for pruning. Topical applications of wound dressings have been shown to be ineffective and may prolong the period of susceptibility to rot organisms. After limb removal is complete, the tree crown should be examined for split crotches or off-balanced limbs that may need artificial supports.

Bark Torn from Trunk. Large wounds such as those caused by vehicle collision (Fig. 8-4A) should be treated as soon as possible after the damage occurs. Any bark that is still moist and can be refitted on cambium should be tacked into place and covered with burlap to prevent drying out during reattachment. The dead bark and wood splinters may be removed after a year. If the loosened bark has already dried out, it should be carefully removed and the wound scribed to make smooth, curved edges (Fig. 8-4 B). In some cases, this may cause removal of cambium and bark tissues that could have aided in wound closure. Leaving even small pieces of bark and cambium on recently exposed sapwood will speed up the wound closing process (Fig. 8-4C). No other action should be taken except water, mulch, and fertilizer application as appropriate to maintain the vigor of the tree. Again, wound dressings are ineffective.

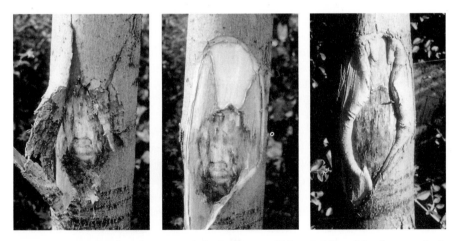

Fig. 8-4. A. Damaged and torn maple bark after a collision. B. Wound has been scribed, but some live cambium has been exposed at top of the wound. C. Wound partially closed some time later.

Lightning Damage. The injured tree should be inspected carefully before any attempt is made to repair the damage. Many trees are severely injured internally or below ground, despite the absence of external symptoms, and will soon die regardless of treatment. Consequently, expensive treatments should not be undertaken until the tree appears to have a good chance of recovery.

Where damage is not great, lightning-struck trees may benefit from immediate treatment. Before they have been allowed to dry out, the long, thin pieces of bark that have been split or lifted from the sapwood should be tacked back on with small nails. Then the area should be covered with burlap for a few days to allow the bark to reattach. Torn bark should *not* be trimmed immediately but should be left for a year, when it may be removed up to the living callus.

Lightning Protection. Trees growing in locations favorable for strikes as well as large, rare, or otherwise valuable trees can be adequately protected at comparatively small cost by the installation of proper equipment. The theories and details of installing lightning-protection equipment are too involved to permit discussion in this book. This information is available in Tree Preservation Bulletin No. 5, which may be obtained from the Superintendent of Documents, U.S. Government Printing Office, Washington, DC 20402, for a small fee. Installation must be entrusted to a specialist in tree care and should never be attempted by the layperson.

The principles governing the protection of buildings and other structures from lightning bolts are employed also in safeguarding trees. The major difference is in the materials used and methods of installation. Tree protection systems must have flexible cables to allow for the swaying of the trunk and branches and adjustable units to allow for the tree's growth.

A vertical conductor is fastened along the trunk with copper nails, from the highest point in the tree to the ground. The ground end is fastened to several radial conductors, which are buried underground and extend beyond the root area. These, in turn, are attached to ground rods 8 feet in length, which are driven vertically into the ground.

The system should be inspected every few years and the air terminals extended to the new growth, and other adjustments necessitated by expansion of the tree should be made. Ground terminals should also be checked periodically.

Cavity Filling

It is apparent that cavity fillings do not add strength to the tree and in most instances are not needed. However, filling can improve the appearance of some trees, so the one method of filling is described here (Fig. 8-5).

First, decayed wood is removed from the cavity. Care must be taken to remove only the softest rotted wood without damaging the surrounding healthy tissues. If the boundary zone separating invaded from healthy tissues is broken, decay will spread rapidly and the tree will suffer. Mechanical devices have been developed to remove the punky wood, but they should not be used or only used with great care.

The cavity may then be filled with any material that will not rub against living tissue. Painting of the interior of the cavity with orange shellac, creosote, tar, or asphalt paint is of no benefit.

It is tempting to bore drain holes to remove water from tree cavities. This practice will not help the tree because the boundary zone will be breached and decay may spread.

Although often used, concrete is not a good material for filling cavities. Concrete will not bend with the swaying of the tree. As a result, the concrete may crack or crumble or the wood may wear away at the point where friction occurs. Inability of the filler to flex with tree movement may result in further spread of decay and possibly breakage of the trunk. Thus concrete should be used only where there is little possibility of movement, for instance, small cavities near the base of the tree.

Polyurethane foam available for filling cavities is more suitable for larger cavities than is concrete. This material is tough, somewhat resilient, and adheres well to both sapwood and heartwood. Moreover, it is light in weight, can be poured into the cavity, expands and solidifies rapidly, is somewhat flexible after setting, and permits callus initiation along the edges of the filled cavity.

Fig. 8-5. Cavity filling.

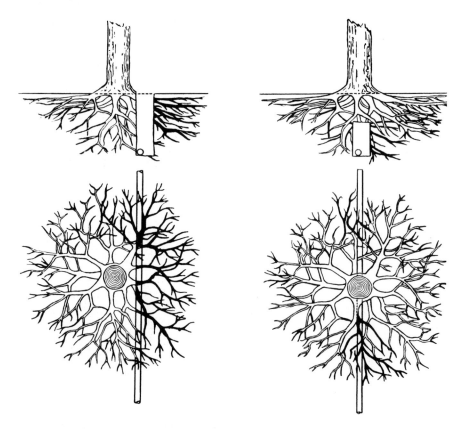

Fig. 8-6. Tunnel beneath root systems. Drawings at left show trenching that would probably kill the tree. Drawings at right show tunneling under the tree, which will preserve many of the important anchorage and feeder roots.

Preserving Trees at Construction Sites

Excavation for buildings, roadbeds, or pipelines is always potentially threatening to nearby trees. Even when care is taken to avoid hitting trees with heavy equipment, many trees die from root damage. Roots may be severed during excavation, suffocated by compaction or grade raising, or suffer from drought or standing water when drainage patterns change. Trees suddenly exposed to direct sun or reflected heat from paved surfaces may suffer from heat stress, and with newly injured roots or branches, sunscald. However, many trees can be saved if proper precautions are observed.

Excavation. The installation of water or sewer pipes or underground utility lines, digging out footings for a building foundation, or cutting out space for a sidewalk or roadway near trees may cause severe root damage. The degree of damage depends largely on how close the excavation comes to the base of a tree. The most critical zone encompasses the main root plate, the area adjacent to the base of the tree containing the largest roots. A trench crossing this area will threaten the life of the tree. It will not only deprive the tree of needed water, stored food, and minerals but will also interfere with anchorage and subject the tree to the threat of wind-throw. Extending out from the root plate to the edge of the tree crown is an area that contains a high density of important conducting roots. Excavations in this area will cause a serious threat to tree health. Since tree roots may extend out as far as twice the tree crown, even trenches that sever roots beyond the edge of the tree crown will cause damage.

The degree of damage may be reduced by careful placement of trenches to avoid large roots and by tunneling under rather than cutting roots greater than $1^1/_2$ inches in diameter (Fig. 8-6). Open trenches should not be routed beneath the crowns of trees that are to be preserved. If larger roots are severed, they should be cut off squarely. If the root system is appreciably reduced, water and mineral nutrient status should be monitored, especially during dry periods. Trenches should not be dug near trees located in wet areas because trees with inadequate anchorage may topple in the wet soil.

Trees can be saved from root destruction by tunneling under their root systems. In one case, only 2 out of 34 silver maples died after an underground utility line was placed alongside their crowns. These were 64-year-old street trees. The utility company tunneled under all roots greater than 2 inches in diameter. The city of Lexington, Kentucky, recently installed a sewer line beneath the root system of a 200-year-old bur oak using a tunneling machine that bored a hole from a pit excavated some distance from the tree. Modern tunneling systems are available that utilize remotely controlled "moles" capable of tunneling long distances entirely

Fig. 8-7. Damage to roots may be minimized by building bridges for walkways.

beneath tree root systems and dragging the cable or pipe behind them. In these systems, deep pits are not needed to site the equipment and initiate the tunnel.

Location of building, roadway, and sidewalk sites needs to be carefully thought out if trees are to be preserved.

There are alternatives to excavations for building foundations. Homes or walkways elevated on pilings will do less physical damage to tree root systems and allow preservation of the existing terrain (Fig. 8-7). Partial foundations to support a structure, much like a bridge over a river, physically destroys roots on only two sides of the building.

Lowering Grade. Leveling building sites often requires removal of soil around the base of trees. Significant root severance is likely if the level is lowered even a few inches. Injury to the tree as a result of even a slight lowering of the grade can cause injury due to soil compaction. Exposed large roots, if cut or broken, should be cut off squarely and covered with peat moss to prevent drying. If many large roots are cut or damaged, the tree will need additional care afterwards similar to the care provided to a newly transplanted tree. After grading, mulch should be applied to prevent desiccation of exposed roots. The remaining undisturbed roots should receive water and mineral nutrients as needed.

Under some conditions injury can be reduced by terracing the new slope around the tree, rather than cutting the soil away in a uniform gradient (Fig. 8-8). The terrace should extend at least to the edge of the tree crown and be sup-

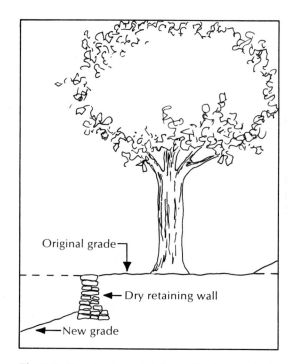

Fig. 8-8. Construct a retaining wall when grade is lowered.

Fig. 8-9. Piling soil over the root zone of trees can be very harmful. Soil compaction and root severance are also likely to cause tree decline in this construction site.

ported by a retaining wall. In this manner, the original soil depth and many of the conducting roots will be preserved. Tree water status should be carefully monitored. After the grade change, tree growth should be monitored for a few years to be sure that the tree has become adapted to the change.

Trees left along the edge of road cuts often continue to die for many years after the initial work was done. Trees should be removed some distance back from the top of the cut in proportion to the depth of cut.

Raising Grade. Extra soil may be placed beneath trees either to deliberately level the construction site or as a convenient way of disposing of excavated soil (Fig. 8-9). In either case, serious damage to trees may ensue. Because most tree species are sensitive to the addition of soil over their roots, excavation soil should never be piled, even temporarily, over the roots of trees which are to be preserved.

The injury to trees as a result of deliberate raising of grade more than a few inches can be reduced by following the procedure illustrated in Figure 8-10. The ground around the base and underneath the branch spread is cleared of all plants and sod, and the soil is then broken up without disturbing the roots. Commercial fertilizers are applied if needed. Four- or six-inch agricultural tile or split sewer pipes are laid in a wheel-and-spoke design with the tree as the hub. The radial lines of tile near the tree should be at least 1 foot higher than the ends joining the circle of tile. A few radial tiles should extend beyond the circle and should slope sharply downward to ensure good drainage. An open-jointed stone or brick well is then constructed around the trunk up to the level of the new fill. The inner

Fig. 8-10. Preventing injury by soil fills. Upper: Bird's-eye view. Lower: Side view showing A, the dry well; B, ground tile; and C, upright bell tile.

circumference of the stone well should be about 2 feet from the circumference of the trunk. Six-inch bell tiles are placed above the junction of the two tile systems, the bell end reaching the planned grade level, and stones are placed around the bell tiles to hold them erect. All ground tiles are then covered with small rocks and cobblestones to a depth of 18 inches. Next, the large rocks are covered with a layer of crushed stone, and these in turn with gravel, to a level of about 12 inches from the final grade. Cinders should never be used as a part of the fill. A thin layer of straw, hay, or synthetic filter material should be placed over the gravel to prevent soil from sealing the air spaces. Then good topsoil, which is the same or a coarser texture than the original soil, should be spread over the entire area except in the tree well and the bell tile. To prevent clogging, crushed stone should be placed inside the dry well over the openings of the radial tile.

The procedure outlined can be followed where a grade is to be raised around several trees in a group. In such cases the tile can run from one tree to another, increasing the air circulation.

The area between the trunk and the stone well should be either covered by an iron grate or filled with a mixture of 50 percent crushed charcoal and 50 percent sand, to make it less hazardous. This will prevent not only filling with leaves and consequent impairment of the efficiency of air and water drainage, but also rodent infestation and mosquito breeding. Bell tiles should be filled with crushed rock and covered with a screen.

The method outlined constitutes an ideal one to ensure the least disturbance by fills, but there are alternative methods. Where water drainage is not a serious problem, coarse gravel in the fill can be substituted for the tile. This material has sufficient porosity to ensure air drainage. Instead of bell tile in the system, stones, crushed rock, and gravel can be added so that the upper level of these porous materials slants toward the surface in the vicinity below the outer branch spread. Despite preparation of a structure to aid in soil aeration where grades are raised, trees sometimes decline and die. Thus, there is some risk involved. Avoidance of grade changes are best for tree preservation.

Compaction. Soil compaction can result from continual foot or vehicular traffic. Sometimes, before a new home is landscaped, a few inches of topsoil, or even subsoil, is placed over compacted soil. Plant roots become established in the upper soil layer. However, the compacted soil beneath can form an artificial hardpan and either restrict needed deep rooting or produce a high perched water table. Eventually the newly planted trees will suffer from insufficient or excess soil moisture and decline.

Compaction damage to existing desirable trees can be prevented on construction sites. Only one roadway should be used for movement of trucks and equipment to the building site. Piling thick layers of wood chip mulch over the tree root zone in the area likely to be compacted may reduce the impact of equipment traffic during construction. Grading should not be done while soil is very moist. Workers should be cautioned against trampling soil near trees. Barriers surrounding valuable trees should be erected at the dripline to keep machinery away.

In areas such as unpaved parking lots or playgrounds where compaction is unavoidable, the impact on trees can be reduced. It has been reported that expanded slate, rotary tilled into the surface soil, provided the best protection against compaction. The slate accounted for 33 percent by volume of the surface soil mixture. Expanded slate is a commercially available material that is made by a heat expansion process so that the slate becomes porous (about 50 percent pore space), inert, and rigid. A layer of wood chips placed to a depth of 3 to 4 inches over the soil surface was also found to reduce compaction in heavily used areas. Vertical mulching and soil fracturing with compressed air may also relieve compaction effects (see Chapter 3).

Paving. Grading and paving with either asphalt or concrete to build driveways or walkways can damage adjacent trees. If larger roots extend entirely across the intended drive or walkway, they may be bridged over. The roots may be wrapped in 2 to 3 inches of polystyrene foam so that as the root grows the foam becomes compressed. This system will allow roots to transport water and minerals from the absorptive roots on the opposite side of the paved walk or drive. In the long term, the covered roots could continue to expand and raise or crack the pavement.

Near valuable trees, paving materials that allow greater air and water penetration should be used in preference to asphalt or concrete. Alternative materials that are less damaging to tree roots include sand-set bricks, stones, or concrete pavers; gravel; decomposed granite; wood chips, blocks, or planks; and stepping stones. When using these materials, one should avoid deep excavation to minimize the number of roots severed. Specialized paving bricks with aeration holes are used in the parking area of the Missouri Botanical Gardens in St. Louis. They make an attractive and useful surface while protecting the trees that shade the parked cars.

Changing a Tree's Environment. Following construction, a tree once growing in partial shade may suddenly be exposed to full sun, perhaps exacerbated by reflected heat from buildings and pavements. Such a tree, often with an injured and reduced root system, may react to this new circumstance with leaf scorch, sunscald, and even death. On a less frequent basis, a tree growing in full sun may be shaded by a new building; trees that cannot tolerate shade will suffer. A change in soil environment to increased or decreased air, water, and mineral availability can be detrimental to tree health.

Tree Response to Construction Injury. One or more of the problems just discussed can result in gradual or sudden decline and death of the tree. The effects may not show up for several years, or they may occur within months. Frequently, tree decline involves leaf growth reduction, twig dieback, epicormic foliar growth, branch and limb dieback, and death. See Chapter 9 for a discussion of diagnosing these decline problems.

Adjusting to Sidewalk Lifting by Tree Roots

As tree roots growing under concrete walks increase in size, the walks may be lifted, creating a hazard to pedestrians. There are several remedies that, depending on circumstances, may help to solve the problem. Walks can be reconstructed to curve around the tree's basal roots, or a thin concrete slab, possibly with a foam cushion beneath it, can be utilized. Paver bricks that adjust easily to changes in terrain can be used. Root pruning and root control barriers can also be used, though drastic root pruning will harm the tree. Prevention of sidewalk damage by tree roots is discussed in Chapter 4.

Selected Bibliography

Anonymous. 1995. American national standard for tree care operations—tree shrub and other woody plant maintenance—standard practices. ANSI A300-1995, New York. 9 pp.

Blair, D. F. 1995. Arborist equipment: A guide to the tools and equipment of tree maintenance and removal. International Society of Arboriculture, Champaign, Ill. 291 pp.

Felix, R. E., and A. L. Shigo. 1977. Rots and rods. J. Arboric. 3:187–90.

Jeffers, W. A., and R. E. Abbott. 1979. A new system developed for guying trees. J. Arboric. 5:121–23.

Kozlowski, T. T. 1985. Soil aeration, flooding and tree growth. J. Arboric. 11:85–88.

Mayne, L. S. 1982. Specifications for construction around trees. J. Arboric. 8:289–91.

Schoenweiss, D. F. 1982. Prevention and treatment of construction damage to shade trees. J. Arboric. 8:169–75.

Shigo, A. L. 1986. A new tree biology. Shigo and Trees, Associates. Durham, N.H. 595 pp.

———. 1991. Modern Arboriculture. Shigo and Trees, Associates. Durham, N.H. 424 pp.

Yingling, E. L., C. A. Keeley, S. Little, and J. Burtis, Jr. 1979. Reducing drainage to shade and woodland trees from construction activities. J. Arboric. 5:97–105.

II

Diagnosis and Management of Tree Problems

9

Diagnosing Tree Problems

The authors are often asked by commercial arborists, county extension agents, city park departments, nurserymen, landscape architects, public utility companies, and homeowners to diagnose the cause of decline or death of shade and ornamental trees. For example, at the request of a public utility company, during the summer of 1956 P. P. Pirone examined and then diagnosed the cause of many abnormalities among more than 300 trees in New York City. Since then he examined several thousand more trees in New York, New Jersey, and Long Island, New York. In addition, the authors have examined thousands of diseased and declining landscape trees in Kentucky, California, Louisiana, Wisconsin, Indiana, England, and France. These activities have provided the experience needed to address the complex subject of tree problem diagnosis.

Successful treatment of any tree abnormality depends primarily on correct diagnosis. Treatment based on incorrect diagnosis may only make the problem worse. Some troubles occur so frequently on certain trees, and have such specific symptoms and signs, that they are readily diagnosed by the competent tree worker. Other troubles are so complex, unresearched, or deficient in reliable background information that even the most expert arborist cannot fathom them.

A reliable arborist will not hesitate to admit an inability to diagnose some abnormalities and will not hesitate to recruit the aid of a specialist to help in diagnosis, much as medical doctors often consult specialists in difficult cases. Corrective treatments of unknown value are to be avoided and would only be considered as a last resort after all attempts at an accurate diagnosis have failed.

Responsibilities of a Tree Diagnostician

Make Careful Observations. The primary responsibility of a good diagnostician is to make careful observations. The observer must not only scrutinize the obvious symptoms, but also look for hidden ones.

Many years ago, P. P. Pirone was asked to diagnose the cause of death of trees growing along the streets of a New Jersey community. One tree, a Norway maple 16 inches in diameter, was purported to have been killed by natural gas escaping from leaking mains. The only basis for this diagnosis was that a gas leak had been repaired in the general vicinity. Dr. Pirone removed the soil around the base of the tree and found that someone had completely girdled the tree below ground at least a year or so earlier. The uneven cuts made by an instrument were still plainly visible! Here was a case, then, where a little digging by the first "expert" might have resulted in an entirely different diagnosis.

Use Good Judgment. The most important qualification of a good diagnostician is plain ordinary common sense; some refer to it as a knack, intuition, or good judgment. Sometimes poor judgment is exercised when one unwisely becomes so absorbed in detailed symptoms that important gross symptoms are overlooked. On the other hand, failure to notice details can result in indefinite diagnoses such as "stress-related decline" when the real problem was something specific such as a scale infestation.

Essential elements of good judgment in diagnosis are a desire to learn more about the problem, to persist until the problem is solved, and to avoid quickly jumping to the wrong conclusion. This may sound like a detective solving a mystery, and in most cases that is exactly what is being done. One should be able to put together seemingly unrelated pieces of information to arrive at a solution. In order to solve a case of suspected phytotoxicity to a tree due to application of a herbicide, one of the authors had to examine a pesticide bill of sale from the previous property owner. In another instance, the observation of aphid honeydew on the rooftop of an automobile parked some distance away from the problem tree led to questioning about the car's usual parking place and the conclusion that the tree problem was caused by a previous heavy aphid infestation.

Understand the Tree. The diagnostician must have a thorough understanding of a normal tree. It is necessary to know the name of the particular tree species in question and its characteristics, such as winter-hardiness, tolerance to dry and wet soils, and reactions to other environmental factors.

Frequently, certain symptom patterns can provide clues as to where the problem lies if the tree's structure and the function of its parts are known. For example, a swelling on a branch coupled with abnormal red leaf color suggests that the phloem just below the swelling is impaired. By knowing that the phloem transports carbohydrates manufactured by the leaves downward to other parts of the tree one would expect that phloem dysfunction would result in carbohydrate accumulation, hence swelling just above that point. In addition, excess carbohydrates are sometimes converted to red pigment-forming sugars, hence red leaf color. Furthermore, when food is not moved to the roots, they starve and the tree may begin to show symptoms of nonfunctioning roots such as mineral nutrient and water deficiency. Therefore, look for a girdling wire, canker, or some other cause of phloem obstruction.

Understand the Causes of Tree Problems. To diagnose and treat a tree problem, one must determine its cause. Nonparasitic causes, insects and mites, and parasitic diseases are described in subsequent chapters. Diagnosticians must know the nature of nonparasitic factors and the biology of the disease or insect causes of disease to identify them and determine their role in the tree problem.

Very often there is no single cause of a tree problem. This is especially true of problems associated with decline and death of the tree. Trees are exposed to long-term factors that predispose them to decline. Such factors include old age, poor adaptation of a particular species, unfavorable climate, stressful urban growing conditions such as soil moisture, fertility, and compaction, or chronic air pollutants. Definite events in the tree's life may incite a more rapid slide towards decline. Such incidents include insect and disease defoliation, acute air pollutants and salts, mechanical and chemical injuries, and weather events such as winter freeze, spring frost, heat, drought, or floods. When in a weakened state due to the foregoing, trees may succumb to diseases and insects that contribute to decline. Insects such as bark beetles, diseases such as cankers, wood decay, and root decay, or other diseases caused by fungi, bacteria, viruses, and phytoplasmas often contribute to tree decline and death. So, just because a microbe or insect has been found on a declining tree, this does not mean that it is the whole cause of the problem. The tree diagnostician must be aware of these complexities to do a good job.

Ask Questions. The good diagnostician will not be afraid to ask questions. The diagnostician must know the history of the tree, such as when and how it was planted. The past climatic history including information regarding drought periods, severity of previous winters, and prevalence of hurricanes or other unusual weather needs to be known. This information is acquired either from the diagnostician's records or from the owner or person in charge of the tree. Critical events such as timing of construction activities and application of deicing salts may not seem important to the tree owner, so this information will not be volunteered and must be obtained through questioning.

The timing of symptom occurrence is also important. Unfortunately, many tree owners do not recognize early symptoms and think the tree is healthy until the "sudden" appearance of severe symptoms. In addition, tree owners, for some mysterious reason, often withhold information essential for proper diagnosis. Thus questions couched in diplomatic language sometimes yield information not otherwise available. After all, if a tree's trouble is the result of someone's mistake, questions asked in an accusing way will not yield much information.

Know the Tree's Environment. The tree diagnostician must also have a thorough understanding of the relationship between the soil and the tree. Is the soil properly drained? Is it well aerated? Does it hold enough moisture? Is it fertile, based on a soil test? The expert must also have a working knowledge of tree physiology, entomology, and plant pathology. Many of these subjects are treated in some detail in various parts of this book. Additional reading and certainly extensive field experience are necessary before one can really qualify as an expert diagnostician. The primary benefit of field experience is that, when confronted with an unknown problem, at least those maladies with which one has had experience can be ruled out, thus narrowing the remaining possibilities.

Importance of Correct Diagnosis

An incorrect diagnosis at the start will naturally lead to improper treatment. For example, a number of years ago P. P. Pirone was called to explain the branch dieback in several large oaks growing on an estate in Westchester County, New

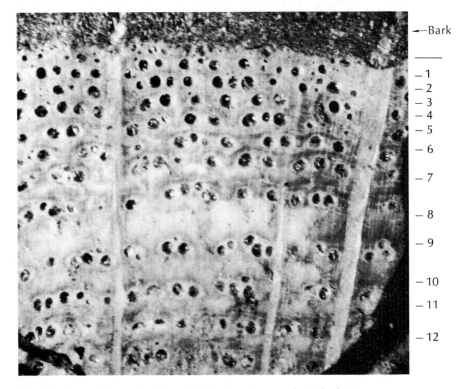

Fig. 9-1. Annual rings of white oak. Note that rings made in the last 6 years are closer to each other than those made 7 to 12 years earlier.

York. The problem was thought to result from a lack of mineral nutrients. The trees were fertilized for several years, but branches continued to die back until the top of one of the trees was nearly bare.

A little digging at the base of the trees and probing into the bark just below the soil line showed that the Armillaria root rot fungus had invaded and killed virtually all of the important tissues beneath the bark. In all probability, this fungus had been present in the roots for years and was the prime cause of the dying back of the top. All the fertilizer in the world applied to trees whose roots were diseased would not have helped.

In another case, Dr. Pirone was asked to determine the cause of death of a 150-year-old white oak (*Quercus alba*) on a private property in Glencoe, Illinois. The tree's owner had planned to sue the local utility on the assumption that the tree had died as a result of a natural gas leak about 45 feet away. The leaves on the tree were wilted and brown when examined in late October. This indicated that the tree had been at least partially alive earlier that growing season. Many trees, including another white oak of approximately the same age located closer to the leak, appeared to be perfectly healthy.

A study of the terrain and of the past history of the tree and the property elicit-ed the following information. A very expensive and beautiful house had been built about 7 years earlier, the front of which was within 25 feet of the tree. The grade had been changed. In the area nearest the house at least 2 feet of soil had been placed over the original soil level. In addition to raising the soil level, a cir-cular driveway of asphalt had been constructed around the tree. This paved area covered 67 percent of the absorptive root area of the tree.

One of the lower branches, about 4 inches in diameter, was removed from the tree and cut into 1-foot lengths for more detailed study. Examination of the annual rings in the laboratory revealed that the tree had begun to do poorly about 6 years earlier, or about a year after the house and the driveway had been con-structed. This is clearly indicated in the enlarged photograph of a cross section of the branch (Fig. 9-1). The rate of growth of a tree can be measured by the spac-ing between the large-holed cells that form in the spring and are known as spring wood; the wider the space between each row of spring wood cells, the more vig-orous the tree's growth.

The tree's demise was actually caused by several factors, including a change in grade, construction of the driveway around the tree, and the use of poor soil fill. When faced with the facts and observations gathered by Dr. Pirone, lawyers rep-resenting the owner of the tree refrained from damage claims against the utility supplying the natural gas.

One of the more difficult diagnoses involves troubles caused by nutrient defi-ciencies. Research on symptoms of and cures for mineral element deficiencies has been done for just a few landscape tree species. Generally speaking, foliage defi-ciencies of the various elements become apparent either at the base or at the tip of the current season's growth, depending on whether the element can move from one part of the plant where it has been used to another part, to be reused. A defi-ciency of nitrogen, phosphorus, potassium, magnesium, or zinc is often apparent first on the older leaves, because the plant moves these needed elements to the newest growing tissues. On the other hand, a deficiency of iron, manganese, sul-fur, calcium, boron, or copper is apparent first on the youngest leaves, because once incorporated into the older leaves these elements remain there and are not mobilized to satisfy the new growth (see Chapter 6).

Several state colleges of agriculture and private laboratories provide leaf nutri-ent analysis services for professional arborists. Increased use of these tests may aid arborists and tree owners in diagnosing and controlling ailments caused by nutri-ent deficiencies and toxicities.

Diagnostic Procedures

Standard procedures used by specialists in diagnosing tree troubles vary slightly with the individual. Some diagnosticians take more stock in symptoms above than below ground. Others, like the authors, feel that the most serious tree trou-bles and those most difficult to diagnose are more frequently associated with below-ground symptoms and factors.

Before examining any part of an ailing tree, one should study the general sur-roundings. Are nearby trees and other plants healthy? Have any special treatments been given prior to the appearance of the abnormal condition? Is the tree under diagnosis so situated that a leaf bonfire beneath it may have played a part in its decline? After these and questions related to patterns of distribution or timing of the problem have been answered, one should then proceed with the direct examination of the tree.

Examination of Leaves. The leaves constitute the best starting point because they are most accessible and are first to show outwardly the effects of any abnormal condition. Here, also, a complete understanding of a normal leaf is essential, because the size and the color of normal leaves vary greatly among the different tree species and even among trees of the same species.

Insect injuries to leaves are rather easily diagnosed, either by the presence of the pest or by the effect of its feeding. The leaves may be partly or completely chewed, or they may be yellowed as a result of sucking of the leaf sap, blotched from feeding between the leaf surfaces, or deformed from feeding and irritation.

Leaf injuries produced by parasitic fungi are not diagnosed as readily, however, because the causal organisms are usually not visible to the unaided eye. In some instances, tiny black pinpoint fungal bodies in the dead areas, visible without a hand lens, are indicative of certain types of causal fungi. Lesions resulting from fungal attack are more or less regular in outline with varying shades of color along the outer edges. They may range in size from tiny dots to spots more than half an inch in diameter. When several spots coalesce, the leaves may become blighted and die.

Atmospheric conditions preceding the appearance of spots on leaves can often be used to advantage in determining the cause of the disorder. For example, when leaf spots appear after a week or 10 days of continuous rains and cloudy weather, there is a greater likelihood that some parasitic organism is responsible, because such conditions are favorable for its spread. When leaves are spotted or scorched following a week or more of extremely dry, hot weather, lack of water (see Leaf Scorch, Chapter 10) may be responsible. Low temperatures in late spring may also result in injury to tender leaves.

Changes in leaf structure, appearance, or function may result from such widely different causes as air pollution, deficient or excessive moisture, lack of available minerals, poor soil aeration, root injuries, or diseases. Some of these can be disregarded if nearby trees of the same species as those under study are perfectly healthy.

Examination of Trunk and Branches. A careful inspection of the branches and trunk should follow examination of the leaves. Sunken areas in the bark indicate injury to tissues beneath. They may have been produced by fungal or bacterial infection or by nonparasitic agents, such as low and high temperatures, or even by improper pruning. The presence of fungal bodies in such areas does not necessarily indicate that the fungus is the primary cause. Only a person with considerable mycological training can distinguish the pathogenic from the nonpathogenic species. The diseased wood beneath the bark, if it is the direct result of fungal attack, shows a gradual change in color from diseased to healthy tissue. It is usually dark brown in the earlier, more severely infected portions, then changes to olive green or light brown, finally reaching a lighter shade of either in the more recently affected parts. On the other hand, injuries resulting from low or high temperatures are usually well defined, an abrupt line of demarcation appearing between affected and unaffected tissues.

The bark of the trunk and the branches should also be examined for small holes, sawdust, frass, and scars or ridges, which indicate borer infestations in the inner bark, sapwood, or heartwood. As a rule, most borers become established in trees of poor vigor; thus it is necessary to investigate the cause of the weakened condition rather than to assume that the borers are primarily responsible. Branches and smaller twigs should always be examined for infestations of scale insects. Although most scales are readily visible, a few so nearly resemble the color of the bark that they are sometimes overlooked.

Branches or twigs with no leaves or with wilted leaves should be examined for discoloration of the sapwood, the usual symptom of wilt-producing fungi. The service of a mycologist is needed for determining the species of fungus involved, because positive identification can be made only by laboratory isolations from the discolored tissue.

The appearance of suckers or water sprouts along the trunk and main branches may indicate a sudden change in environmental conditions, structural injuries, disease, or excessive, incorrect, and ill-timed pruning.

Look for galls, or swellings on the twigs and branches. Many are insect-caused galls, and several, such as rust swellings and crown galls, are caused by fungi and bacteria. Although galls or overgrowths occasionally present on the main trunk may be caused by parasites, many are induced by irritations to the tree not clearly understood by scientists.

The general vigor of a tree usually can be ascertained from the color in the bark fissures and the rapidity of callus formation over wounds. In vigorously growing trees, the fissures are much lighter in color than the bark surface. A rapidly developing callus roll over the wound indicates good vigor; however, it does not necessarily mean that the tree is good at limiting internal decay.

Examination of Roots. Because of the inaccessibility of roots, many arborists rarely inspect them. In diagnosing a general disorder in a tree, however, the possibility of root injury or disease must be carefully considered. More than half of the abnormalities in the hundreds of street and shade trees examined by the authors were found to be caused by injuries to or diseases of the root systems.

The sudden death of a tree usually results from the destruction of nearly all the roots or from the death of the tissues at the trunk base near the soil line. The factors most commonly involved in such cases are infection by fungi such as the Armillaria fungus, the Verticillium wilt fungus, and the Ganoderma fungus, winter injury, rodent damage, lightning strikes, and toxic chemicals like gasoline, oil, salt, and certain weed killers. Trees that become progressively weaker over a period of years may be affected by girdling roots, decay following injury caused by sidewalk and curb installations or road improvement, poor soil type, poor drainage, lack of food, changes in grade, and excessively deep planting from the start.

Nematodes are responsible for the decline of many shade trees. The root-knot nematode *Meloidogyne incognita* produces small swellings or knots on the roots of susceptible trees. Such swellings or knots are not always caused by nematodes;

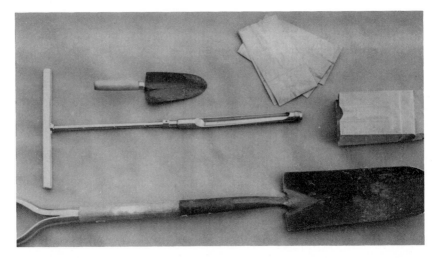

Fig. 9-2. Soil sampling tools: spade, soil auger, trowel, collection bags.

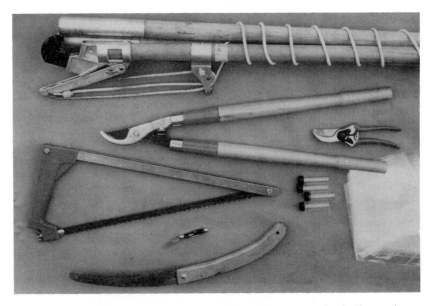

Fig. 9-3. Foliage and limb sample collection tools: folding saw, pocket knife, saw, loppers, pole pruner, collection vials, hand shears, collection bags.

those found on the roots of alder, Australian-pine, and Russian-olive may be natural outgrowths or nodules associated with nitrogen fixation.

Diagnostic Tools

Certain implements are essential in diagnosing tree troubles. Most important of these is a shovel or small spade for removal of soil around the base of the trunk to facilitate detection of girdling roots, rodent damage, winter injury, and infection by the Armillaria fungus or other parasitic fungi. In addition, a spade is used for digging soil in the vicinity of the smaller roots to be examined for the presence of decay. A soil auger or soil sampling tube is essential for collecting representative samples of soil for analytical tests and for determining the nature of the subsoil and the drainage (Fig. 9-2). Pruning shears and a pocket knife are necessary for cutting twigs and small branches to determine the presence of discoloration in the sapwood caused by wilt fungi and other injuries in the inner bark, cambium, and wood. A small saw is helpful where branches larger than can be handled with pruning shears must be cut (Fig. 9-3). The authors find that a portable pole pruner, either telescoping or assembled from short lengths, is excellent for cutting samples.

A most important tool is a chisel or a curved gouge with a heavy handle for tapping the bark to locate dead areas and for making incisions into the bark and sapwood (Fig. 9-4). Bark and sapwood specimens needed for further study are also collected with this instrument with the aid of strokes from a 2- or 3-pound composition mallet. A sharp instrument such as an ice pick is also useful to help determine the extent of soft, decayed wood.

Often, to make an accurate diagnosis, it is necessary to view or visualize what is occurring inside the trunk or branches. Descriptions of tools and instruments used for internal examination of the tree are presented later in this chapter under the section on identifying hazard trees.

A hand lens is valuable for detecting the presence of red spider mites and tiny

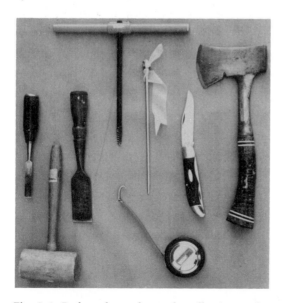

Fig. 9-4. Bark and wood sample collection tools: small wood chisel, mallet, large wood chisel, increment borer, pocket knife, hatchet, diameter tape.

fungal fruiting bodies in leaf lesions. Binoculars are useful for making observations of symptoms where sampling is not possible.

Screw-top vials should be available for collecting insects for identification using reference books in the office or for submission to a professional entomologist for identification. Plastic bags for holding leaf specimens and paper bags or other containers for holding bark and sapwood specimens, or soil samples, should also constitute part of the diagnostician's working kit.

If the diagnostician has a laboratory, additional tools and instruments are helpful for tree problem diagnosis. A soil pH meter and a device for measuring soil-soluble salts are needed to provide preliminary information for diagnosing some soil nutrient and fertility problems (Fig. 9-5). For soluble salts determinations, it is important to follow the instructions provided by the manufacturer. Different instruments express soluble salts levels in different ways, but they should have charts provided that can link numerical values with possible plant injury levels. At the least, such a chart should contain numerical values that would discriminate among soils having low fertility, optimum fertility, high fertility, and injury to most crop plants, including trees. The diagnostician must take into account recent weather and irrigation, soil type, and plant species before concluding that excess soluble salts have caused an injury. Other models of portable pH meters are available, including some that can simply be plunged into the soil on site.

A dissecting microscope ranging in power from 10x to 70x provides enough magnification to observe tree rings, fungal fruiting structures, nematodes, and insects (Fig. 9-6). Dissecting tools such as scalpels, picks, and blades are also needed. A compound light microscope is helpful to people familiar with its use.

Fig. 9-5. Left, conductivity meter, a device for measuring soil soluble salts. Right, pH meter in use.

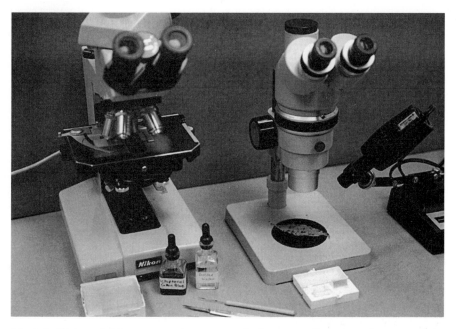

Fig. 9-6. Diagnostic laboratory tools: compound microscope, dissecting microscope, light source, cover slips, scalpel, pick, microscope slides, water, stain.

This microscope, generally ranging in power from 50x to 1000x , can be used to detect fungal spores, bacteria, viral inclusion bodies, and plant abnormalities. Transparent tape can be used to make quick microscope mounts of fungi growing on the surface of leaves and stems. Glass slides, cover slips, and dropper bottles containing water and some stains also are needed.

Portable "laboratories" containing some of the diagnostic equipment listed are commercially available.

A Sample Diagnostic Questionnaire

Following is a sample questionnaire that may be sent to tree owners who desire a tree diagnosis by mail:

Name And Address of Sender

The Problem

What are you observing to be wrong with the tree? Describe any abnormalities you see, with regard to size, color, shape, or death of leaves, twigs, branches, limbs, trunk, and roots.

General Questions

1. Kind, age, and size of tree?
2. Where is the tree situated? Along street, on lawn, in a park? Near body of water, salt or fresh? On level ground or on a slope?

3. How long has the tree exhibited the trouble? If the trouble appeared suddenly, describe the weather conditions occurring just previously. Describe any other unusual conditions.
4. Is the trouble visible all over the tree, on only one side, in the lower or upper branches?
5. Do any other trees of the same species in the near vicinity show the same symptoms? Do other species show it?
6. How much annual growth have the twigs made during the past 3 years?
7. Has the grade around the tree been raised or lowered during the past 7 years? If so, explain amount of change and describe the procedure and type of fill used.
8. Has any construction work been done nearby within the past 3 years—house, road, driveway, curb, garage, or ditches for laying water or sewer pipes?
9. What work has been done on this tree recently? Has it been pruned within the past 2 years? If so, how much? Fertilizer or pesticide treatments—when, what, and how much?
10. If a young tree, how long since it was planted in its present location? How deeply was it planted? (Use a spade to determine depth of roots.) What treatments were given during the first year after transplanting?

Questions about the Soil

Answers to these questions are extremely helpful, particularly if the tree has been dying back slowly over a period of years.

1. What kind of soil surrounds the tree? Sandy, loamy, or clayey? What soil cover—asphalt, cement, crushed rock, cinders, sand, grass, mulch, weeds, or no cover? How much open area is there?
2. What is the depth to subsoil, to rock or shale, to hardpan?
3. What is the pH of the soil?
4. Does water stand on the soil after a heavy rain?

Questions about the Roots

These are the most difficult questions to answer, since considerable digging may be required. The answers are most important, however, especially when a general disorder is involved.

1. Is a girdling root present? Sometimes such roots are well below ground and you may have to dig a foot or so before you can be sure.
2. Are the larger roots normal in color? Do they have rotted bark or discolored wood? If so, submit specimens.
3. What is the appearance of the finer roots? Are root hairs abundant and white? (Dig down a foot or so beneath the outer spread of the branches.)

Questions about the Trunk

1. Are there any long, narrow, open cracks present? If so, which direction do they face?

2. Are cavities present? If so, describe size of opening, condition of interior if unfilled. If filled, give details.
3. Is there any bark bleeding? If so, how extensive? Submit specimens including bark and sapwood.
4. Are there any swollen areas? Describe.
5. Is there a swollen area completely around the trunk? If so, cut into the swelling with a chisel to determine the possible presence of some foreign object such as a wire.
6. Are there any fungi (mushrooms or bracket-type) growing out of the bark? If so, include a few specimens.
7. Are there cankers (dead sunken areas in the bark)? Submit specimens.
8. Are there any borer holes or other evidences of insect activity?
9. Is the bark at or just below the soil line healthy? (Use a curved chisel to determine this.)
10. If wood beneath this bark is discolored, describe color and extent. Submit several pieces of bark and wood.

Questions about the Branches

1. Is the bark cracked for some distance? On what side of the branches?
2. Are there any cankers in the branch?
3. Is there any discoloration in the branches or twigs that have wilted leaves or which are leafless?

Specific Possibilities

Did the trouble appear

1. Immediately after a thunderstorm?
2. After chemicals were injected into the trunk?
3. After sprays were applied? (Name the ingredients used and indicate when applied.)
4. After weed killers were applied in the vicinity?
5. After treating a nearby house for termite control?
6. After any other chemical treatment?

Shipping the Specimens

By the time you have answered the questions that pertain specifically to your particular case, you are ready to collect and ship specimens.

Woody specimens without leaves should be wrapped in newspaper and packed in a cardboard box. Twigs and small branches should be cut to approximately 1 foot in length.

Include with diseased branches or twigs a portion of the adjacent healthy parts. (When submitting specimens be sure to give the name of the tree.)

Leaves should be wrapped dry, preferably individually, in ordinary waxed paper or wrapping paper and placed in a cardboard box. Never wrap leaves in wet paper toweling, wet cotton, or plastic bags because many of the specimens will be decayed or covered with all kinds of organisms by the time they reach the diagnostician.

Fig. 9-7. This silver maple, weakened by decay, toppled during a thunderstorm.

Trunk specimens should include a portion of the dead bark and sapwood. If possible, adjacent healthy tissue also should be included on the same specimens. Bark and sapwood specimens should be about 2 inches long and 1 inch wide and separately packed in waxed paper.

Label all specimens and have this label correspond with the description in your letter or on your questionnaire.

Certain answers to the questionnaire will result in immediate recognition of the problem by most people, especially readers of this book. Many state university colleges of agriculture have plant disease diagnostic laboratories staffed with professional diagnosticians, and diagnostic forms with questions appropriate for their lab. Most county extension offices have access to such laboratories, and specimens taken there are often mailed to the university lab. Some private diagnosticians also provide questionnaires and shipping instructions.

Identifying Hazard Trees

In urban settings, dropping branches and falling trees occasionally pose a hazard to people and property. Healthy trees with their upright and spreading growth have a stable structural design that balances the need for photosynthesis, anchorage, and moisture acquisition. Sometimes, however, due to disease, mechanical injuries, or exceptional mechanical forces such as windstorms, the tree structure breaks down. Trees break down, sometimes with disastrous results, when weakened branches drop off, or the tree falls because the trunk has snapped (Fig. 9-7) or the roots have given way.

It should be the legal responsibility of the tree owner or the tree maintenance person to act prudently to insure that their trees are not a liability under normal circumstances. It is the responsibility of the arborist to recognize hazard trees and make recommendations to have something done about them before a catastrophic event occurs. Although the tree diagnostician cannot predict extreme weather events such as tornadoes, hurricanes, and ice storms, which destroy even healthy trees, most hazard trees that could fall during normal weather can be identified and removed before they cause serious injury, provided there is a will and there are resources to act.

The ISA has published a detailed guide, cited at the end of this section, to the process of hazard tree evaluation. The assessment is based on the likelihood of tree failure due to disease, injury, unstable structure, or location; the size of the part likely to drop or fall; and the importance of and likelihood of there being a target in the place where the tree lands. The target is important in hazard tree evaluation because a falling tree in the middle of a forest is not the hazard that a similar tree would be in a school playground.

External evidence. Much of the job of identifying hazard trees can be done visually. Many of the visual indicators and procedures are the same as those used in diagnosing tree diseases. Following is a partial list of tree hazard indicators that, if present, are cause for concern and a stimulus to look more closely at the problem to determine if a hazard really exists.

1. Tree branch and foliage appearance (indicating a root or lower trunk problem): diminished size, frequency, and health of buds; decreased annual twig growth; reduced canopies; unsatisfactory size, density, and color of foliage; epicormic growth; deadwood and dieback in the crown.
2. Tree structure: poor crown balance; multiple branch attachments; narrow crotch angles between trunks or between branches and trunk, especially with included bark (Fig. 9-8); long, slender nontapering branches; abnormal crooks in branches; and leaning trees.

Fig. 9-8. Ash tree with co-dominant trunks that are prone to eventually splitting apart.

3. Tree diseases and defects: canker diseases; trunk and branch cracks, splits, and bulges (Fig. 9-9); dead bark; bark texture changes; weeping wounds; cavities and hollows (Fig. 9-10); topped trees; branch stubs; flush cuts; dead wood and broken, hanging branches; root decay; deep stem fluting; lack of basal flare; and girdling roots.

4. Biological indicators: fruiting bodies of decay fungi on buttress roots, trunk, or limbs; fungal mycelial mats or fans; fungal rhizomorphs; insect emergence holes; insect frass; bird or mammal nesting holes; and bee colonies inside the tree.

5. Site factors: nearby building construction; trenching through the root zone; changes in water drainage patterns; soil compaction (Fig. 9-11); clearing of a densely wooded site leaving remaining trees exposed to wind; soil erosion; changes in grade such as cuts and fills; extremely light soils providing poor root anchorage.

Hidden evidence. Less accessible, but very important to tree hazard determination, is the partially buried area at the base of the trunk, which includes the buttress or flare roots. The root flare area may need to be carefully excavated to reveal decay, fungal mycelium and rhizomorphs, dead bark, injuries, cracks, and other tree defects. For a tree to be safe and healthy, it is necessary for most of the lower trunk and buttress roots to be free from injury and disease.

Internal evidence. Decay of the wood inside the tree can often be foretold by visual evidence such as fungal fruiting bodies, hollows, and cracks. Nevertheless, to survive, grow, and have healthy foliage, trees need only intact bark and a few outer rings of the wood, so a badly decayed tree might not always appear to be a hazard. Therefore, it is important to determine just how much decay is present when making a hazard assessment. The presence of decay and a cavity (Fig. 9-12) is not necessarily an indication that tree removal is necessary. If the thickness of the healthy shell of the trunk surrounding the decayed center is at least one-third of the trunk radius, the risk of failure is about the same as for a completely sound trunk. Normal wind forces shouldn't cause either trunk to break. An apt analogy

Fig. 9-9. Two indicators of hazard, an abnormal crook in the limb, and a deep crack, are present in this tree.

Fig. 9-10. This cavity at the base of the tree may indicate trunk rot or root and butt rot.

Fig. 9-11. Soil in the tree root zone may become compacted when the area is used often for temporary parking.

Fig. 9-12. The hollow present in this tree is obvious. Less obvious is evidence of how extensive the decay might be in other parts of the tree.

would be a water pipe compared to a solid rod, Most trees with cavities do not have a complete shell, but if the sound shell extends more than three-fourths of the circumference, the shell can probably withstand normal forces.

There are tools that can be used to determine what is inside the tree so that a more accurate picture of the tree strength can be obtained. A simple sounding of the trunk with a mallet to determine the extent of the hollow can be useful for detecting a cavity, but it does not tell how much decay is present. The sound impulse hammer (e.g., Metriguard hammer) is an electronic instrument that measures the velocity as sound waves through wood. Defect-free wood has a characteristic sound velocity pattern that would be disrupted if the tree wood contained cavities or cracks. The travel time of sound between a set of screws a known distance apart and that are affixed to the sapwood is measured, and the instrument calculates sound velocity. The sound impulse hammer makes only small injuries to the tree. Ultrasonic instruments such as the Silvatest and Arborsonic detectors work on the same principle, but with an automatic impulse generator and a system for recording transmitted and reflected waves. Detectors using sound for determining decay take little time to use and will reliably reveal the presence of tree defects, although perhaps not their extent.

An increment borer (Fig. 9-4) facilitates sampling for study of the annual growth rate of the tree (dendrochronology) and for determining the extent of internal wood decay. One must be reasonably careful using this instrument on

Fig. 9-13. The Resistograph 500 bores a tiny hole in the tree while simultaneously recording the resistance to penetration.

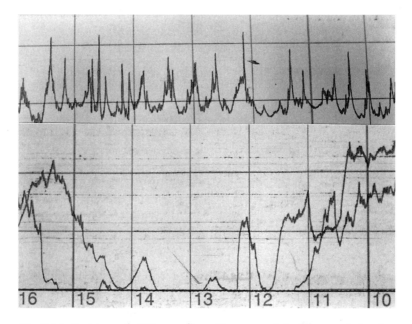

Fig. 9-14. Resistograph 500 recordings. Top: normal, healthy wood showing 3–4 growth ring peaks per centimeter. Bottom: Unsound, decayed wood, as determined with two recordings, is indicated between 11 and 15 centimeters.

living trees, for discoloration and decay could develop around holes made by it. An increment hammer is also available for collecting shallow samples of the bark and wood. A sample taken with an increment borer allows the investigator to examine directly the wood for evidence of decay. Although it is invasive and leaves a hole up to one-fourth inch in diameter, it is useful to withdraw a specimen for culturing the decay fungi or to measure the strength of the core sample withdrawn. The increment borer core sample, when used with an instrument called a Fractometer 150 FE, allows the tree maintenance expert to determine the breaking strength of the wood core sample withdrawn from the tree. This assessment may be needed in cases where there needs to be verification of the wood strength because there is uncertainty as to the exact extent or kind of decay. If it is clear that more than one-third of the tree radius is sound wood, there is no need to do increment corings.

A portable electric drill with a long narrow twist drill bit can be used to determine interior decay and hollow conditions by noting changes in sawdust color, texture, and odor and changes in resistance to penetration at various depths. A more precise variation of this method is provided by the Resistograph F500 (Fig. 9-13) decay detection device. A thin, about $1/16$ inch diameter, probe rotating at high speed is inserted into the tree and penetrates up to a depth of almost 20 inches at constant speed. The changes in power demands on the electric motor, resulting from passing through sound wood and decayed wood, are printed out on a recording device at the same scale as the distance traversed. Compression wood, annual rings, and hollows and cracks may be more accurately located in this way (Fig. 9-14). Because the probe is thin, the wound left in the tree is minute and presumably less attractive to decay organisms. Any drill frass produced is left in the hole. Surface roots, buttress roots, limbs, and trunks can be examined in this way.

Other detectors employing the concept of wood penetration, such as the Densitomat-400, the Decay Detecting Drill (DDD 200), and the Resistograph 1410, operate on the same principle as the Resistograph F500 and also provide a graphic record of the results of drilling. In some cases, results are obtained graphically via a connection between the drill and a portable personal computer with printer. These drilling devices make very small holes in the tree and provide direct, fairly detailed data for the arborist. Interpretation of the results requires much experience and expertise.

Electrical resistance measurements inside the tree will show a difference between decayed and healthy wood. Some diagnosticians are making use of the Shigometer to detect internal discoloration and decay in the tree and also to provide a relative measure of its vitality. Through a tiny predrilled hole, an electrode provides data on the relative resistance or conductivity of the tree tissues being tested. Decaying wood, the cells of which are disrupted, contains higher levels of moisture and salts and therefore conducts electricity more readily than healthy wood. Completely hollowed-out parts of the tree are not electrically conductive and yield a reading reflecting less conductivity than healthy wood. Thus without cutting the tree, one can gain insight into its internal condition. The tiny hole bored for the probe is also less invasive than increment corings. Operating on a similar electrical conductivity basis, the Vitamat uses a needlelike high-pressure probe to obtain internal conductivity readings. This instrument is,

Fig. 9-15. A Shigometer in use for determining relative vitality of cambial tissues in the tree.

however, capable of making deep wounds in the tree. Using another kind of probe with two sharpened pins, electrical resistance/conductivity of the cambium and outer xylem can be measured with the Shigometer (Fig. 9-15). Electrical conductivity of dead tissues examined in this way would be expected to be less than that of healthy tissues. If large numbers of trees are examined at the same time in exactly the same way, the relative vitality of individuals can be determined. For these instruments to work well, they require calibration, and experience is needed for accurate interpretation of the results.

New noninvasive technology such as thermography (measuring a tree's radiant heat), ultrasonic tomography (using sound sensors outside the tree with computer analysis of data), echography (radar), and computed tomography (like the CAT scan used in medicine) are being tried experimentally to detect decay in trees. As tree decay detection technology advances, it may be possible someday to accurately map out the extent of decay inside the tree. If future technology brings computer-aided 3-dimensional models and species-specific tree strength formulas, it should make hazard tree evaluation even more precise. Less sophisticated, but novel, is the idea of identifying trees with decay by using "sniffer dogs" that are sensitive to volatile compounds formed when trees decay. All these internal indicators must be combined with external and biological evidence in order to make a sound judgment about the tree condition.

Managing large numbers of trees for hazards. Municipalities and other managers of public trees should be keenly aware of potential liabilities associated with structurally unstable trees. Conscientious municipal arborists will want to implement a proactive plan that soundly manages risks from tree hazards. A long-range tree hazard risk management plan must begin with an inventory showing species, size, condition, and location of all trees under jurisdiction.

1. Species. In most areas, trees such as silver maple, Siberian elm, willow, poplar, cottonwood, and boxelder have a high potential for breaking or falling and should be monitored closely. A plan for gradual removal of problem species like these can be formulated. Even if the trees are young, it is easier to remove them at that stage than to wait for them to become larger and more dangerous.
2. Size. Large trees present more risk than small trees because they are more massive and have farther to fall. They need routine monitoring.
3. Condition. Any tree recorded in the inventory as being in poor condition or with structural defects must be examined closely. Large trees in poor condition should be removed first. Small trees in poor condition are unlikely to grow up to become healthy and strong, so they should be removed and replaced as resources permit.
4. Location. Trees in schoolyards, along crowded walkways, and at busy intersections pose greater hazards because of the increased chance of damaging property or injuring people. Such trees must be monitored more closely than trees in passive park areas, for example.

The success of communities implementing a plan to reduce risks associated with hazard trees is ultimately dependent on the resources available. Developing a long-range plan and systematically anticipating and mitigating hazard tree problems will help avert a future tragedy. Improved risk assessment also ensures that hazardous trees are removed and nonhazardous trees are not removed unnecessarily.

Other tree hazards. Trees growing in urban areas pose other risks to people and property. In certain clay soils that shrink as they dry out, tree roots can cause a change in soil volume by extracting the water from the soil. When this happens, nearby buildings may settle and crack due to soil subsidence. Reducing risks of subsidence for buildings near trees requires the expertise of both an arborist and a structural engineer.

Tree roots are capable of lifting concrete sidewalk slabs, creating a risk for pedestrians (see Chapter 4). If sidewalks are constructed properly with thick slabs and deep foundations or are underlain with engineered load-bearing soil, tree roots will not lift them.

Tree roots can enter leaky underground drain pipes and block them. Intact pipes are not subject to root penetration.

Selected Bibliography

Anonymous. 1973. A tree hurts too. U.S.D.A. Forest Service. NE-INF. 16-73, Washington, D.C. 28 pp.
Anonymous. 1995. The tree in its various states. Proceedings of the Second European Congress in Arboriculture. Versailles, Sept. 1995. 154 pp.

Bethge, K., C. Mattheck, and E. Hunger. 1996. Equipment for detection and evaluation of incipient decay in trees. Arboricultural Journal 20:13–37.

Drooz, A. T. 1985. Insects of eastern forests. U.S.D.A. Forest Service Misc. Publ. 1426, Washington, D.C. 608 pp.

Farr, D. F., G. F. Bills, G. P. Chamuris, and A. Y. Rossman. 1989. Fungi on plants and plant products in the United States. APS Press, St. Paul, Minn. 1252 pp.

Grogan, R. 1981. The art and science of diagnosis. Ann. Rev. Phytopathol. 19: 333–51.

Hepting, G. H. 1971. Diseases of forest and shade trees of the United States. U.S.D.A. Forest Service Agriculture Handbook 386, Washington, D.C. 658 pp.

Horst, R. K. 1982. Westcott's plant disease handbook. 4th ed. Van Nostrand Reinhold, New York. 803 pp.

Johnson, W. T., and H. H. Lyon. 1988. Insects that feed on trees and shrubs. 2nd ed. Cornell University Press, Ithaca, N.Y. 556 pp.

Lloyd, J. 1997. Plant health care for woody ornamentals. International Society of Arboriculture, Champaign, Ill. 223 pp.

Manion, P. D. 1981. Tree disease concepts. Prentice-Hall, Englewood Cliffs, N.J. 399 pp.

Matheny, N. P., and J. R. Clark. 1994. Evaluation of hazard trees in urban areas. 2nd ed. International Society of Arboriculture, Champaign, Ill. 85 pp.

Nicolotti, G., and P. Miglietta. 1998. Using high-technology instruments to assess defects in trees. J. Arboric. 24:297–302.

Pirone, P. P. 1978. Diseases and pests of ornamental plants. 5th ed. Wiley, New York. 566 pp.

Rose, A. H., and O. H. Lindquist. 1982. Insects of eastern hardwood trees. Can. For. Serv. For. Tech. Rep. 29. 304 pp.

Shurtleff, M. C., and C. W. Averre III. 1997. The plant disease clinic and field diagnosis of abiotic diseases. APS Press, St. Paul, Minn. 245 pp.

Sinclair, W. A., H. H. Lyon, and W. T. Johnson. 1987. Diseases of trees and shrubs. Cornell University Press, Ithaca, N.Y. 574 pp.

Strouts, R. G. 1994. Diagnosis of ill-health in trees. HMSO Publications, London. 308 pp.

Tattar, T. A. 1978. Diseases of shade trees. Academic Press, New York. 361 pp.

Wallace, H. R. 1978. The diagnosis of plant diseases of complex etiology. Ann. Rev. Phytopathol. 16:379–402.

10

Damage Due to Nonparasitic Factors

Poor growth and even death of trees are frequently caused by nonparasitic agents. Nonparasitic disorders generally affect a wide range of tree species, whereas most parasites such as disease organisms and insects usually attack specific tree species. Following are some of the most common nonparasitic problems.

Damage Caused by Grade Changes and Construction Around Trees

The addition or removal of soil may seriously disturb the delicate relationship between roots and soil and thus result in considerable damage or death of the tree. Yet such procedures are done when new homes and highways are built, lawns or terraces graded, or street improvements made.

The authors are often asked to diagnose the cause of the gradual weakening and even the death of beautiful oaks, maples, or beeches on properties in newly developed residential sections (Fig. 10-1). When homes are built on wooded lots, the possibility of soil fill injury is a primary consideration. First the trunk base is examined to see whether it flares out. Where no fill has been applied, the trunk is wider at the soil line than it is a foot or so above. If the trunk has no flare at its base and it enters the ground in a straight line, then a fill has probably been applied. Some digging with a spade will then reveal the depth of fill (Fig. 10-2).

Other visible symptoms are small yellow leaves, premature fall color, the presence of numerous suckers (epicormic growth) along the main trunk and branches, many dead twigs, and in some instances large dead branches. When the owner is questioned as to what precautions were taken to minimize the shock of placing the roots in a new environment, the usual answer is that no precautions were thought necessary. The building contractor dug the subsoil from the cellar location and spread it around the trees. A foot of good topsoil was placed over this and the job was done! When symptoms become visible a few months to several years later, the results are angry and frustrated homeowners who had built the house of their dreams on a costly wooded lot.

Why Trees Suffer from Soil Fills. The moment a blanket of soil is placed over the existing soil, a marked disturbance of soil water and the amount and

Fig. 10-1. Soil fill placed around tree roots during construction caused the death of these trees in a new housing development.

Fig. 10-2. Soil removed to show the depth of fill placed over the tree root system.

kind of air existing in the soil occurs. A decrease in oxygen content of soil yields an anaerobic soil environment. Air (primarily oxygen) and water are essential for normal functioning of roots. Soil microorganisms, necessary to break down soil organic matter that serves as food for the roots, also need air. In addition, when air is lacking, certain gases and chemicals may increase and become toxic to roots. Toxins produced by anaerobic bacteria may be as harmful as the damage resulting from asphyxiation due to lack of oxygen. Fills may raise the water table, increasing the prospect of an anaerobic root environment, or fills may impede water penetration, subjecting the tree to drought. Roots do not readily regenerate in the new soil layer, and as a result the roots die. Symptoms may become visible in the aboveground parts within a month, or they may not appear for several years. The stockpiling of soil around trees for later use may also have a detrimental effect, especially if it occurs during the growing season.

Factors Governing the Extent of Injury from Fills. The extent of injury from fills varies with the species, age, and condition of the tree; the depth and type of fill; drainage; and subsequent exposure to parasites and nonparasitic agents. Sugar maple, beech, dogwood, oak, tuliptree, pines, and spruce are most severely injured; birch, hickory, and hemlock suffer less; elm, poplar, willow, planetree, pin oak, and locust are least affected. Older trees may be more sensitive to root smothering than younger trees, although both are affected. Trees in weak vegetative condition at the time the fill is made are more severely injured than trees in good vigor.

Obviously, the deeper the fill the more marked is the disturbance to the roots and consequently the more serious are the effects. Clay soil fills cause most injury, because the fineness of the soil may shut out air and water almost completely. The application of only an inch or two of clay soil may cause severe injury. Gravelly fills cause least trouble because air and water permeate them more readily. As a rule, the application of a layer several inches deep of gravelly soil, or even of the same kind of soil in which the tree has been growing, will do little harm. Such trees eventually become accustomed to the new situation by producing additional roots near the surface. Some trees are able to survive despite the addition of a considerable amount of soil over the original level. They apparently produce a layer of roots some distance above the original ones.

Treating Soil Fill Injury. Injury caused by soil fills can often be prevented (see Chapter 8). It is much more difficult to rescue trees from the effects of existing fills. All too often, the problem is not diagnosed until severe symptoms of decline are apparent.

Lowering Grades. Depending on how much soil is removed, lowering the grade can be as disastrous as raising the soil level. Since the majority of the tree's roots are in the top 18 inches of soil, and most of the feeder roots are in the top 6 inches, removal of only a few inches can cause serious harm. Lowering the grade removes valuable topsoil, exposes feeder roots to drying out and cold temperatures, severs roots needed for anchorage and water transport, and lowers the water table. Symptoms of grade lowering injury are similar to those for trees having soil fills: leaf stunting and chlorosis, epicormic growth along branches and trunk, twig and branch dieback, and death. The extent of damage done to the tree

Fig. 10-3. Many tree roots were severed to install a curb.

Fig. 10-4. Excavation for building foundation severed many tree roots.

by grade lowering will be greatest when the grade cut is made deep and near the trunk (Fig. 10-3). Some trees will not suffer appreciably from a removal of only a few inches of topsoil in an area well away from the trunk.

Excavations. Excavation for building foundations and utility lines damages trees by severing the roots (Fig. 10-4). Depending on how near the tree the excavation occurs, symptoms of damage like that for grade changes may range from slight, for a trench made outside the tree branch spread, to severe decline and death following trenching close to the trunk. There are virtually no remedial measures for severe decline, although providing adequate air, fertility and water to the remaining root system may help. Insofar as possible, trees should be protected when excavations are made for water and sewer lines. The trenches should be located away from the trees. If they cannot be routed around the trees, the best alternative is to tunnel under them. Power-driven soil augers are available for this job (see Chapter 8).

Soil Compaction. Roots suffer from soil compaction at many construction sites, despite physical protection of roots from cuts, fills, and excavation. Soil compaction involves physical compression of the soil from the surface so that as soil particles are pressed together the spaces between them become smaller. Compacted soil is not easily penetrated by new roots or by water and air, two essential needs for roots. Thus tree growth is affected adversely, and typical decline symptoms set in. Soil compaction results from construction equipment traffic, piles of building supplies, soil piles, and even foot traffic (Fig. 10-5). Compaction occurs more readily in clay than sandy soils and in wet more than drier soils.

Fig. 10-5. Soil compaction during construction results from heavy equipment traffic.

Fig. 10-6. This tree, remaining after a parking lot was constructed, lost nearly all of its effective roots.

Paving. Applying a layer of concrete or asphalt over the root zone of existing trees has much the same effect on the trees as adding soil fill or compacting the soil. Often paving is done following a shallow excavation or grade lowering to establish a foundation for the paving (Fig. 10-6). The extent of tree decline that results will be in proportion to the amount of damage done to the tree's roots. Root systems prevented from getting water and air by the impervious paving layer are not easily rescued from the impending damage.

Injury by Girdling Roots

Many trees are weakened and some are killed by roots that grow closely appressed to the main trunk or large laterals in such a way as to choke the members they surround, much as does a wire left around a branch for a number of years. The most likely effect of this strangulation is to restrict movement of carbohydrates to the roots through the phloem, thus leading to gradual root starvation. Starved roots provide less water and mineral elements to the top. This choking action may also restrict the xylem so that movement of water and mineral elements in the trunk or in the strangled area of the large roots is also impeded. When a large lateral root is severely girdled, the branches that depend on it for water and mineral elements commonly show weak vegetative growth and may eventually die. If the trunk of a tree is severely girdled, the main branch leader may die back. As a rule, trees affected by girdling roots do not die suddenly, but become progressively weaker over a 5- or 10-year period despite otherwise good tree maintenance practices.

Girdling roots that develop below ground level can often be detected by examining the trunk base. If the trunk ascends straight up from the ground, as it does when a soil fill has been made, or is slightly concave on one side instead of showing a normal flare (swelling) or buttress at the soil line, then one can suspect a girdling root. Of course digging the soil alongside the trunk should reveal whether there is a strangling member.

Trees growing along paved streets suffer more from girdling roots than do those in open areas, and middle-aged or old trees suffer more than younger ones. Norway and sugar maples, oaks, elms, and pines are affected most frequently. Girdling roots occur on transplanted trees, whereas trees growing in the wild do not seem to have the problem.

Development of Girdling Roots. In the nursery, when primary roots radiating out from the trunk are pruned to form the root ball for transplanting, subsequent growth of secondary roots may be stimulated. While new roots formed at the cut ends continue to grow outward, many of the secondary roots emerging some distance back from the end of the severed root may develop at right angles to the main root. Over time as the trunk and primary roots expand and as the secondary and even tertiary roots enlarge, the two meet and girdling may occur. When girdling root development is advanced, they can become so intertwined with other roots and embedded into the trunk that it is impossible to correct the problem.

Careless placement or overcrowding of the roots at transplanting time may also lead to root girdling. Container-grown trees in which the roots have been allowed to encircle the growing medium may cause girdling of the stem later. Unless such roots are straightened out or severed at transplanting, later girdling could occur. While transplanting too deeply kills many trees, it by itself is not a cause for girdling root development.

The tendency for expansion of secondary roots perpendicular to the cut main root varies with species and variety. Even within the same variety, there may be differences in root growth habits. Hopefully, those trees with nongirdling tendencies for root growth can find widespread use in the nursery industry in the future.

Diagnosis of Girdling Roots. The best time to determine the presence of girdling roots is in late summer. At that time, the side of the tree above the girdled area will have leaves that are lighter green and smaller than normal and that tend to abscise prematurely. One must always bear in mind, however, that other factors may produce similar symptoms and that the most positive sign is the discovery of girdling roots. Such roots can be uncovered by carefully removing the soil around the trunk on the side showing weak top growth, poor bark development, or a pronounced lack of flare at the base of the trunk. Usually the girdling root is found at or a few inches below the soil surface.

Treatment for Girdling Roots. If the girdling root has not yet seriously impaired the tree's chance of recovery, it should be severed with a chisel and mallet (Fig. 10-7, center) at its point of attachment to the trunk or to the large lateral root. A few inches should be cut from the severed end to prevent its reuniting with the member from which it was severed. It may not be necessary to actually

Fig. 10-7. Treatment of girdling roots. Top: A girdling root. Center: Removing the root with a chisel and mallet. Bottom: The girdling root has been removed. In the example, girdling is fairly advanced. Root removal may or may not be helpful because the tree may already have declined considerably.

remove the offending root segment, especially if removal of a deeply embedded root would cause more trunk or root wounding. Finally, the soil is replaced in its original position. If the tree has been considerably weakened, judicious watering should be practiced and, if needed, fertilizer applied. Be aware that following girdling root removal (or potential girdling root removal on young trees) new secondary roots, positioned to become future girdling roots, may form.

Gas Injury

In urban areas, natural gas is commonly transported through underground pipelines, often near the root systems of landscape trees. Although natural gas (composed mostly of methane) is not directly toxic to trees, natural gas seeping from a leaking gas main can have adverse effects on the environment of the roots, thus indirectly injuring nearby trees. When a leak occurs, gas gradually fills up the air spaces in the surrounding soil, displacing oxygen as it does so. Microbial changes occur in this oxygen-depleted environment; these lead to more anaerobic conditions and eventually to the production of hydrogen sulfide by soil bacteria. Hydrogen sulfide inhibits root respiration and nutrient uptake, causing death of the tree. Other toxins produced anaerobically by soil microorganisms may also be toxic to tree roots.

Diagnosis of Gas Injury. Since the symptoms of gas injury are similar to symptoms caused by other disorders, be certain that in fact there is a gas line in the vicinity. Gas detection meters are available for detecting leaks. Gas company officials want to repair gas leaks, so they are likely to cooperate in the effort to find one. Look for a circular pattern of declining or dying plants of all types in the vicinity of the suspected leak. Plants will show the most injury near the center of the area and will die more rapidly from a large leak than a small one. Dig up soil from the affected area. Soil from a serious gas leak will have a dark or black color and a characteristic sour odor. Roots present in such soil will be dead.

Treatment of Gas Leaks. First the leaky gas main must be repaired. The soil needs to be aerated by drilling auger holes or forcing compressed air through an aerifying or watering needle inserted into the affected soil. If affected plants have enough live tissue to warrant saving, dead branches should be removed and fertilizer and water applied as needed. Where gas such as methane is continually being produced in the soil, as in the case of some reclaimed landfills, trees and shrubs may not be feasibly planted until gas levels decrease to safe levels.

Injuries Caused by Low Temperatures

Trees in either a vegetative or a dormant condition may be injured by low temperatures. Injury is most common in regions where seasonal variations are greatest. For purposes of discussion, low-temperature injuries may be divided according to the season of the year in which they occur, that is, spring, autumn, and winter.

Spring Frosts. Sudden drops to freezing temperature or below result in wilting, blackening, and death of tender twigs, blossoms, and leaves of deciduous trees, or reddening of the needles and defoliation of the newest growth of needle

Fig. 10-8. Spring frost injury to yew. Newly developing shoots are drooped and brown.

evergreen trees (Fig. 10-8). Swollen buds of broad-leaved trees exposed to freezing temperatures may give rise to expanding leaves being tattered or having symmetrically placed holes. Because of cold air settling in local areas, trees growing in valleys and low areas are damaged more often than those on higher ground. Even in individual trees, frost injury may occur on lower branches while the treetop escapes damage.

Warm days in early spring stimulate premature growth, which is readily injured by the cold days and nights that follow. Injury is most severe when the low temperatures occur later in spring when the new growth is well advanced. When late April or May freezes occur following a lengthy mild period, injured flowers, fruits, leaves, and twigs of many trees will be very evident throughout the region within a few days or weeks.

Autumn Frosts. A cool summer followed by a warm autumn prolongs the growing season. Under such conditions twigs, buds, and branches fail to mature properly and therefore are more subject to injury by early autumn frosts. Unseasonable cold waves in the fall may also result in much damage. When such disastrous frosts and freezes occur in October and November, severe injury and death of large numbers of trees and shrubs can be observed, but not until the next growing season, when these trees fail to leaf out, or wilt and die soon after emergence.

Winter Injury. Despite their dormant condition, trees frequently suffer severely during the winter. Their winter-hardiness is influenced by induction of hardiness by gradually cooling weather in the fall, drainage, location, natural protection, species of tree, and character of the root system, as well as by a combination of unfavorable weather conditions.

Although more sensitive to cold temperatures than other plant parts, tree roots are not normally injured by cold temperatures because of the insulating effects of the soil. The roots of trees are more likely to freeze in poorly drained soils than in well-drained soils; maple, ash, elm, and pine roots are most susceptible to freezing. The effects of frozen roots are seldom noted until the following summer, when the aboveground parts wilt and die. Roots are injured most frequently during winters of little snowfall or in soils bare of small plants and other vegetation.

With the advent of container-grown trees in nurseries and the use of aboveground tree planters in urban landscapes, the possibility of tree roots becoming damaged by cold temperatures increases.

Cold winter temperatures are capable of injuring tree tissues either because the species is not locally hardy or because usual hardening off processes have not occurred. Properly hardened off trunk and limb tissues usually withstand lower temperatures than twig, leaf bud, or flower bud tissues. It is not unusual for some marginally adapted flowering trees to fail to bloom following extremely cold January temperatures, which kill the flower buds but not the leaf buds or other tree tissues. However, early winter cold coming before the tree has completely hardened can be more injurious to the trunk and branch tissues than to other parts of the tree. This kind of winter injury often follows a sudden drop in temperature and is potentially more damaging than severe midwinter cold. If the cold has injured the cambium, then the ability to make new xylem and phloem is lost, and the tree could die. Very cold late winter weather following partial dehardening could also injure trunk tissues more than bud tissues.

We have cut into the bark and wood of numerous peach trees following such winters and observed dark, discolored trunk phloem, cambium, and xylem tissues while limb and twig tissues were healthy and green, as were root collar tissues protected from the cold (Fig. 10-9). Such trees were struggling to survive despite producing flowers and leaves in the spring. Some injured trees died by midsummer despite starting the season with healthy roots and foliage because their transport systems had been injured. Valuable trees have been rescued from this kind of damage using a practice known as bridge grafting.

The ability of the tree to withstand cold is also governed by factors influencing the degree of hardiness acquired during the weeks preceding cold weather. Trees still actively growing in late fall because of excessive nitrogen fertilization or because of mild, moist weather may be injured when the first cold weather strikes. However, there are trees that are not affected by late fertilization and mild weather, perhaps because hardening is induced by gradually decreasing hours of daylight. The subject of winter-hardiness and cold temperature injuries needs further research. Aside from damage to flower buds, trunk and branch tissues, and roots, cold temperatures during the winter initiate or aggravate more or less localized injuries on the trunks and branches. These are known as frost cracks, cup shakes, and frost cankers.

Fig. 10-9. Winter injury to trunk cambium and wood of *Prunus*. Discolored and injured wood is above healthy white wood that was protected from cold temperature by proximity to the soil. This contrasts with collar rot disease where lower trunk tissue would be discolored.

Frost Cracks. Longitudinal separations of the bark and wood are known as frost cracks. In extreme cases, these openings may be wide enough to permit the insertion of a hand and may extend in a radial direction to the center of the tree or beyond.

Frost cracks are most likely to form in periods of wide temperature fluctuations. The water in the wood cells near the outer surface of the trunk moves out of the cells and freezes during sudden drops in temperature. The loss of water from these cells results in the drying of the wood in much the same way as green lumber dries and cracks when exposed to the sun. At the same time, the temperature of the cells in the center of the trunk remains much higher, and little drying of these cells or shrinkage of the wood takes place. The unequal shrinkage between the outer and the inner layers of wood sets up great tension, which is released only by the separation of the layers. The break occurs suddenly along the grain of the wood and is sometimes accompanied by a loud report (Fig. 10-10).

Cracks formed in this manner appear principally on the south and west sides of the trunk, since these are heated by the sun's rays and a greater gradient prevails when the temperatures drop suddenly during the night. If all exposures of the trunk were heated evenly, tension and frost cracks would not develop, as all tissues would shrink at the same rate.

Although the cracks are created and observed during cold weather, it is apparent that injuries to the trunk and roots, often made years before, are the reason the crack appears where it does (Fig. 10-11). Such injuries include making flush cuts rather than proper pruning cuts, allowing trunk injuries to occur at planting time, or severing roots during construction.

Deciduous trees appear to be more subject to cracking than are evergreens. Among the former apple and crabapple, ashes, beech, cherry, goldenrain-tree, horsechestnut, linden, London planetree, certain maples, pin oak, tuliptree, walnut, and willow suffer most. Isolated trees are more susceptible than those growing in woodland areas, and trees at their most vigorous age (6–18 in. in diameter)

Fig. 10-10. Frost crack on trunk of a young horsechestnut tree.

Fig. 10-11. Internal extent of frost crack injury. The crack was probably initiated by an injury when the tree was very young.

are more subject than very young or old trees. Probably because of root injury or decay, trees growing in poorly drained sites are more subject to cracking than those growing in drier, better-drained soils.

When warmer temperatures arrive, the frozen tissues thaw and absorb more water, and the crack closes. The cracked zone in the wood never closes completely, even though the surface may be sealed by callus formation. Because of insufficient support, however, the same area again splits open the following winter. The repeated splitting and callusing eventually results in the formation of a considerable mass of tissue over the seam.

The most serious aspect of frost cracks is that they are a potentially hazardous defect and provide conditions favorable for the entrance of wood-decay fungi. Avoiding unnecessary injuries and pruning properly should reduce frost cracking. In extremely cold climates, trunk wraps or shading structures that keep the sun-exposed trunk from becoming too hot may reduce frost crack injury.

Cup Shakes. Wide temperature fluctuations within a relatively short period precipitate a type of injury known as cup shakes. This results from conditions that are the reverse of those responsible for frost cracks. Following low temperatures, sudden heating of the outer tissues of the trunk by sunshine will cause these tissues to expand more rapidly than the inner tissues, resulting in a cleavage or separation along an annual ring. Forceful winds may also trigger cup shakes. In both cases, the cup shakes follow a weakness caused by wounding with temperature and winds activating the problem. Although not visible externally, cup shakes weaken the tree and would make it hazardous.

Frost Cankers. Low-temperature injuries may be confined to small localized areas on the trunk, branches, or in crotches. The lesions or cankers resulting from this limited injury are common on maple and London planetree and are confined principally to the southern and western exposures of the trunk. Scalding by the sun's rays on these exposures may penetrate as deeply as the cambium, resulting in a well-defined lesion.

Winter Drying of Evergreens. Winter injury to such plants as rhododendron, laurel, holly, pines, spruces, and firs, expressed as winter drying, is rarely caused by excessive cold during the winter. The damage is caused, rather, by desiccation.

Evergreen plants transpire continuously. Although water loss is low during winter months, it may increase considerably when the plants are subjected to drying winds or are growing in warm sunny spots. When roots are embedded in frozen or dry soil, water lost through the leaves cannot be replaced by root uptake, and drying out occurs.

In the earlier stages, winter injury on broad-leaved evergreens is evident as a scorching of the leaves at the tips and along the outer margins (Fig. 10-12). The color of the affected parts tends to be brown rather than yellow. On

Fig. 10-12. Winter injury to rhododendrons and other broad-leaved evergreens appears as browning along the edges of the leaves.

narrow-leaved evergreens, the needles are browned entirely or from the tips downward along part of their length. The terminal buds and twigs are brittle and snap readily when bent.

Prevention of Winter Injury. It is obvious from the foregoing discussion that the ability of a plant to withstand low temperatures depends on many factors beyond human control. In some instances, however, precautions can be taken to reduce the possibility of winter damage.

The selection of well-drained soils as sites for trees cannot be overemphasized. Trees considered susceptible always withstand low temperatures better when planted in soil with good aeration and water drainage. Soil fracturing to improve soil drainage around existing trees may also help.

To protect broad-leaved evergreens, such as rhododendron and laurel, wind-breaks of coniferous evergreens, which as a rule suffer less severely, are to be encouraged. Heavy mulches of oak-leaf mold or acid peat moss to prevent deep freezing and thus to facilitate water absorption by the roots are of great assistance in avoiding winter injury. To ensure an ample water supply during the winter months, soils around evergreens should be well watered before freezing weather sets in.

Winter browning can be prevented, or at least greatly reduced, by spraying the leaves with antitranspirants. These materials form a thin, transparent film on leaf and needle surfaces that reduces escape of water vapor and are normally applied with a sprayer on a mild day in late fall or early winter. Some arborists have had success with such treatments, which assist many kinds of evergreens, including rhododendrons, boxwood, and mountain laurel, to come through the winter undamaged. Antitranspirants may do the most good during a midwinter or early spring thaw when the foliage begins to transpire while the roots, still in frozen soil, are unable to meet the tree's water needs. The argument against using anti-transpirants following transplanting, which could inhibit foliar gas exchange needed for photosynthesis, is not relevant for winter protection, because gas exchange is less critical in winter.

Care of Winter-injured Trees. In any attempt to aid a winter-injured tree to regain its vigor, several precautions should be taken.

Drastic pruning of such plants is not advisable, since additional harm is likely to occur. Moreover, pruning should be deferred until the buds open in the spring, when dead wood can easily be distinguished from live. A moderately pruned tree will recover more quickly than one severely pruned or not pruned. It is impractical to prune winter-injured needles or leaves of evergreens, but all dead wood should, of course, be removed.

Some disagreement exists among arborists as to the specific fertilizer practices to be followed on severely winter-injured trees. Those who favor omission of fertilizer for the entire season following injury might claim that heavy spring applications may either further injure already damaged roots or result in the increased production of foliage, whose demands for water might overtax injured root and trunk conducting tissues. These advocates say that a moderate application of fertilizer might be justified about the first of July, since some new water-conducting tissue will have been formed by that time.

Advocates of medium to heavy applications of fertilizers in the spring following winter injury might claim that such treatments encourage the formation of new tissue, thus making it possible for the tree better to withstand summer conditions. After all, it is the foliage that manufactures the raw materials needed for repair of injured tissues.

Sunscald

The bark of trees in some circumstances is subject to injury through heating by the sun. The sunscalded bark will appear discolored and dried out, often in a long strip on one side (usually the southwest, the sun's direction during the hottest part of the day) of the trunk. Discolored, dead tissue is also present under the bark in the cambial tissues so the trunk is no longer expanding in that region. After a time, the injured area takes on a sunken appearance, and the tree begins to produce callus tissue outside the killed area. Frequently the wood under the sunscalded tissue will decay, the decay eventually occupying the entire woody cylinder present at the time of the initial injury. In extreme cases, such weakened trees can break.

Sunscald is more likely to occur on shade-grown thin-barked trees following sudden exposure to direct sun. This occurs after transplanting a young tree from a closely planted site in a nursery to an open site in the landscape. Similarly, young, thin-barked trees on a wooded lot can be injured by sudden exposure to the direct sun following clearing for construction. Again, injuries to roots and branches during these critical operations enhance the likelihood of sunscald damage. Sunscald can occur at any time of year, even in winter, although differential heating and cooling of bark tissues from day to night may be causing the injury in winter.

Light-colored or reflective tree wraps are normally used to protect newly planted trees from direct sun. White latex paint protects otherwise dark, heat-absorbing tree trunks by reflecting sunlight; painted trunks, however, are rarely accepted in the landscape. Any reasonable means of shading tree trunks on the southwest for the first years of exposure should suffice.

Providing adequate water for newly planted and newly exposed trees is also essential. Transplanting inflicts tree root injury, and trees are especially vulnerable during their first two years.

Leaf Scorch

Although the disorder known as abiotic leaf scorch is most prevalent on Japanese red, sugar, silver, Norway, and sycamore maples, it also occurs on dogwood, horsechestnut, ash, elm, and beeches and to a lesser extent on oaks and other deciduous trees.

Symptoms. Scattered areas in the leaf, appearing first between the veins or along the margin, turn light or dark brown (Fig. 10-13). The edges of the discolored areas are very irregular. All the leaves on a given branch appear to be affected more or less uniformly. When a considerable area of the leaf surface is discolored, the canopy of the tree assumes a dry, scorched appearance. In most cases, the leaves remain on the tree, and little damage results. In severe cases, the leaves

Fig. 10-13. Leaf scorch on flowering dogwood, a result of drought. Leaf margins are dry and brown.

dry up completely and fall prematurely. When such defoliation occurs before midsummer, new leaves are formed before fall.

The position of the most severely affected leaves can be used, in many instances, as a means of distinguishing leaf scorch from leaf spot and blight damage caused by fungi. Leaves affected by scorch are usually most abundant on the side of the tree exposed to the prevailing winds or to the most intense rays of the sun, whereas spotted and blighted leaves resulting from fungus attack are usually scattered throughout the treetop, the greatest numbers being on the lower, more densely shaded parts.

Biotic scorch caused by xylem-limited bacteria (see Chapter 12) more nearly resembles abiotic scorch, but patterns of leaf browning may differ. Xylem-limited bacteria frequently cause browning along the leaf margin, and damage appears in any part of the tree. Furthermore, the problem seems limited to the southern and mid-Atlantic regions of the United States.

Cause. Leaves are scorched when the roots fail to supply sufficient water to compensate for that lost at critical periods. Newly transplanted trees are frequently victims of abiotic scorch. They are most severely affected during periods of high temperatures and drying winds. The inability to supply the necessary water is influenced by the moisture content of the soil and by the location and condition of the root system. Even the most extensive root system cannot supply enough water to compensate for the tremendous amounts lost through the leaves if the moisture content of the soil is low because of a prolonged drought.

Though of less common occurrence, scorch may appear in trees growing in excessively wet soils. Under such conditions lack of air greatly impedes the water-absorbing capabilities of the roots.

Trees with diseased roots, those whose root systems have been reduced as a result of transplanting, curb installation, or building construction, and those whose roots are restricted by or covered with impervious materials are most subject to leaf scorch. For example, sugar maples adjacent to paved concrete driveways and near street intersections are commonly observed to have more scorch than those more favorably situated along the same street or those growing on lawns or along country roads.

Where injury to a part of the root system is severe and permanent, or where soil conditions are continuously unfavorable, leaf scorch may occur regularly. In fact, some trees show leaf scorch every year regardless of whether the season is cool and moist or hot and dry.

Heavy infestations of aphids and other sucking insects usually contribute to the severity of leaf scorch. Potassium deficiency is known to cause marginal scorch symptoms. However, potassium-deficient soils are relatively uncommon.

Control. Any practice that increases root development and improves the tree generally will help reduce leaf scorch. Where soil has been tested and potassium has been found deficient, the application of potassium fertilizers is suggested. In addition, insects and fungal diseases should be kept under control.

Where soils are likely to become dry, water should be applied, especially during seasons of low rainfall. Where the soil is heavy and compacted, some improvement will result from breaking up the surface layers, vertical augering, or inserting upright tile to facilitate penetration of water. In the rarer cases where scorch results from the inability of the roots to absorb water because of excessively wet soils, drain tiles may need to be installed.

Leaf scorching in particularly valuable trees can be prevented, or at least reduced, by spraying the leaves with antitranspirant before the regular "scorching" period.

Lightning Injury

Every year thousands of shade and ornamental trees throughout the country are struck by lightning. The amount and the type of injury are extremely variable and appear to be governed by the voltage of the charge, the moisture content of the part struck, and the species of tree involved.

The woody parts of the tree may be completely shattered (Fig. 10-14) and may then burn. A thin strip of bark parallel to the wood fibers down the entire length of the trunk may be burned or stripped off, the internal tissues may be severely burned without external evidence, or part or all of the roots may be killed. The upper trunk and branches of evergreens, especially spruces, may be killed outright, while the lower portions remain unaffected. In crowded groves, trees close

Fig. 10-14. Shattering of the trunk on a lightning-struck tuliptree.

to one directly hit will also die. In many cases, grass and other vegetation growing near a stricken trunk will be killed.

So-called hot bolts with temperatures over 25,000° F can make an entire tree
burst into flames, while cold lightning—striking at 20,000 miles per second—can
make a tree literally explode. On occasion, both may fail to cause apparent damage, but months later an affected tree dies from burned roots and internal tissues.

Tall trees, trees growing alone in open areas, and trees with roots in moist soils
or those growing along bodies of water are most likely to be struck. Though no
species of tree is totally immune, some are definitely more resistant to lightning
strikes than others. Beech, birch, and horsechestnut, for example, are rarely
struck, whereas elm, maple, oak, pine, poplar, spruce, and tuliptree are commonly hit. The reason for the wide variation in susceptibility is not clear. Some
authorities attribute it to the difference in composition of the trees; for example,
trees high in oils (beech and birch) are poor conductors of electricity, whereas
trees high in starch content (ash, maple, oak, etc.) are good conductors. In addition, deep-rooted or decaying trees appear to be more subject to attack than are
shallow-rooted or healthy trees.

It is commonly believed that lightning never strikes twice in the same place.
This is not true, for some trees have been struck by lightning as many as seven
times, judging from the scars on their trunks.

Wires carrying electric current may also cause some injury to trees. Suitable
nonconductors should be placed over wires passing in the vicinity of the trunk
and branches. Much of the injury by wires can be avoided by judicious pruning,
as discussed in Chapter 7. Outdoor evergreens bedecked with lights at Christmas
can be damaged if there are too many poorly placed bulbs or if worn equipment
is used. The bulbs should be placed so they do not come into direct contact with
the needles or twig tips.

Prevention and repair of lightning injury are discussed in Chapter 8.

Hail Injury

Hailstorms can cause considerable damage. They may completely defoliate trees
or shred the leaves sufficiently to check growth sharply. They are most injurious
to young trees or those with undeveloped foliage in early spring. Areas in the bark
and cambium may be severely bruised or even killed from the impact of the hail-
stones. When bark injuries do not callus over rapidly, they often serve as entrance
points for wood decay fungi. Where hail has caused damage, all bark injuries will
be on the same side of the affected twigs and branches.

Mechanical Injuries

Damage to trees, especially the trunks, by motor vehicles, by children, and even
by thoughtless adults occurs commonly, and the cause is usually obvious.

Most mechanical injuries are confined to the outer portions of the trunk (Fig.
10-15). The effect of mechanical injuries on tree health may be negligible if the
injury is small and if the tree is vigorous and good at limiting decay. Or an injury
may be serious enough to cause rapid death due the trunk girdling or decline and
death following invasion of disease organisms such as decay fungi.

Fig. 10-15. Mechanical injury to a linden trunk. An older wound is located just above and to the left of the fresh wound.

Equally obvious to even the casual observer is mechanical injury caused by ice storms and windstorms. Tree damage ranges from broken twigs and branches to uprooting of entire trees.

Accumulations of snow, especially on evergreens such as yews, hemlocks, white pines and junipers, can cause severe branch breakage (Fig. 10-16). Such accumulations should be gently shaken off as soon as possible with a broom or leaf rake. Ice-covered trees should be handled even more carefully. If the temperature is on the rise, rinsing the ice-coated trees with water from a hose will thaw the ice.

Some deciduous trees are prone to cracking and splitting of branches in ice storms and windstorms. Among those most subject are the ashes, catalpas, hickories, horsechestnuts, red maple, empress-tree, Siberian elm, silk-tree, tuliptree, and yellowwood.

Less obvious to many tree owners is mechanical injury caused by girdling

wires and by lawn mowers and string trimmers. Wires and plastic twine are sometimes improperly used for lacing up the burlap of a balled and burlapped tree.
Even when the girdling wire was not originally placed tightly against the root collar, later trunk expansion spells trouble for the tree. Often, as in the case of a
group of white pine trees observed to be in a state of decline 10 years after planting, the wire is deeply embedded into the bark tissues of the root collar a few
inches below the soil surface and is not noticed until too late.

Wires used for tree support at transplanting or even for attaching the plant
label in the nursery have been known to girdle tree trunks and branches some
years later (Fig. 10-17). These aboveground girdles are usually easy to diagnose.
In the early stages, the tree begins to swell just above the constriction. Eventually,
the entire tree above the girdle will begin to decline and die.

The lawn mower or string trimmer is a more destructive instrument for
debarking trees than are automobiles and malicious children. Careless use of
these tools results in severe damage to the inner bark and cambium near the soil

Fig. 10-16. White pine with numerous branch
stubs due to branch breakage by a heavy snow
load.

Fig. 10-17. Wire used to help support this tree caused complete girdling of the trunk.

line. Parasitic fungi, for example, *Ceratocystis*, which infects planetrees, and wood-decay fungi can establish themselves in wounds made by lawn mowers.

Insects such as the dogwood borer are attracted to these injuries. Removal of phloem tissues by this debarking will lead to gradual root starvation and tree decline. If mulch replaces grass planted close to the base of the tree, there is less chance of causing damage. Mulch confers other benefits such as reducing competition and conserving water. Installation of a tree guard also eliminates the need for edging and trimming grass around lawn trees.

Lichens

Lichens often appear as a green coating on the trunks (Fig. 10-18) and branches of trees. They are actually two plants in one, being composed of a fungal body and an algal body, which live together in complete harmony. The alga supplies elaborated food to the fungus and, in turn, receives protection and some food from the fungus.

Lichens do not parasitize trees, but merely use the bark as a medium on which to grow. They are unsightly rather than injurious, although when extensive they may interfere with the gaseous exchange of the parts they cover. Because of their extreme sensitivity to sulfur dioxide released from factory chimneys, lichens seldom appear on trees in industrial cities. They rarely develop on rapidly growing trees, because new bark is constantly being formed before the lichens have an opportunity to grow over much of the surface. Because of this, lichens on certain species may indicate poor tree growth. We have noticed that in some pin oak plantings, those most vigorous have fewer lichens than those of the same age nearby in a state of decline. However, no research has been done to verify any correlation between lichen growth and tree vigor.

Fig. 10-18. Lichens growing over a tree trunk.

As a rule, lichens can be eradicated by spraying the infested parts with Bordeaux mixture or any ready-made copper spray.

Vines

Many kinds of leafy vines grow up the trunks and out of the branches of trees (Fig. 10-19). Several are woody, twining types that have the potential to strangle or girdle the tree. Examples include trumpet vine, Japanese honeysuckle, bittersweet, and domestic or wild grapes. Other vines climb by means of aerial roots or tendrils, some having adhesive disks. Such climbers are a threat to the tree because of their shading tree foliage, bending and possibly breaking branches from their weight, and competing for soil moisture. Examples include winter creeper, English ivy, Virginia creeper, cross vine, Kudzu, and poison ivy. Vines should be kept off young trees. Some of the nontwining types can be tolerated on large tree trunks, but they should not be allowed to grow out on the branches.

Rodent Damage

Young trees with thin, succulent bark are subject to damage by rodents such as field mice, pocket gophers, and rabbits during winters of heavy snowfall, or where

Fig. 10-19. English ivy growing on pin oak.

Fig. 10-20. Mouse damage to the base of a crabapple trunk. The bark was completely stripped from the base of the trunk.

the trees have been mulched with some straw-type material (Fig. 10-20). Placing a collar of a quarter-inch wire mesh or even ordinary window screening around the base of such trees will prevent much damage. Use of straw mulches in which mice are likely to nest should be avoided. The trunk bases of susceptible trees can also be painted with a repellent such as thiram.

Sapsucker Damage

Yellow-bellied sapsuckers have been observed feeding on 26 kinds of trees in the northeastern United States. The birds under observation fed particularly on Atlas cedar, hemlock, red maple, mountain ash, and yellow, paper, and gray birches. Damage due to bird pecking has also been observed on apple, cherry, maple, and Scots pine. Sapsuckers peck holes around the tree to obtain sap (Fig. 10-21), although insects also make up a part of their diet. The bird bores even rows of holes spaced closely together. As these holes fill with sap, the sapsucker uses its brushlike tongue to draw it out. Extensive and repeated pecking may cause cambium and bark injury and branch or trunk swelling. Girdling may kill portions of the tree above the boring injury.

To discourage sapsuckers from feeding on landscape trees, wrap hardware cloth or burlap around the area being injured or apply a sticky repellent material, such as Tree Tanglefoot, on the bark. Decayed aspen trees make good nesting sites for sapsuckers, so their removal may discourage the birds from using the area.

Fig. 10-21. Yellow-bellied sapsucker and feeding holes on cherry.

Fig. 10-22. Larval and adult stages of carpenter ants infesting wood of black birch.

Termites and Ants

Termites are usually associated with damage to wood in buildings, but they are occasionally found feeding on sound heartwood and sapwood of living trees. Several different kinds of termites, including dry wood and the more destructive subterranean (*Reticulitermes* spp.) types, can cause damage. Subterranean termites nest underground and enter the tree through injured roots, or they may build protective earthen tubes along the outer bark connecting the underground shelter to their feeding area of the tree trunk. Serious dieback of individual branches of upright yew can be caused by termite activity. Parting of the branches to expose the earthen tubes along the trunk reveals the cause of the problem. There are reports of termites attacking and severely damaging live trees in some areas of the southern United States.

Ants of several different kinds also infest live trees. Perhaps the most common are carpenter ants (*Camponotus* spp.). Ants generally form galleries in the already decayed heartwood and sapwood of trees (Fig. 10-22). Occasionally the galleries extend into healthy wood, and although the ants do not feed on the wood, removal of chips of wood to extend the galleries weakens the tree. Arborists examining trees for structural weakness need to look for evidence of carpenter ant foraging trails and accumulation of sawdust at the base of the infested tree.

Ants are morphologically different from termites. The principal difference is that ants have a noticeable constriction of the body between the thorax (where legs and wings are attached) and the abdomen, whereas termites have a broad connection between these two sections. Ants can be controlled by blowing Diazinon dust into the colony with a hand duster, or spraying the colony with Diazinon. Termite extermination requires destruction of underground nests, generally by a professional exterminator. Certain termite-proofing materials are toxic to tree and shrub roots and to humans, so precautions must be taken.

Tree Band Damage

Banding trees with a sticky substance to trap insects such as cankerworm moths can be damaging unless means are taken to prevent the inner bark and cambium from absorbing the material. If the banding material is applied directly to the bark, or if the thicker outer bark is shaved or cut away and the material then applied, severe damage may follow.

Sugar maples appear to be more susceptible than other trees to banding substances applied directly to the bark.

Air Pollution Injury

Fossil fuel combustion and a variety of industrial processes produce or induce production of phytotoxic gases, aerosols, and particulates. The major groups of air pollutants are oxidants, sulfur dioxide, and industrial byproducts. Their effects on trees and other plants depend on pollutant concentration, sources and distribution, plant sensitivity, and timing of occurrence.

At one time, air pollution was thought to be a local problem with readily identifiable sources, and for some geographical locations and many industrial by-products this is still the case. However, cars and trucks operate in such large numbers, and smokestacks are built to such heights, that air currents carry pollutants to locations far removed from their sources. Some of these pollutants can become trapped by weather conditions favoring inversion layers in the atmosphere, which prevent upward movement and dilution of the pollutant. Symptoms of air pollution injury sometimes show up after several days of such air stagnation. The effects of air pollution levels below those that cause symptoms on plants are not known, but increased plant stress could occur.

Oxidant Air Pollutants. The burning of fossil fuels should ideally yield nontoxic carbon dioxide and water; however, impurities, additives, and incomplete combustion yield other products as well. When these products are exposed to sunlight, ozone and peroxyacetyl nitrate (PAN), both toxic to plants, are produced. These oxidant air pollutants are important components of photochemical smog, which affects air quality in cities throughout the world. Ozone and PAN are toxic in very low concentrations.

Needle evergreen trees exposed to ozone show chlorotic flecking, banding, or needle tip burn symptoms. Broad-leaved plants show brown, purple, or chlorotic upper leaf surface flecking (Fig. 10-23). Ozone injury does not occur commonly, but when it does happen, symptoms can be striking. Newly expanded leaves

Fig. 10-23. Chlorotic flecking, a symptom of injury caused by oxidant air pollutants.

appear to be most susceptible, so when symptoms are observed, the timing of the most recent episode can be determined. If the timing coincides with air stagnation periods and symptoms are also occurring on ozone-sensitive indicator plants such as grapes, petunias, and cucumbers, then the symptoms of some air pollutant can be suspected.

Symptoms of some diseases, nutritional problems, and feeding of insect pests such as lace bugs and spider mites may resemble ozone injury. Ozone can also be produced by electrical discharges from sparking electric motors and lightning. Levels of the pollutant in these situations are very low, and unless somehow confined, as in a greenhouse, damage does not occur.

The main symptom of exposure of broad-leaved plants to PAN is a patchy silvering or light tan glazing of the lower leaf surface resulting from collapse of the epidermal cells. The affected leaf may exhibit spots or patches of papery-thin, almost transparent tissues. Damage from PAN is less common than ozone damage. Nitrogen dioxide (NO_2), also a product of fossil fuel combustion, is important for ozone and PAN production; NO_2 is less toxic to plants but causes yellowing of leaf margins and interveinal tissue when present at high levels.

Sulfur Dioxide and Acid Rain. Sulfur dioxide (SO_2) is the most common component of smoke from fossil fuel power plants, which is toxic to trees. Coal and oil commonly contain sulfur as an impurity, and it oxidizes to sulfur dioxide during combustion. When emissions are not adequately dispersed, concentrations of the pollutant can become high enough to cause plant symptoms. Conifer trees show a reddish needle tip burn, and reddish brown dead areas appear between the veins on leaves of broad-leaved trees. Mild exposure may only cause interveinal leaf yellowing. Symptoms of exposure to sulfur dioxide resemble scorch resulting from drought or xylem-limited bacteria, and because damage can occur some distance from the pollutant source, diagnosis can be difficult. Finding similar symptoms on a wide range of nearby vegetation is helpful in diagnosis of damage due to SO_2 and other toxicants, but this assumes the diagnostician is familiar with the possible causes of symptoms on plants other than trees.

Power plants are not the only source of sulfur dioxide. One of the authors once diagnosed a very local sulfur dioxide exposure that resulted from the demolition of an old-fashioned electric refrigerator that used the chemical as a refrigerant. Heavier than air, the sulfur dioxide gas migrated to lower elevations in the landscape, killing plant leaf tissue the entire distance.

Acid rain results from an increased sulfuric acid content of rainwater due to high atmospheric sulfur dioxide levels; acid rain can have a pH as low as 4.0. Acid rain has not been shown to directly damage trees, and in fact may be beneficial if trees need sulfur or soil pH needs lowering.

Aerosols of sulfur dioxide, deposited and condensed on leaf surfaces in a process described as acid deposition, may be converted to more concentrated sulfuric acids when dissolved in small amounts of water such as dew. It is thought that concentrated acid is actually causing damage, rather than the more dilute acid rain. The actual effect of acid deposition or acid rain on tree health or productivity is not known; however, these substances may be having effects on freshwater lakes and man-made corrodible structures. Because there is concern about acidic precipitation resulting from sulfur dioxide pollutants, it has been suggested that use of high-sulfur fuels be reduced and improved ways to remove sulfur from fuel and smoke be developed.

Air Pollutants from Industrial Processes. Hydrogen fluoride is a by-product of manufacturing processes that use fluoride-containing ores as raw materials. Copper smelting and cement, glass, fertilizer, and aluminum manufacturing may produce fluorides toxic to trees. Tree foliage injury has been observed near glass etching enterprises that use hydrofluoric acid. Fluoride injury is similar to sulfur dioxide injury. Fluorides cause water soaking, collapse, and drying out of affected leaf tissue, leaving scorched areas. Reddish brown areas on the leaf margins, between the veins, and on needle tips are characteristic symptoms. Severe episodes can cause defoliation. Since the symptoms cannot easily be differentiated from those of other pollutants, knowing the source of fluoride and its proximity to the damage will provide circumstantial evidence of its occurrence.

Injury and death of tree foliage can result from exposure to ammonia, chlorine, hydrogen chloride, and other industrial chemicals. Injury from these substances is likely to follow inadvertent exposures such as leaks and spills.

Particulate Air Pollutants. Some industrial processes, farming practices, and even gravel roads produce dust or smoke particles that coat nearby vegetation. When tree leaves are coated with visible dust or soot, stomata may be blocked and leaves may receive little sunlight; therefore, the leaves photosynthesize less actively. Because the leaves of evergreen trees remain attached throughout the year (in some species for several years), a considerable deposit may accumulate, thus making such trees' survival difficult where particulate air pollutants occur. Some dusts or ash particles may contain substances directly toxic to the foliage; after prolonged soil accumulation, changed soil characteristics may cause toxicity. In one instance oak decline has been attributed to the accumulation of cement dust.

Arborists and others interested in the care of trees should not be too hasty in concluding that leaf abnormalities are necessarily caused by air pollutants. They should bear in mind that unfavorable weather, insects, fungi, bacteria, spray materials, growth regulators, and viruses also produce symptoms that might be confused with symptoms of air pollution.

Over the years, the authors have examined hundreds of cases of alleged air pollution injury to vegetation in highly industrialized areas and have found that only a small percentage of the damage was caused by smoke or gases. Before blaming poor growth of vegetation on air pollutants, therefore, one would do well to call in a competent diagnostician.

Chemical Injuries

Weedkillers. While the chemicals used for weed control are referred to as weedkillers or herbicides, the chemicals themselves cannot distinguish between weeds and desirable plants such as trees. When carelessly used, the various chemical weedkillers can damage or even kill trees. The authors have noted many cases of toxicity to trees caused by such materials. Weedkillers function in a variety of ways. The most commonly used materials are growth regulators, applied to existing weeds in the landscape, and seed germination and seedling inhibitors, applied to soil to prevent weeds from appearing in the landscape. Other materials include

Fig. 10-24. 2,4-D injury to mulberry. Affected leaves are curled, cupped, and malformed.

contact and systemic chemicals applied to existing weeds and soil sterilants applied to soil to kill existing and future weeds.

Herbicides containing plant growth regulators such as 2,4-D (2,4-dichlorophenoxyacetic acid) or dicamba are commonly used for eliminating weeds from lawns. Since their chemical structure mimics plant hormones, when they are taken up by a tree they have an effect on the size and shape of developing parts of the tree. The most general symptom is a distortion or malformation of leaves and twigs (Figs. 10-24 and 25). Leaves also roll upward or downward at the midrib or along the edges, thus becoming cupped. Leaf petioles may curl downward, in some cases forming loops. Although these herbicides are almost always used without incident, we have noted growth regulator chemical injury on many trees, including elm, dogwood, boxelder, hawthorn, Norway maple, tuliptree, various oaks, sassafras, and yew. Unless the trees received an unusually heavy dose of chemical, they have usually recovered.

Applying high concentrations of volatile formulations of 2,4-D near trees during warm spring weather invites injury. Even trees not yet in leaf can absorb enough hormone into their dormant buds to produce distorted leaves. To avoid possible damage, use the less volatile forms of herbicide, avoid windy days, and use a coarse droplet spray at low pressure to reduce drift. Follow herbicide label directions and do not use the herbicide sprayer for other spraying jobs; even minute traces of 2,4-D in such sprayers can cause tree injury. Where a sprayer with a metal tank has been used, the 2,4-D can be removed as follows: Fill the tank with warm water and household ammonia at the rate of 1 gallon of ammonia or 5 pounds of sal soda for each 100 gallons of water. Then pump out a few gallons to wash pump parts, hose, and nozzles, and allow the remainder to stay in the tank for 2 hours or so. Finally, drain the tank and rinse several times with clear water. Activated charcoal can be used to rid spray tanks of 2,4-D residues.

Fig. 10-25. Growth-regulator herbicide injury to yew. Curled needles may remain green or they may turn brown.

Some growth-regulator herbicides are applied as granular formulations, often mixed with a fertilizer. Obviously, these products would not make good tree fertilizers. Furthermore, some formulations contain dicamba. In the soil, dicamba is more persistent than 2,4-D. It can move downward in the soil, be absorbed by tree roots, and move systemically to the foliage where leaf cupping and distortion can occur.

Preemergence herbicides such as crabgrass preventers are applied to the soil and generally do not injure trees.

One of the most common systemic herbicides applied to existing weeds in the landscape is glyphosate (Roundup, Kleenup). Applied to growing herbaceous and woody weeds, it is translocated to the roots, and, soon after, the plants begin to die. If it is mistakenly applied to landscape trees, they too can be killed. Glyphosate applied to a nearby weed can drift to a woody plant, and symptoms might not appear until the next season. In such cases, the new growth may be bunchy and distorted, resembling growth-regulator herbicide injury. We have observed several cases of young flowering crabapple trees suffering injury in this way. Trees may or may not be killed, depending on the dose and species involved. Diagnosing glyphosate injury on trees may require a knowledge of weed control operations the year before. In general, small, thin-barked trees are more vulnerable than large, thick-barked trees. The effect of glyphosate application on exposed tree roots has not been studied, but such roots represent a potential mode of entry.

Contact herbicides such as paraquat kill green tissues on contact. Where such materials are used near trees, avoidance of foliage and of thin green bark is essential. The chemical is not systemic, so a small amount of kill resulting from paraquat drift is seldom fatal to the tree. Small circular brown spots have been observed on leaves of trees exposed to fine drift of this chemical.

The herbicide tebuthiuron kills woody plants. This brush-killer, applied to the soil, kills the tree roots, resulting in gradual decline and death of the plant. Although excellent for control of unwanted brush, this material should not be used where desirable species of trees or shrubs are present.

Soil sterilant herbicides are often used to destroy vegetation in sidewalk cracks, under fences, and along driveways. Many users are not aware that some of these materials are long lasting, can move in the soil, and are very toxic to trees whose roots take up the chemical (Fig. 10-26). Ignorance of just where tree roots grow only compounds the problem.

Herbicides such as Pramitol (prometon) and Hyvar (bromacil) are especially damaging when misused. These chemicals are taken up by the roots and injure foliage on the same side of the tree. This one-sided effect can sometimes be mistaken for vascular wilt diseases. Some conifers such as spruce, having a spiral xylem arrangement, will show damage in a spiral pattern up the tree. Affected branches turn yellow, then reddish brown, and eventually die while the rest of the tree remains green. Some of these materials are so long-lasting that where applied, soil has to be removed and replaced in order to get any plants to grow.

Calcium Chloride. Trees growing along country roadsides may be damaged

Fig. 10-26. Pear leaves injured by soil sterilant. Browning of leaf margins and leaf chlorosis may precede tree death.

by the calcium chloride used to control dust on dirt roads. Heavy rains wash the chemical off the roads and carry it down to the roots, which absorb it and then transport it to the leaves, where it causes injury. Correct grading of the roadside to facilitate removal of flood waters and prevent their being soaked into the soil around roots will greatly reduce the chances of damage.

Tree species vary in their tolerance to calcium chloride. Red oak, white oak, and American elm are most tolerant, whereas balsam fir, white spruce, beech, sugar maple, cottonwood, and aspen are least tolerant.

Damage to the leaves appears as a leaf scorch, which is difficult to distinguish from physiological leaf scorch caused by lack of water in the soil. Leaf cupping symptoms have been observed on emerging linden leaves in spring following a winter of sidewalk calcium chloride deicing.

Salt. Salt is frequently scattered over roads, parking lots, and sidewalks in winter to melt ice or prevent water from freezing. Each year millions of tons of salt (sodium chloride) may be applied along streets and highways in the United States for deicing. In parking lots, snow mixed with previously applied salt often is repeatedly pushed into piles on the landscape strips or ends where trees are planted, thus concentrating all the salt from an entire area into a few locations. When the salty solution finds its way to open soil near roots it can be very toxic. In fact, sodium chloride (ordinary table salt) is five to ten times more toxic to some trees than is calcium chloride. High soil salt concentrations tightly bind soil water so that water becomes unavailable to tree roots. Lacking water, trees essentially undergo drought and respond similarly. Gradual decline with all the typical symptoms may occur, or in extreme cases, leaf scorch and death occur. The presence of sodium and chloride ions in the soil makes it more difficult for the tree to take up potassium and phosphorus, so deicing salts can cause nutrient problems as well. Sand or sawdust should be used to prevent slipping in areas near a tree and other plant roots.

Sand treated with salt and piled along roadsides for use on icy roads may also damage trees. The salt is leached down into the roots and causes burning. Sand piles thus treated should be placed at some distance from the root areas of the trees.

Researchers investigating the association of salt applications and tree mortality have observed that trees within 30 feet of highways were affected, whereas the trees beyond that distance were nearly always healthy. Salt injury appears as a marginal leaf scorch, early fall coloration and defoliation, dieback of twigs and branches, and in severe cases death of the entire tree.

Some trees are more tolerant to salt than others and may be classified as follows:

Very tolerant: Black cherry, red cedar, red oak, and white oak
Tolerant: Black birch, black locust, gray birch, largetooth aspen, paper birch, white ash, and yellow birch
Moderately tolerant: American elm, linden, hop-hornbeam, Norway maple, red maple, shagbark hickory
Intolerant: Beech, birch, hemlock, red pine, speckled alder, sugar maple, white pine

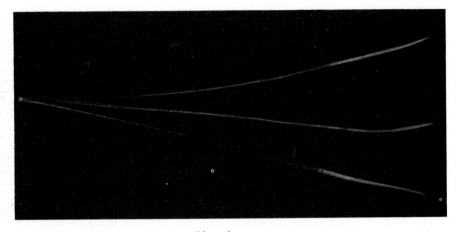

Fig. 10-27. Needle tip necrosis caused by salt spray.

Salt Spray. Trees growing along the seashore are often injured by salt spray blown from the ocean. During hurricanes the spray has actually damaged leaves 50 miles from salt water. Most of the damage, however, is confined to within a few miles of the coast. Salt spray is also damaging to the foliage of needle evergreens planted downwind from salted high-speed roadways. Needle tips on the exposed side of the trees turn brown, principally from dehydration and salt toxicity (Fig. 10-27).

Among the broad-leaved trees, elm, magnolia, Norway maple, sugar maple, tuliptree, and tupelo appear most susceptible to salt spray, whereas red oak and horsechestnut are very resistant. Among the evergreens, white, jack, Balkan, Japanese red, and Swiss stone pines, hemlock, and junipers are most harmed, whereas spruce, Austrian, ponderosa, and Japanese black pines, holly, and yew suffer least.

When salt injury occurs, removal of badly damaged trees and pruning of mildly affected trees are all that can be suggested.

Fungicides and Insecticides. The various sprays recommended in Chapter 13 of this book for disease and insect control may occasionally injure trees. In general, damage from fungicides and insecticides is more severe on undernourished trees and on those growing in poorly drained soils than on vigorous specimens. Insect injury and frost damage also predispose leaves to spray injuries. Cool, damp weather favors the chances of injury from bordeaux mixture and other copper fungicides. High temperatures and humidities increase the chances of injury from lime sulfur and other sulfur sprays. High temperatures also enhance the probability of injury resulting from malathion and oil sprays.

Copper Sulfate. This chemical is frequently mentioned as a control for algae (green scum) in ponds and lakes. The usual concentration is 1 part of the copper sulfate in a million parts of water, by weight. When the concentration is in excess of this figure, there is danger of harming trees growing around the edge of the pond or lake.

Fig. 10-28. Oak leaf chlorosis.

Chlorosis

The uniform yellowing of leaves resulting from a reduction in the normal amount of chlorophyll is termed chlorosis. This loss reduces the efficiency of the leaf in manufacturing food. Chlorotic leaves may result from fungus, virus, or insect attack; low temperatures; toxic materials in the air or soil; excessive soil moisture; surpluses of soil minerals; lack of or nonavailability of soil minerals.

Chlorosis of many shade trees, especially of pin oak, appears to be associated with nonavailability of iron rather than with the lack of iron in the soil. This is especially true in soils containing limestone, ashes, or other alkaline materials, where the pH of the soil ranges from 6.7 to 8.5. Iron may be present in such soils, but in a form that cannot be absorbed by the plant.

Symptoms. The leaves of affected trees first turn uniformly yellowish green, or they may remain green along the veins but turn yellow in the intraveinal areas (Fig. 10-28). The terminal growth of twigs is reduced, and the tree is generally stunted. The tissue between the leaf veins or along the leaf edge may die on trees affected for several years with chlorosis. Eventually whole branches or the entire tree may die prematurely unless the condition responsible is corrected. New growth may be more severely affected than growth produced earlier in the season.

Control. Chlorosis caused by deficiency or nonavailability of iron can often be corrected by special treatments. These treatments may involve soil acidification using sulfur, injection or implantation of iron into the tree trunk, or application of iron to the foliage. These methods are discussed in Chapter 6.

Selected Bibliography

Alden, J., and R. K. Hermann. 1971. Aspects of cold hardiness mechanism in plants. Bot. Rev. 37:37–142.

Evans, L. S. 1984. Acidic precipitation effects on terrestrial vegetation. Ann. Rev. Phytopathol. 22:397–420.

Harris, R. W. 1992. Arboriculture. Prentice-Hall, Englewood Cliffs, N.J. 674 pp.

Hoeks, J. 1972. Effect of leaking natural gas on soil and vegetation in urban areas. Agric. Research Report 778. Centre for Agricultural Publishing and Documentation, Wageningen, Holland. 120 pp.

Jacobson, J. S., and A. C. Hill (eds.). 1970. Recognition of air pollution injury to vegetation: A pictorial atlas. Air Pollution Control Assoc., Pittsburgh, Pa. 100 pp.

Katterman, F. (ed.) 1990. Environmental injury to plants. Academic Press, San Diego, Calif. 290 pp.

Kozlowski, T. T., and S. G. Pallardy. 1997. Physiology of woody plants. Academic Press, San Diego, Calif. 411 pp.

Loomis, R. C., and W. H. Padgett. 1973. Air pollution and trees in the east. U.S.D.A. Forest Service, State and Private Forestry, Northeast Area and Southeast Area, Upper Darby, Pa. 28 pp.

Reinert, R. A. 1984. Plant response to air pollutant mixtures. Ann. Rev. Phytopathol. 22:421–42.

Shigo, A. L., 1991. Modern Arboriculture. Shigo and Trees, Associates, Durham, N.H. 424 pp.

Shurtleff, M.C., and C.W. Averre. 1997. The plant disease clinic and field diagnosis of abiotic diseases. APS Press, St. Paul, Minn. 245 pp.

Sinclair, W. A., H. H. Lyon, and W. T. Johnson. 1987. Diseases of trees and shrubs. Cornell University Press, Ithaca, N.Y. 574 pp.

Strouts, R. G., and T. G. Winter. 1994. Diagnosis of ill health in trees. HMSO, London. 307 pp.

Treshow, M. 1970. Environment and plant response. McGraw-Hill, New York. 422 pp.

Watson, G. W. , and D. Neely (eds.) 1995. Trees and building sites. International Society for Arboriculture, Champaign, Ill. 191 pp.

Wood, F. A., and J. B. Coppolino. 1972. The influence of ozone on deciduous forest tree species. In: Effects of air pollutants on forest trees. VII International Symposium of Forest Fume Damage Experts. Vienna, Sept. 1970, pp. 233–53.

11

Insect and Mite
Pests of Trees

A working knowledge of insect pests, of how they injure plants, and of methods of combating them is essential to any tree preservationist. A detailed discussion on insect identification methods is best left to texts written specifically for that purpose; however, important characteristics of the major insect pests are presented here.

Accurate identification of any insect can be obtained by submitting the specimen to the state entomologist, usually located in the state capital, or an insect identification laboratory located in the state agricultural experiment station. The county extension office is a good place to start.

Structure of Insects

Insects belong to a larger group of animals called Arthropoda. The Arthropoda are characterized by having segmented legs, segmented bodies, and a more or less tough skin that also serves as the skeleton. Besides insects, this group includes spiders, mites, sowbugs, millipedes, centipedes, crayfish, and many other similar animals. The insects differ in structure from other arthropods in that the body segments are organized into three distinct sections: head, thorax, and abdomen.

The head segments are closely fused and appear to be a single segment that bears the feelers, or antennae; the eyes; and the mouthparts. Mouthparts of adult plant-feeding insects are adapted for chewing, sucking, or some combination of both.

The thorax, the section just behind the head, has three segments with a pair of segmented legs on each segment, for a total of six legs. Other arthropods usually have more legs. If wings are present, as they usually are in insects, they are also attached to the thorax. Other arthropods never have wings.

The abdomen is composed of a variable number of segments depending on the species. In the female of some species, a sharp-pointed organ for depositing eggs, the ovipositor, is found at the posterior end of the abdomen.

The major groups of insects, or orders, are separated from each other primarily on the basis of the number and structure of the wings, the type of mouth parts, and the type of metamorphosis they go through in their life cycles.

Life Cycle of Insects

One of the most interesting mysteries of nature is found in the study of the life cycle of the insect as it passes through various stages from egg to adult. The process that embodies all these changes in form is known as metamorphosis.

The life cycle of butterflies, beetles, sawflies, true flies, and moths comprises four separate stages: egg, larva, pupa, and adult. Insects with this life cycle are said to have complete metamorphosis.

After mating, the female deposits her eggs in a safe place that will provide a favorable environment for the usually wormlike larvae when they emerge, or she produces the eggs in such large numbers that survival of only a small percentage will assure abundant offspring. The names of the larval forms vary according to the groups to which they belong. Caterpillars are the larvae of butterflies and moths; grubs are the larvae of beetles (Fig. 11-1 [left]); false caterpillars are the larvae of sawflies; and maggots are the legless larvae of flies.

Most insect larvae have chewing mouthparts, though their parents may not. They do not have their parents' compound eyes and wings, and many do not have six true legs, having in some cases none and in some cases many more false legs on the abdomen. Not only do the larvae look different from their parents, they often live in different places and eat different foods.

The common names of insects often reflect the first or most common plant with which they were associated. The host range of an insect is often much wider than that implied by name. The two-lined chestnut borer, for example, attacks oak, beech, and hornbeam, in addition to chestnut.

As a larva grows it sheds its skin, or molts, several times, each successive skin being larger and better able to accommodate the insect's ever-increasing body. Finally the larva becomes full grown and is ready to enter the third stage of development, the pupa. Larvae of many insects prepare a protective covering for this stage. The cocoon of the moth and sawfly, the pupal case of the true flies, and the pupal cell of the beetle represent such protective coverings.

Although the pupal stage is considered a resting period because the pupa does not feed or move about, it is marked by considerable metabolic activity and many important changes, which culminate in the emergence of the adult insect.

A different form of development, called incomplete metamorphosis, characterizes such insects as grasshoppers, aphids, plant bugs, lace bugs, and leafhoppers. The egg develops into a small wingless creature called a nymph. Often resembling the adult except for its ungainly head and small body, the nymph passes through several molts. Wings become more developed and prominent, and the entire body assumes the appearance of the adult insect with each successive molt. Thus the last stage is attained directly, no pupa being formed in the life cycle.

Types of Insect Pest

The insect pests of plants may be grouped according to their placement in the natural classification system, or according to the habit by which they are pests. For instance, beetles are a natural grouping, but beetles are also included in such habit groups as borers and leaf miners. Because pest control methods are usually more contingent on the pest's habits than on its scientific classification, we use

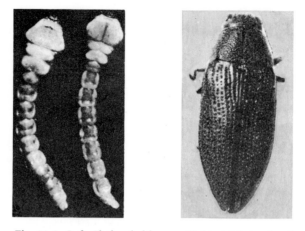

Fig. 11-1. Left: Flatheaded borers. Right: Adult beetle of one of the flatheaded borers. Wing covers are hard.

the habit classification as much as possible. The following sections describe general insect and mite pests. Pest management suggestions presented here may include use of pesticides. Before buying or using pesticides, be sure that the tree and pest are listed on the label. Follow all label directions. New chemical formulations and uses are constantly being developed; these suggestions should not be considered complete. Your Cooperative Extension Service may have up-to-date pesticide information.

Borers. Any insect that feeds inside the roots, trunks, branches, or twigs of a tree is known as a borer. The borer is usually a larva or worm stage of a beetle, moth, or wasplike insect, although some beetles may bore into the tree as adults.

Bark beetles generally mine in the phloem and cambial region between the bark and the wood, resulting in frass production and in conifers, pitch exudation. Adults are small, dark, hard-bodied beetles with strong jaws for boring. Eggs are laid in tunnels bored by the adult. Larvae are grublike, cream or white, often with a dark head. Adult and larval galleries under the bark sometimes form characteristic patterns, which aid in identification. Examples of bark beetles that attack trees include cedar bark beetles, *Phloeosinus* spp.; engraver beetles, *Ips* spp.; European elm bark beetle and shothole borer, *Scolytus multistriatus* and *S. rugulosus*; oak bark beetles, *Pseudopityophthorus* spp.; and red turpentine beetle and western pine beetle, *Dendroctonus valens* and *D. brevicomis*.

Flatheaded borer adults are flattened, oval beetles, often with metallic coloration. Eggs are laid in bark crevices, and the larvae bore winding frass-filled tunnels beneath the bark and sometimes into the wood. Larvae of many species are broad and flat in the front and more tapered and narrow farther back. Examples of flatheaded borers that attack trees include the flatheaded appletree borer and Pacific flatheaded borer, *Chrysobothris femorata* and *C. mali*; and *Agrilus* borers such as bronze birch borer, *A. anxius*; bronze poplar borer, *A. liragus*; flatheaded alder borer, *A. burkei*; oak twig girdler, *A. angelicus*; and two-lined chestnut borer, *A. bilineatus*.

Roundheaded borer adults are medium to large cylindrical beetles with long antennae. They are sometimes called longhorned beetles. From eggs laid in bark crevices, borers tunnel beneath the bark and sometimes into the wood of tree limbs, trunk, and main roots. Borer larvae are white and cylindrical. Examples of roundheaded borers that attack trees include eucalyptus longhorned borer, *Phoracantha semipunctata*; oak root borers, *Prionus imbricornis* and *P. laticollis*; and poplar borer and roundheaded apple tree borer, *Saperda calcarata* and *S. candida.*

Clearwing moth larvae feed beneath tree bark, producing large amounts of frass and often destroying phloem and xylem, leading to tree decline and death. The whitish larvae are cylindrical, with brownish markings around the thorax. Adult moths are recognized by their resemblance to paper wasps with their narrow, clear wings. Female moths emit a pheromone that attracts the males; after mating, eggs are laid in bark cracks and crevices. Some examples of clearwing moths damaging to trees include ash borer, *Podosesia syringae*; dogwood, lesser peachtree, peachtree, and sycamore borers, *Synanthedon scitula, S. pictipes, S exitiosa,* and *S. resplendens*; Sequoia and Douglas-fir pitch moths, *Synanthedon sequoiae* and *S. novaroensis*; and western poplar clearwing, *Paranthrene robiniae.*

Factors Favoring Borer Infestations. Although a few borers can attack vigorously growing trees, most species become established in trees low in vigor. Consequently, any factor that tends to lower the vitality of a tree predisposes it to borer attack. Among the more important of these factors are prolonged dry spells, changes in the environment unfavorable to the growth of the tree (see Chapter 10 for examples), loss of roots in transplanting, repeated defoliation by insect or fungal parasites, and bark injuries caused by frost, heat, or mechanical agents. In some instances, trees of low vitality and injured trees discharge chemical signals that are attractive to borer insects.

Prevention of Borer Attack. Many borer infestations can be prevented by improving, when feasible, any unfavorable situation around the trees. Proper fertilization, adequate watering during dry spells, control of defoliating insects and diseases, and pruning of infested or weakened branches are recommended. All bark wounds should receive immediate attention to facilitate rapid healing and thus reduce the amount of exposed tissue.

Control. The method used to control already established borers depends on the part of the tree infested and the species of borer involved. Twig and root borers are controlled by pruning and discarding infested parts. Borers that make galleries that lead to the bark surface (sawdust-like frass is a telltale sign) can be controlled by injecting a toxic paste into the holes and then sealing them with a small wad of chewing gum, grafting wax, or putty. The pastes are sold under such trade names as Bor-Tox, Borer-Kil, and Borer-Sol. Some borers inside the tree can be killed by crushing with a flexible wire or sharp knife inserted into the opening. A suspension of an insect-parasitic nematode, *Steinernema carpocapsae,* can be similarly injected into the galleries of clearwing moth larvae. The nematode and mutualistic bacteria together kill the insect pest—a biological control.

Application of insecticide to the bark during the adult's egg-laying period or before the eggs hatch and the borers enter the bark is standard procedure for controlling trunk and branch borers. For many clearwinged moth borers, the egg-

Fig. 11-2. Holly leaf miner damage.

laying time can be determined by using pheromone traps, which trap male borers. The presence of males in the trap indicates that mating, egg laying, and larval emergence will occur in due course. In some locations, pest emergence has been correlated with timing of tree seasonal development. For managing most borers and bark beetles, use the insecticides Dursban, Lindane, or Thiodan. The bark on the trunk and larger branches should be sprayed thoroughly. An understanding of the borer life cycle, especially the timing of egg hatching, is needed for best results. Consult the nearby Cooperative Extension office for information on timing of borer sprays. Insecticides injected into the tree xylem rarely work to control borers because the insects are usually in the phloem while the chemical is in the xylem.

Leaf Miners. Leaf miners are insect larvae that feed between the upper and lower leaf epidermis of several kinds of landscape trees. This feeding activity results in unsightly light green to brown blotches or serpentine mines in the leaf (Fig. 11-2). Heavy feeding can cause the entire leaf to die, and a heavy infestation can damage a tree. Interestingly, though several different leaf miner larvae cause similar symptoms, the adults may be completely different kinds of insects.

Birch leaf miner, *Fenusa pusilla*, is a sawfly larva causing blotch mines, which may run together and blight the entire leaf. Boxwood leaf miner, *Monarthropalpus buxi*, is a midge fly larva causing blister mines of boxwood leaves. There are five holly leaf miner species; the two most common, *Phytomyza ilicis* and *P. ilicicola*, feed on English and American hollies. These two species of fly larvae cause blotch or serpentine mines on holly leaves. Pin pricks in leaves are egg-laying and feeding

Fig. 11-3. Holly leaf miner adult.

punctures made by the small, black, robust flies (Fig 11-3). Locust leaf miner, *Odontota dorsalis*, is a beetle larva causing blotch mines that skeletonizes leaves. The adult beetle is wedge shaped, flat, and orange with a black line down the back. Two species of moths, the solitary and gregarious oak leaf miners, *Cameraria hamadryadella* and *C. cincinnatiella*, respectively, attack red and white oaks, causing blotch mines. The former has one and the latter several larvae per mine. Arborvitae leaf miner, *Argyresthia thuiella*, and cypress leaf miner, *A. cupressella*, are important pests of these hosts and of juniper.

Control. Systemic insecticides can be used to kill leaf miner larvae in the mines. Products with systemic action include Cygon, Merit, and Orthene. Protectant sprays are timed generally in springtime to kill adults before they lay eggs in the leaf. Such materials include Astro, Diazinon, Dursban, Dylox, Pounce, and Sevin.

Foliage-Feeding Caterpillars. Caterpillars can be conspicuous consumers of tree leaves. These larvae are relatively large and move freely on the tree. When they are present in sufficient numbers, they destroy a noticeable portion of the foliage and in some cases weaken the tree. Caterpillars have chewing mouthparts, and they have six true legs on the thorax and from four to ten unjointed legs on the abdomen; they may be smooth, hairy, or spiny, and may or may not be brightly colored. The adult phase is a moth or butterfly, a member of the order Lepidoptera, a large and well-known insect group. These insects undergo a complete metamorphosis during their life cycle and may have more than one generation per year. Moth larvae are more numerous than butterfly larvae as tree pests. Caterpillar attacks late in the season are normally not a cause for concern; however, 30 percent defoliation in summer, or even 20 percent defoliation in spring could adversely affect tree health, especially if defoliation occurs annually. Monitor for caterpillars by visual inspection, pheromone traps, and frass collection.

The gypsy moth, *Lymantria* (=*Porthetria*) *dispar*, is a highly destructive pest of

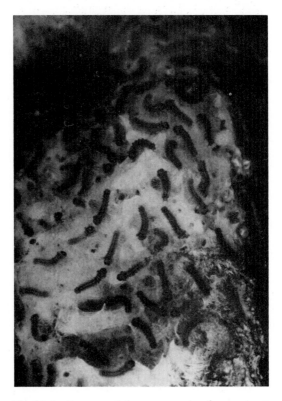

Fig. 11-4. Gypsy moth larvae emerging from egg mass.

trees in the northeastern United States, where it has caused enormous losses and has been the subject of much publicity. The extent of infestation is still increasing, but whether or not it will become a pest nationwide remains to be seen.

The dark and hairy larvae (Fig. 11-4) may grow to a length of $2^{1}/_{2}$ inches after hatching in early spring from egg masses laid the previous season on branches, buildings, or other outdoor objects such as automobiles and camping equipment. Larvae crawl up the tree, begin feeding and, while still small, spin a silken thread from which they are suspended. This allows the wind to carry them several hundred yards, where they can feed on another tree. As the larvae mature, they may feed only at night, crawling down the tree during the day to hide in a protected place.

Larvae of heavily infested trees may continue to feed throughout the day. They feed for about 7 weeks, then pupate. In midsummer the adults emerge and mate; the females lay eggs and then die, without feeding. Oaks suffer from the defoliation more than many other trees, and they may not survive two or more consecutive years of leaf loss. Other trees damaged by gypsy moths include alder, apple, arborvitae, black gum, beech, elm, hawthorn, hemlock, hickory, linden, maple, poplar, pine, sassafras, spruce, and willow.

Other caterpillars damaging to trees include fall and spring cankerworms,

Alsophila pometaria and *Paleacrita vernata*, which have wingless female adults and inch-long yellowish green, green, brown, or black smooth larvae. Orange-striped oakworm, *Anisota senatoria*, and yellow-necked caterpillar, *Datana ministra*, are widespread in the East and feed on several different kinds of trees. Red-humped caterpillar, *Schizura concinna*, and white-marked tussock moth, *Orgyria leucostigma*, also have wide host feeding preferences. The larva of the io moth, *Automeris io*; the linden looper, *Erannis tiliaria*; and elm span-worm, *Ennomos subsignarius*, are all important tree leaf eaters. A few butterfly larvae such as mourning-cloak, *Nymphalis antiopa*, and silverspotted skipper, *Epargyreus clarus*, feed on trees. Several caterpillars construct webbed nests in the tree for protection. Fall webworm, *Hyphantria cunea*; eastern tent caterpillar, *Malacosoma americana*; forest tent caterpillar, *M. disstria*; and California tent caterpillar, *M. californicum*, are common nest makers.

Control. Caterpillars have many natural enemies including birds and parasitic wasps. Sometimes these natural controls are sufficient for pest management. Caterpillars such as web worms and tent caterpillars that form protective nests in the tree can often be physically removed and destroyed, nest and all, and others forming conspicuous egg masses, such as gypsy moth and tussock moth, can be reduced by scraping egg masses off bark or other objects and destroying them. An insecticide specifically toxic to caterpillars is made from the spores of a bacterium, *Bacillus thuringiensis* (Bt), and sold under the names of Thuricide, Biobit, Victory, Larvo-BT, Foray, Dipel. *B. t.* variant *kurstaki* is good for mimosa webworm, which infests honey locusts. Other insecticides used against these pests are Astro, Decathlon, Dursban, Imidan, Orthene, Malathion, Mavrik, Methoxychlor, Scimitar, Sevin, Talstar, and Tempo.

Sawflies. Sawflies are nonstinging wasps that resemble flies. The larval stage (Fig. 11-5) of these insects damages trees. Some sawfly larva resemble butterfly and moth caterpillars, whereas others look like slimy, legless slugs. The larvae of sawflies have six legs on their thoracic segments and prolegs on all abdominal segments.

Many sawfly larvae such as red-headed pine sawfly, *Neodiprion lecontei*; mountain-ash sawfly, *Pristiphora geniculata*; butternut woolyworm, *Eriocampa juglandis*; and pear slug, *Caliroa cerasi*, are defoliators capable of devouring needles or skeletonizing and consuming leaves. Other sawfly larvae are leaf miners, for example, birch leaf miner, *Fenusa pusilla*; elm leaf miner, *F. ulmi*; and European alder leaf miner, *F. dohrnii*. The maple petiole borer, *Caulocampus acericaulis*, is also a sawfly larva.

Adult female sawflies have a sawlike ovipositor at the tip of the abdomen, which is used to slit or cut plant tissue. Eggs are laid in these slits.

Control. Most sawfly larvae are controlled with sprays of insecticidal soaps, Azatin, Dursban, Merit, Orthene, Sevin, Tempo or Turplex. Insecticides should be applied to control pine- and leaf-eating sawfly larvae when they appear in early spring and summer. Larvae of sawflies such as the European pine sawfly tend to feed in groups. The most effective control may be to remove and destroy the entire cluster of insects along with the damaged shoots. Leaf miner control requires more precise timing because adults need to be killed at egg laying time

Fig. 11-5. European pine sawfly larvae.

Fig. 11-6. Elm leaf beetle larva.

or larvae killed before they are protected by the mine. A systemic insecticide such as Cygon can be used to destroy mining larvae.

Foliage-feeding Beetles. Adults of these beetles have durable wing covers, and their larvae have thoracic legs but no abdominal legs. Japanese beetle, *Popillia japonica*, and elm leaf beetle, *Pyrrhalta luteola* (Fig. 11-6), are important tree pests. Adult Japanese beetles are about $3/8$ inch long and a shiny greenish bronze color. They often feed in groups on sunny parts of the tree. Elm leaf beetles are less than $1/4$ inch long, but they have long antennae. They are yellowish green with black stripes. The two represent contrasting life histories: Japanese beetle has a wide host range, whereas elm leaf beetle seeks out various elm species or zelkova, a close relative. Japanese beetle adults feed on and defoliate trees (larvae [grubs] feed on roots of grasses), whereas elm leaf beetle adults and larvae both feed on elm leaves (and the larvae are most damaging). Japanese beetle overwinters as a larva, whereas elm leaf beetle overwinters as an adult. In the United States, the Japanese beetle is found only in the eastern and central states, whereas elm leaf beetle is found wherever elms are grown. Both pests skeletonize leaves and defoliate their hosts. Japanese beetles begin feeding in the tree tops in open sun, perhaps because it is warm there—they do not move about much when temperatures are below 70° F. Beetles are attracted to leaves that have already been fed on, which is why once feeding begins, other Japanese beetles soon arrive, and these aggregates of beetles defoliate their host. Trees that lose their leaves in consecutive growing seasons may have their energy reserves depleted; trees thus stressed have less ability to withstand other potentially harmful pests or adverse environment.

Control. Insecticides such as Astro, Azatin, Carbaryl, Cythion, Decathlon, Dursban, Imidan, Malathion, Talstar, Tempo, M-One, Trident II, Sevin, Methoxychlor, and Orthene can be used to control foliage-feeding beetles. Japanese beetle traps are ineffective for preventing beetles from feeding on susceptible plants in the landscape. Sprays for Japanese beetle are applied in June and July, whereas sprays for elm leaf beetle are applied in May and June.

Weevils. Weevils are a diverse group of beetles, often called snout beetles; many of them feed on trees. The heads of adult weevils are elongated into a snout, and they have bent, clubbed antennae. Many weevils have fused wing covers and do not fly. Larvae are whitish grubs that, depending on the species, feed on roots and pupate in the soil, feed on foliage, or bore into the wood of tree trunks and branches. Adult weevils feed on foliage, causing leaves to have circular to irregular holes, notches, or a ragged appearance. Leaves or needles may be clipped from the twigs. Although foliage damage may be highly visible, most weevils, with the exception of root and trunk-feeding weevils, damage trees only slightly.

Sassafras weevil, *Odontopus calceatus*, is a small weevil that makes numerous holes in leaves of magnolia, sassafras, and tuliptree. Larvae also feed on the leaves, making mines and causing blotches. Fuller rose beetle, *Pantomorus cervinus*, is a weevil that feeds on acacia, camellia, citrus, oak, and *Prunus* species in the southern and western states. The larvae feed on roots, doing little damage, but foliar feeding by adults can be noticed. Black vine weevil, *Otiorhynchus sulcatus*, adults feed on yew and rhododendron, often chewing notches into the leaf edges. Larvae feed on roots and stem bark near the soil surface and are the most damaging stage of this weevil, especially on young plants. Even a few weevils should be a cause for concern and should trigger control measures. Conifers such as pine, fir, Douglas-fir, and spruce are attacked by weevils such as white pine weevil, *Pissodes strobi*; deodar weevil, *P. nemorensis*; and Monterey pine weevil, *P. radiatae*. Although foliage damage by conifer weevils may be slight, trunk and branch damage can be severe enough to kill small trees. Some weevils may vector fungi that cause canker diseases.

Control. Monitoring is an important activity for weevil management. Higher numbers of foliage-feeding weevils such as sassafras weevil and Fuller rose beetles can be tolerated than the root and stem feeders such as black vine weevil and white pine weevil. For small numbers of weevils, monitoring for and destroying them may be sufficient for control. Such methods include dislodging weevils from the foliage and killing them, or using pitfall traps, hiding bands, and trap boards to monitor and kill weevils. Sanitation by pruning out infested twigs and branches to eliminate larvae may help control the conifer weevils. For root- and trunk-infesting weevils, persistent soil or trunk insecticide applications may be needed.

Black vine weevil larvae in container nursery production are controlled using insecticides such as Furadan, Oxamyl, or Turcam, or the biological control nematode, *Steinernema carpocapsae*. Insecticides useful against black vine weevil adults include Astro, Azatin, Cryolite Bait, Dursban, Dycarb, Ficam, Guthion, Isotox, Mavrik, Orthene, Scimitar, Talstar, and Tame. Endocide and Phaser are also available for this pest on yew.

Sassafras or yellow poplar weevil adults may be controlled with early summer applications of Carbaryl, Dursban, or Sevin.

Pines can be protected from northern pine weevils and pales weevils with Asana, Dursban, Dycarb, Ficam, or Turcam insecticides. Stumps harboring the weevils must be destroyed by spring to reduce the infestation. Lindane sprays applied to pines in spring and in late summer are used to protect seedlings and young twigs from these weevils. Lindane is also used on pine or spruce shoot tips in spring to prevent white pine weevil infestations. White pine root collar weevils are controlled with Dursban sprays in May, August, and September.

Plant Bugs and Lace Bugs. These insects have piercing-sucking mouthparts, which in many instances cause a stippling symptom on infested leaves. This stippling may involve patches of chlorotic or necrotic cells, different from the finer chlorotic flecks produced by spider mites, for example.

Tarnished plant bug, *Lygus lineolaris*, and four-lined plant bug, *Poecilocapsis lineatus*, produce small, brown, uniform, circular spots on leaves; these symptoms resemble those caused by parasitic fungi or bacteria. Sycamore plant bug, *Plagiognathus albatus*, injures leaves by producing holes, slits, and tears in the leaves, leaving tattered, chlorotic leaves. Honey locust leaves are discolored, deformed, and stunted by nymphs and adults of the honey locust plant bug, *Diaphnicoris chlorionis*. Plant bugs appear as flattened, shield-shaped insects 1/8 to 1/4 inch long.

In addition to chlorotic stippling, lace bugs produce dark, shiny excrement spots on leaf undersides (Fig. 11-7). Lace bugs are about 1/8 to 1/4 inch long and have a broad thorax with fine, lacelike wings. Important lace bugs attacking trees include sycamore, walnut, and alder lace bugs, *Corythucha ciliata*, *C. juglandis*, and *C. pergandei*, respectively.

Fig. 11-7. Lace bug: adults, nymphs, and tarlike excrement.

Control. Most trees can tolerate lace bug damage, so controls are rarely needed. Sprays of Astro, Carbaryl, Cythion, Decathlon, Dursban, Malathion, Orthene, Sevin, Talstar, or Tempo applied in late spring and summer can control plant bugs and lace bugs.

Aphids. Aphids have mouthparts that pierce and suck the sap from foliage, stems, and roots. They may be found singly or in groups on the affected tissue, often forming very large colonies. Aphids vary in size, ranging from 1/16 to 1/4 inch, and their color ranges from purple, red, or black to yellow, pink, or green. Most aphids overwinter as eggs, hatching in the spring as adults, some winged, some not (Fig. 11-8). Many generations of aphids may occur during the growing season, sometimes all on one host, and sometimes on several different hosts.

Aphid damage may include gradual weakening due to depletion of plant nutrients; cupping, stunting, or distortion of leaves or needles; defoliation; and death of branches. Aphids also cause damage by spreading viruses.

Many aphids secrete large quantities of honeydew, partly digested sweet and sticky plant juices. Honeydew is attractive to bees, wasps, ants, and especially sooty mold fungi. These fungi are black and unsightly, giving a dark, dusty appearance to affected foliage and, in severe cases, interfering with photosynthesis by shading out the sun. In addition, honeydew may drip on people and property and cause dust and dirt to stick to the leaves.

Some aphids produce waxy substances that, when combined with cast molted skins, can produce a white or gray protective covering for the aphid colony. Aphids existing in such colonies are often referred to as woolly aphids.

Some aphids that are tree pests include balsam twig aphid, *Mindarus abietinus;*

Fig. 11-8. Top: winged aphid, greatly enlarged. Bottom: A group of nonwinged aphids.

tuliptree aphid, *Macrosiphum liriodendri*; California laurel aphid, *Euthoracaphis umbellulariae*; apple and cowpea aphids, *Aphis pomi* and *A. craccivora*; giant bark aphid, *Longistigma caryae*; green peach aphid, *Myzus persicae*; green spruce aphid, *Elatobium abietinum*; potato aphid, *Macrosiphum euphorbiae*; and woolly apple and woolly elm aphids, *Eriosoma lanigerum* and *E. americanum*. Other pests resembling aphids include adelgids (phylloxera) such as pine bark adelgid, *Pineus strobi*; and balsam woolly adelgid, *Adelges piceae*. Adelgids, found primarily on conifers, lack the posterior pair of protruding tubelike cornicles of aphids. Psyllids such as the pear psyllid, *Psylla pyricola*, also have piercing-sucking mouthparts and can cause distortion of leaves. Psyllids have longer antennae than aphids, and they commonly jump when disturbed.

Control. Aphids have many natural enemies that are predators, but by the time natural populations of the predators have grown to be an effective force, the damage caused by aphids already may have occurred. Insecticides commonly used to control aphids include horticultural oil during the early spring while trees are still dormant, and, when the aphids are observed to be a problem, summer applications of insecticidal soap, Astro, Azatin, Diazinon, Dursban, Guthion, Malathion, Merit, Orthene, Talstar, Tempo, or Thiodan. Gall-forming adelgids cannot be controlled once the gall has formed.

Scales. Another important group of insects, scales have sucking mouthparts and are wingless (except for the males). High populations of these insects can cause tree decline, and young trees may be killed. There are three kinds of scale or scalelike insects that commonly attack trees: mealybugs, soft scales (Fig. 11-9), and armored scales.

Mealybug injury appears as foliar discoloration, deformation, and wilt and eventual death of affected plant parts caused by loss of sap. These insects congregate in large numbers and may attack roots, stems, twigs. leaves, flowers, and fruits. They are white and motile, deriving their name from the white, waxy secretions that cover their bodies. Important examples include Comstock and long-tailed mealybugs, *Pseudococcus comstocki* and *P. longispinus*; two-circuli mealybug, *Phenacoccus dearnessi*; taxus mealybug, *Dysmicoccus wistariae*; and beech scale, *Cryptococcus fagisuga*.

Soft scales vary widely in appearance, ranging from the cottony maple leaf scale to the shell-like kermes and lecanium scales. The shell or covering of these insects is part of the soft scale exoskeleton. Soft scales such as the lecaniums have

Fig. 11-9. Calico scales infesting Chinese elm. Notice the honeydew droplets accumulating on the twig just below the soft scales.

two periods of motility: the crawler stage, occurring just after egg hatch, which moves to the leaves; and the second instar stage, which migrates from the leaves back down to the twigs in the fall to overwinter or to lay eggs. These scales debilitate the host tree by sucking phloem sap. They often produce large quantities of honeydew, which promotes growth of sooty mold fungi; these can blacken the affected leaf or shoot. Soft scales, which infest many trees in the landscape, include black, *Saissetia oleae*; brown soft and citricola, *Coccus hesperidum* and *C. pseudomagnoliarum*; calico, *Eulecanium cerasorum*; cottony-cushion, *Icerya purchasi*; cottony maple, *Pulvinaria innumerabilis*; Fletcher's, *Parthenolecanium fletcheri*; hickory lecanium, *Lecanium caryae*; magnolia, *Neolecanium cornuparvim*; and tuliptree, irregular pine, and pine tortoise, *Toumeyella liriodendri, T. pinicola,* and *T. parvicornus.*

Armored scales have a hard, shell-like covering composed of a waxy secretion and old shed skins. The covering is not part of the insect; the insect itself can be exposed by carefully lifting the shell. At this stage, the scale insect has no eyes, legs, or antennae, but it has fine sucking mouthparts, which siphon nutrients from the plant. The removal of phloem sap by large numbers of scales debilitates the tree and can cause gradual death of twigs and branches. Armored scales move only in the first instar, which is also called the crawler. Once the crawler settles and begins feeding, it remains in that place until it dies.

The following are examples of armored scales damaging to many landscape tree species: black pineleaf, *Nuculaspis californica*; California red, *Aonidiella aurantii*; European fruit lecanium and frosted, *Parthenolecanuim corni* and *P. pruinosum*; Florida red and dictyospermum, *Chrysomphalus ficus* and *C. dictyospermi*; fiorinia hemlock, *Fiorinia externa*; greedy and latania, *Hemiberlesia rapax* and *H. latania*; hemlock, *Abgrallaspis ithacae*; ivy and oleander, *Aspidiotus hederae* and *A. nerii*; juniper and minute cypress, *Carulaspis juniperi* and *C. minima*; obscure and gloomy, *Melanaspis obscura* and *M. tenebricosa*; oystershell, *Lepidosaphes ulmi*; pine needle and scurfy, *Chionaspis pinifoliae* and *C. furfura*; pit-making, *Asterolecanium* spp.; San Jose and walnut, *Quadraspidiotus perniciosus* and *Q. juglansregiae*; and white peach, *Pseudaulacaspis pentagona.*

Control. Most scale-infested trees can be treated with a dormant horticultural oil spray applied in early spring before bud break. The crawler stages are vulnerable to insecticides such as Diazinon, Dursban, Malathion, Merit, Orthene, Sevin, and Tempo, but the insecticide must be applied in spring or summer when crawlers are active. Monitor crawlers using double-sided sticky tape traps so that insecticides can be accurately timed. It is important to obtain an accurate diagnosis of the scale to know exactly which chemical to apply and when. Scales are often controlled by natural enemies including parasitic wasps and many species of lady beetles. These natural control agents may be killed by insecticides while the scales may be merely suppressed. One scale survivor can produce many eggs, and the outbreak can resume quickly in the absence of natural controls. Thus, unless the scales really threaten tree health, it is usually best to let natural biological controls work.

Spider Mites. Spider mites (family Tetranychidae), which are not really insects—they have eight legs rather than six—feed on trees and cause damage.

There are many species of spider mites, and they damage both conifer and broad-leaved trees. Spider mite mouthparts penetrate leaf cells and extract the sap. A flecking or stippling symptom develops on the foliage as the pests destroy the chlorophyll- containing cells of the epidermis. The heavily infested leaf or needle may take on a yellowish or bronze color, which can sometimes be mistaken for ozone or oxidant air pollutant injury. Some mites produce a weft of fine webbing over the surface of the infested foliage, and eggs, egg shells, and cast skins are found on most infested leaves.

Mites usually overwinter as eggs and develop successively into six-legged larvae, protonymph, deutonymph, and eight-legged female and male adults. Adult mites are tiny, only $1/100$ to $1/25$ inch long! A diagnostic test that works well for infested conifers involves sharply shaking or jarring the foliage over a piece of white paper. Mites that land on the paper can be seen as very fine, dark flecks moving about on the paper. With the aid of a hand lens, eggs, cast skins, webbing, and crawling mites can be observed on the leaves. Reasonable field identifications can be made if their color, host plant, and geographic location are known.

Mite populations can build up rapidly, especially during dry weather. They often become a serious problem when predator mite populations are reduced by the use of insecticides such as Sevin or Lindane. Mites destructive to landscape trees include European red mite, *Panonychus ulmi*; southern red mite, *Oligonychus ilicis*; spruce spider mite, *O. ununguis*; and two-spotted mite, *Tetranychus urticae*.

Control. Rainy weather reduces spider mites; similarly, overhead sprinkling will work. Predatory mites can be purchased and released on infested plants; be sure the right species and host are listed. Most spider mites are sensitive to horticultural oils and soaps. Thorough coverage of the mite infestation is essential. Spider mites of evergreens are usually controlled with sprays of Vendex or horticultural oil in early spring. Late spring and midsummer applications of Avid, Dursban, Hexygon, Kelthane, Malathion, Morestan, Pentac, or Talstar are often used to control deciduous tree spider mites. Pesticide applications are often toxic to predatory mites and are likely to disrupt natural buildup of biological controls.

Eriophyid Mites. These mites are different from spider mites in that they have only two pairs of legs. Their bodies are soft, spindle shaped, and almost wormlike. Eriophyid mites are extremely small, requiring a hand lens for observation. Their life stages consist of egg, nymph, and adult. The latter may take on different shapes, making identification difficult.

Eriophyids, called blister, bud, gall, or rust mites, feed only on plants. They may attack leaves, shoots, twigs, stems, buds, flowers, and fruits. Symptoms include abnormal shape or distortion, blisters, discoloration, erineum (a form of gall or abnormal hair growth from leaf epidermis) galls, stunting, russeting, bronzing, withering, and witches' broom. Most of these mites are host-specific, that is, they feed on only one or a small group of closely related types of plant.

Maple bladder (Fig. 11-10) and spindle galls are familiar pouch galls formed as a result of attack by eriophyid mites *Vasates quadripedes* and *V. aceriscrumena*. The galls are a localized growth reaction of the host following mite feeding. Eggs hatch and young adults live inside the galls.

Fig. 11-10. Maple bladder gall mite damage.

Conspicuous witches' brooms of hackberry result from *Eriophyes celtis* infestation of the tree buds. The infestation interferes with normal growth; numerous buds become clustered together on the stem. When the buds break, thin, stunted, tightly bunched twigs develop, resulting in a witches' broom.

Ash flower galls, caused by *Eriophyes fraxinivorus*, are distinctive because affected flowers are swollen, fused, and distorted.

The erineum of beech, caused by *Acalitus fagerinea*, consists of patches of pale green to yellow felty hairs located in the angles between the lateral veins and the midrib of the lower leaf surface. *Eriophyes elongatus* and *E. calaceris* form bright red erineum patches on the upper surface of maple leaves.

The eriophyid mite that attacks hemlock, *Nalepella tsugifoliae*, causes foliage to have a yellowish or brownish cast without galls being produced. Its feeding is debilitating to infested trees.

Control. Eriophyid mites are not normally damaging enough to warrant control measures. If there is a need, they are usually controlled by an application of horticultural oil in early spring just before budbreak. Insecticidal soap, Carbaryl, Dicofol, Joust, Kelthane, Morestan, Pentac, or Sevin can be applied when leaves or flowers first begin to form.

Selected Bibliography

Barbosa, P., and M. R. Wagner. 1989. Introduction to forest and shade tree insects. Academic Press, San Diego, Calif. 639 pp.

Dreistadt, S. H., J. K. Clark, and M. L. Flint. 1994. Pests of landscape trees and shrubs: An integrated pest management guide. Publication 3359. University of California Division of Agriculture and Natural Resources. 327 pp.

Drooz, A. T. 1985. Insects of eastern forests. U.S.D.A. Forest Service Misc. Pub. 1426, Washington, D.C. 608 pp.

Johnson, W. T., and H. H. Lyon. 1988. Insects that feed on trees and shrubs. 2nd ed. Cornell University Press, Ithaca, N.Y. 556 pp.

Keifer, H. H., E. W. Baker, T. Kono, M. Delfinado, and W. E. Styer. 1982. An illustrated guide to plant abnormalities caused by eriophyid mites in North America. U.S.D.A. Agriculture Handbook 573, Washington, D.C. 178 pp.

Lloyd, J. 1997. Plant health care for woody ornamentals. International Society of Arboriculture, Champaign, Ill. 223 pp.

Pirone, P. P. 1978. Diseases and pests of ornamental plants, 5th ed. Wiley & Sons, New York. 566 pp.

Rose, A. H., and O. H. Lindquist. 1982. Insects of eastern hardwood trees. Can. For. Serv. For. Tech. Rep. 29. 304 pp.

Westcott, C. 1973. The gardener's bug book. Doubleday, Garden City, N.Y. 689 pp.

12

Parasitic Diseases of Trees

Although disease control unquestionably constitutes one of the major phases of successful tree maintenance, it is often the least clearly understood by the arborist. This lack of understanding is due partly to the difficulty of recognizing disease symptoms and partly to the complex nature of the parasitic agents. Insufficient knowledge of the various chemicals used to combat diseases and of their proper applications is also a contributing factor.

Broadly speaking, a tree is considered diseased when its structure or functions deviate from the normal ones described in Chapter 2. Such a definition includes abnormalities caused by drought, excessive moisture, lack of nutrients, and chemical and physical injuries, as described in Chapter 10. In this chapter discussion is limited to diseases caused by infectious agents such as fungi, bacteria, viruses, phytoplasmas, nematodes, and parasitic higher plants. The symptoms they produce, their nature, and the chemicals and arboricultural practices used to combat them are also included in the discussion.

Disease Symptoms

Symptoms are an expression of the tree's reaction or response to disease. Tree disease symptoms occur when a part of the tree fails to function as it should following an attack by a pathogen. Often the symptoms are a direct expression of tissue death caused by the pathogen. Leaf spots, twig blights, cankers, and root rots are good examples (Fig. 12-1). However, even in these cases, the symptoms may be more extensive. Root rot not only causes root death or necrosis, but the entire top of the tree may show stunting, wilt, dieback, or nutrient deficiency symptoms because the roots are not functioning as they should. Table 12-1 presents tree disease symptoms associated with disruptions of tree functions when certain tissues are attacked by pathogens.

In addition to symptoms produced on the plant attacked, certain signs, such as fruiting and vegetative structures of the causal organisms, help in diagnosing the disease. For example, many fungi produce fruiting bodies that are easily recognized and facilitate diagnosis. Such bodies vary in size from pinpoint dots, which are barely visible to the unaided eye, to bodies a foot or more in diameter

Table 12.1 Typical symptoms observed when specific tissues are attacked.

Tissues Attacked	Function Disrupted	Typical Symptoms
Roots	Water and nutrient uptake	Root death, tree wilt, branch dieback, nutrient deficiency
Sapwood, xylem	Upward water movement	Sapwood discoloration, leaf wilt and death
Heartwood	Strong, sound structure	Trunk, branch breakage, cavity formation
Stem phloem	Downward food movement	Discolored, dead inner bark, root starvation, foliar nutrient deficiency, stunting, branch dieback, tree wilt
Stem cambium	Xylem and phloem formation	Discolored, dead cambium, sunken area on stem
Meristematic tissues	Normal cell division	Stem, root, or leaf deformation or galls
Succulent shoots	Normal terminal growth	Dead or thinned shoots, bushy regrowth
Leaves	Photosynthesis	Dead spots, chlorosis, defoliation, tree weakening

(Fig. 12-1), which commonly occur in wood-decay diseases and appear as shelves or brackets adhering to the dead bark.

Additional signs that aid in determining the cause of a disease include droplets of milky ooze that appear on the bark of trees affected by fire blight disease and bright orange telia that appear in spring on rust-infected cedar twigs and branches.

Nature of Parasitic Agents

Plant parasitic fungi and bacteria, the most common tree pathogens, are microorganisms that do not contain chlorophyll, which is necessary for the manufacture of food. As a result, they are dependent on the higher plants for their sustenance. Bacteria and fungi that depend entirely on decaying organic matter for their food source are known as *saprophytes*. These organisms are frequently beneficial, in that they decompose organic materials and liberate minerals for use in future plant growth. Forms that are able to obtain their nourishment by direct attack on living plant cells are known as *parasites* or *pathogens*. It is this group that is involved in disease production.

Bacteria. Bacteria are microscopic, single-celled organisms, which multiply by simple fission, a single individual dividing to form two. Some types of bacteria possess flagellae, whiplike appendages that effect a limited amount of locomotion in liquids. Bacteria have several forms, but those that parasitize plants are usually rod-shaped.

Plant pathogenic bacteria are capable of living on organic food sources in the

Fig. 12-1. Symptoms and signs of fungal diseases. Upper left: Cankers on ash stems. Upper right: Circular spots on maple leaf caused by the fungus *Phyllosticta*. Lower left: Pin-point fruiting bodies of the *Nectria* canker fungus. Lower right: Fruiting bodies of the fungus *Polyporus sulphureus*.

absence of the host plant. Many, such as leaf-spotting and shoot-blighting bacteria, are found on the host plant surface, living epiphytically but not causing disease until conditions are right. Some bacteria, implicated as causal agents of scorch diseases, are limited to the host xylem during pathogenesis.

Fungi. Fungi, as a rule, are much more complex than bacteria. They may possess two distinct stages—vegetative and reproductive. The vegetative stage usually consists of a weft or mass of microscopic, threadlike strands called *hyphae* (Fig. 12-2). These hyphae may be composed of innumerable cells joined together. Certain tissues in this complex are capable of performing highly specific functions.

The reproductive stage may be simple or very complex, either microscopic or visible to the unaided eye, sometimes reaching 12 or more inches in diameter. In all cases, however, the reproductive bodies produce microscopic spores of various shapes and colors, which function much as do the seeds of higher plants. The spores may be ejected from, drop out of, or ooze from the reproductive bodies and are carried by some means to susceptible plant parts. In a moist environment, these spores germinate by pushing forth a tiny strand known as a germ tube. Once inside the host plant, the tube branches into numerous strands, which constitute a new fungus body. Eventually the fungus produces a new crop of spores directly, or it produces fruiting bodies capable of producing spores.

Viruses. Plant viruses are infectious particles composed of protein and nucleic acid. They are introduced into plant cells by specific vectors, most commonly insects and nematodes in the case of trees, and harness the manufacturing capacity of the cell to make more virus particles. Typical virus symptoms on infected leaves include mosaic or mottling patterns, ring spots, flecks, necrotic lesions, chlorosis, line patterns, curling, and dwarfing. Stunting, decline, and excessive branching may also be symptoms. Virus diseases of landscape trees have been studied very little, although viruses of fruit trees are known and presumably occur in their landscape relatives. Mosaic viruses of ash, elm, birch, poplar, and oak have been described. Ash, birch, and poplars are subject to infections by other viruses as well.

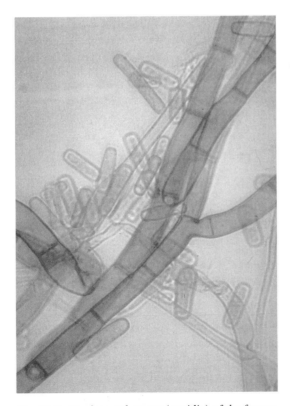

Fig. 12-2. Hyphae and spores (conidia) of the fungus *Thielaviopsis*, cause of black root rot.

Phytoplasmas. These organisms, essentially bacteria without cell walls, are important tree pathogens. Witches' brooming and yellowing are two types of symptoms commonly caused by phytoplasmas. Lethal yellowing of palms and elm yellows (phloem necrosis) are caused by these organisms.

Nematodes. Plant parasitic nematodes are tiny unsegmented roundworms, which have well developed reproductive and digestive systems. Their life stages consist of eggs, larvae, and adults, the latter two stages spent feeding mainly on host plant roots. These nematodes have a needlelike stylet or feeding tube, which allows them to feed on individual plant cells to obtain nutrients for growth. Some feed by grazing here and there on root cells, whereas others settle in one spot, continuing to feed in that location, often causing root galls. Root feeding nematode problems are generally more serious in warm temperate and tropical regions.

Root-knot nematodes feed on roots of woody plants, causing root swellings or galls, which disrupt the uptake of water, producing symptoms of wilt, nutrient deficiency, and poor growth of the top. Boxwood, catalpa, dogwood, maple, peach, pine, and willow are susceptible to root-knot nematode. Root lesion nematode kills portions of the root on which it feeds. The top of the plant becomes stunted or declines in proportion to the amount of parasitism. Landscape trees are not normally heavily damaged by lesion nematodes; however, the nematodes will feed on ginkgo, maple, pine, spruce, sweetgum, and tuliptree roots and seriously damage shrubs such as American boxwood and blue rug junipers. Stubby root nematodes cause reduced top growth because of root tip feeding. They have been associated with pines having littleleaf symptoms. Dagger nematodes feed on ash, elm, maple, and spruce roots and also vector certain viral diseases such as ash ringspot virus. Pine wilt nematode (Fig. 12-3) feeds on cells lining the resin canals of pine branches and stems, causing wilting and death.

Parasitic Higher Plants. Mistletoes and dwarf mistletoes are common parasites of some trees in some areas. These plants grow on twigs and branches of their host, extracting water, mineral elements, and food from the tree by way of a parasite nutrient-uptake organ. The tree branch is often swollen at the point of mistletoe attachment.

The true or leafy mistletoes (*Phoradendron* and *Viscum*) are most frequently associated with hardwoods growing in the southern two-thirds of the United

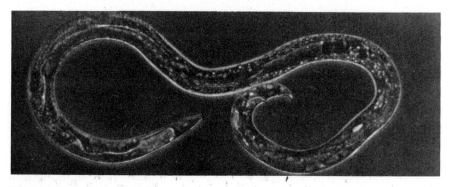

Fig. 12-3. Pine wilt nematode.

Fig. 12-4. Top: Clumps of mistletoe on a black walnut limb. Bottom: Close-up of a mistletoe plant, showing berries, leaves, and stem.

States. These small leafy plants (Fig. 12-4)—the same mistletoe used during holidays—are capable of making much of their own food and depend on the host tree for water and minerals. Mistletoe seeds, spread by birds that feed on the berries, germinate on the tree branch and penetrate the host. The top foliage may be harvested or may die back in extremely cold weather, but the plant grows back in a few years. True mistletoes may be harmful to the tree during times of stress, and although they may not be too damaging in the wild, they would harm trees growing in the already stressful urban environment. They can be controlled by pruning, if necessary.

Dwarf mistletoes (*Arceuthobium*) are more damaging to their tree hosts because they lack true leaves for food manufacture. They are parasites of pine, fir, Douglas-fir, and other conifers in the western United States and of black spruce in the East. The invaded twigs, branches, or trunk tissues may be swollen or produce witches' brooms. The overall effect is a reduction in tree vigor, poor growth, and even death of older trees. The parasitic plant produces sticky seeds that are shot out at some distance, allowing spread of the plant to nearby trees. Dwarf mistletoes should be pruned and destroyed, if possible, and badly infected trees should be taken down.

Pathogenesis, Dissemination and Survival of Parasitic Agents

Bacterial plant pathogens and many fungal pathogens gain entrance into the plant tissue mainly through natural openings or wounds. Natural openings include stomata and hydathodes in leaf tissues or lenticels in the stem, while wounds may be caused by insects or by mechanical means. Many fungi such as those causing scab and powdery mildew are also able to penetrate directly the unbroken epidermis of the leaves.

Once inside the host, growth or multiplication within and at the expense of the host tissue ensues. Parasitic microbes may produce toxins affecting the host, absorb cell contents and cause plant tissue death, or stimulate other host reactions such as gall formation. Plant tissue functions are disrupted. Some soil-inhabiting microbes live as saprophytes in the soil until they come in contact with susceptible tree roots. They enter such roots, even unwounded ones, and become parasites. Wilt fungi grow directly into the conducting tissues, where they cause a partial clogging or secrete toxins that are carried in the water stream to the leaves. There the toxins accumulate and cause a collapse of the leaf tissues.

Parasitic fungi are disseminated primarily in the spore stage, while bacteria are moved as individual cells. They may be splashed by rain or carried by air currents, insects, or animals, on the hands of the tree worker, or on tools, from tree to tree or from place to place.

Wind-blown rain is probably the most important agent in the local dissemination of bacteria and of fungus spores, particularly of those fungi that cause leaf spots, leaf blights, and twig cankers. Spores produced on the surface of infected leaves or in cankers are splashed or washed to nearby healthy leaves and twigs, on which new infections occur.

Air currents are also important agents of dissemination. For example, the spores of some rust fungi are blown by the wind many miles from the tissues on which they are formed.

Insects are well-known agents of dissemination. Bees and flies are known to spread the bacteria that cause fire blight disease. Several species of bark beetle are chiefly responsible for the spread of spores of the fungus causing the Dutch elm disease. Insects are the chief means of spread of viruses and phytoplasmas.

The chestnut blight fungus is known to be spread by birds and squirrels lighting on or traveling over extensive bark cankers on whose surface spores are produced in large numbers. These spores adhere to the talons or feet and may be deposited again on some distant tree.

Spores may be spread on tree workers' hands and clothing, as is the case with the Cytospora canker disease of spruce, especially when diseased branches are pruned during wet weather. The fungus responsible for planetree cankerstain may be disseminated on infested pruning saws, and even in many types of wound dressings. Spores of *Ophiostoma ulmi*, the Dutch elm disease fungus, may be spread on pruning tools.

Soil-inhabiting fungi, bacteria, and nematodes, capable of producing diseases such as wilts and root and collar rots, are carried largely by the transfer of infested soil on tools, machinery, or shoes, or by washing during heavy rains. Winds that lift soil particles and carry them from place to place may also be a means of dissemination of this group of plant parasites.

Overwintering of Bacteria and Fungi

Leaf-spotting and leaf-blighting bacteria and fungi overwinter in fallen infected leaves, on dormant buds, and in small twig cankers. Canker-producing bacteria and fungi overwinter in the margins of the cankers. Wilt fungi pass the winter in the vascular tissues of infected trees and in the soil. Many soilborne fungi overwinter in the form of sclerotia or chlamydospores, fungal bodies highly resistant to destruction.

Factors Required for Disease Development

Three conditions must be met for a tree disease to occur. There must be (1) a susceptible host, that is, a tree in which it is possible for a specific disease to occur; (2) a virulent pathogen, one that is capable of causing the disease in this host; and (3) an environment suitable for disease to develop (Fig. 12-5). If any one of these three factors is missing, no disease will occur.

Although most trees are resistant to most diseases, it is well established that virtually all trees are susceptible to at least one disease. Furthermore, virulent tree

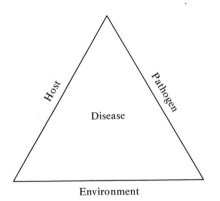

Fig. 12-5. The disease triangle.

Fig. 12-6. Fire blight canker and blight on pear. The bacteria first entered via the flowers and have killed the fruiting spur.

pathogens occur in most areas. However, merely bringing together virulent pathogens and susceptible trees is not enough. Many diseases can occur only under specific environmental conditions, such as the presence of a tree wound, a specific stage of tree growth, a long period of leaf wetness, saturation of the soil with moisture, or stressful growing conditions. Many fungi that normally are weak parasites, which do little damage to trees growing under favorable conditions, can readily destroy trees growing under adverse conditions.

There are some diseases that develop more readily on lushly growing trees. The dense foliage of such trees may provide the most favorable microclimate for disease development; succulent tissues are often more susceptible to infection.

Fire Blight

Fire blight is known primarily as a disease of apple and pear trees, and consequently it is of most concern to orchardists. It also occurs on many ornamental trees and shrubs in the rose family. Commercial arborists, estate superintendents, and even the owner of a single susceptible tree must often cope with this very destructive disease. Besides attacking apple and pear trees, including flowering crabapple and callery pear in the landscape, fire blight appears commonly on several species of cotoneaster, cockspur hawthorn, English hawthorn, and mountain-ash. It occurs less frequently on firethorn, serviceberry, flowering quince, cultivated quince, Christmas berry, flowering plum, spirea, rose, and Stransvaesia.

Symptoms. The leaves near the growing tips and the flowers suddenly wilt, turn brown and black, and look as though they were scorched by fire (Fig. 12-6).

Twigs also are blighted on most of the ornamental hosts. Affected shoots with dead leaves hanging from them often curve near the tip, forming a shepherd's crook. In some, the mountain-ashes, for example, the infection spreads down to involve large branches. Extensive cankers on the trunk and main branches develop on the larger ornamental trees, as well as on apple and pear. The flowering quince is susceptible principally to blossom blight, the disease being rare on woody parts. The presence of bloom infection and the absence of typical twig blight in such cases give the infected tree the appearance of having been injured by frost.

The disease is widely distributed, but its yearly occurrence is sporadic. Occasionally very heavy damage has been observed on mature 'Aristocrat' flowering pears in the Ohio River valley region preceded or followed by several years when few or no disease symptoms were evident.

Arborists and nurserymen wishing to forecast whether or not fire blight symptoms will occur in a given year and needing to obtain better control can use a computer program called MARYBLYT to manage the disease. MARYBLYT enables the user to predict and determine if, and exactly when, primary infections are occurring and when first symptoms are to appear. The computer program, available from pest management suppliers, is easy to use and only requires daily observations of maximum and minimum temperatures, rainfall, and observation of ornamental pear and crabapple springtime growth stages from bud break to dropping fall of their flower petals.

Cause. Fire blight is caused by the bacterium *Erwinia amylovora*. Fire blight disease is thought to occur as follows: The bacteria overwinter in cankers from the previous year's infections. Natural, epiphytic populations of *E. amylovora* develop on apparently healthy shoots, leaves, and buds of the host plant tissues without producing blight symptoms. During the spring, canker-residing bacterial populations increase, and droplets of bacterial ooze are produced in the margins of cankers. Bacteria, spread by insects attracted to the bacterial ooze and by rain and wind, increase in numbers on the plant surface without causing disease. Primary infections begin in the flowers when these epiphytic bacteria build to large numbers on the flower parts and are washed into the nectaries at the base of the flowers. From infected and dying flowers and fruit spurs, more bacteria are produced and disseminated to invade leaves and shoots through stomata, hydathodes, or wounds. They then penetrate and spread within the host tissues, causing death of twigs and branches. Spread to other parts of the tree and to nearby trees by insects, wind, and rain continues until new shoot growth ceases, at which time cankers become dormant.

Fire blight disease is favored by long frost-free periods before bloom, followed by humid weather with 65 to 70° temperatures during bloom, with occasional rains during that period.

Control. Blighted twigs should be pruned in winter just below the infected areas and destroyed. Large limb cankers can also be surgically removed by cutting beyond the margin of the canker. This practice eliminates an important potential source of inoculum for subsequent epidemics, although it is difficult to thoroughly prune large landscape trees. The pruning must be done carefully, so that all infected branches are removed. It is not necessary to sterilize the pruning tools for dormant pruning. Trees that are badly infected should be removed, as should old, neglected pear trees, which can be sources of inoculum.

While trees are dormant, apply copper sulfate (4-5 lb/100 gal) to the twigs and branches to help reduce overwintering bacterial inoculum. This chemical may color nearby objects blue, so care must be taken if this is done in urban landscapes. Thoroughly wash the spray tank following use, since the chemical is corrosive.

In a nursery or similar planting, spray the antibiotic streptomycin (50–100 ppm solution, or 4 oz of 21% streptomycin sulfate/100 gal) at 4 to 5-day intervals during bloom. Sprays can be timed for best effectiveness using the fire blight computer program described above. Streptomycin is best applied late in the day, when the air is still and when intense light will not rapidly break the chemical down. Repeated applications help the streptomycin penetrate the flower tissues. During a severe outbreak of fire blight, just one or two spray applications will not be effective.

We prefer not to see streptomycin used in urban landscapes. Resistance of *E. amylovora* to streptomycin has been reported, and the possibility of resistance transfer to human pathogens, though extremely remote, is real. Fixed copper such as Basic Copper Sulfate (not copper sulfate) can be used instead of streptomycin, although it is not nearly as effective.

Sucking insects such as aphids, leafhoppers, plant bugs, and pear psylla vector the bacteria and create wounds that bacteria can enter. Control these insects in the nursery before bloom and throughout the summer, especially if the planting has had a history of an insect problem. In the landscape, insecticide use for fire blight control should not be needed but may provide a benefit if it is being applied for insect pest control in any case.

Excess nitrogen fertilization, which encourages rapid tree growth and increased susceptibility to fire blight, should be avoided. Removal of suckers and water sprouts may reduce disease.

Susceptible trees should be inspected frequently during the growing season and infected spurs and terminals removed by breaking them out. Again, arborists and nurserymen could be assisted by the MARYBLYT computer program by knowledge of which day symptoms would be occurring. Breaking out infected flower spurs and shoots when symptoms first appear is more effective than wait-ing a few days or weeks to prune. If pruning tools are used, sterilize between cuts using a solution of 1 part Lysol concentrated disinfectant or sodium hypochlorite bleach to 5 parts water. Prune or break back at least 12 inches from active fire blight. This job will be very difficult to do for large landscape trees. If a thorough job of blight removal cannot be accomplished, it is better to leave this task until winter.

Disease-resistant flowering crabapple cultivars are available and should be planted in areas where fire blight is likely to be a problem.

Crown Gall

Crown gall is a common disease of trees in the rose family, such as apple, cherry, pear, plum, and flowering almond. The disease also occurs on such widely differ-ing species as chestnut, Arizona cypress, European juniper, incense cedar,

Fig. 12-7. Crown gall on walnut stem.

sycamore maple, oleander, poplar, English walnut, hickory, willow, English yew, and, in Hawaii, on the macadamia tree.

Symptoms. Roughened swellings or tumorlike galls of varying sizes appear at the base of the tree or on the roots (Fig. 12-7). On poplar, hickory, and willow, these galls may also appear on limbs and branches. Growth of the tree may be retarded, and the leaves may turn yellow, or the branches and roots may die as a result of water and nutrient transport interference caused by the presence of these galls. Although established trees are rarely damaged appreciably, young trees may be killed. Other agents such as *Phomopsis* fungi can cause symptoms resembling crown gall. We have observed mature hickory trees showing extensive phomopsis galling on twigs and branches.

Cause. Crown gall is associated with the soilborne bacterium *Agrobacterium tumefaciens*. The bacterium does not kill the parts attacked but carries a separate genetic factor known as a plasmid, sometimes called tumor-inducing principle, which stimulates abnormal cell growth. It enters plants through wounds and is primarily a parasite of young nursery stock.

Control. Sanitation is important for crown gall control. Nurserymen should discard all young fruit trees and ornamental trees showing galls on the stem or main roots. They should follow practices that minimize wounding the stems and roots of young trees in the nursery. They should also be aware that infected plants, infested soil, insects, contaminated tools, and irrigation water are capable of carrying the crown gall bacteria.

A biological control agent, Galltrol A or Norbac 84, consisting of a certain

Fig. 12-8. Slime flux oozing from a pruned branch stub of an American elm.

strain of *Agrobacterium radiobacter*, shows promise for reducing crown gall in nursery plantings. This bacterium competes with the crown gall bacterium for attachment to wound sites and is effective when used as a root dip or spray for cuttings and seedlings.

Gallex is a product used as a crown gall treatment after infection has already occurred. The active ingredients are 2,4-xylenol and *meta*-cresol. The material is applied directly to the galls and causes them to shrink. Gallex may need to be reapplied and is limited to those galls that can be exposed and reached with treatment. Follow manufacturer's directions for use and necessary precautions.

Tree species and cultivars not susceptible to crown gall include bald cypress, beech, birch boxwood, catalpa, cedars (*Cedrus*), ginkgo, goldenrain-tree, hemlock, holly, hornbeam, larch, linden, magnolia, pine, spruce, tuliptree, tupelo, and zelkova.

Bacterial Wetwood and Slime Flux

The foul-smelling and unsightly seepage from wounds in the bark or wood of various shade trees is known as slime flux. It occurs most commonly on bacterial wetwood-infected trees, such as elm, poplar, oak, birch, and maple. Although slime flux development is seasonal, evidence of wetwood and slime flux-stained bark is visible anytime (Fig. 12-8). This slime is different from the liquid associated with alcoholic flux. Alcoholic flux emanates from shallow wounds and persists only for a short time.

Symptoms and Cause. Wetwood seepage originates from infections of the heartwood and inner sapwood by common soil-inhabiting bacteria such as *Enterobacter cloacae* (*Erwinia nimmipressuralis*). There are several other bacterial species also associated with wetwood. Wetwood bacteria are capable of growing anaerobically (without oxygen) in the internal wood tissues. Methane and osmotic or metabolic liquids, two by-products of the bacterial activity, accumulate under pressure and are forced out of the tree through the nearest available opening, usually a trunk wound or branch stub. Pruning a branch or taking a core with an increment borer can sometimes release the materials under pressure, squirting the worker with foul-smelling liquid and gas.

Normally flowing to the wounded bark surface, the wetwood fluid is a clear watery liquid containing several nutrients. On the surface it soon changes to a brown, slimy ooze, as a result of feeding by fungi, yeasts, bacteria, and insects. This surface slime flux may kill injured cambium and bark surface organisms as well as grass growing near the base of the tree,

Wetwood-infected trees have an internal core of wood that is wet but not decayed. These infected branch, trunk, and root tissues also have a high pH. Wetwood-infected wood is resistant to decay by fungi. The extent of wetwood spread in the tree may be limited by tree defenses; however, wetwood can spread into new tissues as new injuries occur. Thus deep injection holes and pruning can expand wetwood infection. Take care to avoid pruning live branches on infected trees.

Control. Thus far, no effective preventive or curative measure is known. If the bark is being stained it may be helpful to drain the slime flux away from the branch or trunk so that it drips on the ground. Drilling a hole into the tree and inserting a copper or semirigid plastic tube has helped in some cases; however, this results in additional wounding and the threat of expanded wetwood or decay should be considered. Loose dead bark should be carefully cut away so that the area can dry.

Bacterial Leaf Scorch

Landscape trees have long been afflicted with leaf scorch symptoms caused by environmental factors such as root damage, road salt, and drought, and by wilt diseases caused by fungi. A major cause, bacterial leaf scorch, has been reported in coastal U.S. states from New York to Texas, and in Kentucky and Tennessee. Hosts include bur, pin, red, and shingle oak; elm; mulberry; sycamore; sugar and red maples; and sweetgum.

Symptoms. In oak, scorch symptoms first appear in late summer in individual branches where leaves show dead margins with green tissues near the main veins and leaf petiole. Often there is a fine yellow or reddish zone between brown and green tissues (Fig. 12-9). Many affected leaves drop prematurely. In succeeding years, the annual late summer leaf scorch progresses to all parts of the tree. Gradually, infected trees suffer a chronic decline with branch dieback affecting more of the tree each year. Secondary factors can contribute to tree demise, and eventually the tree needs to be removed. Tree decline, from first discovery of the disease to removal, may take place over a period of five to ten or more years. The bacterium apparently causes leaf scorch and defoliation of landscape trees

Fig. 12-9. Shingle oak with symptoms of bacterial leaf scorch. Leaf tips and margins are brown while leaf tissues nearest the midrib and base are still green. There is a yellow band between brown and green tissues.

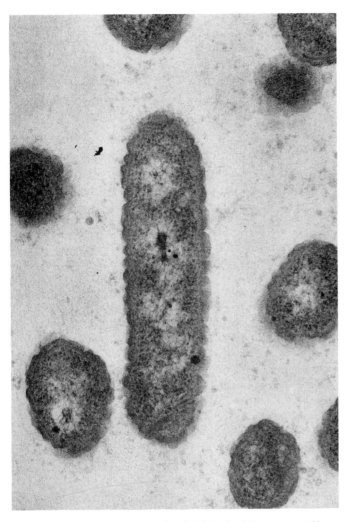

Fig. 12-10. Electron micrograph of *Xylella fastidiosa*, cause of bacterial leaf scorch.

because of water stress due to xylem occlusion. Bacterial leaf scorch symptoms may be more severe when trees are suffering stresses related to adverse growing conditions.

Cause. Bacterial leaf scorch is caused by the xylem-limited bacterium *Xylella fastidiosa*. It can be detected in infected xylem tissues with an enzyme-linked immunosorbent assay (ELISA) laboratory test. The rod-shaped bacteria associated with the xylem of diseased trees have been observed microscopically (Fig. 12-10). The causal bacteria are vectored by leafhoppers, and the disease seems to spread slowly within a population of susceptible trees.

Control. There are no cures for this disease. Provide good growing conditions

Fig. 12-11. Witches' broom symptoms caused by the ash yellows phytoplasma. Note that some twigs have died.

for affected trees. Since the disease develops relatively slowly, tree owners may plant substitutes nearby so that they will be well established by the time the infected trees must be removed a few years later. Micro-injections with the antibiotic oxytetracycline (Mycoject) have been shown to suppress symptoms. The effect is temporary, and additional injections are needed to maintain tree health. Where repeated injections are needed, there is a concern that long-term tree health might be jeopardized from the injections. More research is needed to find the balance between suppressing bacterial leaf scorch disease and enduring numerous small injuries from injection applications.

Yellows Diseases

Ash yellows and elm yellows are two important diseases found in landscapes. Although these diseases only occur occasionally, they are usually lethal when they occur. These two diseases are caused by phytoplasmas, bacteria-like microbes lacking cell walls and which live in the phloem tissues of trees. Phytoplasmas infect trees systemically and are be vectored by certain leafhoppers and possibly meadow spittlebugs.

Ash yellows. This disease is found primarily in white ash growing in the northern parts of the United States from the Great Plains to the East Coast.

Symptoms. Infected ash trees show a variety of symptoms, including premature spring growth, premature fall color, stunted and folded leaves, reduced twig and branch elongation, reduced trunk diameter growth, epicormic shoot growth at the base of the trunk, loss of terminal shoot apical dominance in branches, witches' brooms (Fig. 12-11), sometimes chlorotic leaves, branch dieback, decline and death. Trees may survive several years with the disease, but highly susceptible trees often die in just a few years.

Cause. Ash yellows has been diagnosed primarily by field symptoms. In the laboratory, the disease can be confirmed by microscopic examination of stained phloem tissues, or observation of the phytoplasma in infected tissues using an electron microscope.

Control. There have been no control measures developed to prevent or cure this disease. Although green ash is reported to be more tolerant than white ash, little is known about the susceptibility of others such as blue ash.

Elm yellows. Formerly called elm phloem necrosis, elm yellows has sometimes been mistaken for Dutch elm disease because, like that fungal disease, yellows disease also kills infected trees within a few months or a year.

Symptoms. Infected elms display leaf yellowing, leaf petiole epinasty (bending downward), premature defoliation, branch death, and tree death. These symptoms are probably caused by root deterioration associated with the disease. Roots of infected trees die, and discoloration and death of phloem tissues at the base of the tree is evident. Tan to brown discoloration of the inner bark, the phloem tissue, can be observed, and samples of freshly exposed diseased inner bark may yield the odor of wintergreen. Slippery elms may show witches' broom symptoms before other yellows symptoms develop.

Cause. The elm yellows phytoplasma is systemic in the phloem tissues.

Control. Elm yellows, like Dutch elm disease may be spread from infected to nearby healthy trees via root grafts, so root graft breakage is a useful tool for preventing disease spread in a cluster of elm trees. Otherwise, there are no treatments for prevention or cure of elm yellows. American, slippery, and winged elms are susceptible, and Asiatic elms are resistant.

Leaf Spot Diseases

The most obvious and visible diseases of trees are those affecting the leaves. The effect of each leaf spot disease on tree health varies with the disease and tree affected. In general, those leaf spots that result in premature defoliation will cause the most harm to the tree, and those that appear late in the season or that cause very little death of leaf tissue will only slightly harm the tree. Thus, it is important to diagnose leaf spots correctly and to know how a particular tree is going to respond to the disease.

Symptoms. Distinctions can be made as to the different kinds of spots and their general symptoms. Leaf spots are usually thought of as well defined lesions or dead areas on leaves or needles. They may be circular or angular on broadleaves, or bandlike on needles. Scab disease spots are circular, superficial and sometimes somewhat roughened lesions while leaf blisters are swollen or raised blisterlike spots on the leaf surface. If an individual leaf spot tends to be spreading into surrounding tissues, it may be referred to as a blotch. Anthracnose diseases often appear first on leaves as a spot that becomes an irregular blotch and finally as extensive dead areas that involve the whole leaf and shoot. Shot hole symptoms develop in leaves when the dead tissue of the leaf spot drops out. Because there are so many different fungi and bacteria that cause leaf spots, and

Fig. 12-12. Apple scab disease of flowering crab-apple. In the early stages, scab spots (3 in this photo) are relatively indistinct. Infected leaves turn yellow and drop prematurely.

there are so many different ways that tree species and cultivars react to the diseases, the symptoms will vary considerably. Symptoms of individual diseases are described in the last section of this book.

Cause. Many fungi, most of them ascomycetes or imperfect fungi, are parasites of tree leaves and cause spots. Fungi such as *Ascochyta*, *Cercospora*, *Cylindrosporium*, *Elsinoe*, *Marssonina*, *Microsphaera*, *Mycosphaerella*, *Phyllosticta*, *Rhytisma*, *Septoria*, *Taphrina*, and *Venturia* can each cause leaf spot diseases of several different trees. Leaf spot symptoms may also be caused by anthracnose fungi, powdery mildews, and rust fungi, discussed elsewhere in this chapter. Bacteria such as *Pseudomonas* and *Xanthomonas* also cause leaf spots. Determining the exact pathogen causing the the leaf spot by microscopy or special laboratory tests is sometimes necessary to develop effective control strategies.

Control—biological and cultural.

1. Use sanitation to reduce pathogen levels. For the vast majority of leaf spot diseases, raking up and destroying or thoroughly composting diseased leaves is sufficient for control. Some of the leaf spot fungi, for example, Septoria leaf spot of poplar, also cause cankers on the twigs, and pruning out cankered wood is needed to reduce inoculum.

2. Manage the growing environment. Avoid sprinkler irrigation that wets the foliage. Thin out crowded branches and prune away overhanging vegetation from nearby trees to improve ventilation and sunlight penetration.
3. Provide good growing conditions for trees in the landscape. Some leaf spot diseases such as Actinopelte leaf spot of oak attack trees under stress.
4. When available, use disease resistant cultivars. There are many excellent cultivars of flowering crabapple that are resistant to scab (Fig. 12-12), for example.

Control—chemical. If the tree is a valuable specimen and despite the use of good cultural practices the leaf spot disease continues to be detrimental to the health of the tree, fungicide sprays should be considered. Use protectant fungicides such as Captan, Chipco 26019, Daconil 2787, fixed copper, Fore, Manzate, Syllit, and Ziram to control a wide range of leaf spot diseases. Use systemic fungicides such as Banner Maxx, Bayleton, Benomyl, Cleary's 3336, Eagle, Immunox, or Rubigan to target certain specific diseases. Some of the active ingredients can be purchased as fungicide mixtures (see Chapter 13) having the advantage of both protectant and systemic activity. Bacterial leaf spot diseases are managed with fixed copper chemicals such as Kocide, Champion, or Bordeaux mixture. Choose fungicides and bactericides that have the tree and the disease listed on the label. The timing of sprays for leaf spots is critical, and applications often will need repeating. Be sure that the spray equipment that is being used thoroughly covers all surfaces of the foliage.

Powdery Mildews

The leaves and twigs of many deciduous trees and shrubs are occasionally covered with a grayish white dusty material known as powdery mildew. The mildew is made up of delicate, cobweblike strands of fungus tissue covered with microscopic colorless asexual spores. Many minute, spherical, black bodies—fruiting structures of the sexual stage of the fungus—may be visible to the unaided eye in the grayish white areas. The spores of the fungus may overwinter in these bodies. Some powdery mildews overwinter vegetatively in dormant buds of the infected tree.

Symptoms. Although mildew fungi grow mainly over the surface (Fig. 12-13), they also penetrate the leaf surface with numerous fine filaments, which extract nutrients from the host plant. Heavily mildewed leaves may turn yellow, dry up, and fall prematurely. Mildews appear most commonly toward the end of summer and in late fall and are most prevalent on trees or branches growing in shaded and damp locations.

Cause. There are a number of distinct species of powdery mildew fungi, which can be distinguished only by microscopic examination. The following list includes most of the species, together with the trees they parasitize:

Brasiliomyces trina on oak.
Erysiphe aggregata on alder.
E. cichoracearum on aspen, eucalyptus, and smoke-tree.
E. lagerstromiae on crapemyrtle.
E. liriodendri on tuliptree.

Fig 12-13. Powdery mildew of dogwood. Note white fungal signs on leaf surfaces.

E. polygoni on acacia, black locust, and serviceberry.

Microsphaera spp. on alder, ash, beech, birch, buckthorn, catalpa, chestnut, dogwood, elm, hackberry, hickory, holly, honey locust, hornbeam, hop hornbeam, linden, magnolia, maple, oak, pecan, planetree, sycamore, and walnut.

M. diffusa on black locust.

Phyllactinia guttata on alder, ash, beech, birch, black locust, boxelder, boxwood, catalpa, chestnut, chinaberry, crabapple, dogwood, elm, hawthorn, hickory, holly, hornbeam, horsechestnut, linden, magnolia, maple, mountain-ash, mulberry, oak, pear, planetree, sassafras, serviceberry, sycamore, tuliptree, walnut, willow, and yellowwood.

Pleochaeta polychaeta on sugarberry.

Podosphaera leucotricha on apple.

P. clandestina on almond, apricot, cherry, hawthorn, mountain-ash, peach, pear, persimmon, plum, and serviceberry.

Sphaerotheca lanestris on oaks.

S. pannosa on flowering *Prunus*.

S. phytophila on hackberry.

Uncinula spp. on buckeye, elm, hackberry, linden, maple, poplar, and willow.

Control. Beneficial management practices such as pruning out infected twigs, raking up infected leaves, improving air movement around the trees, and increasing the amount of sunlight reaching them can often reduce severity of the disease. Dusting or wettable sulfur fungicides control powdery mildew. Other fungicides such as Banner Maxx, Bayleton, Benomyl, Cleary's 3336, Eagle, Immunox, Rubi-

gan, and Strike are effective for powdery mildew. As a rule, however, only particularly valuable large specimens and small nursery trees are sprayed to control
mildew. In most cases, the disease appears so late in the growing season that it
does little real damage. Resistant cultivars of flowering dogwoods and crabapples
are available and offer an excellent means of control.

Rust Diseases

Most of the rust diseases of landscape trees involve one of the cedar rust diseases,
or one of the pine rust diseases. The rust diseases have complex life cycles, often
involving more than one host plant.

Cedar Rust Diseases. Three cedar rust diseases occur on landscape trees in
the eastern United States: cedar-apple rust, cedar-hawthorn rust, and cedar-
quince rust. All three are caused by different species of the fungus *Gymnosporangium*, each of which must spend a phase of its life cycle as a parasite on
Juniperus species, such as native red cedars or ornamental junipers, and another
phase on roseaceous hosts such as apple and hawthorn.

Fungus Life Cycle. In general, cedar rusts go through distinct growth stages
on their different hosts. Beginning in springtime on the diseased cedar, usually in
a gall, the fungus produces a highly visible bright orange telial stage, with
teliospores. From the teliospores, microscopic basidia and basidiospores develop
and are conveyed by air currents to the nearby apple, hawthorn, or related host
where, during moist periods, they infect leaves, fruit, or twigs. Some weeks later,
the fungus produces fruiting structures called pycnia with specialized spores that
are important for sexual reproduction. After cross fertilization, the fungus continues to develop in the same leaf tissue, forming an aecial stage with aeciospores. In
summer, aeciospores carried by air currents then infect nearby cedars or junipers,
the alternate host of the fungus. After about 18 months in the cedar, the rust fungus life cycle finally returns, in the second springtime, to the orange telial stage.

Symptoms. Cedar apple rust (*Gymnosporangium juniperi-virginianae)* causes orange-colored leaf spots and defoliation of crabapple, apple, and occasionally hawthorn. Yellow leaf spots appear in spring on the upper leaf surface, and by
late summer the spots enlarge to the leaf underside, becoming more orange with
whitish tubular spore-bearing structures (aecia). Cedar apple rust only occasionally affects twigs or fruit. On cedar and juniper, rust infections result in brown,
spherical galls an inch or so in diameter, and sometimes twig dieback. Gall surfaces are marked by regular, circular depressions. In springtime, galls become
bright orange with spore horns of the fungus.

Cedar-hawthorn rust (*G. globosum*) appears on hawthorn, crabapple, apple,
pear, quince, serviceberry, and mountain-ash. Leaf spots are similar in appearance to cedar-apple rust, but few of the tubular aecia form within them. Cedar-
hawthorn rust occasionally affects twigs and fruits of their hosts. On cedar and
juniper, galls produced by cedar-hawthorn rust are similar in appearance to those
of cedar-apple rust, but are smaller, more irregular in shape and do not develop
the regular arrangement of circular depressions. Spore horns, too, are shorter,
generally fewer in number and are wedge or club-shaped.

Cedar-quince rust (*G. clavipes*) is the most damaging of the cedar rusts, and it

Fig. 12-14. Cedar-quince rust symptoms on hawthorn. Twig cankering and swelling are associated with the aecial stage of the fungus, visible as white projections.

Fig. 12-15. Cedar-quince rust of juniper. Infected twig is slightly swollen at the base; bark is cracked and peeling.

affects many hosts, including hawthorn, crabapple, apple, serviceberry, mountain-ash, quince, flowering quince, pear, photinia, medlar, chokeberry, and cotoneaster. Cedar-quince rust normally does not cause leaf spots, but twig and fruit infections are common and sometimes damaging. Young, green hawthorn and crabapple twigs and thorns infected by the cedar-quince rust fungus can become swollen, cankered, and die (Fig. 12-14). Most cedar-quince-rust-infected fruits are enlarged with protruding white, tubular aecia emerging from them. Infected apple fruits develop a corky texture. Cedar-quince rust also affects cedars and junipers, but it does not form rounded galls. Instead, this rust forms perennial, spindle-shaped swellings on the twigs (Fig. 12-15), which bear spore-producing gelatinous orange "cushions" in the spring.

There are many other cedar rusts with similar life histories, including juniper-broom rust, *G. nidus-avis;* medlar rust, *G. confuscum;* and others. Some are characterized by witches' brooms, which form on infected junipers.

Control.

1. Grow resistant apples, crabapples, hawthorn, mountain-ash, or junipers. Lists of some of these rust-resistant plants are available through Cooperative Extension Service offices.
2. Destroy nearby wild, abandoned, or worthless apples, crabapples, cedars, or junipers. When practical, prune out and destroy rust galls found on ornamental junipers and cedars. Although landscape plants may occasionally become infected by spores produced up to several miles away, most infections result from spores produced on infected *Juniperus* within a few hundred feet of the landscape.
3. Fungicide spray programs may be used to control rust of apples, crabapples, or hawthorns. Apply Banner Maxx, Bayleton, Fore, Eagle, Immunox, Manzate, or Rubigan fungicides when the trees show immature green foliage just before they begin to flower, again during bloom, and finally, just after bloom. Spray timing can also be based on carefully monitoring the cedars and spraying when the orange gelatinous telial stage of the fungus is active. Choose fungicides that have the tree and the disease listed on the label and be sure to obtain thorough coverage of the twigs, blooms, and foliage.

Pine Rust Diseases. Some rust diseases can be very destructive to pines. Pine rust diseases are found more commonly in forests or Christmas tree plantings than in landscape environments. Like most rusts, pine rusts have alternate host plants, dicots such as oaks, currants, and asters. Often, these alternate hosts are little damaged by the rust disease they host.

Fungus Life Cycle. In general, pine rusts go through distinct growth stages on their different hosts. Beginning on the diseased dicot host plant, the fungus produces a telial stage with teliospores. From the teliospores, basidia and basidiospores develop and are conveyed by air currents to the nearby pine. Some weeks, months, or years after infecting the pine, the fungus produces spermagonia and specialized spores called spermatia, which are important for sexual reproduction. Still in the pine, the fungus then develops an aecial stage with aeciospores. This stage is sometimes characterized by swelling of affected pine tissues. Aeciospores carried by air currents then infect the nearby alternate host plant. On the dicot, uredia and urediospores that can reinfect the dicot are produced before the rust fungus life cycle finally returns to the telial stage.

Symptoms. Pine needle rust (*Coleosporium asterum*) affects Austrian, Japanese black, jack, loblolly, longleaf, mugo, pitch, red, Scots, shortleaf, and Virginia pines. Needle rust is not as destructive as some of the other rusts. It takes one year for this fungus to complete its life cycle. Alternate hosts include aster, goldenrod, and other composites.

Eastern gall rust (*Cronartium quercuum*) affects Austrian, jack, loblolly, mugo, pitch, red, Scots, shortleaf, and Virginia pines. When a stem or branch is infected with the fungus, it typically produces a spherical swelling and kills the stem or branch and all distal branches within a few years. The life cycle of this fungus takes two years or longer. Alternate hosts include many oak species.

Western gall rust (*Endocronartium harknessii*) affects Austrian, jack, mugo, and Scots pines, with symptoms similar to those of eastern gall rust. There are no alternate hosts; the fungus lives its entire life on pine, becoming reinfected by aeciospores produced on pine.

Fusiform rust (*Cronartium quercuum* f. sp. *fusiforme*) affects Austrian, loblolly, and longleaf pines. The disease is very destructive, causing swollen cankers on the affected stems and branches. Fusiform rust is an important disease in southern pine plantations. The two-year or longer life cycle of this rust also has many oak species as alternate hosts.

White pine blister rust (*Cronartium ribicola*) affects eastern white pine, producing girdling cankers that are very destructive. The alternate hosts—currant, gooseberry, and other *Ribes* spp.—often grow in the same regions as white pines. Where the disease is common, regulations prohibit the growing of the alternate hosts near white pine.

Control.

1. Avoid growing pines near alternate hosts, or remove the alternate hosts, if possible.
2. Remove and destroy pine galls, or, if necessary, whole trees with galls to break the disease cycle.
3. The fungicides Bayleton or Strike may be used for pine needle rust.

Anthracnose Diseases

Anthracnose diseases are common on the foliage and sometimes the twigs of dogwood, ash, sycamore, maple, and oak. The anthracnose fungi can also attack walnut, hickory, elm, birch, catalpa, linden, planetree, tuliptree, and horsechestnut.

Symptoms. Early in spring, anthracnose fungi may kill twigs and newly expanding leaves, causing symptoms that resemble frost injury. Small, sunken dead areas, cankers that can girdle the branch, form on infected twigs and branches (Fig. 12-16). Later infections of the leaves or leaflets cause dead blotches along the leaf veins and sometimes distortion. Infected leaves and leaflets may drop from the tree, causing extensive defoliation in severe cases. Trees normally develop another crop of leaves following this defoliation. The dogwood anthracnose fungus may infect leaves and then invade and kill twigs and branches.

Fig. 12-16. Sycamore anthracnose: A, twig canker; B, leaf infection.

Cause. Anthracnose of landscape trees is caused by several species of *Discula* (imperfect) or *Apiognomonia* (perfect) formerly referred to as *Gloeosporium* or *Gnomonia*. The fungus overwinters in fallen leaves and twigs as well as in branch cankers. In spring, the fungus produces spores, and in wet weather it infects leaves and twigs. In sycamores, cool, wet weather favors twig and branch attack, whereas warmer wet weather favors leaf infection. The disease normally does not continue building up during the summer.

Control. Normally, little is done to control anthracnose, except to rake up and destroy fallen leaves and twigs to reduce fungal inoculum. Even that activity may not be effective unless combined with pruning all infected twigs and branches.

Fungicides such as Banner, Benomyl, Cleary's 3336, Fore, Manzate, or Basic Copper Sulfate can be used to reduce infections. They need to be applied first at budbreak in spring, again when first leaves are partially expanded, and finally two weeks later. Spraying might be warranted for young or particularly valuable specimens or following several successive severe anthracnose years. Arbotect 20-S injections can be used to manage sycamore anthracnose, but injection-induced trunk injuries could cause long-term problems. Fertilizing and watering trees to promote replacement foliage following defoliation may also help.

Verticillium Wilt

Verticillium wilt, one of the most common fungal diseases, is found on more than 300 kinds of plants, including shade and ornamental trees, food and fiber crops, and annual and perennial ornamentals. Among its more important hosts are the many species of maples planted along city streets. In fact, this disease has caused the death of more maples over the last half-century than any other disease. A survey in New Jersey by P. P. Pirone revealed that silver maple (*Acer saccharinum*) is the most susceptible of the streetside maples, with the Norway maple (*A. platanoides*), red maple (*A. rubrum*), and sugar maple (*A. saccharum*) decreasingly susceptible in that order. In other parts of the United States, however, it is the sugar maple that appears most susceptible.

The following additional trees are known to be susceptible to Verticillium wilt: hedge, black, Japanese, Japanese red, Schwedler's, sycamore, Drummond, tatarian, and trident maples; boxelder and California boxelder; horsechestnut; tree-of-heaven; Spanish chestnut; southern and western catalpa; carob; redbud; camphor-tree; yellowwood; flowering dogwood; smoke-tree; quince; Japanese persimmon and Texas persimmon; persimmon; Russian-olive; white, green, black, and European ash; Kentucky coffee tree; China tree; goldenrain tree; tuliptree; osage-orange; southern, saucer, and star magnolia; sourgum or tupelo; olive; tea olive; avocado; cork tree; pistachio; aspen; almond; sour, sweet, Mahaleb, or Morello cherry; garden and cherry plum; apricot and Japanese apricot; prune; pin oak; black locust; sassafras; pepper-tree; Japanese pagoda-tree; American, little-leaf, and Crimean linden; American, English, slippery, Scotch, and Siberian elm.

Symptoms. Wilt symptoms vary somewhat with the kind of tree, its age, its location, and other factors, but there are some symptoms common to all situations. A sudden wilting of the leaves on one limb, on several limbs, or even on the entire tree is a general symptom (Fig. 12-17). The severity of this symptom depends less on the species of tree than on the amount of infection in the belowground parts. Long before wilting becomes visible, the fungus has been at work below ground. If most of the roots are infected, the tree will die quickly. On the other hand, if infections in the roots are confined to one side of the tree, only the parts above that side will show wilt symptoms. Periods of drought frequently accentuate the wilting symptoms and speed the death of the infected tree.

The wilted leaves may fall after they have dried completely, or they may hang

Fig. 12-17. This maple is heavily infected by the fungus *Verticillium albo-atrum*.

on for the remainder of the season. Sometimes a few branches or even the entire top of the tree will die during the winter and fail to leaf out the following spring. Cases of this sort are usually attributed to winter injury.

Because many other agents cause similar symptoms, positive diagnosis of Verticillium wilt can be made only by culturing sapwood tissue in the laboratory. The cultures are taken from discolored sapwood, which occasionally appears in the branches but more commonly and more reliably near the base of the main trunk or in the main roots. When limb discoloration occurs, it frequently shows up on the undersides of infected branches nearest the trunk, reflecting direct movement of the fungus from the trunk to limbs via the xylem. Verticillium is a soil-inhabiting fungus primarily, thus its most common avenue of entrance is through the roots. One would expect, therefore, that the discoloration of the sapwood would be more marked and more common closer to the point of entrance. This is actually the case, as we have discovered in our studies on the cause of landscape and streetside tree death over many years.

When the Verticillium fungus is present, the outer sapwood rings (those nearest the bark) show a discoloration. The color varies with the species of tree and the duration of the infection. In Norway maple, it is a bright olive green; in American linden, dark gray; in tree-of-heaven, yellowish brown; in black locust, brown to black; in northern catalpa, purple to bluish brown; in American elm, brown (like the Dutch elm disease fungus); and in Japanese pagoda-tree, greenish black. But the presence of discolored sapwood tissue cannot be used as an exact diagnostic symptom; it merely indicates that a fungus may be involved. For example, P. P. Pirone isolated at least six different kinds of fungi from sapwood of Norway, silver, and red maples that showed the bright green discoloration! Moreover, he reproduced the same discoloration in young, healthy maples by inoculating them with each of these fungi.

It is well to reiterate, therefore, that a positive diagnosis for Verticillium wilt, as for Dutch elm disease, can be made only by laboratory isolation tests.

Cause. Wilt is caused by the fungi Verticillium dahliae and V. albo-atrum. Many strains of these fungi are known to exist. The fungus can live in soil for some time without parasitizing roots. Although it is thought to enter roots and stems only through injuries, infection may occur in young, uninjured roots. Some research workers believe air currents may spread spores over long distances. When such spores lodge in tree wounds, aboveground infection may occur. Sapwood wounds are usually necessary for the start of an infection in the aerial portions of the tree. Contaminated pruning tools may promote aerial infections.

Once inside the water-conducting tissues, the Verticillium fungus grows upward longitudinally; that is to say, the infection spreads upward from the point of inoculation. The actual discoloration that results may be due to the presence of the fungus itself or to a chemical change resulting from the action of the fungus.

Control. Trees showing general and severe infection by the Verticillium fungus cannot be saved. Such trees should be cut down at once and destroyed quickly and completely. As many of the roots as possible should also be removed. A species of tree satisfactory for the particular location but not susceptible to wilt may be planted.

Narrow- and broad-leaved evergreens appear to be safe replacements for trees that have died from the wilt disease. The following deciduous trees are not known to be susceptible to wilt and might be used as replacements: beech, birch, chestnut, crabapple, ginkgo, hackberry, hawthorn, hickory, holly, honey locust, hop-hornbeam, hornbeam, Katsura-tree, mountain-ash (European), mulberry, oak (white and bur), pawpaw, pear, pecan, planetree, serviceberry, sweetgum, sycamore, walnut, willow, and zelkova.

Where it is absolutely necessary to plant a Verticillium-susceptible tree in an infested site, soil fumigation with Vapam drenches, methyl bromide soil injections, and even formaldehyde drenches have been suggested. These materials are extremely toxic to the user, and only experienced, certified fumigators can be trusted with such a task. All previously existing roots have to be removed from the soil and the soil has to be well worked up beforehand to allow good fumigant penetration. Finally, a wait of several weeks between fumigation and planting is needed to allow the soil time to air out. Even with all this effort, the reliability of such a treatment is questionable, especially if the dead tree is large. Crop rotation to noninfested soils is essential for managing Verticillium wilt in nurseries and tree farms.

We and other investigators have noted recovery of trees from mild cases of Verticillium wilt after liberal applications of a relatively soluble nitrogen fertilizer. However, it should be noted that Verticillium wilt is often fatal, and fertilization can be a wasted effort that could in some circumstances promote spread of the fungus in the tree. The material must be used early in the growing season on trees that show only one or two wilted branches. The heavy fertilization apparently stimulates leaf growth, which in turn enables the rapid formation of a thick layer of sapwood that seals in the infected parts beneath (Fig. 12-18).

Fig. 12-18. Greenish dark sapwood discoloration may be caused by *Verticillium* fungus. Several layers of healthy sapwood now cover the infected area in this branch.

Fig. 12-19. Top: Wood decay visible in old pruning wound. Bottom: Fungal fruiting bodies at the base of an extensively decayed tree.

We and other investigators have tried injecting fungicides into infected trees or applying them to the soil around the roots as a curative measure. Such treatments are not reliably effective and hence must be considered as experimental at present.

Wood Decay

Living landscape trees are subject to internal decays of branches and trunk and of lower stem and roots. These decays are called heart rots, canker rots, and root and butt rots. They are sometimes hard to detect because the decaying tree may appear sound from the outside. The wood decay diseases are important to arborists because they afflict living trees, affecting the structural integrity and therefore the strength and relative hazard of the tree.

Decays of dead trees, of dead branches of living trees, or of wood products are sometimes called sapwood rots. Sapwood rots also represent a hazard, but an obvious one, because dead branches or dead trees are apparent. This discussion concentrates on decay in the living tree rather than in the dead one, though both are important.

Symptoms. Decay in living trees appears as a softening or weakening of the woody xylem tissues of the sapwood and the heartwood (Fig. 12-19). Although the extent of decay is not normally visible from the outside, the decay can sometimes be viewed through completely rotted branch stubs, wounds, and cracks in the bark.

If the tree were cut open, decayed wood might show brown rot or white rot symptoms. Brown rot fungi often produce a dry, crumbly, brown residue, which breaks up into cubical blocks, whereas white rot fungi usually produce a white, stringy decay. In either case, the wood is obviously weakened. A hollow where no wood is present can occur where decay has progressed extensively. An abnormal dark or light discoloration of the wood may be found in areas bordering the decay. The pattern of wood decay in the tree varies depending on the fungus, tree species, and growing site. Some fungi cause mainly canker rots, when wood and bark are both invaded, sometimes in sequence, so that large concentric callus ridges are formed in response to infection. Others cause root and butt rots, so that trees failing to generate roots faster than they are destroyed may show gradual decline or be easily uprooted. Some decays are primarily heart rots of the main trunk and branches of the tree.

Often, fruiting bodies of the decay-causing fungus are present on the outside of the tree. These fruiting bodies, called conks, are indicators of advanced decay (Fig. 12-19). Conks may be leathery to woody and perennial, or they may be soft, fleshy, and ephemeral. They range in size from an inch to several feet. The color is usually gray or brown, but it may be white, black, orange, or yellow. The fruiting body may protrude from the tree or develop flat against the tree surface. The downward-facing surface of the fruiting body, where the fungal spores are released, may be covered with tiny pores, slits, or teeth. Some conks produce no spores.

Cause. Wood decay in trees is caused by the growth of one of several fungi, mainly of the class Basidiomycetes, although some Ascomycetes and bacteria may also cause decay. The disease begins when a windblown fungal spore comes in contact with a tree wound and, given the right conditions (possibly involving other nondecaying microorganisms), germinates. Wounds such as branch stubs, lower trunk or root injuries, branch crotches, and pruning wounds are typical locations for this activity. The germinating spore produces a germ tube and then branched hyphae and mycelia, which invade fiber, vessel, tracheid, and ray cells of the wood. The fungal hyphae release enzymes that break down cellulose (brown rot) or cellulose and lignin (white rot), thus providing food for the fungus and loss of rigidity for the tree.

However, trees are not passive victims; they are able to respond to wounding and fungal invasion. This response has been studied by Dr. Alex Shigo, formerly of the U.S. Forest Service, and co-workers. They developed the concept of compartmentalization of decay in trees (CODIT). There is considerable literature on this subject, and the interested reader should consult Dr. Shigo's work for details.

When an injury occurs, the invading wood-decay fungi encounter several tree defense mechanisms. Vertical movement of the decay fungus is impeded by plugging of xylem tracheids and vessels with resins and tyloses above and below the wound. The horizontal movement of decay fungi is impeded in an inward direction by the already existing growth rings and in a radial direction by toxic substances produced by the ray cells. Decay is prevented from moving outward to new growth by barriers laid down by the vascular cambium after injury occurs.

Of these various defenses the strongest is that formed by the new growth. Consequently, decay will not proceed into subsequent yearly growth increments unless they are reinjured.

In circumstances where wounding is minimal and the tree is quick to respond to wounding and invasion, the tree suffers some internal discoloration and little or no decay, thus retaining its structural integrity. However, where wounding is extensive and when the tree is not capable of producing a quick, effective response, the decay fungus spreads and the tree is weakened.

The decay fungi gradually digest wood upward and downward as well as inward toward the center of the tree. Meanwhile, outside the original wound, callus cells and new wood may have formed. Decay fungi do not invade wood laid down after the wound has closed. Therefore, after a period of years, the outer cylinders of wood may be sound despite the internal decay. Canker rot fungi, able to invade both bark and wood tissues, cause internal decay while keeping the wound open and visible from the outside. Because most decay fungi primarily utilize the central portion of the tree, the term heart rot is used to describe this disease, even though both sapwood and heartwood (where present) are rotted.

The fungal mycelium continues to develop within the stem until, when conditions are right, the fungus begins to grow following a path of least resistance to the stem's outer surface. When it reaches the surface, the fungus produces a fruiting body, or conk. This often occurs near a dead branch stub or other injury. These fruiting bodies are capable of producing large number of spores, which then are available to start new infections of nearby wounded trees. The fruiting bodies of heart rot decay fungi produce spores in such great numbers that there are generally spores present at any wound site.

The presence of a fungal fruiting body signifies that internal decay has progressed extensively and, perhaps more significantly, that the tree has failed to limit decay and that additional decay is likely. Since the outer rings of sapwood, the cambium, and phloem carry out most of the tree's vital functions, the tree can appear healthy despite the presence of extensive internal decay.

Wood decay fungi are usually identified by their conks. The conks of some of the common wood decay fungi important to tree health are described here. The first listed have slits, or gills, on the underside of the conk or mushroom cap, and the others have pores on the conk underside. Several of them are characterized as having a broad host range; this usually includes most or all of the following trees: acacia, ash, birch, catalpa, cherry, chestnut, elm, eucalyptus, fir, magnolia, maple, oak, pine, poplar, spruce, sweetgum, tuliptree, walnut, and willow. Other decay fungi are specialized, attacking just one or a few tree species.

- *Armillaria mellea*, the shoe string root rot or oak root rot fungus, has such a wide host range and is so important in shade tree pathology that it is discussed by itself in the next section of this chapter.
- *Flammulina velutipes* (*Collybia velutipes*) annually produces clusters of mushrooms with stout stalks and moist, smooth, reddish orange to reddish brown caps. The cap has white gills and is 1 to 3 inches across, and the dark, hairy stalk is 1 to 3 inches long. This fungus produces white rot of a wide range of tree species.
- *Pleurotus ostraetus*, the oyster mushroom, causes a white rot of the heartwood and sapwood of many landscape trees, including those listed above. The fungus produces an off-white fleshy shelflike mushroom up to 8 inches across. The mushroom may have a short stout stalk, and the underside of the conk has gills.
- *Schizophyllum commune* annually forms clusters of small, leathery, gray, fan-shaped conks 1 or 2 inches across. The gilled cap of this fungus is attached directly to the tree without a stalk and causes a white rot. The fungus may be found causing white rot on a very wide range of declining and dead trees.
- *Fomes fomentarius* produces gray, woody, perennial hooflike fruiting conks. The conks have pores on their underside and may be 8 inches wide and several inches thick. The fungus causes white rot on a wide range of trees.
- *Ganoderma applanatum* (*Fomes applanatus*) forms a thin shelflike gray-brown conk up to 1 foot across. The underside of the conk consists of minute white pores, and the surface turns brown when bruised or scratched—giving it the name artists' conk. This fungus causes a white rot of a broad range of tree species.
- *Ganoderma lucidum* also produces a large conk, usually at the base of the infected tree. The upper surface has a reddish brown, smooth, varnishlike appearance, while the undersurface with pores is white to tan. This fungus also has a wide host range.
- *Laetiporus sulphureus* (*Polyporus sulphureus*), also called the sulfur fungus, causes a brown rot of the butt and heartwood of living and dead trees. This fungus produces clusters of annual bright yellow or yellow-orange conks up to a foot across. The undersides of the conks are covered with minute pores. It has the wide host range mentioned above.

- *Phellinus robiniae* (*Fomes rimosus*), a pathogen of black locust, produces yellowish brown to gray conks with a brown pore-bearing surface. The conks are several inches to a foot across and protrude shelflike away from the trunk of affected trees.
- *Stereum gausapatum,* a major oak pathogen, and other related *Stereum* species cause white rot of many different trees. Conks appear as clusters of thin, brownish, shelflike structures a little more than two inches across. Injured conks may leak a red fluid.
- *Trametes versicolor* (*Coriolus versicolor, Polyporus versicolor*) fruiting bodies annually form dense overlapping clusters on decaying branches and trunks. The caps, 1 to 2 inches across, are leathery and colorfully arrayed with zones of white, yellow, red, brown, gray, green, and bluish green. Tiny white pores can be found on the cap underside. Generally causing white rot on stressed trees, it can in some circumstances cause heart rot of wounded, but not otherwise stressed trees.
- *Trametes hirsuta* (*Coriolus hirsutus, Polyporus hirsutus*), somewhat larger than *T. versicolor,* produces leathery gray to brownish caps without zones. This fungus also causes a white rot of many tree species.

Control. Unnecessary wounding or injuries to the wood and bark should be avoided. Trees should be maintained in a vigorous condition and examined periodically for symptoms of decay and signs of the fungus-causing decay (conks). Structurally weakened trees may not be able to support their own weight and should be removed before they break and injure people or damage property.

Armillaria Root Rot

Some of our most valuable shade and ornamental trees are susceptible to Armillaria root rot disease. This disease is also called shoestring root rot disease because of the long, round, black strands of tissues, closely resembling shoestrings, produced by the fungus underneath infected bark, over infected roots, or in the soil. Among shade and ornamental trees, oaks and maples appear to be the most commonly infected, although the disease is occasionally destructive on apple, birch, black locust, cherry and other *Prunus* species, chestnut, empresstree, eucalyptus, goldenrain-tree, Katsura-tree, larch, mountain-ash, pine, plane-tree, poplar, redbud, rhododendron, and spruce. In all probability the disease may occur on almost any tree or shrub grown, if the necessary conditions for infection are present.

Symptoms. The aboveground symptoms cannot be differentiated from those produced by many other diseases or agents that cause root or trunk injuries. Probably the most striking external symptom is a decline in vigor of a part or the entire top of the tree (Fig. 12-20). Where the progress of the disease is slow, branches die back from time to time over a period of several years.

The most positive signs of this disease are found at the base of the trunk at or just below the soil line or in the main roots in the vicinity of the root collar. Here, fan-shaped, white wefts of fungal tissue closely appressed to the sapwood are visible when the bark is cut away or lifted. Scraping or lifting the white wefts of

Fig. 12-20. Branch dieback and general decline in this oak are symptoms of Armillaria root rot.

mycelia, which have a strong mushroom odor, will reveal water-soaked sapwood. Where the entire top has wilted, the fungal tissue will be found completely around the trunk. Where a large branch has died back, or only one side of the tree shows poor vigor, the fungus will be found on one or two main roots or on one side of the trunk base. Where the tree has been dead for some time, the dark brown to black "shoestrings" may occur beneath the bark or in the soil near the infected parts. Clusters of light brown mushrooms, called honey mushrooms, may appear in the vicinity of the rotted wood in late autumn (Fig. 12-21). These rarely occur near infected street and shade trees, however, inasmuch as conditions

Fig. 12-21. The mushroom of the Armillaria root rot fungus.

are usually unfavorable for their development or the infected trees are removed before the mushrooms have a chance to form.

Cause. Armillaria root rot is caused by the fungus *Armillaria mellea*. Infection of living trees is thought to occur almost exclusively by means of root contact with the fungus shoestrings or by root grafts with infected trees. The fruiting stage, the honey mushroom, somewhat resembles the mushroom commonly sold in stores but is slightly larger, yellowish brown, and the top of the cap is dotted with dark brown scales. The mushrooms usually grow in tight clusters with their stem bases pressed together. Spores released from the underside of the cap are blown by the wind to bark injuries at the base of living trees where infection can occur.

Because the disease is almost consistently associated with trees previously in poor vigor, several investigators believe that the fungus is only weakly parasitic and cannot attack vigorously growing trees. Others believe the causal fungus is capable of attacking living roots and can penetrate sound, healthy bark. The fact is that once the fungus becomes established in trees, it usually kills them in a relatively short while.

A similar and important root rot disease is caused by the fungus *Clitocybe tabescens*. The fungus kills the root collar cambium, producing mycelial fans under the bark. It also rots the roots, thus having a pathology similar to that of *Armillaria*. The main difference is that the mushroom lacks an annulus and, furthermore, that *Clitocybe* does not produce the flat, black rhizomorphs or shoestrings of *Armillaria mellea*. Clitocybe root rot generally occurs in the South, whereas Armillaria occurs, often on the same tree species, in the North and West.

Control. Because much of the evidence at hand indicates that Armillaria root rot is associated with stressed trees, probably the best precautionary measure is to provide good growing conditions for the tree.

An infected tree whose entire root system or trunk is diseased cannot be saved. The large roots in the vicinity of the trunk as well as the trunk itself should be removed and destroyed. Soil in the immediate vicinity should also be removed.

Attempts to retard disease development on partly infected valuable trees are occasionally successful, especially where only one or two main roots or one side of the trunk are involved. In such cases, the soil within a radius of about 2 feet around the trunk should be removed in the spring or summer to expose the root collar and the large roots. Roots whose bark is completely rotted should be removed and the affected bark on partly diseased roots and on the trunk carefully cut away. All visible fungal tissue and water-soaked wood should be removed. The bark edges should be trimmed. Branches should be thinned out and plants growing nearby removed to allow as much sunlight and dry air as possible in the vicinity of the treated parts.

Where the disease has killed one tree in a group, spread to nearby trees can occasionally be delayed or even completely checked by removing the affected and adjacent trees and their roots. All soil from the affected area should be carted away. This treatment is expensive and will not necessarily check the spread of the disease. The remaining trees should, of course, be watered and given all possible care to ensure continued vigorous growth.

The problem of planting another tree in the site on which one has previously died from this disease occasionally confronts the shade-tree commissioner, arborist, or property owner. We know of cases where three successively planted trees have died from Armillaria root rot under such conditions. Where a tree must be planted in the same place, the replacement should not be the same species as the one removed. Oaks, maples, and other highly susceptible species should be avoided.

Other Root Rot Diseases

Two other root diseases, black root rot and Phytophthora root rot, are often found on trees in landscapes and nurseries.

Phytophthora Root, Crown, and Collar Rot. The fungus *Phytophthora*, found in soils and plant debris worldwide, is capable of causing root, crown, and collar rot diseases of landscape trees, especially those in moist sites. *Phytophthora* produces zoospores that are motile in water, allowing the fungus to swim from infected to healthy roots in flooded or waterlogged soils. In addition to aiding dispersal, wet soils stress the root systems, making them more susceptible to Phytophthora diseases. The fungus also produces resistant oospores enabling the fungus to survive in soil and plant debris for long periods of time. Susceptible trees include apple, beech, birch, boxwood, camellia, cherry, citrus, cypress, dogwood, elm, fir, Franklin-tree, horsechestnut, juniper, maple, oak, palm, pear, pine, plum, rhododendron, strawberry-tree, sweetgum, tuliptree, and yew.

Symptoms. Root and lower trunk symptoms include death of absorptive and transport roots, and death of phloem and cambial tissues on the bark at the base of the trunk and on buttress roots. Dead bark may be scaly or peeled back; bleeding cankers sometimes occur. Infected inner bark and cambium tissue typically turns a cinnamon brown or dark brown color. Often, when root or crown infections are well advanced, the tops of affected trees show symptoms. They include undersized, chlorotic, folded, epinastic leaves; thinning, tufting, or browning of the foliage; premature fall color; stunting; dead branches; and tree decline and death. Not all *Phytophthora* infections occur at the base of the tree; some result in trunk and branch cankers, shoot blight, and even fruit rots.

Cause. Root, crown, and collar rots are caused by many species of the fungus *Phytophthora*, including *P. cactorum*, *P. cambivora*, *P. cinnamomi*, *P. citricola*, *P. citrophthora*, *P. cryptogea*, *P. drechsleri*, *P. lateralis*, *P. megasperma*, *P. nicotianae* var. *parasitica*, and *P. syringae*. To accurately diagnose a Phytophthora root rot disease requires microscopic examination of laboratory cultures made from infected plant tissues, and to differentiate one species from another requires a specialist.

Control. Improved soil drainage is essential for management of Phytophthora root rots. This can be done before planting by installing drain tiles, creating planting mounds and berms or drainage ditches, and amending heavy soils to promote internal drainage. Choose tree species and rootstocks that are resistant to Phytophthora diseases. Fungicides such as Aliette, Banol, Subdue, Subdue Maxx, Terrazole, or Truban can be used in the nursery and sometimes in the landscape to suppress Phytophthora.

Black Root Rot. This fungal disease does not often receive much notice as an important tree disease, perhaps because its symptoms are not very dramatic and its fungal fruiting bodies are microscopic. Black root rot has been problem for hollies in landscapes and nursery production facilities. It is most frequently observed on Japanese holly, blue holly, and inkberry, but can also be found on American holly. Other trees reported to be susceptible to black root rot include black locust, catalpa, elm, lilac, red and Scots pines, and yews. Flowering plants in the landscape such as allium, begonia, bergenia, geranium, nicotiana, pansy, phlox, snapdragon, sweet pea, verbena, and viola are also hosts and could introduce or harbor the fungus near susceptible trees. Black root rot is favored by high soil pH conditions.

Symptoms. The first symptoms of black root rot include yellowing and marginal scorch of the foliage. Later, twigs or stems and eventually the entire plant may die back and die. American hollies may only show thinning of the foliage without a noticeable decline. The root system of the declining plant is stunted and decayed. Black lesions appearing on the tips or elsewhere along the length of absorptive roots contrast sharply with the adjacent healthy white portions.

Cause. Black root rot is caused by *Thielaviopsis basicola* (*Chalara elegans*). This fungus can persist indefinitely in the soil or it can survive as a saprophyte on plant debris.

Control. If susceptible trees are to be used, examine their root systems prior to planting them in the landscape. If blackened roots are evident, the presence of the fungus can be confirmed through laboratory assay. Avoid planting susceptible trees into high pH soils. Declining, badly infected trees should be removed and the site replanted with a nonsusceptible host. Good cultural practices may enable some trees to continue to grow in spite of the disease. Trees in the early stages of infection can be kept looking good if they are well fertilized and watered. Fungicides such as Cleary's 3336 or Domain applied to the root zone soil may suppress this disease during nursery production, but they have not been shown to control black root rot disease in the landscape.

Selected Bibliography

Blanchard, R. O., and T. A. Tattar. 1981. Field and laboratory guide to tree pathology. Academic Press, New York. 285 pp.

Carter, J. C. 1961. Illinois trees: Their diseases. Illinois Nat. Hist. Surv. Circ. 46. 99 pp. (second printing with alterations)

Farr, D. F., G. F. Bills, G. P. Chamuris, and A. Y. Rossman. 1989. Fungi on plants and plant products in the United States. APS Press, St. Paul, Minn.

Hepting, G. H. 1971. Diseases of forest and shade trees of the United States. USDA Forest Service, Agriculture Handbook 386, Washington, D.C. 658 pp.

Hickman, G. W., and E. Perry. 1997. Ten common wood decay fungi on California landscape trees. Western Chapter, ISA, Sacramento, Calif. 27 pp.

Horst, R. K. 1979. Westcott's plant disease handbook. 4th ed. Van Nostrand Reinhold, New York. 803 pp.

Manion, P. D. 1981. Tree disease concepts. Prentice-Hall, Englewood Cliffs, N.J. 399 pp.

Pirone, P. P. 1978. Diseases and pests of ornamental plants. 5th ed. Wiley, New York. 566 pp.

Schwartz, F. W. M. R., D. Lansdale, and S. Fink. 1997. An overview of wood degradation patterns and their implications for tree hazard assessment. Arboricultural Journal 21:1–32.

Shigo, A. L. 1986. A new tree biology. Shigo and Trees, Associates, Durham, N.H. 595 pp.

Sinclair, W. A., H. H. Lyon, and W. T. Johnson. 1987. Diseases of trees and shrubs. Cornell University Press, Ithaca, N.Y. 574 pp.

Stipes, R. J., and R. J. Campana (eds.). 1981. A compendium of elm diseases. American Phytopathological Society, St. Paul, Minn. 96 pp.

Tattar, T. A. 1978. Diseases of shade trees. Academic Press, New York. 361 pp.

Van der Zwet, T., and H. L. Keil. 1979. Fire blight: A bacterial disease of rosaceous plants. U.S.D.A. Agricultural Handbook 510, Washington, D.C. 200 pp.

13

Coping with Tree Pests and Diseases

In this chapter, the term pest will be used to mean all types of tree pests, whether insect, mite, fungus, bacterium, phytoplasma, virus, nematode, parasitic plant, weed, or animal. For the most part the discussion of pest management will focus on arthropod (insect and mite) pests, which cause tree damage, and microbial pests (pathogens such as fungi, bacteria, and nematodes), which cause tree diseases. Some of the chemicals used to manage pests are called pesticides and include fungicides, bactericides, nematicides, insecticides, miticides, herbicides, and rodenticides.

Once an actual or anticipated tree pest problem has been identified, the decision of what to do remains. There was a time when finding the right pesticide to kill the offending pest was the only decision required. Although arborists are still dependent on chemical pesticides, their use in landscapes has been mitigated by laws, societal and environmental pressures, new scientific discoveries and pest management technology, and common sense. Several important general steps should be followed in dealing with pest problems in the landscape:

1. Anticipate pest problems. The circumstances that promote their occurrence and their potential for damage should be well known. Key weather parameters should be monitored, as should the condition of the trees and the effect recent weather is having on them and their potential pests. The development of most fungal and bacterial pathogens is favored by specific temperature and moisture conditions. The life stages of many insect pests can be keyed to the stage of development of landscape plants or to growing degree days. Many pest problems can be prevented if accurately anticipated.
2. Monitor trees in the landscape regularly. Discovery of pests before they cause serious problems and ability to locate the exact point of infection or infestation will make a rescue action, if needed, more successful.
3. Accurately determine the cause of the problem. If you do not know the answer, seek out competent help.
4. Determine a course of action. In some cases the best decision will be to do nothing at the present time because the damage has already been done, or because there will be little or no damage.

5. Carry out the control decision properly. Pest management results will not be satisfactory if improperly implemented.

A tree maintenance firm should have a pest management adviser who knows where to obtain good information about managing tree pests. Most states have County Cooperative Extension offices with staffs competent in pest identification and management or with access to this kind of information through land-grant university plant disease clinics, insect clinics, and university publications on tree pest control. Local, state, and federal foresters often have pest management information which relates to landscape trees. Meetings and workshops on tree care and tree pest management are held in many areas. Knowledge gained from textbooks is helpful, but it should be supplemented with real pest encounters. Practice and experience are essential.

Plant Health Care

The concept of plant health care (PHC) has been developed as a holistic approach for coping with pests and for maintaining tree health. PHC not only incorporates the precepts of integrated pest management (IPM), discussed in the next section, but also includes other maintenance activities that affect the health of a tree (or population of trees and other plants) over its lifetime. It has been said that most pest problems result from poor tree maintenance practices, beginning with poor tree and site selection and continuing with poor watering, training, and pruning practices. If tree health can be maintained optimally and in balance with its surroundings using PHC, perhaps the tree will be more resistant to pest attacks and more amenable to treatments using IPM approaches.

Using PHC under ideal conditions, the arborist or horticultural consultant would be involved in decisions relating to site and plant selection, safeguarding existing trees, soil preparation, planting, post-planting to mature plant care, and the maintenance level desired for the landscape; these are discussed in earlier chapters. Unfortunately, arborists are often not involved in early decisions that ultimately affect plant health; nevertheless, the PHC approach can provide benefits even when applied to existing trees. The International Society of Arboriculture has published an excellent manual describing the principles and practices of PHC.

Integrated Pest Management

When an integrated pest management (IPM) program is implemented, important pests are monitored, all suitable pest management methods are considered, and decisions on what methods to use are based on ecological, economic, and sociological values. The task is to integrate cultural practices, plant resistance, biological control, and pesticide application to best manage the pest problem. Several methods might be used to manage tree pests; all IPM practices need to be evaluated and need to be compatible with tree maintenance and pest management objectives. Thus, IPM is compatible with the objectives of PHC.

Landscape pest management objectives. Deciding what to do about a pest problem will depend on one's idea of what the landscape should look like. A person expecting a formal, high maintenance landscape will have different needs than one desiring an informal landscape where only the fittest trees are expected to survive. In addition, a person revering all forms of life might seek a different approach to tree pest management than one with entomophobia. One IPM approach is to educate the public (or one's clients) to accept higher pest levels or damage thresholds than they otherwise might. Obviously one would consider this approach only when the perceived pest does not threaten the health or life of the tree.

Pest and tree health monitoring. The landscape is too complex a place to look for every possible pest. Even when an arthropod or pathogen is found, it does not always mean that a problem exists. For efficient monitoring, select the most important tree and pest combinations, based on experience and advice. Monitor specifically for those pests regularly and systematically. In addition, it is very important to look for trees in trouble. If the tree does not look normal, seek out the cause of the problem! As mentioned, early detection allows timing and location of control measures to be more precise. In addition, any adverse effects on potential natural controls such as beneficial insects can be minimized.

Monitoring procedures vary from one landscape to another, but the following suggestions may be helpful:

1. Examine trees in the landscape routinely. Begin by seeking information about species, cultivar, age, and current growth rate. A knowledge of soil type, pH, fertility, drainage, and recent maintenance activities such as planting, pruning, and fertilization is important.

2. Identify the key potential pests and understand their biology so that control measures can be aimed at their most vulnerable stage. Sources of infestation or inoculum should be sought in the landscape or in the neighborhood. This will allow the destruction of pest insect egg masses and pupae as well as overwintering phases of fungal and bacterial pathogens. For Dutch elm disease, oak wilt, and other devastating problems that can move in from nearby landscapes, the surrounding community should be inspected so that preventive action to preserve a valuable tree can be taken, or a community-wide control strategy can be developed if necessary.

3. Use pheromone traps, if available, for the important insect pests that require precise timing of control measures. A trap in one landscape should be adequate for the needs of other landscapes nearby. (This is discussed further in the next section.)

4. Keep track of degree days and the seasonal development of plants. The activity of certain insects can be predicted each season based on the timing of certain plant phenology events such as bud break, bloom, or petal fall. These predictive aids work best if they have been created locally. Nearby Cooperative Extension offices or university extension specialists may be of help. The rate of development of some diseases and many insect pests can be predicted by keeping track of the accumulated degree days. The pest manager can thus determine when key stages in insect or disease development will occur and time control activities more accurately.

5. Use effective detection methods such as observing pest damage directly; shaking needle evergreen branches over white paper to detect dislodged

mites; noting honeydew episodes for aphids; discovering frass droppings for certain caterpillars; monitoring sticky traps for whiteflies, thrips, and leafhoppers; using two-sided sticky tape to catch and count scale crawlers; and placing burlap bands on the tree trunk to detect gypsy moth larvae.

6. Record pest-monitoring observations so that comparisons can be made and trends detected.

One of the inherent difficulties in relying on an IPM approach to managing landscape tree pests is the cost of monitoring. Our experience with a residential landscape IPM program implemented in Kentucky showed that most tree owners were unwilling to pay the cost of monitoring, relying instead on attempting pest control after the problem became obvious. Similarly, IPM programs run by a tree maintenance company may cost the customer more than a preventive spray schedule without monitoring. So although IPM may make biological, environmental, and sociological sense, some clients will not accept the increased cost. Nevertheless, some corporate, governmental, or institutional clients may be willing to afford regular monitoring. Pesticide use was reduced in the National Capitol Region, National Park Service following use of an IPM program, a benefit certainly worth the cost in this heavily used public area.

Although IPM and PHC provide good approaches for tree pest management, there still are some uncertainties associated with these approaches. For many diseases and insect infestations, little is known about pest threshold levels that affect tree health, and aesthetic injury levels for trees could vary widely. In many instances the biology of the pest is not well enough known to make an informed decision about its control. Accurate diagnosis could also be a problem unless the program is backed by an unbiased pest diagnostic service. Too often, the tree care expert making the diagnosis is also the person who stands to gain monetarily from some kind of treatment, a situation that might make a tree owner uneasy.

Cultural Practices to Combat Tree Pest Problems

Cultural methods for reducing tree pest problems should be integrated into good tree maintenance practices. Cultural practices are aimed at increasing host resistance or reducing attractiveness to pest attack, reducing pathogen inoculum or insect infestation levels, and making the environment for disease or insect activity less favorable.

Maintaining tree vigor. Improvement of vigor and avoidance of plant stress by proper tree maintenance is almost always warranted. Attention must be given to planting, watering, mulching, pruning, aerifying soil, and other best practices for maintaining trees. Optimum levels of vigor will vary with the species; too much growth stimulation of flowering pears or crabapples may increase susceptibility to fire blight disease, for example. In general, however, stressed trees are more attractive to various wood-boring insects and often become insect breeding sites. Dogwood borers are attracted to flowering dogwoods growing in more stressful sunny sites; elm bark beetles breed in dying elm trees; and many canker-causing fungi are only active on trees growing in stressful circumstances.

Sanitation. Sanitation measures should be applied routinely to destroy pest sources and to keep healthy trees from becoming contaminated with insects and parasitic microbes. Rake up and destroy infected leaves and prune out infected and infested twigs and branches to reduce sources of further insect and disease outbreaks. Diseased and insect-infested abandoned trees should be removed and destroyed to benefit the healthy trees remaining. One of the most important measures used to stop the spread of Dutch elm disease is the community-wide removal and destruction of dead and dying elms and elm firewood. Sanitation in this case is often mandated by the local governing body.

Changing the environment. Modifying the tree habitat or environment to produce an unfavorable situation for pests is also appropriate where practical. For example, shade favors powdery mildew and many leaf spot organisms. Thinning a tree or removing overhanging branches from one nearby could reduce disease by increasing sunlight penetration and air movement into the canopy. Improving soil drainage or planting trees in raised beds will help to reduce the occurrence of Phytophthora root rot disease.

Pest-Resistant Cultivars

Many problems can be avoided simply by selecting trees prone to fewer pest problems. Unless the tree owner is prepared to invest heavily in pest management efforts, trees such as flowering plums, crabapples, flowering cherries, and most fruit trees are regarded as poor choices in the landscape because they often suffer multiple pest problems. In certain locations, depending on pest pressure, many pines, flowering dogwoods, white birches, and certain junipers are at risk from debilitating pest attacks. Trees that are relatively free from serious problems such as oaks, maples, ginkgo, and yews, are less likely to require frequent treatments for pest management. Insect- and disease-resistant cultivars are the most cost-effective means of pest management and should be considered wherever landscape trees are being replaced.

Disease-resistant trees. Selection of disease-resistant tree cultivars is a form of biological control. The development of flowering crabapple cultivars has progressed to the point that cultivars are available that resist several important diseases: scab, fire blight, powdery mildew, rust, and black rot. New horticulturally pleasing disease-resistant cultivars are available and should be requested when flowering crabapples are purchased for the landscape. Other trees with disease resistance include elms resistant to Dutch elm disease, maples and other trees resistant to Verticillium wilt, trees not susceptible to Armillaria and Phytophthora root rots and crown gall, a sycamore resistant to anthracnose, poplars resistant to Cytospora canker, and junipers resistant to tip blight. Specific suggestions of cultivars with reduced disease problems are considered in other parts of this book. Lists of best locally adapted disease-resistant cultivars are often available from nearby Cooperative Extension offices.

Insect-resistant trees. Insect-resistant trees or trees less attractive to insects are an important component of an IPM approach to managing tree pests in the landscape. Certain trees such as tuliptree, white ash, and red maple produce toxic alkaloids, chemicals that make these trees less likely to be fed on by gypsy moths. There are flowering crabapple cultivars known to be less attractive to Japanese beetles, acacia cultivars resistant to acacia psyllids, and pines resistant to

Nantucket pine tip moths. There may soon be transgenic trees, perhaps some genetically engineered with the *Bacillus thuringiensis* toxin gene, that will be used to manage insect pests. Pests have ways of mutating to overcome host plant resistance, so genetic resistance must be broadly based, involving several host plant genes, to be long lasting. Pest-resistant trees are living organisms and as such could be considered a form of biological control.

Physical and Mechanical Methods

Physical and mechanical controls may be considered as substitutes for chemical control strategies. Insects such as bagworms and tent caterpillars can often simply be removed by hand and destroyed. Even Japanese beetles can be hand-picked from small trees and crushed or drowned. Barrier bands encircling the trunks of trees under attack by gypsy moth larvae can be used to prevent movement of larvae from the ground to the top of the tree. Larvae stuck in the band can be destroyed. A strip of burlap encircling the tree can be used as a hiding band where gypsy moth larvae seek refuge. The hiding band can be opened up routinely and the larvae removed and destroyed. Weed pests are managed mechanically by mowing, pulling, and cultivating and are suppressed physically with application of mulches.

Biological Control of Insects and Diseases

Biological control involves the use of living agents to manage pests. Although their use is not yet widespread and many agents are still under development, this technology could become a major pest management approach, providing a safe and effective way to control pest problems. Biological control of insects includes the use of parasites, predators, and bacterial and viral organisms. More broadly, it also includes the use of insect-resistant cultivars as well as the use of microbial insecticides, sterilizing agents, and attractants or lures. Biological control of diseases includes the use of microbial antagonists of pathogens and disease-resistant cultivars. Biological control is usually long lasting, environmentally sound, and cost-effective, but some plant damage by pests might be expected if buildup of the pest's parasite, for example, lags behind pest buildup.

    Insects—introduction of parasites and predators. One of the outstanding examples of biological control was the introduction of the vedalia ladybird beetle from Australia in 1887 into citrus orchards in California to control the cottony-cushion scale. Within two years of its release into the orchards, the larval and adult stages of this predator had annihilated the scale, a pest that had threatened to wipe out California's citrus industry.

    A more recent successful experience with introduced natural enemies is the project done in California on management of obscure scale, *Melanaspis obscura*. Following the discovery of a severe outbreak of this pest on several oaks in a Sacramento park, a parasitic wasp called *Encarsia aurantii* was imported from Texas and released in the infested trees. Researchers monitored the populations of the scale and the parasitic wasp for establishment of the parasite and, eventually, reduction of the scale insect and found levels approaching complete biological control. From first release to confirmation of enduring establishment of successful biological control in 1997 took most of a decade. The use of introduced parasites, even in this limited area, takes time and a sound knowledge of pest and parasite biology. There are many benefits of using a natural enemy in this example. Oak trees in the park and presumably in surrounding neighborhoods will not be subject to the stresses associated with heavy scale infestations. The environmental, sociological, and monetary costs associated with using pesticides for scale management in the park have been avoided. The spread of the scale infestation may be retarded. In this case, control of an introduced pest was achieved by introducing its natural enemy; it is thought that augmenting a population of parasitic wasps in an area where they already exist might also be beneficial.

In the urban landscape, parasitic wasps, natural enemies of certain aphids, have been used to reduce aphid problems of street trees in California. Other stingless wasps attack whiteflies such as the ash whitefly, and certain wasps are being tried for managing the giant whitefly infesting trees in Southern California. Each wasp lays a single egg inside a whitefly nymph, the immature stage of the whitefly pest. The tiny wasp larva hatches from its egg, feeds on the inside of the whitefly nymph, and kills it. Later, an adult wasp emerges from the dead whitefly to start the cycle again.

Some beneficial invertebrates that are natural enemies of tree pests are commercially available. They include lacewings, ladybird beetles, minute pirate bugs, parasitic nematodes, parasitic wasps, and predatory mites. Other insects in the landscape such as assassin bugs, hover flies, gall midges, predaceous ground beetles, and soldier beetles also feed on pest species. Praying mantids are also available, but although these predators feed on some pest species such as aphids, mealybugs, and scales, they also feed on harmless insects. Tree owners cannot assume that the natural enemies they release will remain in the vicinity of their release, and they also run the risk of losing their released predators to insecticides used by their neighbors. Sometimes populations of predators do not build up to sufficient levels to affect the pest population until the pest has seriously damaged the tree.

Given the diversity of trees in the urban landscape and their uneven distribution, predators and parasites will not be widely used for landscape tree pest control without much more research and development. Managing pests of landscape plants is very complex, and there is little information on the use of the few products what are available. During the twentieth century the U.S. Department of Agriculture has imported hundreds of natural enemies of pests. Some of these have survived and become established, but only a few have substantially controlled the pest they were imported to combat. However, because there are a few cases where parasites and predators have controlled destructive insects successfully, there is hope that these biological means can become an important tool, along with other methods for managing tree insects.

Insects—use of microbes. The bacterium *Bacillus thuringiensis* (Bt) produces a toxin that kills leaf-chewing insects. The toxin from one Bt strain, 'kurstaki,' acts against larvae of butterflies and moths such as gypsy moth, and the toxin from another strain, 'tenebrionis,' acts against larvae of certain beetles such as elm leaf beetle. When a larva ingests Bt, it is poisoned by the toxic crystals, which affect the gut lining and cause the insect to stop feeding. Death often occurs within a few days. Bt insecticides are most effective when the larvae are young; if they reach more than half size, it is often too late for effective control. The Bt insecticide must be present on the surface of the tree leaves being eaten by the insects and in some cases will need to be reapplied. The Bt toxin is not harmful to humans and other mammals.

Bacillus popillae, another bacterium, is the causal agent of milky disease of Japanese beetle grubs. Once bacterial spores become distributed in the soil, milky disease provides effective control of turfgrass grubs for many years. If adult Japanese beetle control is to benefit trees, however, the bacterium would need to be used on a very wide scale over an entire community, an expensive proposition.

Insect viruses that infect and sicken or kill insects are being used to control certain caterpillars on trees. The gypsy moth nucleopolyhedrosis virus, formulated commercially as Gypcheck, specifically affects gypsy moth, causing population reductions. Treatments are best applied where gypsy moth populations have built to high levels. Other insect viruses have been used to manage infestations of the great basin tent caterpillar and the European pine sawfly larvae. Insect viruses are normally specific to just one pest and are usually applied to forest trees by airplane to control a pest outbreak.

A fungus, *Entomophaga maimaiga*, a gypsy moth pathogen, has been associated with population declines of this insect from one year to the next. In some areas, the fungus appears naturally a few years after the initial invasion of the insect pest. The fungus produces spores that may persist in the soil under infested trees, providing inoculum for continual suppression of the gypsy moth.

While some of these treatments are useful in vast plantings, the widely scattered nature of specific urban tree species sometimes makes it difficult for these microbes to become permanently established on insect pests in the diverse urban forest.

Insects—pheromones. Pheromones are chemical sex attractants which enable male insects to find and mate with receptive females. Each pheromone is species-specific, and males of that species are sensitive to extremely small amounts of the chemical in the air. Other scented lures may attract insects with odors related to plant odors or to feeding smells. Pheromones can be used to monitor populations of male insects in the area; to attract, trap, and kill males so that mating does not occur; or to fill the air with enough scent to confuse the males.

To monitor insect pests in the landscape, pheromone traps (Fig. 13-1) are designed with the chemical mixed with a poisonous bait or placed near a sticky surface to attract males, which then die. Pest populations are monitored in this way so that insecticide treatments can be accurately timed. They are not normally used to reduce insect populations in the landscape. Pheromone traps are used to monitor populations of gypsy moth and clear-winged moth borers such as dogwood borer, lilac borer, and ash borer. Also available are pheromone traps for monitoring the population of codling moth, oblique-banded leafroller, peach tree borer, oriental fruit moth, lesser peach tree borer, and red-banded leafroller in fruit trees.

Trapping and killing Japanese beetles directly using floral lures and pheromones has been successfully accomplished with commercial Japanese beetle

Fig. 13-1. Pheromone trap used to monitor flights of certain male insects.

traps that can attract huge numbers of adult beetles into a sack from which they cannot escape. University of Kentucky entomologists found, however, that the traps do not really reduce Japanese beetle populations in the landscape. Host plants near the trap may actually suffer a greater beetle infestation than they would have had the traps not been present. Although many beetles are destroyed, others that normally would not have come are attracted from greater distances into the landscape.

Mating disruption occurs when high levels of pheromone in a tree confuse the males so they cannot find a receptive female. Disparlure, the gypsy moth pheromone, can be placed in dispensers throughout the tree canopy for mating disruption. Males are unable to find the female gypsy moths, and the resulting eggs laid by the females will be sterile. An approach to disseminating very small Disparlure dispensers to the tops of trees in the forest might be to seed them into the tree tops by airplane. Apple codling moth pheromones are impregnated into twist ties which are attached by the hundreds to apple trees for mating disruption

Insects—miscellaneous biological control approaches. Sterilization of the males of certain harmful pests may have potential as a form of biological control. The male insects are exposed to gamma radiation or chemical sterilants, rendering them sterile. After their release, females mating with treated males produce no offspring; thus the insect population is greatly reduced.

Diseases. Biological control of crown gall disease (Chapter 12) has been achieved commercially by using a related nonpathogenic strain of bacteria. *Agrobacterium radiobacter* strain 84 inhibits the pathogen, *A. tumefaciens*, by competing for infection sites around wounds. It is available commercially as

Galltrol A. The product must be used as a preventative and will not cure already existing crown gall.

Chestnut blight disease control has been accomplished in European chestnut plantings by introducing hypovirulent strains of the pathogenic fungus *Endothia parasitica* into developing cankers. Hypovirulent strains of the fungus are not highly pathogenic to chestnuts, in contrast to the normal pathogenic strains. Hypovirulent strains contain a double-stranded RNA, similar to that found in plant-infecting viruses. Normal strains exposed to hypovirulent strains become weakened and the chestnut tree may then overcome the canker. In the United States, research is being done to find ways to easily and rapidly transmit the "virus" to diseased trees.

Chemical Pesticides

Chemical pesticides are convenient to use. Some are toxic to only one kind of pest, providing precise control as needed. Some are broad spectrum, thus allowing many kinds of pests to be controlled with one application. Some are systemic and can be applied to the soil for root uptake or by injection directly to the tree with little hazard to its surroundings. The decision to use a chemical for tree pest control should be based on sound biological, economic, aesthetic, and practical considerations. Preventive applications for tree diseases and insect pests are warranted in the following circumstances:

1. The tree has very high value. Of course the control measure should not cost more than the replacement value of the tree!
2. The disease or insect pest is potentially highly destructive. Even in such instances sprays should not be applied unless the threat is real, for example, an insect infestation that threatens to reach a level that could cause defoliation or a wilt disease such as Dutch elm disease where inoculum is actually present in the area. If the pest has minimal impact on tree health, sprays are unnecessary, other than for aesthetic reasons.
3. There is a history of the problem on that individual tree or on that species in the area. In this case it is assumed that the problem will recur yearly, like scab disease of susceptible flowering crabapple. However, even in this case, treatments may be unnecessary in a dry spring. One should be aware that if cultural practices that readily reduce inoculum have not been attempted, spraying to overcome intense pest pressure could be futile.
4. All reasonable cultural and biological control efforts have been implemented.
5. Effective, legal treatments are available.
6. Special equipment, e.g., tree injection devices, is available where appropriate.
7. Only a few treatments are necessary. One should consider alternatives if a weekly season-long spray program is needed to maintain landscape tree health.
8. The tree is already temporarily under stress from other problems such as transplanting or defoliation.
9. The potentially serious disease or insect pest cannot be monitored or predicted. For diseases, it is generally too late to attempt control when the symptoms are already showing.

Unanticipated pest problems should be dealt with on a case-by-case basis, with the decision to apply chemicals on a rescue basis lying with the tree expert.

There are obvious problems with pesticide dependence. Pest resurgence, a buildup of the pest population to even higher levels than before, or secondary pest outbreaks can occur, especially if pest predators are killed by the pesticide. Resistance to pesticides can also develop so that when a pesticide is really needed, it may not be effective. Misapplication of the pesticide can result in phytotoxicity and drift of the pesticide to non-target locations. If misused, some chemical pesticides are hazardous to people, pets, and wildlife. Some pesticides present environment problems by contaminating soil and water supplies or by killing non-target organisms. Despite these shortcomings, there are some kinds of serious tree pest problems that can be controlled only by using chemicals.

Chemical Application Methods and Equipment

Successful control of diseases and of insect pests is dependent on correct placement of the properly selected and prepared pesticide on the infested or susceptible parts of the tree. This is best accomplished with good spraying or injecting equipment and techniques.

Injections and Implants. Techniques and materials have been developed for controlling specific insects or diseases in certain trees by injecting insecticides or fungicides into the sapstream at the base of the trunk. The chemicals then move throughout the tree to provide protection and control. These systems are useful where sprays cannot be used. Much less chemical is needed, and it is all delivered into the tree and not into the surrounding environment. Problems of uneven distribution of chemical in the tree crown do sometimes occur, however, and each injection creates a wound that is a potential colonization site for decay fungi. In this regard, the smaller the injection the better. Injections would not normally be used on small trees and shrubs.

Trunk injections using the Mauget system of injector units containing a small amount of insecticide or fungicide have provided many arborists with excellent results (Fig. 13-2). Similar injection systems using the fungicide Alamo have been used to control oak wilt and Dutch elm disease. A small plastic cylinder containing the pesticide is attached to a short plastic tube inserted into a predrilled hole in the lower trunk or basal root flare. Injectors are normally left on the tree for several hours until the capsule is emptied. Installing the injector units requires knowledge and practice. Accordingly, these are used by arborists, nurserymen, and horticulturists who have received special training.

The Arbor Systems company has developed an injection system that more closely resembles the hypodermic needle approach familiar to human medicine. Through a tiny slit in the bark, a wedge-shaped needle is forced into the sapstream and the chemical is then injected into the tree. Using small volumes of chemical, a tree can be injected in just a few minutes.

The Acecap implant technique uses gelatin capsules filled with the insecticide Orthene that are placed in holes drilled in the trunk. The capsules dissolve in the sapstream liquid, releasing the pesticide to be distributed in the tree crown. Implants are simpler to use than injections, but they require a larger trunk wound.

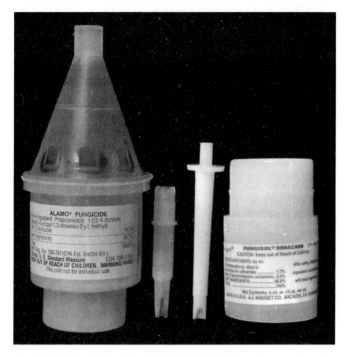

Fig. 13-2. Capsules and tubes for injecting low volumes of pesticides into trees.

Apparatus is also available for injecting liquid insecticides and fungicides into the tree under high pressure. Small holes are first bored into the tree trunk to the depth of the outer layers of sapwood. A special injector screw is then inserted into the hole. A high-pressure hose is connected at one end to the screw and at the other end to a hydraulic sprayer capable of building pressures to 400 pounds per square inch. The insecticide and/or fungicide is then pumped into the tree at high pressure. Only experienced professional arborists should use this apparatus. The Elm Research Institute provides information on use of a low-pressure system for injection of fungicide into elm trees to control Dutch elm disease. The arrangement of a series of T-shaped nozzles connected to one another with tubing would be similar to the apparatus used for the high-pressure system described above.

Injection systems do cause wounds to trees. If the problem being treated is only temporary, for example, Dutch elm disease threats from nearby infected trees that will eventually be destroyed, the wounds from one or two years of injection may not impact tree health. Where circumstances require repeated injections, the possible consequences should be considered. The long-term effects of multiple injection wounds on tree health are not known. The tree maintenance expert must explore all options before committing the tree to a program of injections. One option may be to replace the tree with one that is better suited for a problem-prone site.

Fig. 13-3. Hand-pumped backpack sprayer. Small trees are easily treated with portable spray equipment.

Spraying Equipment. Small trees can be sprayed with hand-pumped, bucket, hose-end, backpack, or small power sprayers. For low-growing trees, the "trombone" sprayer is very efficient and easy to operate. Similarly, a hand-pumped backpack sprayer (Fig. 13-3) is useful for small trees. Large trees can be properly sprayed only with large spraying machines. Such machines are expensive and are owned principally by commercial arborists and by park and shade tree departments that do considerable tree work.

Hydraulic Sprayers. The modern spray machine consists of a large steel or plastic tank, ranging in capacity from 100 to 1,000 gallons, for holding the spray material; an agitator inside the tank, which revolves to keep the spray material uniformly dispersed; and a gasoline or diesel-driven pump, which forces the material from the pump through the spray hose (Fig. 13-4). These parts are usually mounted on a truck or trailer chassis. Certain accessories, such as adequate hose and spray guns with special nozzles, complete the equipment.

Spray machines are available with pumping capacities ranging from 5.5 gallons a minute at 300 pounds pressure to 55 gallons at 800 pounds. Specially made hose of $^3/_4$- or 1-inch size must be used to withstand some of the higher pressures. Several companies manufacture large spraying machines. Your local pesticide dealer has information on prices and specifications of spray equipment. The manufacturers whose names are listed in some of the trade magazines (e.g., *Arbor Age; American Nurseryman; Weeds, Tree Care Industry; Trees, and Turf;* and *Grounds Maintenance*) also will provide information. Some of the magazines have special product directory issues. National and state tree industry trade shows are good places to see demonstrations of some of the equipment.

The stream of spray material and the height it will reach are governed by the size and construction of the nozzle, the amount of material discharged, and the pressure behind it. There are five principal types of spray nozzles: the disk nozzle, which throws a hollow cone-shaped spray; the Vermorel nozzle, which throws a misty, more or less cone-shaped spray; the bordeaux nozzle, which throws a flat, fan-shaped driving spray; the spray gun, which throws a cone-type to solid stream spray; and the park nozzle, which throws a solid stream spray. Each type has its own particular advantages and uses.

Fig. 13-4. Stream of spray suspension delivered by a hydraulic sprayer.

The pressure of the spray material at the nozzle opening, as well as the type of nozzle, governs the success of a spraying operation. Other factors (e.g., the size of the nozzle opening) remaining the same, any increase in pressure breaks up the spray material into finer particles, forces them to a greater distance from the nozzle, and results in the application of a greater volume of the insecticide or the fungicide.

To spray a tree approximately 20 feet high, it is necessary to have a nozzle with a 1/8-inch opening in the disk, which will deliver 6 gallons of spray a minute at 180 pounds pressure. A tree 90 feet high can be adequately sprayed with a nozzle having a $5/16$-inch opening, releasing 44 gallons per minute at a pressure of 300 to 400 pounds. Because of friction as the spray moves through the hose, the pressure at the nozzle is much lower than that developed by the pump at the spray tank. For each 100-foot increase in length of 1-inch hose from the tank to the nozzle in the example cited above, the pressure in the tank must be increased by 33 pounds to provide sufficient pressure at the nozzle to enable coverage of the 90-foot tree.

Sprays are applied to tall trees in a so-called solid stream; that is, the material leaves the nozzle much as water issues from a fire hose. This stream, forced out under great pressure, soon reaches a height where it breaks into a mist, which drifts onto the leaves and stems.

In contrast to solid-stream spraying, hydraulic mist spraying is used on trees less than 20 feet high. The spray breaks into a fine mist as soon as it leaves the nozzle, giving rapid and complete coverage. Most handheld spray guns have nozzles that can be adjusted to accomplish both types of spraying. Hydraulic applications are intended to apply pesticide suspensions just up to the point of runoff.

Hydraulic sprayers must be handled properly to maintain peak efficiency. To achieve this objective, lubricate all moving parts daily; pour chemical mixtures through a good strainer before they enter the tank; rinse out the tank, hose, and nozzles with clean water at the end of the day; and carry, rather than drag, the spray hose over rough terrains.

Mist Blowers. Mist blowers, or air sprayers, use blasts of air to propel the droplets of pesticide, in contrast to hydraulic sprayers, which use water as the vehicle for the pesticide. With mist blowers, it is possible to cover more trees in shorter time and at far less cost. The use of highly concentrated materials not only speeds up the refilling time but sharply reduces runoff or drip waste, a major component of lost pesticide with hydraulic applications.

Some mist sprayers can disperse spray particles for relatively long distances, often up to 100 feet, against a mild wind. One should bear in mind, however, that if the pesticide vaporizes before it reaches its target, the dry chemical will not adhere to the leaf surface. In other words, to deposit a coating of pesticide properly, the liquid-pesticide mixture must reach the leaves and twigs before the liquid phase evaporates.

There are several sizes and types of mist blowers on the market. Some are more suitable for use on large shade trees (Fig. 13-5), and others are better adapted for watershed, park, and forest plantings. Small motor-powered backpack mist blowers (Fig. 13-6) are available for low-growing plant materials. One disadvantage of mist blower use is ease of pesticide drift through and past the target tree. Even the slightest breeze can carry fine droplets to unintended targets.

Fig. 13-5. Mist blower producing fine particles that drift into the tree. These workers may be applying biological control suspensions and perhaps do not perceive a need to wear protective equipment.

Helicopters are now being used to apply insecticides and fungicides to large uninhabited areas such as woodlands and golf courses. Biological control materials and insecticides have been applied from helicopters to control gypsy moth caterpillars in woodland areas.

Spraying Practices. Although correct spraying technique can be perfected only through practice, the following suggestions will help in its acquisition.

The intensity and direction of the wind must be considered before spraying operations begin. Poor coverage and considerable waste of material are bound to occur on windy days; therefore, pesticide application should never be attempted on windy days. Even in slight breezes, sprays will drift. Maximum coverage is obtained when the spray is directed toward the leeward sides of the leaves and twigs. As the spray is caught by the breeze, it is blown back in the opposite direction, thus ensuring complete coverage. The wind factor is of even greater importance when using mist blowers; that is to say, there should be little to no wind.

Another factor to be considered is the proximity of the trees to buildings and other structures. The operator should determine beforehand how to get the best coverage of the tree while getting as little spray as possible on these structures. Where mist blowers are used near parked automobiles, the operator must take into account the possibility of coating the automobiles with the spray. The problem can be solved by enforcing regulations that prevent parking of automobiles under trees that are to be sprayed.

There is no ironclad rule as to the amount of spray that should be applied to any particular tree. The object of spraying is to cover every leaf, twig, and branch that is infested or that might later become infested or infected by an insect or

Fig. 13-6. Backpack mist blower. Applicator is not properly protected for pesticide application.

parasite. Once complete coverage is obtained, the operator should move on to another part of the tree or to another tree. Usually it is not necessary to continue the spray once the material begins to drip from the leaves.

Thorough coverage is essential when using protectant pesticides. For protection against microbes that cause infectious diseases, both leaf surfaces usually need to be covered. Systemic pesticides are transported throughout the plant, hence complete coverage is less important.

Timing of pesticide applications is also important. To protect against infection by fungi or bacteria, the pesticide must be present on the plant before infection occurs. Once the pathogen has entered the plant, it is usually too late for the spray treatment to be of any value. Certain diseases such as scab of flowering crabapple can be controlled a few days after infection if the right systemic fungicide is used, but examples such as this are rare.

Pesticide Labels

The pesticide label is a legal document, and all pesticide users are expected to understand the information printed on it. The label states the brand name of the product, the type of formulation, the common chemical name with its percentage as active ingredient indicated, and the weight or volume of the material as packaged. The signal words and symbols indicating relative toxicity; description of potential human, environmental, and physical and chemical hazards; antidotes to counteract accidental exposure; and a statement of whether it is a restricted or general use material are also included. The Environmental Protection Agency registration number and manufacturer's name and address are stated. Directions for use, label violation penalties, a reentry statement, if needed, and any possible restrictions for use of the product are presented. Finally, storage and disposal directions are given.

Pesticide Toxicity. Pesticide labels indicate the pesticide's degree of toxicity to people. The important signal words to look for are *Danger/Poison*, with skull and crossbones, *Warning*, and *Caution* (Table 13-1). The most hazardous pesticides (*Danger/Poison*) generally are available to certified and licensed commercial applicators. All pesticide labels state that the pesticide must be kept out of reach of children. Ill effects or injury resulting from pesticide exposure depends as much on how the pesticide is used as on the toxicity of the chemical. A high-hazard situation with even a low-toxicity pesticide can occur when the material is used constantly in a careless manner at a high concentration. A low-hazard situation with a pesticide can exist when a trained and knowledgeable applicator wearing protective clothing (Fig. 13-7) uses a dilute, infrequently needed material.

Fig. 13-7. Operator wearing appropriate safety clothing for applying hazardous pesticides, including hat, gloves, long sleeves and pants, and respirator.

Table 13-1. Relationship of pesticide label signal words to pesticide toxicity

Signal Words	Toxicity	Approximate Oral Dose of Pesticide Concentrate Needed to Kill the Average Person*
Danger/Poison	High	A taste to a teaspoonful
Warning	Moderate	A teaspoonful to an ounce (two tablespoons)
Caution	Low	More than an ounce (two tablespoons)

*Pesticides may also be toxic when inhaled or in contact with the skin. These kinds of exposures are more likely to occur than ingestion. A pesticide labeled Danger/Poison may be lethal if only small amounts are inhaled or contacted.

Some additional safety precautions are:

• Choose the correct pesticide.
• Always read the label before using pesticides. Respect warnings and precautions listed.
• Keep pesticides out of reach of children, pets, and irresponsible adults.
• Transport pesticides safely.
• Store pesticides in their original containers.
• Do not store pesticides near food.
• Use appropriate application equipment.
• Wear protective clothing.
• Do not smoke or eat while spraying.
• Avoid inhaling sprays.
• Do not spill pesticides on skin or clothing. Should some be accidentally spilled, remove contaminated clothing quickly and wash the skin thoroughly.
• Wash hands and face thoroughly with soap and water and change to clean clothing after spraying pesticides.
• Dispose of empty containers so they pose no hazard to any form of life.
• If symptoms of illness occur, call a physician or local poison control center.

Preparing Small Quantities of Pesticides. Pesticide labels frequently provide mixing instructions geared to the commercial user. Individuals wishing to use small quantities are sometimes put off by having to calculate teaspoons per gallon from pounds or pints per 100 gallons listed on the label. Table 13-2 can be used to make these calculations.

Specific Pesticides for Insect, Mite, and Disease Control

Most of the materials mentioned in the following lists are generally available for tree pest control. Some pesticides can be used only by those people who have obtained certification for application of restricted use materials from their state regulatory agencies. In addition, people who are hired to apply pesticides need to obtain a license from their state regulatory agency.

The pesticides that are discussed here are listed both by brand name and by

Table 13-2. Amount of wettable powder and liquid pesticide to use for various quantities of water*

Gallons of Spray	Quantity of Pesticide	
	Wettable Powder	Liquid
100	1 lb	1 pt
50	8 oz	1 cup
5	3 tablespoons	5 teaspoons
1	2 teaspoons	1 teaspoon

*The density of wettable powders varies but is about half that of liquids.
†If the pesticide label calls for more pints or pounds per 100 gallons, simply multiply the number of cups, tablespoons, or teaspoons proportionally.

common chemical name. Different formulations of the same chemical may have different toxicities. Pesticide use patterns do change; by all means read the label before you buy, making sure that the material you choose is currently legal and safe for the intended use. The miticides listed primarily control mites; however, many of the insecticides control both insects and mites.

Some materials may be safely combined with others to control fungal or bacterial diseases and insects in a single operation. Because improper mixing of these materials may cause injury to the sprayed plants or reduce the efficiency of one or several of the materials in the mixture, one should seek advice from state entomologists and plant pathologists as to the best and safest combinations. A spray compatibility chart is also available from the Meister Publishing Company, Willoughby, OH 44094, for a small charge.

Fungicides and Bactericides

Many fungicides are available for controlling tree diseases, but there are few chemicals available for control of diseases caused by bacteria. Following are some of the materials approved for use in the United States as of this writing. Not all fungicides listed can be used on all trees in the landscape; some are intended only for nursery, fruit tree, or Christmas tree plantation use. It is the responsibility of the user to be sure that the plant to which a chemical is applied is actually listed on the label of the product to be used. Common chemical names are listed first, followed by examples of corresponding brand names. Brand names followed by an asterisk (*) are mixtures of more than one fungicide and appear under more than one category.

Surface Protectants. Used for treatment of plant, plant parts, or soil, these fungicides are used to prevent the pathogen from entering the plant and causing disease. Chemical control must be in place before the pathogen is poised to enter the plant. Because protectants have a broad spectrum mode of action, they can be combined with systemics to reduce fungicide resistance to the latter type of fungicide.

Inorganics, mainly formulations of copper and sulfur, historically are among the first fungicides developed. Bordeaux mixture, first used in France in the 1800s is still useful today as a broad spectrum fungicide. Inorganic fungicides may be made from minerals mined from the earth or may be manufactured. Inorganics containing copper are used to control some fungal leaf spots and certain bacterial diseases. Sulfur-based fungicides are useful for powdery mildew and apple scab diseases. Lime sulfur controls powdery mildew and aids in management of some insect and mite pests. In general, inorganics must be used in fairly large quantities. They have the potential to be phytotoxic to new growth and, for most diseases, are less effective than the more modern synthetic fungicides. They have the advantage of being inexpensive and persistent.

> copper hydroxide—Champ, Champion, Kocide 101
> copper sulfate + hydrated lime—Bordeaux mixture
> fixed copper—Basic Copper Sulfate, Copper Oxychloride Sulfate, and variants
> copper sulfate with organic component —Phyton 27
> sulfur—Microthiol Special, wettable sulfur, dusting sulfur
> calcium polysulfide—Lime sulfur, Dormant Disease Control

Dithiocarbamates and similar compounds are good examples of effective surface protectants. They are used for protection against a wide range of fungal leaf spots, scab, blights, gray mold, anthracnose diseases, rust diseases, and downy mildews, but not powdery mildews.

> mancozeb—Dithane, Fore, Manex II, Manzate, Penncozeb, Protect T/O, Zyban*, Duosan*, Pace*
> metiram—Polyram
> ferbam—Carbamate, Ferbam
> ziram—Ziram
> thiram—Spotrete, Thiram

Carboximides are primarily surface protectants, but some have locally systemic activity, penetrating the tissues to which they are applied. They are used for protection against a wide range of fungal leaf spots, scab, blights, gray mold, and downy mildews, but not powdery mildews or rusts. Some, such as fludioxonil, are used to manage root or stem rots caused by *Rhizoctonia.*

> captan—Captan, Captec
> iprodione—Chipco 26019, Rovral, Benefit*
> vinclozolin—Curalan, Ornalin, Ronilan, Touche, Vorlan
> terraneb—Cloroneb
> fludioxonil—Medallion

Miscellaneous protectant fungicides do not fit a particular chemical category. Some, such as chlorothalonil protect foliage against fungal leaf spots, blights, gray mold, and downy mildews, but not powdery mildews and rusts. Others, such as PCNB, applied to the soil, are most effective for control of root rots or stem rots caused by *Rhizoctonia* or *Sclerotium.* Etridiazole protects specifically against water mold fungi such as *Pythium* and *Phytophthora.* Neem oil, extracted from a tropical tree, is used for fungal leaf spot diseases.

dodine—Syllit
clorothalonil—Bravo, Daconil 2787, Ortho Daconil, Thalonil, Ultrex, Weather
  Stik, Broadway*, ConSyst*, Ridomil/Bravo*, Two Some*
Pentachloronitrobenzene (PCNB)—Terrachlor, Turfcide
etridiazole—Terrazole, Truban, Banrot*
Neem oil—Triact

Antibiotics and other bactericides are used for bacterial disease control.
Streptomycin may be used for fire blight control in nurseries. Gallex and Galltrol
are used for management of crown gall disease.

streptomycin—Agri-Strep, Agrimycin, Phytomycin
oxytetracyclene—Mycoshield, Mycoject, Terramycin
2,4 Xylenol, meta cresol, and penetrants—Gallex
Agrobacterium radiobacter strain 84—Galltrol-A

Systemics (Protectants and Eradicants). Systemic fungicides kill the
pathogen on the plant surface, as it enters the plant, or inside the plant, sometimes even after infection is established (eradicant activity). Some indirectly kill
the pathogen by inducing plant resistance mechanisms. Because most systemics
have a narrow mode of action, they are vulnerable to fungicide resistance and are
therefore often combined with protectants.
  Benzimidazoles such as benomyl and thiophanate-methyl are systemic fungicides, but are often also used as surface protectants. They are used for protection
against, and limited eradication of, several fungal leaf spots, scab, blights,
anthracnose diseases, gray mold, some twig blights, and powdery mildews, but
not downy mildews and rusts. Some, such as Arbotect or Elmsafe, are injected
systemically into trees for control of Dutch elm disease.

benomyl—Benomyl
thiophanate-methyl—Cleary's 3336, Domain, Fungo-Flo, Topsin-M, Banrot*,
  Zyban*, Duosan*, ConSyst*, Benefit*
thiabendazole, MBC phosphate, and related compounds —Arbotect, Correx,
  Elmsafe, Fungisol, Lignasan, Mertect, Thiabendazole

Acylalanine fungicides are systemic, and when applied to the plant are active
against water mold fungi such as *Pythium* and *Phytophthora*.

mefenoxam—Subdue Maxx
metalaxyl—Apron, Pythium Control, Ridomil, Subdue, Ridomil/Bravo*, Pace*

Phenyamide fungicides specifically control rust diseases.

oxycarboxin—Plantvax

Sterol biosynthesis inhibitor (S.B.I) fungicides, some of which are also
referred to as demethylation inhibitor (D.M.I.) fungicides, are systemic in plants
following application and for some diseases have curative effects after infection
has begun. They are effective at low doses against a few fungal leaf spots, scab and
black spot, some anthracnose diseases, rust diseases, and powdery mildews.
Some, such as Alamo, are used via tree injection for control of oak wilt and Dutch

elm disease. S.B.I. fungicides include general chemical groups such as azoles (triazoles), imidazoles, pyrimidines, and piperazines.

propiconazol—Banner, Banner Maxx, Alamo, Orbit
tebuconazol—Elite
fenbuconazol—Indar
myclobutanil—Eagle, Immunox, Nova, Systhane
tiradimefon—Bayleton, Strike
triflumizole—Procure, Terraguard
fenarimol—Rubigan AS, Two Some*, Broadway*
triforine—Triforine, Funginex
cyproconazole—Sentinel

Fungicides that induce plant disease resistance are the newest group of fungicides and are just becoming available for plant disease control. Although these compounds are not actually fungicidal or bactericidal, they can, when applied, induce host plant resistance to some bacterial and fungal diseases. Additional plant disease resistance inducers are under development.

anilinopyrimidine—Actiguard

Strobilurin fungicides are another new group of fungicides derived from *the Strobilurus tenacellus* mushroom. They may be capable of controlling several different fungal diseases. The various strobilurin fungicides differ slightly in their properties and disease control spectrum, so it is important to check the labels in advance of use for the disease to be controlled. Trifloxystrobin is reported to be mesosystemic, meaning that it is absorbed into the leaf cuticle and redistributed to nearby cells and tissues by vapor action.

azoxystrobin, kresoxim-methyl, trifloxystrobin—Heritage, Compass, Abound, Sovran, Quadris

Miscellaneous systemics include two materials, Aliette and Banol, which are specific for control of root rot diseases caused by water molds such as *Phytophthora* and *Pythium*. Pipron is specific for powdery mildews. Vanguard is both a protectant and a systemic fungicide effective against scab.

cyprodinil—Vanguard
fosetyl-Al—Aliette, Prodigy
piperalin—Pipron
propamocarb hydrochloride—Banol

Fungicides for tree diseases. If it becomes necessary to employ chemical management for tree diseases, the following list indicates chemical groups best used for specific types of diseases. Check the labels of specific individual plant disease control chemicals to be sure that the tree and disease to be controlled are listed, and follow all label directions for safe and effective use.

Fungal leaf spots—mancozeb, thiophanate-methyl, benomyl, SBI fungicide, chlorothalonil, or fixed copper
Rosaceous plant scabs—SBI fungicide, thiophanate-methyl, benomyl, mancozeb, captan, or sulfur

Tip blights—thiophanate-methyl or benomyl

Anthracnose diseases—SBI fungicide, thiophanate-methyl, benomyl, or mancozeb

Downy mildews—metalaxyl, mancozeb, or fixed copper

Rust diseases—SBI fungicide, or mancozeb

Gray mold—iprodione, vinclozolin, mancozeb, or chlorothalonil

Powdery mildews—SBI fungicide, thiophanate-methyl, benomyl, or sulfur

Water mold root rots—metalaxyl, etridiazole, or fosatyl-Al

Rhizoctonia root rot—PCNB, thiophanate-methyl, or benomyl

Vascular wilts (some)—MBC phosphate, or SBI fungicide

Bacterial blights—fixed copper, streptomycin, or oxytetracyclene

Crown gall—Galltrol or Gallex

Insecticides and Miticides

The following list of chemical insecticides and miticides can be used as a guide for
finding products for controlling tree pests. The materials listed have been
approved for use in the United States as of this writing. Not all insecticides listed
can be used on all trees in the landscape; some may be intended only for nursery,
fruit tree, or Christmas tree plantation use. It is the responsibility of the user to be
sure that the plant to which a chemical is applied and the intended pest are actually listed on the label of the product to be used. Common chemical names are listed first, followed by examples of corresponding brand names. The use of insecticides followed by an asterisk (*) are restricted to commercial use or to licensed
applicators. Some of the restricted chemicals, formulated at lower concentrations,
are less toxic, and may be available for general use. Other restricted insecticides
may not be particularly toxic but they are formulated in specialized application
devices such as injection capsules that are only available to commercial applicators.

Natural Product-Based Insecticides and Miticides. Sometimes, naturally
occurring materials are our best sources of materials toxic to insects and mites.
The number of chemicals available that are derived from plant or microbial
sources has been increasing in recent years. In some cases, chemical manufacturers have been able to synthesize some of these natural materials creating more
effective chemical variants. Also included in this section are inorganics, soaps,
and oils.

Botanical insecticides are derived from natural plant products. Azadirachtin,
an insect-feeding deterrent and growth regulator, comes from the neem tree
grown in India. Pyrethrins, extracted from a species of chrysanthemum, are irri-
tating to insect nervous systems and are sometimes mixed with a synergist and
rotenone to make them more effective. Rotenone, extracted from the roots of two
different plants grown in South American or Asia, is toxic to leaf-feeding caterpillars and beetles. Although neem is not very toxic to people or animals, rotenone is
highly toxic to fish, and pyrethrum can cause severe allergic reactions. Thus, just
because these insecticides are "natural," they are not necessarily more safe.

azadirachtin (azatin, neem)—Azatin, Bioneem, Margosan-O, Neemisis, Turplex

pyrethrins—Pyrellin, Pyrenone, Pyrethrin

rotenone—Prentox, Rotenone

Microbial insecticides are toxic compounds derived from bacteria or fungi. Bacterial spores plus crystalline delta-endotoxin are found in formulations of cultures of a microbe, *Bacillus thuringiensis*. When young caterpillars or beetle larvae feeding on tree leaves ingest the bacteria and toxin, they stop feeding and soon die. Different strains of the bacteria may be specific for different pest insect species. The fungus *Streptomyces* produces avermectin toxins, which are formulated into pesticides that can be toxic to humans as well as to spider mites. Naturally occurring soilborne microbes are capable of synthesizing chitin-degrading enzymes that can damage soilborne insect and nematode chitin exoskeletons. Amending soil with chitin is said to enhance these enzyme activities, thus affecting the pests. *Steinernema*, a nematode, is parasitic to certain weevils.

Bacillus thuringiensis (Bt) var. *kurstaki*—Bactospeine, Biobit, Caterpillar Attack, Dipel, Foray, Javelin, Larvo-BT, MVP II, Thuricide, Victory
Bacillus thuringiensis (Bt) var. *tenebrionis*—M-One, Trident II
Bacillus thuringiensis (Bt) var. *san diego*—M-Trak
abamectin—Avid*, Inject-a-Cide AV, Vivid
chitin—Clandosan
spinosad—Conserve SC
Steinernema carpocapsae—Ecomask, Savior, Vector TL

Horticultural oils are highly refined petroleum or plant oils often used during the dormant season to suffocate overwintering insects and mites. Some lighter weight oils can also be used in summer to control insects and mites. They are sometimes combined with other insecticides at reduced rates to enhance the toxic effect of both. It is essential to read and follow label directions when using horticultural oils, because when they are used improperly they can cause phytotoxicity. Citrus oils can also be insecticidal.

petroleum oils—Dormant Oil, Summer Oil, Superior Oil
citrus oil

Pyrethroid insecticides are synthetic versions of the botanical pyrethrin insecticides.

bifenthrin—Talstar*
cyfluthrin—Decathlon*, Tempo
esfenvalerate—Asana XL*
fenpropathrin—Tame
fluvalinate—Mavrik Aqua Flow
lambda-cyhalothrin—Scimitar*
permethrin—Ambush*, Astro, Pounce*
resmethrin—Resmethrin

Soaps can be formulated to have insecticidal properties by disrupting respiration and cell membranes. More toxic forms may be formulated with citrus oil or pyrethrins. Insecticidal soaps applied directly to the arthropods control soft bodied insects such as aphids, beetle larvae, caterpillars, mealybugs, soft scales, thrips, and whiteflies, and also spider mites.

pesticidal soap—Aphid-Mite Attack, Insecticidal Soap, M-Pede, Safer's Insecticidal Soap

Inorganic insecticides may consist of minerals mined from the earth, or they may be manufactured. Kryocide, consisting of sharp mineral particles, punctures the gut cells of leaf-feeding beetles, caterpillars, and sawflies. Sulfur is toxic to many insects and mites.

cryolite—Kryocide
sulfur—Lime-Sulfur

Synthetic Insecticides. Most of the synthetic insecticides are organic compounds synthesized from petroleum products. Some of the compounds listed here often have specific modes of action and target only a few pests, but most are fairly broad spectrum.

Organophosphates make up the largest group of synthetic insecticides. They are mostly neurotoxins and last only a short time in the environment. Some can be very toxic to humans.

acephate—Dendrex, Isotox IV, Orthene, Orthenex, Pinpoint
azinphos-methyl—Guthion*
chlorpyrifos—Dursban*, Pageant*
diazinon—Diazinon*, Knox Out, Spectracide, D-Z-N
dicrotophos—Bidrin, Inject-a-cide B*
dimethoate—Cygon, Dimethoate
disulfoton—Di-Syston*
fenetrothion—Pestroy
isophenphos—Discus*, Oftanol*
malathion—Cythion, Malathion
methyl parathion—Methyl Parathion*
phosmet—Imidan
naled—Dibrom
oxydemeton-methyl—Harpoon*, Inject-a-cide*, Metasystox-R2*
parathion—Parathion*
trichlorfon—Dylox*, Proxol*

Carbamate insecticides are mostly neurotoxins.

bendiocarb—Dycarb, Ficam*, Turcam*
carbaryl—Carbaryl, Sevimol, Sevin
carbofuran—Furadan
fenoxycarb—Precision
methiocarb—Grandslam*, Mesurol
oxamyl—Oxamyl*, Vydate*

Chlorinated hydrocarbon insecticides include banned materials such as DDT that lasted a long time in the environment and found their way into the food chain. Less persistent members of that group, used for tree insect management are listed here.

dicofol—Dicofol, Kelthane*
dienochlor—Pentac*
endosulfan—Endocide*, Phaser*, Thiodan*
lindane—Lindane*, Lindand Borer Spray
methoxychlor—Marlate, Methoxychlor

Miscellaneous synthetic chemicals such as propargite, a sulfite ester, and oxythioquinox, a dithiocarbamate, have specific uses for spider mite management. Metacetaldehyde compounds are often formulated as baits to manage slugs on herbaceous and garden plants, and could have utility in certain tree nursery circumstances. Insect growth regulator (IGR) chemicals like diflubenzuron mimic insect hormones and modify the metabolism, growth, and developmental cycle of the insect pest. Imidacloprid is a systemic chloronicotinyl insecticide that can be applied to the foliage or to the soil for plant uptake. Soil application must be done in advance of the pest appearance due to delays in root uptake and systemic distribution of the insecticide in the plant.

diflubenzuron—Dimilin*
hexythiazox—Hexygon
imidacloprid—Imicide, Marathon, Merit, Pointer
metaldehyde—Bug-Geta, Deadline, Slug-Geta
oxythioquinox—Joust, Morestan
propargite—Ornamite
cryomazine—Citation

Insecticides and miticides for tree pests. If it becomes necessary to use chemicals to manage arthropod (insect and mite) pests of trees, the following list indicates the chemical groups best suited for the control of specific types of pests. Check the labels of specific individual pesticides to be sure that the tree and pest to be controlled are listed, and follow all label directions for safe and effective use.

Caterpillars (foliage feeding)—azadirachtin, *Bacillus thuringiensis*, carbamates, imidacloprid, organophosphates, pyrethroids
Sawflies—azadirachtin, carbamates, imidacloprid, organophosphates, pyrethroids
Foliage feeding beetles—carbamates, imidacloprid, organophosphates, pyrethrin, pyrethroids
Bagworms—azadirachtin, carbamates, organophosphates, pyrethroids
Plant bugs, leaf bugs, and lace bugs—carbamates, imidacloprid, organophosphates, petroleum oils, pyrethroids, pesticidal soaps
Aphids and adelgids—carbamates, imidacloprid, organophosphates, petroleum oils, pyrethrin, pyrethroids, pesticidal soaps
Scales and mealybugs—carbamates, imidacloprid, organophosphates, petroleum oils, pyrethrin, pyrethroids, rotenone, pesticidal soaps
Spider mites—carbamates, chlorinated hydrocarbons, hexythiazox, organophosphates, oxythioquinox, petroleum oils, propargite, pyrethroids, pesticidal soaps
Eriophyid mites—carbamates, chlorinated hydrocarbons, organophosphates, oxythioquinox

Thrips—carbamates, imidacloprid, organophosphates, pyrethroids
Leafhoppers—carbamates, imidacloprid, organophosphates, pyrethrin, pyreth-
   roids, rotenone, pesticidal soaps
Whiteflies—carbamates, imidacloprid, organophosphates, petroleum oils,
   pyrethrin, pyrethroids, rotenone, pesticidal soaps
Weevils—carbamates, organophosphates, pyrethroids
Leaf miners—carbamates, imidacloprid, organophosphates, pyrethroids
Bark beetles—carbamates, chlorinated hydrocarbons, organophosphates
Borers—carbamates, chlorinated hydrocarbons, organophosphates
Snails and slugs—carbamates, metaldehyde

Nematicides

Many nematicides are also insecticides and are found in the previous list. Some
nematicides, used as soil fumigants, are also general biocides and kill soilborne
insects, fungi, and weed seeds. These soil-applied biocides are toxic to trees and
can only be used prior to planting.

   Soil-applied insecticides/nematicides. These materials normally kill
nematodes by contact action. Although some may be systemic in the plant,
nematicidal activity is usually limited to the roots and the soil. These pesticides
can be used in soils where plants already exist, as in a nursery or landscape.
Chemical names are followed by trade names.

   ethoprop—Mocap
   fenamiphos—Nemacure
   oxamyl—Vydate

   Soil fumigants. Soil fumigants are applied to nursery or plant bed soil where
trees are to be planted in the future. The fumigants would kill the roots of existing trees if they were used in a landscape or park. The soil fumigant Vapam is used
to kill grafted roots between adjacent trees to prevent spread of Dutch elm disease. Methyl bromide and chloropicrin are often applied as mixtures. Most fumigants are biocides capable of killing fungi, nematodes, insects, and weed seeds in
the soil and are highly toxic to the user; thus their uses are therefore restricted primarily to nursery operations by professional applicators. Methyl bromide is a
potent stratospheric ozone destroyer, and its use is being gradually phased out
worldwide. The current target date for elimination of methyl bromide in the
United States is the year 2005.

   chlorinated hydrocarbons (dichloropropene) Telone, Telone C-17, Telone II
   chloropicrin—Brom-O-Gas, Chlor-O-Pic, Larvacide, MC-2, MC-33, Telone C-
   17, Terr-O-Gas
   metam-sodium—Busan, Vapam
   methyl bromide—Brom-O-Gas, MC-2, MC-33, Meth-O-Gas, Terr-O-Gas

   Biological controls. Certain soilborne fungi are capable of suppressing plant
pathogenic nematodes. A special strain of the fungus *Myrothecium* has been formulated into a nematicide that can be incorporated or surface broadcast before,
during, or after tree planting. Amending soils with chitin can create an environment unfavorable for plant parasitic nematodes.

Myrothecium verrucaria—DiTera
chitin—Clandosan

Spreading and Sticking Agents

Any material that helps spread spray in an even, unbroken film over the surface of leaves, bark, or insects' bodies is known as a spreader; this action is usually accomplished by lowering the interfacial tension of the spray. Materials that help the spray adhere firmly to the leaves are known as stickers.

Formerly, soaps, household detergents, and some oils were recommended as spreaders. We now know that some, particularly soaps, are alkaline and may completely detoxify certain insecticides. Others, such as oils, may not be compatible with fungicides containing sulfur. Where a spreader-sticker is required, it is best to depend on specially prepared products.

Commercial spreaders and stickers widely available include du Pont Spreader-Sticker (SS3), Triton B1956, Filmfast, Nu-Film, Ortho Dry Spreader, and Spray Stay.

Selected Bibliography

Anonymous. 1975. Apply pesticides correctly: A guide for commercial applicators. USDA and USEPA., Washington, D.C. 45 pp.

Anonymous. 1998. Turf and ornamental reference for plant protection products. C&P Press, New York. 792 pp.

Dreistadt, S. H., J. K. Clark, and M. L. Flint. 1994. Pests of landscape trees and shrubs: An integrated pest management guide. Publication 3359, University of California, Division of Agriculture and Natural Resources. Oakland, Calif. 327 pp.

Hartman, J. R., J. L. Gerstle, M. Timmons, and H. Raney. 1986. Urban landscape IPM in Kentucky: A case history. J. Environ. Hortic. 4:120–24.

Jones, R. K., and R. C. Lambe (eds.). 1982. Diseases of woody ornamental plants and their control in nurseries. North Carolina Agriculture Extension Service, Raleigh. 130 pp.

Lloyd, J. 1997. Plant health care for woody ornamentals. International Society of Arboriculture. Champaign, Ill. 223 pp.

Meister, R. T. (ed.). 1997. Farm chemicals handbook. Meister Publishing Co. Willoughby, Ohio.

McGrath, H., J. Feldmesser, and L. Young (eds.). 1986. Guidelines for the control of plant diseases and nematodes. U.S.D.A. Agriculture Handbook Number 656, Washington, D.C. 274 pp.

Potter, D. A. 1986. Urban landscape pest management. In: Advances in urban pest management, G. W. Bennett and J. M. Owens (eds.). Van Nostrand Reinhold, New York, pp. 219–51.

Quist, J. A. 1980. Urban insect pest management for deciduous trees, shrubs, and fruit. Pioneer Science Publications, Greeley, Colo. 169 pp.

Raup, M. J., and R. M. Noland. 1984. Implementing landscape plant management programs in institutional and residential settings. J. Arboric. 10: 161–69.

Thomson, W. T. 1977. Pesticide guide: Tree turf and ornamental. Thompson, Fresno, Calif. 134 pp.

III

Abnormalities of Specific Trees

Abnormalities of Specific Trees

This section deals primarily with the diseases and arthropod pests of trees. The trees are listed alphabetically by common name. The cultural requirements of some of the trees are discussed briefly before the common diseases and pests and their control are described. For diseases and insects, the scientific name of the insect or disease causal agent is given after the common name of the insect or disease.

Control of many pests and problems of trees depends on maintenance of good growing conditions so that the tree has energy reserves to respond to and ward off pest problems. Failure to recognize the importance of integrating tree maintenance practices into pest control operations can lead to unsatisfactory pest control.

The chemical control measures suggested here should be used as a guide. Since pesticide registrations are constantly changing, it is important for the user to read the pesticide label before purchasing and using the product. In addition, most Cooperative Extension offices have information relating to control of landscape tree problems, with control measures that are suited to local conditions. For further information about cultural, biological, and chemical control and application methods, refer to Chapter 13. More detailed information about important selected abiotic diseases, insect pests, and parasitic diseases is found in Chapters 10, 11, and 12. Approaches to diagnosis of problems are presented in Chapter 9.

ACACIA (*Acacia*)

Acacias typically produce bright yellow flowers and are grown outdoors in the Southwest and in greenhouses elsewhere. This legume is a relatively fast grower.

Diseases

The fungal diseases of acacia are not important. Anthracnose, caused by *Glomerella cingulata*, occurs on acacia. Twig cankers caused by *Nectria ditissima* and *Fusarium lateritium (Gibberella bac-* cata) have been reported from California. Leaf spotting by the fungus *Physalospora fusca*, a species of *Cercospora*, and the alga *Cephaleuros virescens* occasionally develops in the South. In California, powdery mildew caused by the fungus *Erysiphe polygoni* is common. Root and trunk rots caused by the fungi *Phymatotrichum omnivorum, Armillaria mellea, Clitocybe tabescens,* and *Ganoderma applanatum* have been reported from the South and the West.

Insects

COTTONY-CUSHION SCALE. *Icerya purchasi.* This ribbed-scale bark louse, which has been a serious pest of acacia and many other trees in California and throughout the southern United States, has been troublesome in greenhouses in the eastern states for a number of years. The mature insect, which may be up to $1/4$ inch long, has a ribbed, cottony covering through which its soft body may be seen. The white, fluted mass that extends from the small, usually inconspicuous, scale is composed largely of eggs covered by a protective waxy material. At its earliest stage the scale is greenish and not contained in a woolly excretion.

Several other species of scale occasionally infest acacia. These are California red, which is round and reddish in color; dictyosperm; greedy, a small gray species; latania, similar to greedy; lesser snow; oleander, a pale yellow kind; and San Jose, which is small and gray with a nipple in its center.

• *Control.* See under scales, Chapter 11.

CATERPILLARS. *Argyrotaenia citrana* and *Sabulodes caberata.* The former, known as the orange-tortrix, a dirty-white, brown-headed caterpillar, webs and rolls the leaves of many trees and shrubs on the West Coast. The latter, known as the omnivorous looper, is a yellow to pale pink or green caterpillar, with yellow, brown, or green stripes on the sides and back, which also feeds on a wide variety of plants in the same region.

• *Control.* See under caterpillars, Chapter 11.

FULLER ROSE BEETLE. See under Camellia.

ACACIA PSYLLID. *Acizzia uncatoides.* This is an important pest of some species of acacia in California. The feeding causes yellowing, and enough honeydew may be produced to make this pest bothersome.

OTHER INSECTS. Acacia is host to a thornbug, *Umbonia crassicarnis*; a leafhopper, *Kunzeana kunzii*; a species of spittle bug, *Clastoptera arizonana*; and giant whitefly (see under Citrus).

AILANTHUS
See Tree-of-Heaven.

ALDER (*Alnus*)

Most species of alder grow best in moist soils. Those grown in ornamental plantings are less subject to diseases than to insect pests. The few diseases that occasionally appear produce little permanent damage. It is a relatively fast growing and short-lived tree.

Diseases

CANKER. A number of fungi cause cankers and dieback of the branches. Among the most common are *Nectria galligena, N. cinnabarina, Botryosphaeria obtusa,* and *B. dothidia.* Trunk cankers and rots may also be caused by *Cerena unicolor, Ganoderma applanatum, Hypoxylon* spp. and other fungi.

• *Control.* Cankered branches should be pruned to sound wood and the trees fertilized and watered to increase their vigor.

LEAF CURL. *Taphrina macrophylla.* The leaves of red alders grow to several times their normal size, are curled and distorted, and turn a decided purple when affected by this disease. Infection is apparent on the leaves as soon as they appear in the spring.

Three other species, *T. amentorium, T. occidentalis,* and *T. robinsoniana,* are known to cause enlargement and distortion of the scales of female catkins. The scales resemble curled, reddish tongues and are soon covered with a white glistening layer of fungal tissue.

• *Control.* Spray the trees with Ferbam in late fall or while they are still dormant in the spring. This will destroy most of the spores, which overwinter in the bud scales and on the twigs and are largely responsible for early infections. Sprays are suggested only for valuable specimens.

POWDERY MILDEW. Several species of powdery mildew fungi attack the female catkins of alder. The most common, *Erysiphe aggregata,* develops as a white powdery coating over the catkin. Two other species, *Microsphaera penicillata* and *Phyllactinia guttata,* occur less frequently.

• *Control.* These fungi rarely cause enough damage to warrant control measures. If sprays are needed for valuable specimens, see under powdery mildew, Chapter 12.

LEAF RUST. *Melampsoridium alni.* Some damage is done to alder leaves by this rust, which breaks out in small yellowish pustules on the leaves during the summer. The leaves later turn dark brown. It is suspected that the alternate host of this rust is a conifer.

• *Control.* The disease is never severe enough to justify control measures.

Insects

WOOLLY ALDER APHID. *Prociphilus tessellatus.* Large woolly masses (Fig. III-1) covering bluish black aphids are found in downward folded leaves. Eggs are deposited and overwinter in bark crevices. Maples are also infested by this species.

• *Control.* See under aphids, Chapter 11.

ALDER LEAF BEETLE. *Crysomela mainensis.*

Both adults and larvae of this beetle feed on alder leaves, sometimes leaving nothing but the midribs and main veins. This pest is found mainly in the northern states.

• *Control.* See under beetles, Chapter 11.

ALDER FLEA BEETLE. *Macrohaltica ambiens.* Dark brown larvae with black heads chew the leaves during July and August. The adult, a greenish blue beetle $1/5$ inch long, deposits orange eggs on the leaves in spring.

• *Control.* Sevin sprays when the larvae begin to feed will control this heavy feeder.

ALDER LACE BUG. *Corythucha pergandei.* This species infests not only alder but occasionally birch, elm, and crabapple.

• *Control.* See under lace bugs, Chapter 11.

ALDER PSYLLID. *Psylla floccosa.* This sucking insect is common on alders in the northeastern

Fig. III-1. Woolly alder aphid on alder.

United States. The nymphal stage produces large amounts of wax. When massed on the stems, the psyllids resemble piles of cotton.

• *Control.* Spray with Carbaryl or Dursban when the nymphs or adults are seen.

ALDER SPITTLEBUG. *Clastoptera obtusa.* The insect produces a white frothy mass on alders, but does little damage. Birch and hickory are also attacked.

FLATHEADED ALDER BORER. *Agrilus burkei.* This borer feeds on weakened trees in western states. Trunk holes made by emerging beetle adults are typically D-shaped. Italian alder is resistant to flatheaded borer.

ALDER BORER. *Saperda oblique.* This borer also attacks alder.

• *Control.* These borers rarely attack vigorous alders (see under borers, Chapter 11).

EUROPEAN ALDER LEAF MINER. *Fenusa dohrnii.* This insect can cause conspicuous blotch mines in alder leaves. See under leaf miners, Chapter 11.

OTHER INSECTS. Alder is a preferred host of gypsy moth (see under Elm) and a host of fall webworm (see under Ash); a leafroller, *Archips negundanus*; elm calligrapha (see under Linden); mottled willow borer (see under Willow); adult bronze birch borer (see under Birch); birch aphid, *Euceraphis betulae*; and an eriophyid mite.

ALMOND, FLOWERING
(*Prunus triloba*).
See under Cherry.

ARBORVITAE
(*Thuja*)

In the wild, the American arborvitae, also called white cedar, inhabits moist, sometimes swampy areas along banks of streams. When under cultivation, it prefers a moist but well-drained soil. The unthrifty condition of many trees in ornamental plantings is often associated with waterlogged soils.

Many horticultural varieties of American arborvitae are now available from nurseries. Most of these are subject to the same troubles that affect the parent tree.

Diseases

LEAF BLIGHT. Although most destructive on the western red cedar of the Northwest, leaf blight occasionally appears on arborvitae. The disease is most prevalent on trees under 4 years old, but it may attack trees of all ages.

• *Symptoms.* From one to four irregularly circular, brown to black cushions appear on the tiny leaves in late spring. The leaves then turn brown, and the affected areas appear as though scorched by fire. Toward fall, the leaves drop, leaving the branches bare.

• *Cause.* The fungus *Didymascella thujina* causes leaf blight. The brown to black cushions on the leaves are the fruiting bodies of this fungus. They discharge spores into the air during rainy periods in summer to initiate new infections.

• *Control.* Small trees or nursery stock can be protected by several applications of bordeaux mixture or other copper sprays in midsummer and early autumn.

TIP BLIGHT. This disease resembles leaf blight but occurs primarily on varieties of *Thuja orientalis*, the golden biota or golden arborvitae being the most susceptible. The leaves near the branch tips turn brown in late spring or early summer. Tiny black bodies, which are visible through a magnifying lens, occur on infected leaves.

• *Cause.* The fungus *Coryneum thujinum* causes tip blight. Several other fungi, including *Cercospora thujina*, *Pestalotiopsis funerea*, and *Phacidium infestans*, are also associated with tip and twig blights.

• *Control.* Spray with a copper fungicide two or three times at weekly intervals starting in late May.

JUNIPER BLIGHT. The fungus *Phomopsis juniperovora*, which is the cause of the blight of red cedar and other species of *Juniperus*, also attacks arborvitae.

• *Control.* See Phomopsis twig blight, under Juniper.

Abiotic Diseases

BROWNING AND SHEDDING OF LEAVES. The older, inner leaves of arborvitae turn brown and drop in fall. When this condition develops within a few days or a week, as it does in some seasons, many persons feel that a destructive disease is involved. Actually it is a natural phenomenon similar to the dropping of leaves of deciduous trees. When the previous growing season has been favorable for growth, or when pests such as red spider have not been abundant, the shedding occurs over a relatively long period through fall and consequently is not so noticeable.

WINTER BROWNING. Rapid changes in temperature in late winter and early spring, rather than drying, are responsible for browning of arborvitae leaves.

Insects and Related Pests

ARBORVITAE APHID. *Cinara thujafilina.* This reddish brown aphid with a white bloom infests roots, stems, and leaves of arborvitae, Italian cypress, and Hinoki cypress.

- *Control.* See under aphids, Chapter 11.

BAGWORM. See under Juniper.

CEDAR TREE BORER. See under Redwood.

LEAF MINER. *Argyresthia thuiella.* The leaf tips turn brown as a result of feeding inside the leaves by the small leaf miner maggot. The adult stage is a tiny gray moth with a wingspread of $^1/_3$ inch. The maggots overwinter in the leaves. This pest also feeds on juniper. *Thuja plicata* is less susceptible to leaf miner than is *T. occidentalis.*

- *Control.* Trim and destroy infested leaves. See under leaf miners, Chapter 11.

ARBORVITAE WEEVIL. *Phyllobius intrusus.* White to pink larvae with brown heads feed on the roots of arborvitae and junipers. The adult, covered with greenish scales, feeds on the upper parts of the plants from May to July.

- *Control.* See under weevils, Chapter 11.

MEALYBUG. *Pseudococcus ryani.* This pest also attacks incense cedar, Norfolk Island pine, and redwood.

- *Control.* See under scales, Chapter 11.

SCALE. *Lecanium fletcheri.* This dark brown, flat, more or less hemispherical scale belongs to the so-called naked or unarmored scales. Newly hatched young are amber in color, then change to pale orange-yellow. The insect overwinters as a partly grown scale on the stems, branches, and

leaves. The pest is relatively unimportant on this host but is rather serious on yews. Crawlers, vulnerable to insecticide sprays are active in June and September.

Several other species of scales also attack arborvitae: European fruit lecanium; Glover, a juniper scale relative; and San Jose.

- *Control.* See under scales, Chapter 11.

SPIDER MITE. See spruce spider mite, under Spruce.

OTHER INSECTS. Arborvitae is also host to a bark-feeding caterpillar, *Dioryctria abietella;* the cypress webber, *Epinotia subvirdis;* a weakened tree-attacking bark beetle, *Phloeosinus canadensis;* and gypsy moth (see under Elm).

ASH (*Fraxinus*)

Ash are popular landscape trees. Several cultivars of native ash are available. Some, such as blue ash, can survive several centuries and attain large size. White, green, and black ash are unusually free from attacks by fungal parasites but are frequently attacked by several insect pests. Other species of ash native to Europe and Asia Minor are also available.

Diseases

ANTHRACNOSE. *Discula sp. (Gloeosporium aridum).* Brown spots occur over large areas of the leaves, especially among the veins (Fig. III-2). In wet seasons, infected leaflets drop prematurely.

- *Control.* See under anthracnose, Chapter 12.

LEAF SPOTS. Several leaf-spotting fungi occur on ash, including *Cercospora fraxinites, C. lumbricoides, C. texensis, Septoria besseyi, S. leucostoma,* and *S. submaculata.* Mycosphaerella leaf spots of ash involve several stages of the causal fungi *Mycosphaerella effigurata (Marssonina fraxini, Asteromella fraxini)* and *Mycosphaerella fraxinicola (Cylindrosporium fraxini, Phyllosticta virdis).*

- *Control.* See under leaf spots, Chapter 12.

RUST. *Puccinia sparganioides.* The leaves of green and red ash are conspicuously distorted and the twigs are swollen by this fungus. Severe cases can cause some twig and branch dieback.

Fig. III-2. Anthracnose of white ash caused by the fungus *Gloeosporium aridum.*

The spores of the fungus, yellow powder in minute cups, appear over the swollen areas in May. The spores produced on ash are incapable of reinfecting ash but can infect marsh and cord grasses (*Spartina* spp.).

• *Control.* The disease is rarely destructive enough to warrant special control measures. See rust diseases, Chapter 12.

POWDERY MILDEW. *Microsphaera fraxini* and *Phyllactinia guttata.* These fungi cause powdery mildew of ash.

• *Control.* See under powdery mildew, Chapter 12.

ASH YELLOWS. This is a phytoplasma-caused disease which is fairly widespread in the northern United States. Bunching, proliferating shoot growth, sometimes called witches' broom, and gradual decline characterize the disease. It is most damaging to white ash and green ash.

• *Control.* See yellows diseases, Chapter 12.

CANKERS. *Cytospora annulata, Diplodia infuscans, Dothiorella fraxinicola, Botryosphaeria dothidea, Nectria cinnabarina, N. coccinea,* and *Sphaeropsis sp.* At least six fungi cause branch and trunk cankers on ash. A species of *Dothiorella* is associated with death of white ash.

• *Control.* Prune out infected branches. Maintain trees in good condition by proper feeding, watering, and spraying for foliar diseases.

HEART ROT. Some aggressive heart rot fungi attack older ash trees, causing decay and hollowing. These include species of *Fomes, Pleurotus, Polyporus, Perennipora,* and *Phellinus.* See Chapter 12.

OLIVE KNOT. *Pseudomonas syringae* pv. *savastanoi.* This bacterial disease causes galls (knots) and cankers (Fig. III-3) on twigs, branches, and limbs of European ash, *F. excelsior.*

WILT. *Verticillium dahliae.* See under Verticillium wilt, Chapter 12.

DIEBACK. In the northeastern United States, white ash has been affected by a branch dieback and tree decline disease. Distinct fungal cankers often appear on affected trees. The primary cause of the disease has not been clearly established although tobacco ringspot and tobacco mosaic viruses have been associated with it. The disease could be transmitted both by grafting and by nematodes. The ash decline may be a complex problem also involving drought, phytoplasmas, and opportunistic canker-causing fungi such as *Cytophoma pruinosa*.

• *Control.* Control measures have not been developed.

Abiotic Diseases

AIR POLLUTION. Ash is relatively sensitive to ozone injury. See Chapter 10.

Borer Insects

LILAC BORER. *Podosesia syringae.* Rough, knotlike swellings on the trunk and limbs and the breaking of small branches at the point of

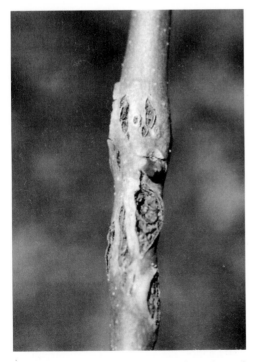

Fig. III-3. Olive knot, a bacterial canker of European ash.

injury are indications of the presence of the lilac borer, a brown-headed, white-bodied $^3/_4$-inch-long larva. The adult female, which appears in late spring, is a moth with clear wings having a spread of $1^1/_2$ inches. Larvae pass the winter underneath the bark.

ASH BORER. *Podosesia aureocincta.* This borer attacks ash and mountain-ash in the Prairie states. It burrows into the lower part of the tree trunk.

CARPENTER WORM. *Prionoxystus robiniae.* Large scars along the trunk, especially in crotches, and irregularly circular galleries about $^1/_2$ inch in diameter, principally in the wood, are produced by this 2- to 3-inch pinkish white caterpillar. The adult moth, with a wingspread of nearly 3 inches, deposits eggs in crevices or rough spots on the bark during June and early July. A period of 1 to 4 years is necessary to complete the life cycle.

RED-HEADED ASH BORER. *Neoclytus acuminatus.* This pest can be a problem of weakened and newly transplanted trees in the Midwest.

OTHER BORERS. Many other borers infest ash. Among the more common are brown wood, California prionus, flatheaded apple tree, Pacific flatheaded, and leopard moth (see under Maple).

• *Control.* See under borers, Chapter 11. Treatments for many of the borers require precise information on timing, which can be supplied by local Cooperative Extension offices.

Other Insects and Related Pests

LILAC LEAF MINER. *Caloptila syringella.* Light yellow, $^1/_4$-inch larvae first mine the leaves of ash, deutzia, privet, and lilac, and then roll and skeletonize the leaves. Small moths emerge from overwintering cocoons in the soil in May to deposit eggs in the undersides of the leaves. A second brood emerges in July.

• *Control.* See under leaf miners, Chapter 11.

FALL WEBWORM. *Hyphantria cunea.* In August and September, webworms chew leaves and cover branches with webs or nests that enclose skeletonized leaves and pale yellow or green caterpillars 1 inch long. The adult moth has white- to brown-spotted wings with a spread of $1^1/_2$ inches.

Fig. III-4. Ash flower gall caused by eriophyid mites.

• *Control.* Sprays applied when webs are well established are ineffective. There are many natural predators that invade the nests of fall webworms and, if left alone, will often eventually control the infestation. See under caterpillars, Chapter 11.

BROWN-HEADED ASH SAWFLY. *Tomostethus multicinctus.* Trees may be completely defoliated in May or early June by yellow sawfly larvae. The adult is a beelike insect that lays eggs in the outer leaf margins. Winter is passed in the pupal stage in the ground.

• *Control.* See under sawflies, Chapter 11.

ASH FLOWER GALL. *Eriophyes fraxinoflora.* This abnormality is caused by small mites that attack the staminate flowers of white ash. The flowers develop abnormally and form very irregular galls up to $1/2$ inch in diameter (Fig. III-4). These galls dry out, forming clusters that are conspicuous on the trees during winter.

• *Control.* See under eriophyid mites, Chapter 11.

EASTERN ASH PLANT BUG. *Tropidosteptes*

amoenus. Infested leaves take on a whitish cast, and dark excrement spots can be found on the leaf undersides.

• *Control.* See under plant bugs, Chapter 11.

OYSTERSHELL SCALE. *Lepidosaphes ulmi.* Masses of brown bodies, shaped like oyster shells and about $1/10$ inch long, covering twigs and branches, are characteristic of this insect. The pests overwinter in the egg stage under the scales. The young crawling stage appears in late May. Oystershell scale has a wide host range. The pit-making pittosporum scale, *Asterolecanium arabidis,* and walnut scale (see under Walnut) also feed on ash.

• *Control.* See under scales, Chapter 11.

ASH WHITEFLY. *Siphoninus phillyrea.* This pest, present in the Southwest on ash citrus and redbud, was first found in California in 1988. Large numbers may build up on leaf undersides and may appear as swirling snowflakes when disturbed.

• *Control.* Natural enemies were introduced to the infested areas soon after the pest first appeared. Because it is under good biological control with introduced natural enemies, ash whitefly is now rarely a pest.

OTHER INSECTS. Ash is subject to attack by the foliar-feeding spiny io moth caterpillar (see io moth, under Sycamore); forest, eastern, and California tent caterpillars (see under Maple, Willow, and Strawberry-Tree); a southern California bagworm, *Oiketus townsendi*; Japanese weevil (see under Holly); a western woolly aphid, *Prociphilus californicus*; cambium miner (see under Holly); buffalo treehopper, *Stictocephala bubalus*; and ash midrib gall midge, *Contarinia candensis*.

ASPEN. See Poplar.

ATHEL TAMARISK (*Tamarix*)

This Mediterranean native is a medium-sized tree, tolerant of salt but not tolerant of cold. It can be used in arid regions, and is relatively tolerant of pests.

AUSTRALIAN-PINE (*Casuarina*)

This tree has been widely planted as an ornamental and as windbreaks in southern Florida. It is fast growing, pinelike, with spreading, drooping branches. *Casuarina* does not tolerate freezing weather.

Diseases

In Florida the Australian-pine is subject to a root rot caused by *Clitocybe tabescens* and to a species of root-knot nematode. In California it is subject to the root-rot fungus *Armillaria mellea*.

• *Control.* Effective control measures are not available.

Insects

AUSTRALIAN-PINE BORER. *Chrysobothris tranquebarica.* This flatheaded borer also attacks red mangrove trees. The adult females are greenish bronze beetles, $1/2$ to $3/4$ inch long, which deposit eggs on the bark in April.

• *Control.* See under borers, Chapter 11.

OTHER INSECTS. The citrus and long-tailed mealybugs and seven species of scales—barnacle, brown soft, cottony-cushion, Dictyospermum, latania, long soft, and mining—also may infest Australian-pine.

• *Control.* Control measures are rarely used.

AVOCADO (*Persea*)

Avocado is sometimes considered a highly desirable ornamental for subtropical areas.

Diseases

ANTHRACNOSE. *Colletotrichum gloeosporioides.* Greenhouse plants in the northern states and garden plants in the South are subject to attack by this fungus. It causes a general wilting of the leaves at the ends of branches as well as cankers on the stem and spots on the leaves and flowers.

• *Control.* See under anthracnose, Chapter 12.

ROOT ROT. *Phytophthora cinnamomi.* Root rot, also known as decline, is the most destructive disease of avocado in California. The causal fungus is soilborne and seedborne and thrives when soil drainage is poor.

• *Control.* Treat seed in hot water at 120° to 125° F (49° to 50° C) for 30 minutes to eliminate seedborne infections. The avocado rootstock cultivar Duke is said to be rather resistant to attack by the causal fungus.

SCAB. *Sphaceloma perseae.* This is a serious disease of avocado in Florida and Texas.

• *Control.* See under leaf spots, Chapter 12.

OTHER DISEASES. Other important diseases of avocado are canker, caused by *Botryosphaeria dothidea*, root rot by *Armillaria mellea*, and wilt by *Verticillium albo-atrum*.

• *Control.* Measures for the control of the latter two are discussed in Chapter 12.

Insects

Golden mealybug, *Nipaecoccus nipae*; mealybugs, *Pseudococcus* sp.; thrips, *Heliothrips haemorrhoidalis*; and tuliptree, ivy, latania, and dictyospermum scales attack avocado.

BALD CYPRESS (*Taxodium*)

This genus, native to the eastern United States, has both evergreen and deciduous members. Bald cypress thrives on wet sites and also tolerates dry soil, forming knees in wet locations. It is relatively free of fungal parasites and insect pests.

Diseases

TWIG BLIGHT. This disease, also present on a number of other conifers, including arborvitae, junipers, cypress, and yews, is rarely serious. It causes a spotting of leaves, cones, and bark and, in very wet seasons, a twig blight.

• *Cause.* The fungus *Pestalotiopsis funerea* is frequently associated with twig blight. It is not considered a vigorous parasite but becomes mildly pathogenic on trees weakened by mites, dry weather, sunscald, or low temperatures.

• *Control.* Control measures are rarely applied, although a copper fungicide would probably be effective.

WOOD DECAY. A number of fungi belonging to the genera *Echinodontium, Fomes, Lenzites, Poria,* and *Polyporus* are associated with wood decay of bald cypress. The last is sometimes found on living trees.

• *Control.* See under wood decay, Chapter 12.

OTHER FUNGAL DISEASES. Canker caused by species of *Septobasidium* and heart rot caused by *Fomes geotropus, F. extensus,* and *Ganoderma applanatum* have been reported.

• *Control.* No control measures are known for these diseases.

Insects and Other Pests

BALD CYPRESS RUST MITE. *Epitrimerus taxodii.* This eriophyid mite can cause premature bronzing and defoliation when population levels become high in the fall. Adult females overwinter in bark crevices protected by a white waxy covering.

• *Control.* See under eriophyid mites, Chapter 11.

BALD CYPRESS WEBWORM. *Coleotechnites apicitripunctella.* The larval stage of this moth mines the leaves of bald cypress and hemlock, then webs them together in late summer. The female adult is a small yellow moth with black markings and fringed wings. The pest is also called green hemlock needleminer.

• *Control.* See under leaf miners, Chapter 11.

CYPRESS LEAF GALL. *Itonidae taxodii.* Small, cone-shaped blue-green to white galls appear on leaves in summer. These galls are caused by a midge. No controls are suggested.

SOUTHERN CYPRESS BARK BEETLE. *Phloeosinus taxodii.* This beetle attacks bald cypress stems.

BANANA (*Musa*)

Plants of the cultivated banana, *Musa paradisiaca* var. *sapientum,* are subject to a number of serious bacterial and fungal diseases and have many insect pests. Plants grown in northern greenhouses for display and educational purposes are subject to mealybugs, whiteflies, and scales. These can be controlled with malathion sprays.

In Florida the dwarf banana, *M. nana,* is subject to anthracnose caused by the fungus *Gloeosporium musarum;* leaf blight by the bacterium *Pseudomonas solanacearum;* and the southern root-knot nematode, *Meloidogyne incognita.* The burrowing nematode, *Radopholus similis,* occurs in Louisiana on the roots of ornamental banana trees.

• *Control.* Plant pathologists at agricultural experiment stations will provide information on control measures.

BANYAN (*Ficus benghalensis*)

The only pests reported on this tree are insects: the long-tailed mealybug and several species of scales—black, Chinese obscure, California red, lesser snow, green shield, and mining.

• *Control.* See under scales, Chapter 11.

BASSWOOD. See Linden.

BEECH (*Fagus*)

Beeches are among our most beautiful trees; their dormant shapes and winter bark contribute much to any landscape.

The American beech (*F. grandifolia*) is a handsome, low-branched, slow-growing tree. Its dense foliage and shallow rooting habit, however, make the maintenance of a lawn underneath difficult. It grows best in cool, moist soils and can withstand neither constant trampling over the soil nor common city conditions very well. It is generally extremely difficult to transplant.

The European beech (*F. sylvatica*) is similar to the American species except that its bark is darker gray and its foliage more glossy. Its many horticultural varieties include purple beech, *atropunicea,* with deep purple leaves; weeping beech, *pendula,* one of the world's most beautiful weeping trees; weeping purple beech, *purpureopendula*; and cutleaf, *laciniata,* with very beautiful, narrow, almost fern-like leaves. Other new or interesting kinds are Darwyck beech, *F. sylvatica fastigiata*; golden-leaved beech, *aurea*; a red fern-leaved beech, *Rohanni*; and the tricolor beech, which has pink, white, and green leaves that turn a coppery bronze.

Diseases

LEAF MOTTLE. A large number of American beeches in the eastern United States have died over the years due to drought and subsequent death of the roots. Associated with the progressive weakening of some of the trees, however, is a disorder known as leaf mottle or scorch.

• *Symptoms.* In spring, small translucent spots surrounded by yellowish green to white areas appear on the young, unfurling leaves. These spots turn brown and dry, and by the first of June the mottling is very prominent, especially between the veins near the midrib and along the outer edge of the leaf. Within a few weeks the entire leaf presents a scorched appearance. A considerable part or, in some instances, all of the leaves then drop prematurely. Where complete defoliation occurs, new leaves begin to develop in July. The second set of leaves in such cases appears quite normal and drops from the trees at the normal time in the fall. Trees of different genetic backgrounds may differ in their response, some developing only leaf mottle and others leaf scorch under the same soil conditions.

The extensive loss of foliage early in the season depletes carbohydrate reserves and exposes the branches to the direct sun, causing considerable scalding of the bark. Such injury predisposes the tree to attacks by the two-lined chestnut borer and possibly by other insects.

• *Cause.* The cause of leaf mottle is not known. Disease incidence is not correlated with site differences. The reactions of individual trees vary, suggesting a genetic basis for susceptibility to leaf mottle.

• *Control.* Provide good growing conditions to reduce the impact of disease stress.

BLEEDING CANKER. *Phytophthora cactorum.* This disease, described under maple, occurs occasionally on beech. As the name suggests, the most common symptom visible on beech, maple, and elm is an oozing of a watery light brown or thick reddish brown liquid from the bark.

The fungus also causes crown canker on dogwood and collar rot of fruit trees.

• *Control.* No effective preventive measures are yet known. Severely infected specimens should be cut down and destroyed to prevent spread to nearby trees. Mildly affected trees have been known to recover. Avoid bark wounds near the base of the tree.

BEECH BARK DISEASE. The natural stands of beech in eastern Canada and northeastern United States have been severely damaged by this disease since the early 1930s. Although it is primarily a disease of forest beech, it is potentially dangerous to ornamental beech.

The attack of the woolly beech scale, *Cryptococcus fagisuga,* and possibly other scales, combined with the fungus *Nectria coccinea* var. *faginata,* and possibly other *Nectria* fungi, causes this disease.

Infestations of the scale on the bark always precede those of the fungus. During August and September, countless numbers of minute, yellow, crawling larvae appear over the bark. By late autumn they settle down and secrete a white fluffy material over their bodies. This substance is very conspicuous, and the trunks and branches appear to be coated with snow. Through the feeding punctures of the insect, the *Nectria* fungus then penetrates the bark and kills it. The insects soon die because of the disappearance of their source of food.

When the infected bark dies, it dries and forms areas on the trunk that are depressed and cracked. Eventually, deeply sunken cankers are formed, which assume a more or less circular or oval shape. The destruction of the bark and other tissues of the tree leads to a progressive dying of the top. Late in the season the leaves usually curl and turn brown, the twigs die, and new buds fail to form. The dead leaves remain attached throughout the winter. In spring, affected branches fail to produce foliage. Other branches, lacking reserve food materials, produce small yellow leaves that usually die during summer. Eventually the entire tree dies.

• *Control.* Because infestations of the woolly beech scale must precede fungal penetration, the eradication of the insect pest will prevent the start of the disease. Malathion sprays applied to the trunks and branches of valuable ornamental trees growing in the infested areas in early August and in September should control the young scales. A dormant lime sulfur spray applied to trunks and branches will control overwintering adult scales. Oil sprays are also effective but are not reliably safe on beech trees.

CANKERS. *Asterosporium hoffmanni, Cytospora* sp., *Strumella coryneoidea, Endothia gyrosa, Nectria galligena,* and *N. cinnabarina.* Canker and branch dieback of beech may be caused by any one of these fungi.

• *Control.* Prune and destroy infected branches.

LEAF SPOTS. *Gloeosporium fagi* and *Phyllosticta faginea.* Leaf spots develop late in the growing season.

• *Control.* See under leaf spots, Chapter 12.

POWDERY MILDEW. *Microsphaera erineophila* and *Phyllactinia guttata.* Two species of powdery mildew fungi occasionally develop on beech leaves in late summer.

• *Control.* See under powdery mildew, Chapter 12.

WOOD DECAY. *Hericium erinaceus, Fomes fomentarius,* and *Phellinus ignarius.* Three species of fungi have been observed in decayed wood. See Chapter 12.

Insects

BEECH BLIGHT APHID. *Prociphilus imbricator.* The bark is punctured and juices are extracted by the beech blight aphid, a blue insect covered with a white cottony substance. The pests also feed on the leaves.

WOOLLY BEECH APHID. *Phyllaphis fagi.* Leaves are curled and blighted by the woolly beech aphid, a cottony-covered insect, the cast skins of which adhere to the lower leaf surface. The purple beech is more commonly infested than the American beech. The giant bark aphid (see aphids, under Sycamore) also feeds on beech.

• *Control.* See under aphids, Chapter 11.

BEECH SCALE. *Cryptococcus fagisuga.* White masses of tiny circular scales, each $1/40$ inch in diameter, on the bark of the trunk and lower branches are the common signs of beech scale infestation. The eggs are deposited on the bark in late June and early July, and the young crawling stage appears in August and September. The pest overwinters as a partly grown adult scale. Beech scale is primarily a pest of forest trees and is mentioned here only because of its association with the beech bark disease. It is present in the northeastern United States.

SCALES. In addition to the *Cryptococcus* scale, beech trees are subject to the following scales: black, cottony-cushion, cottony maple, hickory lecanium, European fruit lecanium, oystershell, Putnam, and San Jose.

• *Control.* See under scales, Chapter 11.

EASTERN TENT CATERPILLAR. See under Willow.

DATANA CATERPILLAR. See yellow-necked caterpillar, under Oak.

GYPSY MOTH. See under Elm.

CATERPILLARS. Many other caterpillars chew the leaves of beech, including the hemlock looper, saddled prominent, walnut, fall and spring cankerworms, elm spanworm, and green mapleworm. The larval stage of the following moths also infests the leaves of this host: imperial, io, luna, and the rusty tussock.

• *Control.* See under caterpillars, Chapter 11.

BROWN WOOD BORER. *Parandra brunea.* Winding galleries in the wood, made by the white-bodied, black-headed borers, 1¹/₄ inches long, and tiny holes in the bark, made by emerging shiny brown beetles ³/₄ inch long, are typical signs of brown wood borer infestation. The eggs are deposited in bark crevices or in decayed wood. Leopard moth larvae (see under Maple) also invade beech wood.

• *Control.* See under borers, Chapter 11.

TWO-LINED CHESTNUT BORER. See under Oak.

ASIATIC OAK WEEVIL. See under Oak.

OTHER INSECTS. Beech leaf miner, *Brachys aeruginosus*; maple trumpet skeletonizer; and maple leaf cutter (see under Maple) attack beech. Mites, including oak mite (see under Oak), and an eriophyid mite, *Acalitis fagerina*, infest beech.

BIRCH (*Betula*)

The birches, graceful as well as beautiful, are highly susceptible to ice and snow breakage. Extremely particular about soil conditions, they are unable to adapt themselves as street trees and are used principally on lawns.

The canoe birch (*B. papyrifera*) has attractive bark and a single trunk. Gray birch (*B. populifolia*) is gray-barked, and many specimens have several trunks.

Ornamental birches are susceptible to several fungal parasites and insect pests. Of these, the bronze birch borer is mainly responsible for the death of trees used in ornamental plantings.

River birch (*B. nigra*) with its shaggy reddish brown bark likes moist sites and is resistant to bronze birch borers.

The Monarch birch (*B. maximowicziana*), introduced into the United States from Japan in 1893 by Professor Sargent of the Arnold Arboretum, is highly resistant to the bronze birch borer. Its leaves are larger than those of native species and more distinctly heart-

shaped. Monarch birch does better in urban environments than the other birches. It should be more widely planted as a lawn or park tree.

European white birch (*B. pendula*) is also widely planted but is very susceptible to borers.

Most native birches, except river birch, are best grown in cold climates.

Diseases

LEAF BLISTER. Two fungi cause leaf blister on many species of birch. *Taphrina carnea* produces red blisters and curling of the leaves on many of the birch species, and *T. flava* forms yellow blisters on gray and canoe birches.

• *Control.* Gather and destroy all fallen leaves. Spray with Ferbam or Daconil 2787 just before buds open in spring.

LEAF RUST. *Melampsoridium betulinum.* Leaves of seedlings and mature trees are sometimes attacked by a rust that causes spotting and defoliation. The rust pustules are bright reddish yellow. The spores from these pustules spread the infection from leaf to leaf. The alternate or sexual stage causes a blister rust on larch. In mixed forest plantings both hosts may be seriously injured.

• *Control.* In ornamental plantings the disease rarely becomes destructive enough to warrant special control measures.

LEAF SPOT. Several leaf spot fungi attack birch. The fungus *Colletotrichum gloeosporoides* (= *Glomerella cingulata*) produces brown spots with a dark brown to black margin. The fungus *Cylindrosporium betulae* forms smaller spots with no definite margin. Both fungi may become sufficiently prevalent to cause some premature defoliation. *Marssonina betulae* causes a leaf blotch.

• *Control.* See under leaf spots, Chapter 12.

CANKER. Canker is caused by the fungus *Nectria galligena.* Small, globose, dark red fruiting bodies of the fungus are barely visible to the unaided eye on dead bark. Young cankers are not readily visible. Upon close examination, however, they are seen to be darker in color than

Fig. III-5. Canker on black birch.

the adjacent healthy tissue and appear water-soaked. Later, the edge of the diseased area cracks and exposes the canker (Fig. III-5). Callus tissue forms over the cracked area but becomes infected and dies. The process is repeated annually, forming concentric rings of dead callus. The bark within the cankered zone falls away, leaving the wood exposed. When the canker completely girdles the stem or trunk, the distal portion dies. Black, paper, sweet, and yellow birches are particularly susceptible.

• *Control.* In thick stands it is advisable to remove trees with trunk infections. Trees having cankers or galls on the branches may be saved by pruning out and destroying the cankers. Since the cankers originate on young growth, inspection of the plantings and early destruction of the cankered young trees are advisable. Trees in ornamental plantings should be fertilized and watered to maintain good vigor.

DIEBACK. *Melanconium betulinum.* Trees weakened by drought may be attacked by this fungus, which causes a progressive dieback of

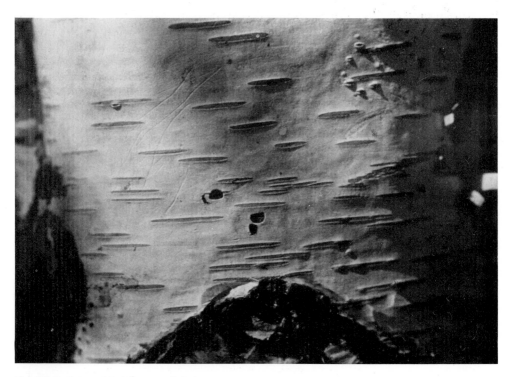

Fig. III-6. Bronze birch borer exit holes.

the upper branches. Infestations of the bronze birch borer, mentioned below, may cause similar symptoms.

• *Control.* Prune affected branches to sound wood and fertilize and water heavily to help revitalize the tree.

WOOD DECAY. Several wood-decay fungi attack birches. One, *Piptoporus betulinus,* attacks dying or dead birches and produces shelf- or hoof-shaped, gray, smooth conks along the trunk. Others, such as *Torula ligniperda, Fomes fomentarius, Inonotus obliquus, Phellinus laevigatus,* and *Ganoderma applanatum,* are associated with decay of living trees.

• *Control.* See under wood decay, Chapter 12.

OTHER FUNGAL DISEASES. Birches are subject to several other fungal diseases: powdery mildews caused by *Microsphaera ornata* and *Phyllactinia guttata*; leaf spots by *Discula betulina* and *Septoria betulicola*; and stem cankers by *Botryosphaeria obtusa* and *Diaporthe alleghaniensis. Gloeosporium* and *Diaporthe* cause sunken, black stem cankers on young branches; distal leaves turn brown.

• *Control.* Control measures are rarely necessary for these diseases.

VIRAL DISEASE: LINE PATTERN MOSAIC; APPLE MOSAIC VIRUS. Decline, dieback, and death of both white and yellow birches in forest plantings may be virus-induced. White birch in ornamental plantings may also be affected but to a lesser extent. Yellow to golden line and ringspot patterns on the leaves are the most striking symptoms.

• *Control.* Control measures have not been developed for this viral disease.

Abiotic Disease

AIR POLLUTION. Birch is relatively sensitive to sulfur dioxide. See Chapter 10.

Insects and Related Pests

BRONZE BIRCH BORER. *Agrilus anxius.* Varieties of birch grown in parks as ornamental shade trees, especially where the soil is poor, and trees grown elsewhere under adverse conditions become prey to this borer. The grub is from $1/2$ to 1 inch long, flatheaded, and light-colored. The adult stage is a beetle $1/2$ inch long. The bee-

tles, which feed on foliage for a time, deposit their eggs in slits in the bark. The borers make flat, irregular, winding galleries just beneath the bark of the main trunk (Fig. III-6). Heavy infestations usually kill the trees (Fig. III-7). Poplar, willow, and dogwood borers also attack birch.

• *Control.* See under borers, Chapter 11.

BIRCH LEAF MINER. *Fenusa pusilla.* Gray, canoe, and cut-leaf birches are especially susceptible to attacks by this leaf miner, a small white worm which causes leaves to turn brown in late spring or early summer (Fig. III-8). The adult is a small black sawfly that overwinters in the soil as a pupa. The first brood begins to feed anytime from very early to late May, depending on seasonal conditions and the location of the trees. The first brood causes the most damage because it attacks the tender spring foliage. Other broods hatch during the summer but cause less damage since they confine their feeding to leaves on sucker growths and to newly developing leaves in the crowns of the trees and do not attack mature foliage.

• *Control.* See under leaf miners, Chapter 11.

WITCH-HAZEL LEAF GALL APHID. *Hamamelistes spinosus.* This insect, which causes the formation of cone galls on witch-hazel, migrates from this host to birches in the summer. It feeds on the undersides of the leaves and resembles nymphs of whiteflies.

APHIDS. *Euceraphis betulae* and *Calaphis betulaecolens.* The former, known as the European birch aphid, is yellow and infests cut-leaf and other birch varieties. The latter, the common birch aphid, is a large green species that produces copious quantities of honeydew, which encourages the growth of sooty mold. Giant bark aphid (see aphids, under Sycamore) is also found on birch.

• *Control.* See under aphids, Chapter 11.

CASE BEARER. *Coleophora fuscedinella.* Leaves are mined and shriveled and small cases are formed under the leaves by the case bearer, a light yellow to green caterpillar, $1/5$ inch long with a black head. The adult is a brown moth with a wingspread of $2/5$ inch. The larva overwinters in a case attached to the bark.

• *Control.* See under caterpillars, Chapter 11.

Fig. III-7. These birches are dying back because of infestations of the bronze birch borer.

Fig. III-8. The leaf miner causes brown blotches on birch leaves. The tiny adult flies and the larvae are also shown.

BIRCH SKELETONIZER. *Bucculatrix canadensisella.* The lower leaf surface is chewed, and the leaf is skeletonized and may turn brown as a result of feeding by this yellowish-green ¹/₄-inch-long larva. The adult moth has white-lined, brown wings with a spread of ³/₈ inch. The apple and thorn skeletonizer (see under Hawthorn) also infests birch.

• *Control.* See under caterpillars, Chapter 11.

DUSKY BIRCH SAWFLY. *Croesus latitarsus.* Larvae with shiny black head capsules and gray-green to yellow-green bodies with black spots feed on birch in spring, often in colonies or groups.

• *Control.* See under sawflies, Chapter 11.

BIRCH LACE BUG. *Corythucha pallipes.* Nymphs feed on birch causing a stippling symptom on the leaf surface. Controls are not normally needed.

SEED MITE GALL. *Eriophyes betulae.* Another conspicuous gall on paper birch and other species is caused by the seed mite. The galls are about 1 inch in diameter and are made up of many adventitious branches and deformed buds. They may resemble witches' brooms.

• *Control.* This pest is never serious enough to require control measures.

SCALES. Several types, including hickory lecanium, terrapin, frosted, and walnut, infest birch.

OTHER INSECTS. Yellow-necked caterpillar (see under Oak), cambium miner (see under Holly), red-humped caterpillar (see under Poplar), white-marked tussock moth and mourning-cloak butterfly (see under Elm), elm spanworm (see under Elm), cecropia moth (*Hyalophora cecropia*) and eastern tent caterpillar (see under Willow), forest tent caterpillar (see under Maple), and leaf roller (*Archips* sp.) feed on birch leaves. Japanese beetle (see under Linden), potato leafhopper (*Empoasca fabae*), planthoppers (see under Cherry), spider mites, and giant hornet wasp (see under Franklin-tree) also infest birch.

BLACK GUM. See Tupelo.

BOXWOOD (*Buxus*)

Boxwood requires a well-drained, neutral soil with an occasional light application of ground limestone. In poorly drained soil it is susceptible to winter injury and fungal diseases. A few kinds, particularly the tree box, grow into small trees and hence are discussed here.

Fig. III-9. Volutella canker of boxwood. Notice the dead shoot tip and the fungal pustules on the affected stems.

Diseases

CANKER. One of the most destructive diseases of boxwood is canker.

• *Symptoms.* The first noticeable symptoms are that the new growth on certain branches or certain plants in a group does not start as early in spring as others, and that the new growth is less vigorous than that on healthy specimens. The leaves turn light green and then various shades of tan. Infected leaves turn upward and lie close to the stem instead of spreading out like the leaves on healthy stems. The diseased leaves and branches show small, rose-colored, waxy pustules, the fruiting bodies of the fungus. The bark at the base of an infected branch is loose and peels readily to expose the gray to black discolored wood beneath. Infection is frequently found to take place at the bases of small dead shoots or in crotches where leaves have been allowed to accumulate.

• *Cause.* Canker is caused by the fungus *Pseudonectria rousselliana,* which is found on the stem cankers. Another stage of the same fungus, known as *Volutella buxi,* attacks the leaves and twigs (Fig. III-9), and produces pale rose-colored spore masses on yellowed leaves. A species of *Verticillium* is occasionally associated with the dieback of twigs. The fungus *Nectria desmazierii,* whose imperfect stage is known as *Fusarium buxicola,* is also capable of causing canker and dieback of boxwood.

• *Control.* Dead branches should be removed as soon as they are noticeable, and cankers on the larger limbs should be treated by surgical methods. The annual removal and destruction of all leaves that have lodged in crotches is recommended. Four applications of fixed copper, Bordeaux mixture 3-3-50, or lime sulfur 1-50 have been shown to be very effective in preventing canker. The first application should be made after the dead leaves and dying branches have been removed and before growth starts in the spring; the second when the new growth is half completed; the third after spring growth has been completed; and the fourth after the fall growth has been completed. The boxwood should be fertilized with occasional applications to the soil of well-rotted manure or commercial fertilizers, and ground limestone.

BLIGHT. Several fungi are associated with the blighting of boxwood leaves. The most common are *Phoma conidiogena* and *Hyponectria buxi.* The exact role of these fungi in this disease complex is not clear.

• *Control.* The fungicides recommended for canker control will also control blight.

LEAF SPOTS. *Macrophoma candolei, Phyllosticta auerswaldii, Fusarium buxicola,* and *Collectotrichum* sp. Leaves turn straw-yellow and are thickly dotted with small black bodies, which are the fruiting structures of the first fungus listed above. The others also cause leaf spotting. All are apparently limited in their attacks to foliage weakened by various causes.

• *Control.* See under leaf spots, Chapter 12.

ROOT ROT. *Phytophthora cinnamomi.* "Off-color" foliage followed by sudden wilting and death of the entire plant is characteristic of this disease. Yews and a large number of other woody ornamental plants are also subject to this disease. Another species, *P. parasitica,* also causes a root rot and blight of boxwood. Root infection by the fungus *Paecilomyces buxi* was found to be associated with boxwood decline.

• *Control.* See under root rots, Chapter 12.

Nonparasitic Diseases

WINTER INJURY AND SUN SCALD. Most boxwood problems in the northeastern United States are due to freezing and sun scalding, which primarily injure the cambium of unripened wood. Several distinct types of symptoms are exhibited by winter injury. Young leaves and twigs may be injured when growth extends far into fall or begins too early in spring. Leaves may turn rusty brown to red as a result of exposure to cold, dry winds during winter. A dieback of leaves, twigs, and even the entire plant may occur on warm winter days when the aboveground tissues thaw rapidly and lose more water than can be replaced through the frozen soil and roots. Another type of winter injury is characterized by the splitting and peeling of the bark. The bark becomes loosened and the stems are entirely girdled, resulting in death of the distal portions.

Fig. III-10. Cupping of boxwood leaves (left and right) caused by the boxwood psyllid (*Psylla buxi*). Leaves in the center are normal.

• *Control.* Fertilizers should be applied in late fall, preferably, or very early in spring. Adequate windbreaks should be provided during winter, especially in the more northern latitudes. Spraying with an antitranspirant on a mild day in December, and again on a mild day in the following February, may provide as much protection from winter winds as do burlap windbreaks. A heavy mulch consisting of equal parts of leaf mold and manure should be applied to prevent deep freezing and to provide a continuous water supply.

Insects and Related Pests

BOXWOOD LEAF MINER. *Monarthropalpus buxi.* Oval, water-soaked swellings on the lower leaf surface result from the feeding inside the leaves by the leaf miner, a yellowish white maggot, 1/8 inch long. The larvae spend the winter in the blister-like yellow or brown leaf mines; the larvae pupate in early spring. The adult is a tiny midge, 1/10 inch long, which emerges in April or May, leaving behind a tan pupal case extending from the mine.

• *Control.* See under leaf miners, Chapter 11.

BOXWOOD PSYLLID. *Psylla buxi.* Terminal leaves are cupped and young twig growth is checked by the boxwood psyllid, a small, gray, sucking insect covered with a white cottony or waxy material (Fig. III-10). The adult is a small green fly with transparent wings having a spread of 1/8 inch.

• *Control.* Spray as soon as new growth appears and again 2 weeks later with Azatin, Carbaryl, Dursban, horticultural oil, insecticidal soap, Sevin, or Tempo.

BOXWOOD WEBWORM. *Galasa nigrinodis.* This pest chews leaves and forms webs on boxwood.

• *Control.* See under caterpillars, Chapter 11.

GIANT HORNET WASP. See under Franklintree.

YELLOW-NECKED CATERPILLAR. See under Oak.

MEALYBUGS. Comstock mealybug is discussed under Catalpa. The ground mealybug, *Rhizoecus falcifer,* feeds on the roots of boxwood, citrus, pine, and spruce. High populations on the roots can cause decline and death of small trees.

• *Control.* A dilute solution of Diazinon applied to the soil around the base of the boxwood should control the ground mealybug.

SCALES. Many species of scales, California red, cottony maple, cottony-cushion, lesser snow, oystershell, Japanese wax, ivy, Glover, and European fruit lecanium, may infest boxwood.

• *Control.* See under scales, Chapter 11.

Other Pests

BOXWOOD MITE. *Eurytetranychus buxi.* A light mottling followed by brownish discoloration of the leaves is caused by infestation of these $1/64$-inch-long mites. They overwinter as eggs, which hatch in April, and the young mites begin to suck out the leaf juices. By June or July, considerable injury may occur on infested plants. As many as six generations of mites may develop in a single season.

• *Control.* See under spider mites, Chapter 11.

NEMATODES. *Pratylenchus pratensis.* Leaf-bronzing, stunted growth, and general decline of boxwood may result from invasion by meadow nematodes. They enter the roots, usually near the tips, and move through the cortical tissue. Invaded portions soon die, and the plant forms lateral roots above the invaded area. These laterals in turn are infested. Repeated infestations and lateral root production result in a stunted root system resembling a witches' broom. Even heavy rains may fail to wet such densely woven root bundles. Boxwoods are also subject to several other parasitic nematodes, including the southern root-knot nematode, *Meloidogyne incognita,* an endoparasite which lives inside the roots; various ring nematodes, *Criconema, Criconemoides,* and *Procriconema*; and the spiral nematode, *Helicotylenchus.*

• *Control.* Chemical treatments for root-infecting nematodes of boxwoods established in the landscape are not available. The life of infested but untreated plants may be prolonged by providing good care and by soaking the soil thoroughly during dry spells. Before boxwood is replaced in infested soil, the planting site should be fumigated with any one of several materials available for that purpose. See Chapter 13.

BUCKEYE. See Horsechestnut.

BUCKTHORN (*Rhamnus*)

This small tree or tall shrub is sometimes used in the landscape.

Insects

Buckthorn may be attacked by black, gloomy, and ivy scales.

CAJEPUT (*Melaleuca*)

This native of Australia grows to medium size in Florida and California, where it is used as a street tree and in landscapes. It is not troubled by serious diseases in the United States.

Insects

At least fourteen species of scale as well as the citrus mealybug have been found on this host in Florida.

• *Control.* See under scales, Chapter 11.

CALIFORNIA LAUREL (*Umbellularia*)

This California and Oregon native has high aesthetic value in many of its native situations and is widely used in home and park landscaping. It is also called Oregon myrtle and pepperwood, the latter because of a strong, pungent, sneeze-inducing camphorlike odor emitted from crushed green bark and foliage.

Leaf Diseases

A bacterium, *Pseudomonas lauracearum,* and the fungi *Kabatiella phoradendri* f. sp. *umbellulariae* and *Collectotrichum gloeosporioides* occasionally cause serious leaf blight in California. *Mycosphaerella arbuticola* and several sooty mold fungi also appear on leaves. Control measures have not been worked out.

CANKER. *Nectria galligena.* This fungus also causes cankers on this tree. Unnecessary wounds should be avoided.

Insect Pests

California laurel aphid, *Euthoracaphis umbellulariae,* sometimes mistaken for an immature whitefly, is found on this host. See Chapter 11.

CAMELLIA (*Camellia japonica* and *C. sasanqua*)

Commonly a garden shrub, camellias can grow into small, broad trees. Camellias are grown outdoors in the warmer parts of the country and indoors in the colder parts. The *sasanqua* varieties are said to be more winter hardy than the *japonica* varieties in the northeastern part of the country along the Atlantic Coast.

Fungal and Algal Diseases

BLACK MOLD. *Meliola camelliae.* The abundant black fungal growth of the *Fumago* stage covers the leaves and twigs of this host.

• *Control.* Control insects such as aphids and scales (See Chapter 11) since the fungus grows on their honeydew. Promptly pick off and destroy infected leaves and discard all debris from infected plants.

CANKER. *Glomerella cingulata.* A canker and dieback of camellias is widespread and frequently destructive in the southern states; it also occurs on greenhouse-grown plants in the North. The fungus enters only through wounds. In nature, the usual entrance points are scars left by the abscission of leaves in spring. In Florida, a species of *Phomopsis* causes somewhat similar symptoms.

• *Control.* Prune and destroy cankered twigs. When the cankers occur on the main stem of large plants, surgically remove the diseased portions and follow with application of a fungicide. Copper fungicides applied periodically to the leaves and stems may help prevent new infections.

FLOWER BLIGHT. *Ciborinia camelliae.* This blight is confined to the flowers, which turn brown and drop. It occurs in the Pacific and Gulf Coast states as well as other southern states from Texas to Virginia. All species and varieties of camellias appear equally susceptible to the blight.

Another flower-blighting fungus, *Sclerotinia sclerotiorum,* has been reported from North Carolina. A bud and flower blight is occasionally caused by *Botrytis cinerea,* particularly after the plants have been subjected to frost.

• *Control.* To control the *Ciborinia camelliae* blight, pick off and discard all old camellia blossoms before they fall. Benomyl, Ferbam, sulfur, or mancozeb sprays help prevent infection. Infections can also be prevented by placing a 3-inch mulch of wood chips or other suitable material around the base of each plant. Such a barrier will prevent the fungal bodies in the soil from ejecting their spores into the air and onto the leaves. Soils heavily infested with sclerotia (which later produce ascocarps) may be treated with Ferbam or Captan. PCNB (Terraclor) provides even more effective control but must be used in soils free of plants. No special controls have been developed for the *Botrytis* bud and flower blight or for *Sclerotinia sclerotiorum.*

LEAF BLIGHT. *Cephaleuros virescens.* The epidermal cells are attacked by this alga, which spreads rapidly over the leaf and causes it to blacken and die.

• *Control.* Remove diseased leaves. Badly infested specimens may be sprayed with a copper fungicide.

LEAF GALL. *Exobasidium camelliae.* The leaves and stems of new shoots are thickened and distorted by this fungus.

• *Control.* Spray once before the leaves unfurl with Ferbam.

LEAF SPOT. *Cercospora theae.* This leaf spot, first reported from Louisiana, develops under conditions of overcrowding, partial shade, and high humidity of a lath or shade house.

• *Control.* No controls have been developed.

ROOT ROT. *Phytophthora cinnamomi.* This disease is common not only on camellia but also on avocado, maple, pine, and rhododendron as well as many other woody plants. Excessive moisture and poor soil drainage favor its development.

• *Control.* See under root rots, Chapter 12.

SPOT DISEASE. *Pestalotiopsis maculans.*

More or less irregular round blotches run together, causing a silvery appearance of the upper surface of the leaves. The diseased area is sharply marked off from the healthy portion. The pycnidia or fruiting bodies of the fungus are visible as black dots. Leaf fall sometimes results. Several other fungi produce leaf spotting: *Phyllosticta camelliae, P. camelliaecola,* and *Sporonema camelliae.* A species of *Sphaceloma* causes scabby spots on the leaves.

- *Control.* See under leaf spots, Chapter 12.

Viral Diseases

Leaf and flower variegation is caused by camellia yellow mottle virus. The condition may be transmitted by grafting from variegated *Camellia japonica* to uniformly green varieties of *C. japonica* and *C. sasanqua.* Some yellow variegation, however, may be due to genetic changes rather than viral infection. Such variegations usually follow a uniform and rather typical pattern, which is more or less similar on all leaves. Many greenhouse-grown camellias in the North have typical ringspot patterns on the leaves (Fig. III-11).

- *Control.* Plants suspected of harboring a virus should be discarded, or at least isolated from healthy plants.

Abiotic Diseases

BUD DROP. Camellias grown in homes, in greenhouses, and even outdoors frequently lose their buds before opening (Fig. III-11), or the tips of the young buds and edges of young petals turn brown and decay. Bud drop from indoor-grown plants usually is due to overwatering of the soil or to some other faulty environmental condition such as insufficient light, excessively high temperatures, or a potbound condition of the roots. Bud drop in the Pacific Northwest may result from a severe frost in September or October, severe freezing during the winter, or an irregular water supply. In California it may result from lack of adequate moisture.

CHLOROSIS. Deficiency of some elements in the soil may result in chlorosis.

EDEMA. Frequently brown, corky, roughened swellings develop on camellia leaves grown in greenhouses. The condition is associated with overwatering of the soil during extended periods of cloudy weather.

SUNBURN. This condition appears on leaves as faded green to brown areas with indefinite margins. It occurs on the upper exposed sides of bushes, particularly those transplanted from shady to very sunny areas.

SALT INJURY. Camellias cannot tolerate high soil salinity and grow best in the acid soils and temperate climate of our eastern and Gulf Coast areas. Salt levels above 1800 parts per million in the soil solution were fatal to camellias in greenhouse tests. Azaleas were found to be equally susceptible to high salt concentrations.

Insects and Other Animal Pests

FLORIDA RED SCALE. *Chrysomphalus aonidum.* This scale insect, common on citrus and other plants in the greenhouse, has been found to live on camellia leaves. The scales are dark brown and more or less circular. It is easy to remove the scale with a needle; this exposes the

Fig. III-11. Upper: yellow mottle virus symptoms. Lower: Bud drop of camellia.

Fig. III-12. Florida red scale on camellia.

very light yellow body of the insect, firmly attached to the leaf by its sucking organ. The leaf shown in Fig. III-12 was photographed after several of the scales had been removed to reveal the insects, which appear as white spots.

Many other species of scales infest camellias. They include black, California red, camellia, chaff, cottony taxus, degenerate, dictyosperm, euonymus, Florida wax, Glover, greedy, hemispherical, Japanese wax, latania, Mexican wax, oleander, oystershell, peony, and soft scale. Two other scales known as the camellia parlatoria and the olive parlatoria also attack this host.

TEA SCALE. *Fiorinia theae.* The most serious pest of outdoor camellias in the South, this scale also infests greenhouse-grown camellias in the North as well as ferns, palms, orchids, figs, and several other plants. It can be distinguished superficially by its oblong shape and the ridge down the center parallel to the sides.

• *Control.* See under scales, Chapter 11.

MEALYBUGS. *Planococcus citri* and *Pseudococcus adonidum.* These two white insects are usually found in the leaf axils and shoot buds.

• *Control.* See under scales, Chapter 11.

FULLER ROSE BEETLE. *Pantomorus cervinos.* This weevil occasionally infests camellias, roses, palms, and many other plants.

A number of beetles also infest this host. The most common are rhabdopterus, flea beetle, and the grape colaspis.

• *Control.* See Chapter 11.

SPOTTED CUTWORM. *Amathes c-nigrum.* This cutworm has been found feeding on the

blasted buds of camellias in greenhouses. Apparently it is able to climb up among the branches, leaves, and flower buds.

• *Control.* See under caterpillars, Chapter 11.

THRIPS. The browning of the tips of buds, followed by decay and dropping, is frequently caused by the attacks of a species of thrips. This should be clearly distinguished from the bud drop caused by overwatering (see above).

• *Control.* Spray with Carbaryl, Diazinon, Dursban, Malathion, Merit, or Sevin.

WEEVILS. *Otiorhynchus sulcatus* and *O. ovatus.* The black vine weevil and the strawberry root weevil feed on the leaves; the larval stages feed on the roots and the base of the stem. The Japanese weevil, *Callirhopalus bifasciatus,* feeds on leaves like other weevils.

• *Control.* See under weevils, Chapter 11.

OTHER INSECTS. A great number of other insects infest camellias in greenhouses, homes, and outdoors. These include the following aphids: black citrus, melon, green peach, and ornate. The following caterpillars chew the leaves: omnivorous looper, orange-tortrix, and western parsley. The fruit tree leafroller and the greenhouse leaf tier also chew the leaves and roll or tie them together. The greenhouse whitefly is common on greenhouse-grown plants.

• *Control.* See Chapter 11.

NEMATODES. Camellias are unusually resistant to the root-knot nematode, *Meloidogyne incognita,* although these pests have been recorded on camellias in Texas. Another species of nematode, *Hemicriconemoides gaddi,* has been reported on the roots of camellia in Louisiana.

• *Control.* Treatments for established landscape plants are not available.

MITES. Southern red mite (see under Holly) attacks camellia as does *Cosetacus camelliae,* an eriophyid mite.

CAMPHOR-TREE (*Cinnamomum*)

An Asian native, this tree is popular in the southern, southwestern, and western United States but does poorly in alkaline, clay soils. This tree has escaped from cultivation from Florida to Louisiana and has become naturalized in Florida.

Diseases

WILT. Verticillium wilt has been found on this tree in the San Francisco Bay area of California. See Chapter 12.

ROOT ROT. Clitocybe and Armillaria root rots have also been reported for this host. See Chapter 12.

ANTHRACNOSE. *Glomerella cingulata (Gloeosporium).* Leaf spot, canker, and shoot blight are symptoms of this disease.

POWDERY MILDEW. This disease is caused by *Microsphaera alni.* See Chapter 12.

THREAD BLIGHT. *Ceratobasidium stevensii.* This disease occurs in Louisiana.

Insects

SCALES. Sixteen species of scale insects infest camphor-trees in Florida. The most destructive one, which can be fatal to this host, is the camphor scale, *Pseudaonidia duplex.* It is circular in shape, convex, dark blackish brown, and $1/10$ inch across.

MITES. Other pests of the camphor-tree are three species of mites—avocado red, plantanus, and southern red—and camphor thrips, *Liriothrips floridensis.*

• *Control.* See under scales and spider mites, Chapter 11.

CAROB (*Ceratonia*)

This tree grows in Mediterranean climates, and its pods are made into a chocolate substitute. Used as a screen or shrubby hedge, carob is somewhat messy and its male flowers are foul-smelling. Carob is susceptible to Verticillium wilt and is attacked by oleander scale.

CATALPA (*Catalpa*)

Two species of catalpa, common and western (or northern), are used in ornamental plantings. Common catalpa (*C. bignonioides*) is a messy tree and should be planted only where its flowers and fruits are not objectionable. Western catalpa (*C. speciosa*) is hardy and grows rapidly. Several leaf spots and powdery mildews, as well as the Verticillium wilt disease (Chapter 12), occur on these hosts. Insects are usually more of a problem than diseases.

Diseases

LEAF SPOT. A common disease of catalpa, leaf spot generally appears during rainy seasons.

• *Symptoms.* Tiny water-soaked spots, scattered over the leaf, appear in May. The spots turn brown and increase in size until they attain a diameter of about $1/4$ inch. Holes in the leaves are common as a result of dropping out of the infected tissue. Where the spotting is unusually heavy, the leaves may drop prematurely.

• *Cause.* Three fungi, *Phyllosticta catalpae, Gloeosporium catalpae,* and *Cercospora catalpae,* are frequently associated with these spots (Fig. III-13). Injury by the catalpa midge, discussed below, and infection by bacteria are believed to increase the susceptibility to leaf spots. *Alternaria catalpae* may be a secondary invader.

• *Control.* See under leaf spots, Chapter 12.

POWDERY MILDEW. Two species of mildew

Fig. III-13. Leaf spot of catalpa caused by the fungus *Phyllosticta catalpae.*

fungi, *Microsphaera elevata* and *Phyllactinia guttata,* attack catalpas.

• *Control.* See under powdery mildew, Chapter 12.

WILT. Verticillium wilt is an important problem. See Chapter 12.

WOOD DECAY. The western catalpa is extremely susceptible to heartwood decay caused by the fungus *Trametes versicolor.* The heartwood becomes straw-yellow, light, and spongy. Another fungus, *Polyporus catalpae,* causes decay of the trunk near the soil line. The wood becomes brown, tough, brittle, and full of cracks.

• *Control.* See under wood decay, Chapter 12.

OTHER DISEASES. Among other diseases of catalpa are twig dieback, caused by *Botryosphaeria dothidea,* and root rot by *Armillaria mellea* and *Phymatotrichum omnivorum.*

• *Control.* Control measures have not been developed.

Abiotic Disease

AIR POLLUTION. Catalpa is relatively sensitive to sulfur dioxide. See Chapter 10.

Insects

COMSTOCK MEALYBUG. *Pseudococcus comstocki.* This small, elliptical, waxy-covered insect attacks catalpa primarily but is also found on apple, boxwood, holly, horsechestnut, magnolia, maples, osage-orange, poplar, and Monterey pine. After hatching in late May, the young crawl up the trunk to the leaves, where they suck out the juices and devitalize the tree. Twigs, leaves, and trunks may be distorted as a result of heavy infestations. The eggs overwinter in bark crevices or in large masses hanging on the twigs.

• *Control.* See under scales, Chapter 12.

CATALPA MIDGE. *Cecidomyia catalpae.* Leaves are distorted, and circular areas inside the leaves are chewed, leaving a papery epidermis, as a result of infestation by tiny yellow maggots. The adult, a tiny fly with a wingspread of $^1/_{16}$ inch, appears in late May or early June to lay eggs on the leaves. Winter is passed in the pupal stage in the soil.

• *Control.* Cultivate the soil beneath the trees

to destroy the pupae, and spray in late May with malathion.

CATALPA SPHINX. *Ceratomia catalpae.* Leaves may be completely stripped from a tree by a 3-inch-long pale yellow and black caterpillar with a black horn on its posterior end. The adult female is a grayish brown moth with a 3-inch wingspread. The winter is passed as the pupal stage in the ground.

• *Control.* See under caterpillars, Chapter 11.

WHITEFLY. *Tetraleurodes* sp. Infestation is often followed by the appearance of sooty mold.

SCALE. See white peach scale, under Cherry.

CEDAR (*Cedrus*)

This genus contains several very beautiful evergreen trees such as the Atlas cedar (*C. atlantica*) and the cedar of Lebanon (*C. libani*). Cedars are relatively free of pests and diseases. These "true cedars" produce cones, unlike juniper, false cypress (also called white cedar), and arborvitae, which are sometimes called cedars.

Diseases

TIP BLIGHT. The fungus *Sphaeropsis sapinea,* formerly called *Diplodia pini,* occasionally causes canker and dieback of branch tips in the South.

• *Control.* The same as for tip blight of pine.

ROOT ROT. Several fungi, including *Armillaria mellea, Clitocybe tabescens,* and *Phymatotrichum omnivorum,* are associated with root and trunk decay. The latter attacks a great variety of trees, shrubs, and ornamental and food plants in the South.

• *Control.* No effective, practicable control measures are known.

Insects

BLACK SCALE. *Saissetia oleae.* This dark brown to black scale is primarily a pest of citrus on the West Coast. It attacks a wide variety of trees and shrubs in the South and West, including the Deodar cedar (*C. deodara*). Besides extracting juice from the plant, it secretes a substance on the leaves and stems on which the sooty mold fungus grows. Scale crawlers are vulnerable to treatments in late spring. Cottony-cushion, greedy, and latania scales have also been reported on cedar.

• *Control.* See under scales, Chapter 11.

DEODAR WEEVIL. *Pissodes nemorensis.* Beginning in spring, this brownish snouted weevil feeds on the cambium of leader and side branches of Deodar, Atlas, and Lebanon cedars. It deposits eggs in the bark, and the $1/3$-inch-long white grubs that hatch from the eggs burrow into the wood. Eventually the leaders and terminal twigs turn brown and die. Small trees may be killed by this pest.

• *Control.* See under weevils, Chapter 11.

OTHER INSECTS. Red-headed pine sawfly, *Neodiprion lecontei*; bagworm (see under Juniper); and mealybug, *Pseudococcus longispinus,* attack cedar.

CEDAR, INCENSE (*Libocedrus*)

This handsome columnar or narrow, pyramidal tree grows mainly in the northwestern United States, although it can be grown in warmer parts of the country. Incense cedar is useful as a tall hedge or windbreak, and the foliage is aromatic when crushed.

Diseases

Bacterial crown gall caused by *Agrobacterium tumefaciens*; blight by the fungus *Herpotrichia nigra*; branch canker by *Coryneum cardinale*; needle cast by *Lophodermium juniperinum*; root rot by *Phymatotrichum omnivorum*; and rust by *Gymnosporangium libocedri* are recorded on incense cedar. The latter forms conspicuous witches' brooms, killing small sprays of foliage.

• *Control.* Control measures are rarely necessary.

Insects

The cypress bark beetle, cypress tip moth, cypress mealybug, and four species of scales—cypress, juniper, pine needle, and Putnam—infest incense cedar.

• *Control.* Effective controls for the cypress bark beetle have not been developed. For mealybugs and scales, see Chapter 11.

Other Pests

The mistletoe *Phoradendron juniperinum* f. *libocedri* infests incense cedar in the western states.

• *Control.* Pruning infested branches will reduce local spread of mistletoe.

CEDAR, WHITE. See False Cypress.

CHAMAECYPARIS. See False Cypress.

CHASTE-TREE (*Vitex*)

This low-growing tree or shrub produces showy purple clusters of blooms in late summer in the North and earlier in the South.

Diseases

LEAF SPOT. *Cercospora viticis.* This disease occurs on chaste-tree along the Gulf Coast.

• *Control.* The disease is rarely severe enough to warrant control measures.

ROOT ROT. *Phymatotrichum omnivorum.* Root rot is present on chaste-tree in Texas.

• *Control.* Control measures are not practicable.

Insects

SCALES. Pit-making pittosporum and oleander scales attack chaste-tree.

CHERRY, JAPANESE FLOWERING, BLACK, AND CHOKE (*Prunus serrulata, P. yedoensis, P. serotina, P. virginiana,* and related species); also FLOWERING ALMOND (*P. dulcis*), FLOWERING PEACH (*P. persica*), and FLOWERING PLUM (*P. cerasifera*).

Flowering *Prunus* species are used extensively as ornamentals. In northern latitudes they are occasionally damaged by low winter temperatures, as is evidenced by longitudinal cracks (Fig. III-14) on the south or west side of the trunk or in the branch crotches. Flowering *Prunus* species are best planted in fall. Native black cherries, which develop into large trees,

Fig. III-14. Winter injury to flowering cherry. This species is particularly susceptible in areas with very cold winter temperatures.

are adapted throughout the eastern United States.

Landscape cherries, peaches, plums, and almonds are susceptible to some of the fungi and insect pests that attack the *Prunus* species grown for fruit. These parasites, however, have not been studied in detail on landscape varieties, and any treatment must be based on the control measures suggested for fruit crops. Information on the diseases and insects of fruiting *Prunus* is readily available from the state agricultural experiment stations.

Bacterial and Fungal Diseases

BACTERIAL SPOT. *Xanthomonas pruni.* The bacterium that attacks peaches and cherries in orchards is also known to attack Japanese cherries, causing a familiar "shot-hole" appearance. The infected tissue dries up and falls out, leaving a hole about 1/8 inch in diameter. This bacterium is also capable of causing stem canker and gummosis, although other bacteria such as *Pseudomonas syringae* are more often involved. Shot-holes in cherry leaves may also be caused

Fig. III-15. Black knot of cherry caused by the fungus *Apiosporina morbosa*.

by the fungus *Coccomyces hiemalis,* discussed below, and by viral infection.

LEAF SPOT. *Blumeriella jaapii,* formerly called *Coccomyces hiemalis.* This disease is widespread during rainy springs. The reddish spots on the leaves drop out, leaving circular holes. Complete defoliation may follow.

SHOT-HOLE. *Wilsonomyces carpophilus.* This fungus causes a leaf and fruit spot disease of several *Prunus* species in the landscape. The disease is sometimes referred to as coryneum blight.

• *Leaf spot control.* See under leaf spots, Chapter 12.

POWDERY MILDEW. *Podosphaera clandestina* and *Sphaerotheca pannosa.*

The Japanese cherry is subject to the same powdery mildew that attacks edible cherries. The leaves and twigs become coated with a mat of fungus growth, which causes dwarfing and death of these branches. The disease is uncommon.

• *Control.* See under powdery mildew, Chapter 12.

BLACK KNOT. *Apiosporina morbosa.* Black, rough cylindrical-shaped galls (Fig. III-15) develop on the twigs of apricots, cherries, and plums. Neglected trees appear to be especially subject to this disease. Wild black cherries are susceptible to black knot and, when growing near the landscape, may be an important source of inoculum for disease of domestic *Prunus.*

• *Control.* Prune knotted twigs and excise knots on large branches during the winter. Spray with Ferbam, lime-sulfur, or Tribasic Copper Sulfate when the trees are dormant, at pink bud stage, at full-bloom stage, and 3 weeks later.

WITCHES' BROOM. *Taphrina wiesneri.* Japanese flowering cherry seems to be quite susceptible to this disease. Large branches may become deformed when many irregular dwarfed branches grow to form a witches' broom. Blossoms develop and leaves emerge on the brooms earlier than on the normal branches. Sometimes large numbers of very small brooms develop all over the tree, killing the end branches and eventually the whole tree. *T. deformans* causes peach leaf curl, and *T. communis,* plum pockets.

• *Control.* Cut off and destroy the brooms. Spray Ferbam in fall or in early spring.

CROWN GALL. *Prunus* species are susceptible. See Chapter 12.

SHOOT AND TWIG BLIGHT. *Monilinia*

fructicola and *M. laxa.* These fungi cause flowers to turn brown and rot in moist weather and also cause leaf blight and fruit brown rot.

• *Control.* Prune and destroy infected twigs. To prevent disease, spray with Benlate or Captan as flowers open, and repeat in 10 days.

WILT. Verticillium wilt affects this genus. See Chapter 12.

PERENNIAL CANKER. *Leucostoma cincta* and *L. persoonii.* These fungi, formerly called *Cytospora,* are opportunistic pathogens, attacking trees injured by cold temperature or drought. Symptoms include twig, branch, and limb cankers, which can seriously damage *Prunus.* As with insect injuries and other cankers, perennial cankers also cause gummosis.

• *Control.* Prune out cankered twigs and branches.

BACTERIAL LEAF SPOT AND TWIG CANKER. *Pseudomonas syringae* pv. *syringae.* This common epiphyte infects *Prunus* species and is also involved in ice nucleation. Symptoms include dormant bud death, flower blast, shoot tip dieback, blackened twigs, stem cankers, and leaf spots or vein discoloration. Spring frosts and wet weather favor the disease.

• *Control.* Copper or Streptomycin sprays are suggested, although in some areas bacteria have developed resistance to these chemicals.

ROOT ROT. *Prunus* species are susceptible to attack by *Armillaria mellea.* See Chapter 12. *Xylaria mali,* the dead man's finger fungus, also decays cherry roots.

WOOD DECAY. *Phellinus tuberculosus.* See Chapter 12.

Viral Diseases

Prunus species are susceptible to mosaic, ringspot, green ring mottle, and stem pitting viruses. Viral infection of fruit trees can be prevented by careful sanitary practices during propagation. Once landscape trees are infected, there is no control.

Phytoplasmal Diseases

Peach yellows and X-disease are important phytoplasmal diseases of *Prunus.* Yellowing, tufted growth, leaf roll and shot-hole, and

decline are some symptoms of these diseases. Control measures are not available.

Abiotic Disease

YELLOWING. Very often, yellowing and premature defoliation of flowering cherry occur without previous spotting of the leaves. These symptoms are associated with excessively wet or dry soils or with low-temperature injuries to the crown and roots. As a rule, a second set of normal-looking leaves is formed after the premature defoliation in spring or early summer.

Insects and Other Animal Pests

PEACH TREE BORER. *Sanninoidea exitiosa.* The grubs of the peach borer cause a great amount of damage to flowering peach as well as to related forms, frequently causing death of the trees. The damage is marked by profuse gummosis at the crown and on the main roots just below the surface of the soil. The trees fail to grow properly, and the leaves turn yellowish. The frass, or borings, of the grubs becomes mixed with the gum. If one follows down the burrows with a chisel or a penknife, white flat grubs from $1/2$ to 1 inch long with brown heads may be found.

LESSER PEACH TREE BORER. *Synanthedon pictipes.* This borer attacks growing tissue anywhere in the trunk from the ground to the main branches. The adult, a metallic, blue-black, yellow-marked moth, emerges in May in the South and in June in the vicinity of New York City to deposit eggs on the bark higher up in the tree than the peach tree borer.

BORERS. Shot-hole and peach bark beetle borers, *Scolytus rugulosus* and *Phloeotribus liminaris,* attack weakened trees; leopard moth larvae (see leopard moth borer, under Maple) tunnel into wood. American plum borer (see under Planetree) and a clear-winged moth, the dogwood borer, also attack *Prunus* (see borers, under Dogwood).

• *Borer Control.* See under borers, Chapter 11.

ORIENTAL FRUIT WORM. *Graphiolitha molesta.* Wilting of the tips of twigs may be due to boring by the oriental fruit worm, a small pinkish white larva, about $1/2$ inch long. The adult

Fig. III-16. Pear slug sawfly larvae skeletonizing a cherry leaf.

female is about ¹/₂ inch long and is gray with chocolate brown markings on the wings. Larvae overwinter in the soil. Oak twig pruner and twig girdler each attack *Prunus.*

• *Control.* Where only a few trees are involved, removal and destruction of wilted tips as they appear is usually sufficient. See under caterpillars, Chapter 11.

PEAR SLUG. *Caliroa cercasi.* These so-called slugs, olive green, semitransparent, and slimy, are the larvae of a sawfly. They are about ¹/₂ inch long, swollen at the front, and shaped somewhat like a tadpole (Fig. III-16). They occasionally infest *Prunus* and may completely skeletonize the leaves. There are two generations per year in the northern states and three in the southern.

• *Control.* See under sawflies, Chapter 11.

PLANTHOPPERS. *Metcalfa pruinosa* and *Ormensis septentrionalis.* These planthoppers injure shrubs and trees by sucking the juices of the more tender branches, which they cover with a woolly substance. The former, also called mealy flata, is ¹/₄ inch long with purple-brown wings and is covered with white woolly matter. It also attacks mulberry. The latter has a beautiful blue-green color and is covered with white powder, as illustrated in Fig. III-17. These insects are about ¹/₂ inch long and are very narrow. They also feed on hawthorn. The rose and other leafhoppers and buffalo tree hoppers also injure *Prunus* foliage.

• *Control.* The problem is rarely serious enough to warrant control measures.

EASTERN TENT CATERPILLAR. See under Willow. Wild cherries are the natural hosts of the tent caterpillar. See Fig. III-18.

OTHER CATERPILLARS. *Dichomeris ligulella,* a juniper webworm relative; yellow-necked caterpillar (see under Oak); red-humped caterpillar (see under Poplar); fall webworm (see under Ash); forest and California tent caterpillars (see under Maple and Strawberry-Tree); and ugly nest caterpillar, *Archips cerasivoranus,* are all *Prunus* foliage feeders.

CAMBIUM MINER. See under Holly.

LEAF MINERS. California casebearer, *Coleophora sacramenta*; and cherry leaf miner, *Phyllonorycter crataegella,* an apple pest, have been reported.

MITES. Spindle gall and marginal fold-forming eriophyid mites are found on cherry leaves.

SAN JOSE SCALE. *Aspidiotus perniciosus.* Gray, closely appressed masses of circular scales, ¹/₁₀ inch in diameter, each with a raised nipple in the center, are characteristic of the San Jose

Fig. III-17. Planthoppers (*Ormensis septentrionalis*) on a cherry branch.

Fig. III-18. Eastern tent caterpillars on wild cherry. Egg masses on twigs.

scale. The winter is passed in the immature stages on the bark.

WHITE PEACH SCALE. *Pseudaulacaspis pentagona.* The presence of this scale is indicated by white encrustations on the bark, which are masses of circular white scales $1/10$ inch in diameter, which suck juices from below the bark. Other *Prunus* scales include cottony-cushion, cottony maple, black, hickory lecanium, terrapin, European peach, globose, greedy, frosted, oystershell, oleander, walnut, euonymus, and Comstock mealybug.

• *Control.* See under scales, Chapter 11.

GREEN PEACH APHID. *Myzus persicae.* This insect can be very injurious during the summer to a number of varieties of flowering peach and related ornamentals.

BLACK PEACH APHID. *Brachycaudus persicae.* This aphid is common on commercial plantings of peaches.

WATERLILY APHID. *Rhopalosiphum nymphaeae.* Ornamental species of *Prunus* grown near waterlily ponds are seriously attacked by the waterlily aphid. The insects migrate to the trees in May and June and in autumn. Apple aphid, *Aphis pomi,* also attacks *Prunus.*

• *Control.* See under aphids, Chapter 11.

ROOT NEMATODE. *Pratylenchus penetrans.* Research at the New York State Experiment Station revealed that this nematode attacks the roots of edible cherry trees. It is possible that the roots of Japanese flowering cherries are also susceptible to the same nematode.

ASIATIC GARDEN BEETLE. *Maladera castanea.* The leaves are chewed at night by a brown beetle $1/4$ inch long. During the day the insect hides just below the soil surface.

• *Control.* See foliage-feeding beetles, Chapter 11.

JAPANESE BEETLE. See under Linden.

OTHER BEETLES. Fuller rose beetles (see under Camellia) also attack *Prunus.*

APPLE AND THORN SKELETONIZER. See under Hawthorn.

GIANT WHITEFLY. See under Citrus.

CHESTNUT (*Castanea*)

Although the majestic American chestnut is no longer a part of the natural landscape in the eastern United States, substitutes such as Chinese chestnuts are occasionally planted as landscape and backyard nut trees.

Diseases

BLIGHT. The rapid disappearance of the American chestnut, one of our best forest, ornamental, and nut trees, is the result of infection by one of the most virulent tree parasites and is too

well known to warrant much discussion in this book.

Though blight is essentially a disease of the American chestnut, it commonly occurs on the chinquapin (*Castanea pumila*) as well. The causal fungus has been found growing on red maple, shagbark hickory, live oak, and staghorn sumac (*Rhus typhina*), and on dead and dying white, black, chestnut, and post oaks.

• *Symptoms.* The fungus attacks the trunk and branches, causing cankers that girdle the stems and cause the leaves on one or more branches or even on the entire young tree to suddenly wilt, turn brown, and hang dry on the branches. The cankers are discolored, slightly sunken areas in the bark (Fig. III-19). Layers of flat, fan-shaped, buff-yellow wefts of fungus tissue are usually revealed between the bark and the sapwood when the bark in the cankered area is lifted or peeled carefully.

• *Cause.* Blight is caused by the fungus *Cryphonectria parasitica,* formerly referred to as *Endothia parasitica.* The fungus produces tiny, pinpoint fruiting bodies, the tips of which barely protrude from the bark. In damp weather, sticky yellowish orange masses of spores ooze from the openings in these bodies. The spores are splashed by rain or are carried by birds and crawling and flying insects to wounds in the bark below or to nearby trees, where they cause new infections. A second type of fruiting structure embedded in the bark produces ascospores, which are shot into the air and are blown for many miles by the wind. The windblown spores lodge in open wounds on chestnut trees and produce new infections when conditions are favorable. Chestnut blight has been unchecked because it is impossible to prevent the dissemination of spores by wind, birds, and insects.

• *Control.* Cankers can be restricted by the introduction of a hypovirulent (weak) parasitic strain of *Cryphonectria parasitica* into the canker. Ways are still being sought to transmit the hypovirulent strains to cankered trees so that the fungus causing the disease can become weakened too. A moist soil compress prepared from soil beneath the tree and affixed to cankers

Fig. III-19. Left: Chestnut blight on young sprouts. The leaves hang dry and brown on girdled stems. Notice the stump of the original tree. Right: Chestnut blight canker with fungal fruiting bodies dotting the infected area.

on stems and branches will cause canker remission. The extensive labor involved and possible retreatment needed makes this method impractical on a large scale.

The best long-term solution is the current development and eventual use of disease resistant American chestnuts, but it will be many years before a resistant American chestnut becomes available. Asiatic chestnuts (*C. japonica* and *C. mollissima*) are resistant to chestnut blight. Most of them are best suited for ornamental plantings and nut production rather than as forest trees. Many have a shrublike growth habit, with multiple trunks arising near the ground level. They do not grow as tall and straight as the American chestnut and cannot compete in wooded areas when interplanted with other trees.

TWIG CANKER OF ASIATIC CHESTNUTS. The increased use of Asiatic species of chestnuts because of their resistance to blight, has revealed that these species are susceptible to a less destructive disease, twig canker.

• *Symptoms.* Cankers on trunks, limbs, or twigs result from fungal invasion of the bark, cambium, and sapwood. Cankers on the branches may cause complete girdling, followed by death within a single season. When complete girdling does not take place in one season, the canker may be callused over temporarily; however, the infection may continue the following season to cause death of the parts above the canker.

• *Cause.* Several fungi have been associated with this disease, the most common and virulent one being *Cryptodiaporthe castanea.* It has been found to kill young trees in nurseries and older trees in permanent plantings, but more commonly it kills individual branches resulting in decreased growth and deformed trees.

• *Control.* Twig canker is most prevalent on trees in poor vigor. Maintaining good vigor will do much to ward off attacks. Carefully select planting sites and then fertilize and water to ensure vigorous growth. All unnecessary injuries to the trees should be avoided, since the causal organisms penetrate and infect most readily through bark wounds. Trees should be carefully inspected in early summer, when symptoms are most apparent, and affected twigs should be pruned to sound wood and destroyed. Large cankers on the trunk or larger branches should be removed by surgical methods.

OTHER DISEASES. The fungus *Monochaetia kansensis* causes a leaf spot of *Castanea mollissima* in Kansas and Mississippi. Powdery mildews, *Microsphaera americana* and *Phyllactinia guttata,* also attack chestnut.

Insects

The insect pests that attack American chestnut are of no importance because of the destruction of the trees by blight. Following, however, are some insects that can attack living American and Asiatic chestnuts.

WEEVILS. *Curculio auriger* and *C. proboscideus.* Two species of weevils native to the United States may seriously damage the nuts of the Asiatic species of chestnuts. Asiatic oak weevil (see under Oak) feeds on chestnut.

• *Control.* See under weevils, Chapter 11.

CATERPILLARS. Yellow-necked caterpillar (see under Oak) and spring cankerworm and whitemarked tussock moth (see under Elm) feed on chestnut foliage.

BORERS. Brown wood borer (see under Beech), leopard moth borer (see under Maple),

and two-lined chestnut borer (see under Oak) all attack chestnut.

JAPANESE BEETLE. See under Linden.

TWIG PRUNER. See oak twig pruner, under Hickory.

SCALES. Obscure and lecanium scales (see under Oak) attack chestnut.

GIANT BARK APHID. See aphids, under Sycamore.

SPIDER MITES. See oak mite, under Oak.

CHINABERRY (*Melia*)

This southeast Asia native, which thrives in difficult growing conditions, is planted in the southeastern United States for quick, dense shade.

Diseases

Chinaberry is susceptible to *Peronospora parasitica* downy mildew, *Phyllactinia guttata* powdery mildew, and *Cercospora meliae* and *C. subsessilis* leaf spots. Controls for these diseases are normally not needed.

Insects

Two scales, greedy and white peach, feed on this host.

CHINESE JUJUBE (*Ziziphus*)

This low-growing tree is adapted to dry, alkaline, salty soils.

CHINESE PISTACHIO (*Pistacia*)

This tree is widely adapted in the United States and tolerates summer heat. Chinese pistachio has no serious disease or insect problems.

CHINESE TALLOW TREE (*Sapium*)

This medium-sized tree, adapted to mild climates, grows best in moist, acid soils. It grows rapidly and is relatively free of diseases and insect pests.

CITRUS (*Citrus*)

Citrus trees are grown in backyard landscapes as well as in commercial orchards in the frost-free areas of the southern and western United States.

Diseases

BACTERIAL DISEASES. Citrus foliage is susceptible to two foliar bacterial diseases: citrus blast, *Pseudomonas syringae,* in California; and the dreaded citrus canker, *Xanthomonas citri.* In the latter case, even patio citrus trees are subject to intense scrutiny from plant industry officials because of citrus's commercial importance.

LEAF SPOTS. Foliar diseases caused by fungi include scab, *Elsinoe fawcetii;* melanose, *Diaporthe citri;* leaf spot, *Septoria citri* or *S. limonum;* grease spot, *Cercospora citri-grisea;* tar spot, *C. gigantea;* and anthracnose, *Glomerella cingulata.*

TRUNK AND ROOT DISEASES. Stems and roots are affected by brown rot gummosis or foot rot caused by several *Phytophthora* species. *Clitocybe tabescens* and *Armillaria mellea* also attack citrus roots.

OTHER DISEASES. Citrus is also susceptible to a phytoplasma, several viral diseases, and to deficiencies of the mineral nutrients copper, zinc, manganese, magnesium, boron, and iron.

Insects

BAGWORM. *Oeketus abbotii.* Bagworm feeds on citrus.

WEEVILS. Fuller rose beetle (see under Camellia) and citrus root weevil attack citrus.

APHIDS. Several types including black citrus, spirea, potato, and cowpea feed on citrus, often producing sufficient honeydew for growth of sooty mold fungi.

WHITEFLIES. Greenhouse whitefly, *Trialeurodes vaporariorum;* citrus whitefly, *Dialeurodes citri;* and woolly whitefly, *Aleurothrixus floccosus,* withdraw sap from leaves and produce honeydew. The giant whitefly, *Aleurodiscus dugesii,* found in coastal southern California, feeds on leaves, weakening the trees. This insect also secretes long waxy threadlike filaments on leaf undersides.

SCALES. The following attack citrus: longtailed and Comstock mealybug, cottony-cushion, black, hemispherical, ivy, brown soft, citricola, California red, Florida red, dictyosper-

mum, citrus, snow, and tea. For control, see under scales, Chapter 11.

OTHER INSECTS. A treehopper, leaf-footed bug, and greenhouse and Cuban laurel thrips attack citrus.

CORK TREE (*Phellodendron*)

Cork trees grow well in cities and are unusually free of pests. The only insects recorded on this host are the lesser snow scale, *Pinnaspis strachani,* and the pustule scale, *Asterolecanium pustulans.*

Malathion or Sevin sprays applied in late spring or early summer will control the crawler stages of these pests.

COTTONWOOD. See Poplar.

CRABAPPLE, FLOWERING (*Malus*)

Over 600 species and varieties of flowering crabapple are grown in North America. Flowering crabapple, which needs some winter chilling, is widely adapted to most climates except those of southern California and other mild weather areas. Trees used for ornamental purposes are subject to attack by many of the fungi and insects that occur on commercial fruiting apples.

The five most common diseases of flowering crabapple are scab, fire blight, cedar-apple rust, powdery mildew, and black rot. Some species and cultivars that are unusually resistant to all five diseases include 'Albright', 'Baskatong'*, 'Christmas Holly', 'Cotton Candy', 'David', 'Dolgo', 'Henry Kohankie', 'Jewelberry'*, 'Prof Sprenger', *Malus x robusta* 'Persicifolia', *sargentii* cv. 'Tina'*, 'Sentinel', and 'Sugar Tyme'. There are several other excellent crabapples such as 'Beverly', 'Coralburst', 'Liset', *rocki,* and 'Snowdrift' that are susceptible to only one of the important diseases. They would be good choices in areas where the disease in question is rare. There are additional cultivars such as 'Anne E.'*, 'Bob White'*, 'Candymint', 'Golden Raindrops', *Malus baccata* cv. 'Jackii', 'Louisa'*, 'Mary Potter', 'Molten Lava', 'Ormiston Roy'*, 'Prairiefire'*, 'Prairie Maid', 'Redbud', 'Red Jade', 'Red Jewel'*, 'Sinai

Fig. III-20. Flowering crabapple thread blight; fungal sclerotia and rhizomorphs are clearly visible.

Fire', 'Strawberry Parfait', 'Tea'*, and 'White Angel'* that resist scab, the most common of the crabapple diseases. Cooperative Extension offices have information on the best cultivars to grow locally. (Cultivars marked * are also less damaged by Japanese beetle.)

Bacterial and Fungal Diseases

SCAB. *Venturia inaequalis.* Olive-drab spots ¼ inch in diameter appear on the leaves, which drop prematurely, and the fruits are disfigured. The imperfect stage of the fungus, *Fusicladium dendriticum,* overwinters on the twigs and fallen leaves. Many flowering crabapples such as 'Almey', 'Hopa', 'Radiant', 'Red Baron', 'Royalty', 'Velvet Pillar', and 'Weeping Candied Apple' are so susceptible to scab that in wet seasons they are nearly defoliated by midsummer. Susceptible trees provided with good scab control produce more profuse blooms the following season than those allowed to become defoliated. The disease resistant cultivars listed earlier should retain their leaves the full season and should not suffer from reduced bloom due to scab.

• *Control.* See under leaf spots, Chapter 12.

FIRE BLIGHT. *Erwinia amylovora.* In most seasons, this disease will not be serious on flowering crab unless commercial orchards of pears and apples are nearby.

• *Control.* See under fire blight, Chapter 12. Use disease-resistant crabapples listed above.

RUST. *Gymnosporangium juniperi-virginianae.* When common red cedar with cedar-apple rust galls is grown near the landscape, a number of brown or orange spots bearing the aecial stage of the rust may later appear on the leaves of nearby flowering crabapples. Much defoliation may follow heavy infection. Crabapple twigs are sometimes killed by a related disease, cedar-quince rust.

• *Control.* See under rust diseases, Chapter 12.

POWDERY MILDEW. *Podosphaera leucotricha.* Leaves and terminal shoots of many crabapple cultivars are attacked by this fungus.

• *Control.* See Chapter 12.

BLACK ROT. *Botryosphaeria obtusa* (imperfect stage, *Sphaeropsis malorum*). This fungus causes cankers on the trunks of crabapples. It often gains entrance through wounds made by lawn mowers and other maintenance equipment, or through fire blight cankers. This fungus also causes a leaf spot called "frogeye" and a fruit rot disease.

• *Control.* Avoid wounding trees. Increase the tree's vigor by fertilizing and watering. Fungicidal sprays will control leaf spot.

OTHER DISEASES. Flowering crabapples are also subject to thread blight (Fig. III-20), *Cerato-*

Fig. III-21. Female periodical cicadas lay their eggs in young twigs and branches.

basidium stevensii; canker, *Botryosphaeria dothidea*; collar rot, *Phytophthora cactorum*; and crown gall, *Agrobacterium tumefaciens*. Viral diseases causing leaf mosaic and stem and limb abnormalities are also found on flowering crabapple.

Abiotic Disease

AIR POLLUTION. Crabapples are sensitive to sulfur dioxide. See Chapter 10.

Insects and Related Pests

Management of most insects that infest the leaves and twigs, such as aphids, alder lace bug, leafhoppers, and several kinds of caterpillars, is discussed in Chapter 11.

PERIODICAL CICADA. *Magicicada septendecim.* This pest (Fig. III-21), also known as the 17-year locust, damages branches of crabapple, apple, oak, and many other trees by making deep slits in the bark during the egg-laying period. Such branches are easily broken during windy weather. Following hatch, nymphs drop to the ground, enter the soil, and begin feeding on roots.

Tree decline may be observed after many years.

• *Control.* A spray containing Carbaryl, Dursban, or Sevin gives excellent control if applied in spring and summer at the time the cicadas are in the trees.

FRUIT TREE LEAF ROLLER. *Archips argyrospilus.* Young caterpillars are green with a shiny black head. They tie or roll leaves together with silken threads and feed inside the nest. This pest feeds and produces webbing on flowering crabapple in the spring. It also feeds on cherry, poplar, oak, hawthorn, and elm.

• *Control.* Apply Imidan or Guthion sprays in late May and early June.

FOLIAR-FEEDING CATERPILLARS. Crabapple leaves are food for yellow-necked caterpillar, tussock moth, eastern tent caterpillar, and many other lepidopterous larvae.

APPLE AND THORN SKELETONIZER. See under Hawthorn.

JAPANESE BEETLE. The insect is described under Linden. There are good, disease-resistant crabapple cultivars that are not attractive to Japanese beetles. See under Crabapple.

BORERS. Leopard moth and flatheaded borers (see under Maple), dogwood borer (see borers, under Dogwood), as well as several other insects feed on the wood of flowering crabapple.

APHIDS. Apple aphid, *Aphis pomi*, feeds primarily on crabapple leaves; *Eriosoma rileyi* is a bark-feeding species. Woolly apple aphid, *Eriosoma lanigerum*, also attacks crabapples. See Chapter 11.

SCALES. Cottony-cushion, cottony maple, black, European fruit lecanium, calico, frosted, oystershell, greedy, California red, and San Jose scales infest crabapples. See Chapter 11.

EUROPEAN RED MITE. *Panonychus ulmi.* This species also infests the leaves of black locust, elm, mountain-ash, and rose, in addition to many kinds of fruit and nut trees.

• *Control.* See under spider mites, Chapter 11.

PEAR SLUG. See under Cherry.

CRAPEMYRTLE (*Lagerstroemia*)

This beautiful shrubby tree is reliably hardy only in the South, although fine specimens are growing in protected places in the latitude of New York City. Crapemyrtle grows best in a hot, dry climate and in a fertile, well-drained loam.

Diseases

POWDERY MILDEW. *Erysiphe lagerstroemiae.* This mildew is most serious in the spring and fall months and causes the leaves and shoots to be distorted and stunted. If the inflorescences are attacked, the flower buds may fail to open. The shoots and leaves may be coated with the white fungal mycelium, and the leaves may assume a reddish color. Two other species, *Phyllactinia corylea* and *Uncinula australiana,* occasionally infect this host. Japanese crapemyrtle has more mildew resistance.
* Control. See under powdery mildew, Chapter 12.

OTHER FUNGAL DISEASES. Black spot caused by a species of *Cercospora;* tip blight by *Phyllosticta lagerstroemia;* leaf spot by *Cercospora lythracearum* and *Pestalotiopsis maculans;* and root rot by *Clitocybe tabescens* are other diseases of this host. Thread blight, caused by *Ceratobasidium stevensii,* also affects crapemyrtle.
* Control. For leaf spot control, see Chapter 12.

Insects

CRAPEMYRTLE APHID. *Sarucallis kahawaluokalani.* This species attacks only crapemyrtle. It exudes a great amount of honeydew on which the sooty mold fungus, *Capnodium* sp., thrives.
* Control. See under aphids, Chapter 11.

FLORIDA WAX SCALE. *Ceroplastes floridensis.* This reddish or purplish brown scale covered with a thick, white, waxy coating tinted with pink attacks a wide variety of shrubs in the South.
* Control. See under scales, Chapter 11.

JAPANESE BEETLE. This pest does extensive damage to crapemyrtle. See under Linden.

LEAFFOOTED BUG. *Leptoglossus* sp. This insect feeds on crapemyrtle flowers.

CRYPTOMERIA (*Cryptomeria*)

Japanese cedar (*C. japonica*), an Asiatic native, is a good park tree, as it is tall and pyramidal.

This tree is practically free of fungal parasites and has no important insect pests.

Diseases

LEAF BLIGHT. Leaves and twigs may be blighted by a species of *Phomopsis* during rainy seasons.
* Control. Pruning out affected leaves and twigs is usually sufficient. Copper fungicides can be used on valuable specimens. Dormant oil sprays may cause serious damage and should never be used on this host.

LEAF SPOT. Two fungi, *Pestalotiopsis cryptomeriae* and *P. funerea,* are frequently associated with a leaf spotting of this host. Infection probably follows winter injury or some other agent.
* Control. Same as for leaf blight.

CUCUMBERTREE (*Magnolia*)

This tree is widely but not frequently distributed in cool, moist forests south of New York and Michigan. It makes a handsome ornamental with a similarity to tuliptree, a close relative.

Diseases

Nectria galligena causes target cankers on stressed trees, and *Phyllosticta cookei* causes a leaf spot.

CYPRESS* (*Cupressus*)

Arizona cypress (*C. glabra*) and Monterey cypress (*C. macrocarpa*) are well adapted to their native regions. The former is grown as a Christmas tree in the South. Italian cypress (*C. sempervirens*) is widely planted in the western United States, it tolerates dry soils.

Diseases

CANKER. Twigs and branches may be girdled by cankers, and the entire tree may be killed. Cypress, particularly the Monterey cypress, junipers, and Oriental arborvitae, are all subject to this disease.

---

*Certain species of *Chamaecyparis* and *Taxodium* are also called cypress.

• *Cause.* The fungus *Seiridium cardinale* forms lethal cankers, especially on cypress growing in hot, dry climates; it has devastated cypress growing in the Mediterranean region. *Botryosphaeria dothidea* is another canker-causing pathogen of cypress.

• *Control.* Control is difficult because the fungal spores are spread by wind-splashed rain, pruning tools, and perhaps insects and birds. Remove and destroy severely infected trees, drastically prune mildly infected ones and spray periodically with a copper fungicide, starting at the beginning of the rainy period.

CYTOSPORA CANKER. The columnar form of Italian cypress, *C. sempervirens,* along the California coast is most subject to this disease. Occasionally it also affects the horizontal form of Italian cypress and the smooth cypress.

• *Symptoms.* Smooth reddish brown cankers, from which resin flows, develop on young branches. Diseased bark on older branches becomes cracked and distorted with a more abundant flow of resin.

• *Cause.* The fungus *Cytospora cenisia* f. *littoralis* causes this canker.

• *Control.* Prune and destroy dead or dying branches. Trees with cankers on the trunks should be removed and destroyed.

OTHER DISEASES. Cypresses are subject to several other diseases: needle blight caused by *Asperisporium sequoiae* (*Cercospora sequoiae*); crown gall by the bacterium *Agrobacterium tumefaciens*; and Monochaetia canker by *Monochaetia unicornis.*

• *Control.* Control measures are rarely adopted.

TWIG BLIGHT. See Phomopsis twig blight, under Juniper.

ROOT ROT. *Clitocybe tabescens* and *Armillaria mellea.* These two fungi attack cypress. See Chapter 12.

Insects

CYPRESS APHID. *Siphonatrophia cupressi.* This large green aphid infests blue and Monterey cypress.

• *Control.* See under aphids, Chapter 11.

CYPRESS MEALYBUG. *Pseudococcus ryani.* Primarily a pest of Monterey cypress in California, this species also infests arborvitae, redwood, and other species of cypress.

• *Control.* Control measures have not been developed.

CATERPILLARS. The caterpillars of the following moths feed on cypress: tip moth, *Argyresthia cupressella*; webber, *Epinotia subviridis*; imperial, *Eacles imperialis*; and white-marked tussock moth (see under Elm).

• *Control.* See under caterpillars, Chapter 11.

CYPRESS BARK MEALYBUG. *Ehrhornia cupressi.* Sometimes called a scale, this pink mealybug covered with loose white wax primarily infests Monterey cypress. It occasionally attacks Guadalupe and Arizona cypress and incense cedar. Leaves of heavily infested trees turn yellow, then red or brown.

• *Control.* See under scales, Chapter 11.

OTHER SCALES. Several other scales, including the cottony-cushion and the juniper, occasionally infest cypresses.

CYPRESS TIP MOTH. *Argyresthia* and *Recurvaria* spp. Both species attack and mine foliage tips.

BAGWORM. See under Juniper.

CYPRESS SAWFLY. *Susana cupressi.* The grayish-green larvae damage cypress in southern California.

DESERT WILLOW (*Chilopsis*)

A catalpa relative, desert willow is a small tree native to the arid southwestern United States. Few pathogens or insect pests are important on this host. *Phyllosticta erysiphoides* causes a leaf spot disease.

DOGWOOD (*Cornus*)

The dogwoods, among our best ornamental low-growing trees, are subject to several important fungal diseases and insect pests. Flowering dogwood (*C. florida*) is native to the eastern United States, and cultivars adapted to various parts of the range are available. Pagoda dogwood (*C. alternifolia*), a little hardier, and Kousa dogwood (*C. kousa*),

Fig. III-22. Top: Leaves killed by dogwood anthracnose. Bottom: Anthracnose fungus fruiting bodies, much enlarged, extruding conidia on plant surface.

which blooms later, are also good landscape trees.

Diseases

ANTHRACNOSE. In the northeastern United States, a leaf spotting, blighting, and twig dieback disease (Fig. III-22) caused by the fungus *Discula destructiva* was first found devastating *C. florida* in 1979 by P. P. Pirone. Discula anthracnose has since spread throughout the range of dogwood from New York to Florida and has been especially devastating in cool, moist Appalachian Mountain areas, where it has resulted in nearly complete loss of native dogwoods in some areas. This disease also occurs on *C. nuttallii* in the Pacific Northwest.

• *Control.* Provide good growing conditions, including adequate water, light, and borer controls. Prune out dead and dying branches as well as limb and trunk sprouts. See under anthracnose, Chapter 12.

POWDERY MILDEW. *Microsphaera pulchra* and *Phyllactinia guttata.* Powdery mildews may attack dogwood, entirely covering the leaves

Fig. III-23. Effects of powdery mildew on dogwood foliage.

with a thin white coating of the fungus. Infection may result in distortion of emerging leaves, dead patches and reddish splotches on leaves, leaf scorching, and poor leaf color and vigor. Often the fungus grows so sparsely on the leaf surface that it is not easily seen. In some landscapes, leaves are so severely curled and scorched that they may be nearly dead by late summer (Fig. III-23). While *C. florida* is very susceptible to powdery mildew, other species such as *C. kousa* and *C. mas* are resistant.

• *Control.* See under powdery mildew, Chapter 12.

CROWN CANKER. An unthrifty appearance is the first general symptom. The leaves are smaller and lighter green than normal and turn prematurely red in late summer. At times, especially during dry spells, they may curl and shrivel. Later, twigs and even large branches die. At first the diseased parts occur principally on one side of the tree, but within a year or two they may appear over the entire tree.

The most significant symptom, and the

cause for the weak top growth, is the slowly developing canker on the lower trunk or roots, at or near the soil level. Although the canker is not readily discernible in the early stages, it can be located by careful examination. Cutting into it will reveal that the inner bark, cambium, and sapwood are discolored. Later, the cankered area becomes sunken, and the bark dries and falls away, leaving the wood exposed. When the canker extends completely around the trunk base or the root collar, the tree dies.

• *Cause.* Crown canker is caused by the fungus *Phytophthora cactorum*. The same parasite apparently causes the so-called bleeding canker disease of maples and canker of American beech. This fungus also parasitizes a large number of other plants. *P. cactorum* can survive in the soil in partly decayed organic matter, and its spores may be washed to nearby uninfested areas. It appears to gain entrance primarily through wounds and then invades the tissues in all directions. Thus far this basal canker is found only on transplanted trees in ornamental plantings. Dogwoods subjected to periodic flooding and *P. cactorum* may sustain root and collar rot damage whether or not they are wounded.

• *Control.* See under root rots, Chapter 12.

DIEBACK. *Botryosphaeria dothidea.* Dieback of dogwood branches, particularly the pink-flowering kinds, is frequently caused by this species of *Botryosphaeria.* The dieback can be erroneously attributed to dogwood borers.

• *Control.* No effective control measures have been developed.

CANKER. A canker disease of unknown cause affects the trunk and lower branches (Fig. III-24) of flowering dogwoods in landscapes, nurseries, and in the wild. The disease causes a one-sided flattening of the stem and scaly, rough bark. These cankers are preferred sites for dogwood borer egg laying.

FLOWER AND TWIG BLIGHT. *Botrytis cinerea.* In rainy seasons the white flower bracts fade and rot. The fungus develops on the rotting bracts and infects leaves on which they fall (Fig. III-25). In some cases twigs are also blighted.

• *Control.* Spray the entire tree lightly with Benomyl or Cleary's 3336 early in the flowering period.

LEAF SPOTS. *Ascochyta cornicola, Cercospora cornicola, Colletotrichum gloeosporioides, Elsinoe corni* (Fig. III-26), *E. floridae, Phyllosticta cornicola, Ramularia gracilipes, Septoria cornicola* (Fig III-27), and *S. floridae.* Many species of fungi cause leaf spots on this host.

• *Control.* See under leaf spots, Chapter 12.

TWIG BLIGHTS. *Myxosporium everhartii, Cryptostictis* sp. and *Sphaeropsis* sp. Cankers and blighting of dogwood twigs may be caused by these species of fungi.

• *Control.* Prune and destroy infected twigs. Fertilize and water to increase vigor of the tree.

OTHER DISEASES. Other diseases of dogwood include root rots caused by *Armillaria mellea* in the North, its counterpart *Clitocybe tabescens* in the South, and *Phymatotrichum omnivorum.* The Pacific dogwood (*C. nuttallii*) is subject to canker caused by *Nectria galligena,*

Fig. III-24. Dogwood canker.

collar rot by *Phytophthora cactorum,* and leaf disease by *Placosphaeria cornicola.*

• *Control.* Control measures are not available.

The flowering dogwood, *C. florida,* is especially sensitive to dry soils. In prolonged dry spells its leaves wilt at the margins and turn brown and appear as though scorched by fire. Applying water to specimens growing in lawns during dry spells will help to prevent this condition.

Insects

BORERS. At least seven kinds of borers attack dogwoods. The most serious are the flatheaded

Fig. III-25. Blight of flowering dogwood leaves caused by the fungus *Botrytis cinerea.*

borer (see under Maple) (Fig. III-28) and the dogwood borer, *Synanthedon scitula.* The dogwood borer adult (Fig. III-29) is a clear-winged moth which resembles a wasp. Eggs are laid on the bark, and borer larvae enter through wounds and scars. They feed in the cambium and produce sawdust-like frass, which may appear on the bark surface. Infested trees eventually become weakened. This insect is attracted to injured trees and trees growing in full sun. Dogwood twig borers, *Oberea tripunctata* (see under Elm) and *O. ulmicola,* also attack dogwood.

• *Control.* See under borers, Chapter 11.

DOGWOOD CLUB GALL. *Mycodiplosis clavula.* Club-shaped galls or swellings, $^{1}/_{2}$ to 1 inch long, on twigs of flowering dogwood are caused by a reddish brown **midge** which attacks

the twigs in late May. Small orange larvae develop in the galls, which drop to the ground in early fall (Fig. III-30).

• *Control.* Pruning and destroying twigs with the club galls during summer will provide control.

Spraying with Carbaryl or Sevin as leaves are expanding in the spring will also control this pest.

LEAF MINER. *Xenochalepus dorsalis.* The flat yellowish white larvae, up to $^{1}/_{4}$ inch long, attack black locust leaves, making blisterlike mines on the underside. The adult beetles skeletonize dogwood leaves by feeding on the underside.

• *Control.* See under leaf miners, Chapter 11.

SCALES. Several species of scales may infest dogwood: dogwood, cottony maple (Fig. III-31),

Fig. III-26. Small, slightly raised reddish gray spots on dogwood leaves are caused by the fungus *Elsinoë corni*. Spots on the white bracts are also caused by *E. corni*.

obscure, oystershell, San Jose, ivy, mealybug, calico, white peach, and tea.

• *Control.* See under scales, Chapter 11.

WHITEFLY. *Tetraleurodes mori.* Leaves of dogwood, mulberry, holly, maple, and planetree infested by whitefly are usually sticky from honeydew secreted by the scalelike young, which are black with a prominent white border. The adults are tiny white flies, which dart away when the leaves are disturbed.

• *Control.* Spray with malathion in midsummer and again in late summer.

OTHER INSECTS. Among other insects that occasionally infest dogwood are melon and potato aphid, psyllid, pitted ambrosia beetle, redhumped caterpillar, green maple worm, giant

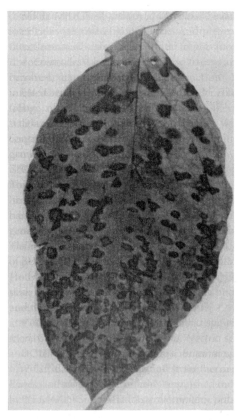

Fig. III-27. Septoria leaf spot of dogwood.

Fig. III-28. Flatheaded apple tree borer and galleries in dogwood branch.

Fig. III-29. Adult stage of the dogwood borer *Synanthedon scitula*.

Fig. III-30. Dogwood club gall opened to show larvae, *Mycodiplosis alternata.*

Fig. III-31. Cottony maple scale on dogwood leaves.

hornet, leafhoppers, greenhouse thrips, leafroller, and the dogwood sawfly. The latter is especially damaging to gray dogwood. Control for most of these can be achieved with Sevin sprays.

JAPANESE WEEVIL. See under Holly.

TWIG GIRDLER. See under Hackberry.

DOUGLAS-FIR
(*Pseudotsuga menziesii*)

Although native to western North America from Alaska to Mexico, this tree is fairly well adapted elsewhere and is used as a landscape tree in eastern North America.

Diseases

LEAF CAST. *Rhabdocline pseudotsugae.* Yellow spots first appear near the needle tips in fall. The spots enlarge in spring, then turn reddish brown, contrasting sharply with adjacent green tissues. With continued moist weather, the discoloration spreads until the entire needle turns brown. When many groups of needles are so affected, the trees appear brown and scorched when viewed from a distance.

Two other fungi produce leaf cast of Douglas-fir: *Phaeocryptopus gaeumannii* and *Rhabdogloeum hypophyllum.* The former produces symptoms closely resembling those of *Rhabdocline pseudotsugae* and both may be present in the same tree.

• *Control.* See under leaf spots, Chapter 12.

GALL DISEASE. The cause is unknown. Bacteria have been implicated but have not been proven to be the cause.

Galls are formed on the twigs of Douglas-fir and big-cone spruce in California. Trees up to 15 years in age are most susceptible. When galls develop on the main stem girdling and death of the upper portion follow.

• *Control.* Control measures have not been developed.

CANKERS. *Valsa abietis, Leucostoma kunzei, Dasyscypha ellisiana, D. pseudotsugae, Phacidiopycnis pseudotsugae,* and *Phomopsis lokoyae.* A number of fungi cause cankers on this host, most of them being prevalent in the Pacific Northwest.

• *Control.* Control measures are rarely practiced except on important ornamental specimens.

OTHER DISEASES. Douglas-fir is susceptible to a leaf and twig blight caused by the fungus *Botrytis cinerea,* which is serious in wet springs; rust by the fungus *Melampsora medusae,* the alternate stage of which occurs on poplar; needle blight by *Rosellinia herpotrichioides*; and witches' broom by the mistletoe *Arceuthobium.* The fungus *Dermea pseudotsugae* causes a dieback and death of young Douglas-firs in northen California. Controls have not been developed.

Insects

APHID. *Cinara pseudotsugae. Cinara* is one of a complex of three aphids which seriously damage young Douglas-fir.

• *Control.* See under aphids, Chapter 11.

COOLEY SPRUCE GALL APHID. See Cooley spruce gall adelgid, listed under Spruce.

TUSSOCK MOTH. *Orgyia pseudotsugata.* Douglas-fir tussock moth may occasionally completely defoliate landscape Douglas-fir in the western United States.

SCALES. Two scale insects, black pineleaf and pine needle, infest Douglas-fir. See Chapter 11 for suggested treatments effective against the young crawling stage of these pests.

OTHER INSECTS. This host is also subject to the spruce budworm, Douglas-fir beetle, white pine weevil, pales weevil, pitch moth, spruce and eriophyid mites, and cambium miners. The controls for these pests are discussed under more common hosts such as pine, spruce, and yew.

ELM (*Ulmus*)

Elms are important landscape trees worldwide. In 1981 it was estimated that the 136 million landscape elms throughout the world had a value of at least 13.6 billion dollars (*Compendium of Elm Diseases*). Almost three dozen species of elms are thought to exist worldwide, and many of them are important landscape trees. Elms grow fast and are tolerant of poor soils, heat, and drought. Unfortunately, Dutch elm disease has decreased the

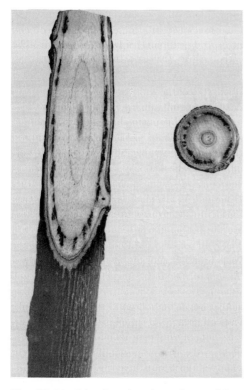

Fig. III-32. Discoloration in twig resulting from infection by the Dutch elm disease fungus, *Ophiostoma ulmi.*

value of American elm (*U. americana*), while Siberian elm (*U. pumila*), an undesirable tree, has tarnished the reputations of Asiatic elms. Chinese elm (*U. parvifolia*) is actually an excellent medium-sized disease-resistant tree with good form, although not the vaselike form of American elm.

Diseases

DUTCH ELM DISEASE. Much has been written both in the United States and in Europe on the Dutch elm disease, the most destructive fungus disease attacking elms. The misleading name given the disease merely refers to the place where it was first identified in 1919, the Netherlands. The disease is believed to have entered the United States in the late 1920s on burled elm logs

from Europe. After killing a majority of the 70 million landscape elms throughout the United States, it now is known to be present in almost every state where elms are grown.

• *Symptoms.* Wilting leaves on one or more branches followed by yellowing, curling, and dropping of all but a few of the leaves at the branch tips about midsummer are the first outward symptoms.

The disease is often recognizable in winter by the tuft of dead brown leaves adhering to the tips of curled twigs. When diseased twigs or branches are cut, spots or flecks are visible in the sapwood near the bark (Fig. III-32). A longitudinal section of the diseased twig shows long brown streaks following the grain of the wood.

Discoloration in the wood also occurs with several other less destructive elm diseases. Consequently, a positive diagnosis can be made only after the fungus responsible for the discoloration has been isolated. Plant pathologists at state agricultural experiment stations are equipped to make a diagnosis, provided adequate twig specimens showing internal discolorations are submitted to them.

• *Cause.* The Dutch elm disease is caused by the fungus *Ophiostoma ulmi* (*Ceratocystis ulmi*), the asexual stages (Fig. III-33 and 34) of which are *Graphium* and *Sporothrix.* An aggressive, highly pathogenic strain of the fungus called *Ophiostoma novo-ulmi* now exists throughout North America and in many parts of Europe.

Several species of insects are largely responsible for the spread of the fungus from diseased to healthy trees. Among the most common vectors in this country are the smaller European elm bark beetle, *Scolytus multistriatus,* and the native elm bark beetle, *Hylurgopinus rufipes.* A number of other boring insects are also known to transmit the causal fungus or are suspected of transmitting it. The fungus may occasionally be spread by rainwater and through grafting of roots between a diseased and a healthy elm. Other suspected means of spread are windblown spores, birds, and pruning tools.

Fig. III-33. The Dutch elm disease fungus as it appears in pure culture when isolated from infected elm twigs.

The fungus penetrates the tree mainly through bark beetle feeding wounds. Once inside the wood, it can spread rapidly either by spores developed in the wood vessels and carried in the sap stream or by growth of the fungal hyphae. Death of affected branches is believed to be due to toxins produced by the fungus or to lack of water as a result of the plugging of the vessels by materials formed in the process of fungus invasion. It has been proposed that compartmentalization of the fungus by the tree also results in the walling off of starch reserves at a time when energy reserves may already be low, leading to tree death by starvation.

• *Control.* The most effective controls are against the insects (bark beetles) that are the principal disseminators of the fungus and against sources of fungal inoculum. Sanitation—removing and destroying dead and dying elm trees—is extremely important in control of Dutch elm disease. Sources of fungal inoculum are reduced when infected trees are destroyed. In addition, removing dead and dying elms eliminates breeding sites for the bark beetle vectors. Thorough sanitation requires removal of all dead, dying, or devitalized elm material such as sick or hurricane-damaged trees, broken limbs, elmwood piles, and elm fenceposts. Such material must be removed from a relatively large area because the smaller European elm bark beetle, the principal disseminator in the United States, can fly several miles in search of suitable breeding places; these same beetles may also feed on living trees. Beetles have been found to carry viable spores of the Dutch elm disease fungus

Fig. III-34. Spore-bearing coremia of the asexual stage of the Dutch elm disease fungus.

for more than 2 miles. The elm bark beetles breed in and feed on all species of elms that grow in this country.

Injection of the herbicide cacodylic acid into diseased elm trees can also be used to reduce beetle populations. Treated trees are readily colonized by bark beetles, but herbicide-induced drying of the bark prevents maturation of the larvae. This "trapping" of bark beetles reduces the probability of Dutch elm disease spread in a community.

Another way to prevent the movement of these beetles is to use a residual-type contact insecticide on elm trees in March or April. Methoxychlor is the insecticide of choice for this use.

Pruning diseased branches (therapeutic pruning) has often been suggested as a means of checking this disease. Although costly, therapeutic pruning done when symptoms first begin to occur will preserve most infected trees. The infected tree has a better chance of survival if the infection in the excised branch has not penetrated too far down in the branch. A minimum of 10 feet of healthy wood must be removed along with the discolored wood. The extent of the infection can be determined by peeling back the bark and seeing how far down the tree the xylem discoloration has progressed. Chain saw blades and lubricating oil should be disinfested regularly. It has been observed that healthy trees that are pruned in late July, August, and September are more likely to contract the disease than those pruned at other times of the year.

The disease may be spread by root grafts between a diseased and a healthy tree. Soil treatments with Vapam, a general purpose soil fumigant, will kill a narrow zone of roots between trees, thus breaking grafts. The Vapam label provides instructions for use. Digging a narrow trench between trees will also break the grafts between healthy and diseased trees.

The fungicide Alamo, injected into the lower trunk or root flares, can be used to prevent Dutch elm disease or to cure trees with new infections. Benzimidazole-derived fungicides such as Arbotect 20-S, Lignasan BLP, and Fungisol are also available for injection. Another product, Phyton 27, may be injected for Dutch elm disease control.

None of the chemicals or injection systems is 100 percent effective. All of the injection systems cause injury to the tree; the larger the hole, the greater the possibility of discoloration and decay.

For best results, a thorough sanitation program involving a community-wide program of dead tree and elm firewood removal is needed. Where a valuable elm tree is being considered for a Dutch elm disease control program, the surrounding neighborhood in an area of several square blocks must be scouted thoroughly for sources of disease inoculum, the already infected dead and dying trees. If a source of inoculum is discovered, a fungicide injection program may be initiated. After the inoculum source has been destroyed, injections are no longer needed.

The greatest hope for eventual control lies in the discovery of American elms that have natural immunity to the disease. American elms with resistance to Dutch elm disease include the cultivars 'Delaware #2', '8630', 'Washington', and 'Liberty'. The latter selection is derived from seedlings, so disease resistance could vary from tree to tree. Two new selections include 'Valley Forge' and 'Trenton'.

There are many disease-resistant hybrid elms, 'Urban Elm' is a cross involving three different elm species, *Ulmus hollandica* cv. *vegeta* (Dutch elm), *U. carpinifolia* (smooth-leaf elm), and *U. pumila* (Siberian elm). It has an upright branching form as compared with the very familiar umbrella shape of the American elm. Other elms such as *U. parvifolia* (Chinese Elm) and *U. davidiana* may be used for creating hybrids. Other disease-resistant hybrids include *U. japonica* x *U. wilsoniana* 'Accolade'; the cultivars 'Danada', 'Vanguard', and 'Charisma'; and *U. carpinifolia* x *U. parvifolia* 'Frontier'. The cultivars 'Dynasty', 'Homestead', 'Jacan', 'Pioneer', 'Regal', 'Sapporo Autumn Gold', and 'Thompson' are reportedly resistant.

The 'Christine Buisman' elm, formerly thought to be highly resistant to the disease, has been found to be susceptible to both Dutch elm disease and the phytoplasmal disease, phloem necrosis. Other Dutch selections such as 'Dodoens', 'Lobel', 'Groenveld', and 'Plantyn' are reportedly resistant. The Hanson Manchurian elm, like most Asiatic elms, is decidedly resistant to the disease. A close relative of the American elm, Japanese keaki (*Zelkova serrata*), is much more resistant to the disease than is the American elm. Although it is vase-shaped, has bark resembling beech, and elmlike foliage that turns red in the fall, it tends to produce excessive twigs, making it less graceful than the American elm.

Occasionally one reads that well-fertilized elms are less likely to contract Dutch elm disease. Actually, the reverse is true. Fertilizers increase the vessel group size, which makes the trees more susceptible to the disease. Unless the trees show a severe nutrient deficiency, they should not be fertilized more often then once every 3 or 4 years.

WILT. *Dothiorella ulmi*. Early symptoms are the drooping and yellowing of the leaves, which are more or less mottled and later become brownish and rolled. The foliage on trees whose trunks are infected is very dwarfed. Much dieback of twigs and branches occurs. This fungus is spread by wind, rain, insects, and birds. The parasite enters through wounds on leaves and tender shoots and develops in the water-conducting system. Branches having dieback develop cankers with associated pycnidia.

• *Control.* Severely infected trees should be removed and destroyed. Mildly infected ones should be pruned heavily to remove as much of the diseased wood as possible. Because the fungus may develop internally well beyond the area of the external symptoms, pruning does not always produce the desired results. Heavy fertilization may help mildly diseased trees to recover. The number of leaf infections can be reduced by applying combination sprays containing a fungicide and an insecticide. Fertilization is suggested as a general precautionary measure despite the fact that there seems to be no correlation between the vigor of the tree and its susceptibility to the disease.

VERTICILLIUM WILT. *Verticillium albo-atrum*. The symptoms of this disease are so much like those of the Dutch elm disease and Dothiorella wilt that culturing of the fungus is necessary to distinguish them. This wilt may in time cause the death of large elms but usually does not become epidemic. See Chapter 12.

ELM YELLOWS. This disease, also called phloem necrosis, is caused by a phytoplasma. It is deadly; thousands of elms in the Midwest have died from its effects since the early 1940s. The disease occurs from the Great Plains to the East Coast.

• *Symptoms, Cause, and Control.* See under yellows diseases, Chapter 12.

BACTERIAL LEAF SCORCH. This bacterial disease occurs on the mall of the nation's capital in Washington, D.C. Foliar necrosis and a gradual crown deterioration and eventual death are the principal symptoms.

See Chapter 12.

WETWOOD. This disorder occurs in American, Moline, Littleford, English, Siberian, and slippery elms.

• *Symptoms.* Elms affected by wetwood have dark, water-soaked, malodorous wood. The condition is usually confined to the inner sapwood and heartwood in trunks and large branches. Little or no streaking occurs in the outer sapwood, and no discoloration is seen in the cambial region or phloem.

• *Cause and Control.* See Chapter 12.

CANKERS. *Botryosphaeria dothidea, Coniothyrium sp., Cytospora ambiens, Nectria galligena, N. cinnabarina, Phoma sp., Phomopsis sp., Schizoxylon microsporum,* and *Sphaeropsis ulmicola.* Many species of fungi cause cankers and dieback of twigs and branches of elms. *Botryodiplodia hypodermia* and *Tubercularia ulmea* cause Siberian elm cankers in the Great Plains.

• *Control.* Many small cankers can be eradicated by surgical means. The cuts should extend well beyond the visibly infected area to ensure complete removal of fungus-infected tissue. If

Fig. III-35. Leaf spot of elm caused by the fungus *Stegophora ulmea.*

the canker has completely girdled the stem, prune well below the affected area and destroy the prunings.

BLEEDING CANKER. *Phytophthora inflata.* This disease, also called pit canker, is characterized by seepage of reddish brown fluid from the margins of perennial cankers produced on the trunk or limbs of trees growing under adverse conditions.

• *Control.* See under Maple.

LEAF BLISTER. *Taphrina ulmi.* Small blisters that lead to abnormal leaf development follow an attack by this fungus. Infection usually takes place soon after the leaves unfold.

• *Control.* Valuable specimens subject to this disease should be sprayed with concentrated lime sulfur or bordeaux mixture in spring just before growth starts.

LEAF SPOTS. *Cercospora sphaeriaeformis, Cylindrosporium tenuisporium, Coryneum tumoricola, Gloeosporium inconspicuum, G. ulmicolum, Monochaetia desmazierii, Phyllosticta confertissima, P. melaleuca, Mycosphaerella ulmi, Septogloeum profusum,* and *Coniothyrium ulmi.* There are so many fungi that cause leaf spots of elms that only an expert can distinguish one from another. Probably the most prevalent leaf spot, however, is that caused by *Stegophora ulmea,* the first symptom of which appears early in spring as small white or yellow flecks on the upper leaf surface. The flecks then increase in size, and their centers turn black

(Fig. III-35). If infections occur early and are heavy, the leaves may drop prematurely. Usually, however, the disease becomes prevalent in late fall about the time the leaves normally drop, and consequently little damage to the tree occurs.

• *Control.* See under leaf spots, Chapter 12.

POWDERY MILDEWS. *Microsphaera neglecta, Phyllactinia guttata,* and *Uncinula macrospora.* These species of powdery mildew fungi develop their mycelia on both sides of the leaves and cause a yellowish spotting.

• *Control.* Damage is so slight that spraying is usually unnecessary.

WOOD DECAY. A number of fungi are associated with decay of elm wood. They cannot be checked once they have invaded large areas of the trunk. Many can be prevented from gaining access to the interior of the tree, however, by avoiding bark injuries, properly treating injuries that do occur, and keeping the tree in good vigor by mulching and watering.

MOSAIC. This viral disease, which causes yellow mottling of leaves, is rare and relatively harmless. The causal agent, cherry leafroll virus, can be transmitted from diseased to healthy trees through pollen.

• *Control.* No control is known for mosaic.

Abiotic Disease

AIR POLLUTION. Elms are sensitive to sulfur dioxide. See Chapter 10.

Insects and Other Animal Pests

ELM LEAF BEETLE. *Pyrrhalta luteola.* Two distinct types of injury are produced by this pest. Soon after the leaves unfurl in spring, the adult beetles, brownish yellow insects ¹/₄ inch long (Fig. III-36), chew rectangular holes in them. The beetles deposit eggs on the lower leaf surface, and later in the season the leaves are skeletonized and curl and dry up as a result of the feeding on the lower surface by the larvae, which are black grubs with yellow markings.

• *Control.* Spray with *Bacillus thuringiensis tenebrionis.* See also foliage feeding beetles, Chaper 11.

GYPSY MOTH. *Lymantria dispar.* The leaves of a large number of forest, shade, and ornamental trees are chewed by the gypsy moth

Fig. III-36. Adult elm leaf beetle, which occasionally becomes a household pest.

Fig. III-37. Gypsy moth adult and egg mass.

Fig. III-38. White-marked tussock moth larva.

larva, a hairy, dark gray caterpillar ranging up to 3 inches in length, with pairs of blue and red dots down its back. Among the most susceptible trees are apple; speckled alder; gray, paper, and red birches; hawthorn; linden; oaks; poplars; and willows. Trees that are also favored as food include ash, balsam fir, butternut, black walnut, catalpa, red cedar, flowering dogwood, sycamore, and tuliptree. The average annual damage caused by feeding of this insect amounts to several million dollars. More than a million acres of woodlands have been 25 to 100 percent defoliated in a single year. See Chapter 11.

• *Control.* For control options, see under caterpillars, Chapter 11. Gypchek, formulated from a virus that attacks gypsy moth, may also be used. A naturally occurring fungal parasite of gypsy moths may provide control in some circumstances. See Chapter 13.

Destroying the egg masses (Fig. III-37) during winter or early spring also helps to protect valuable ornamental trees.

A synthetic sex lure, Disparlure, was developed by United States Department of Agriculture scientists. Male gypsy moths are lured into special traps by this material, which enables federal and state officials to determine the presence and density of gypsy moths in any particular area. Disparlure also has potential for mating disruption. See Chapter 13.

MOURNING-CLOAK BUTTERFLY. *Nymphalis antiopa.* The leaves are chewed by the caterpillar stage, which is 2 inches long and spiny, with a row of red spots on its back. The larva is sometimes called spring elm caterpillar. The adult has yellow-bordered, purplish brown wings. It overwinters in bark cavities and other protected places and deposits masses of eggs around small twigs in May.

• *Control.* Normally this pest does little harm to the tree. See Chapter 11, under caterpillars.

WHITE-MARKED TUSSOCK MOTH. *Orgyia leucostigma.* The leaves are chewed by the tussock moth larva, a hairy caterpillar 1¹/₂ inches long. It has a red head, longitudinal black and yellow stripes along the body, and a tussock of hair on the head in the form of a Y (Fig. III-38). The adult female is a wingless, gray, hairy moth that deposits white egg masses on the trunk and branches. Larvae emerge in May and can devour large amounts of foliage before being discovered.

On the West Coast, the larvae of the western tussock moth, *O. vetusta,* feed on almond, apricot, cherry, hawthorn, oaks, pear, plum, prune, walnut, and willows in addition to elms.

• *Control.* See under caterpillars, Chapter 11.

SPRING CANKERWORM. *Paleacrita vernata.* Spring cankerworms, also called inchworms or measuring worms, are looping worms of various colors and about 1 inch in length that chew leaves. The adult female moth, which is $^1/_2$ inch long and wingless, climbs up the trunk to deposit eggs in early spring.

FALL CANKERWORM. *Alsophila pometaria.* The leaves are chewed by the fall cankerworm, a black and green worm about 1 inch in length. The adult female is a wingless moth that deposits eggs in late fall on the twigs and branches. This insect also attacks oak and maple.

ELM SPANWORM. *Ennomos subsignarius.* This pest, whose adult stage is known as the snow-white linden moth, can cause heavy defoliation of deciduous trees. It feeds mostly on beech, elm, hickory, horsechestnut, linden, maple, oak, pecan, and yellow birch. The larvae are about $1^1/_2$ to 2 inches long, brownish black with bright red head and anal segments. Eggs are laid in midsummer in groups on branches and hatch the following spring. The moths appear in late July in such great numbers that they resemble a snow shower. The larvae of Io moth, *Automeris io,* also feed on elm.

• *Control.* See under caterpillars, Chapter 11.

ELM CASE BEARER. *Coleophora ulmifoliella.* When elms are infested by the case bearer, small holes are chewed in the leaves and angular spots mined between the leaf veins by a tiny larva. The adult is a small moth with a $^1/_2$-inch wingspread. The pest overwinters in the larval stage in small cigar-shaped cases made from leaf tissue.

• *Control.* See under caterpillars, Chapter 11.

APHIDS. Infestations of the woolly apple aphid, *Eriosoma lanigerum,* result in stunting and curling of the terminal leaves. The growing tip may be killed back for several inches. This bluish white aphid spends two generations on elm, migrates to roots of rosaceous hosts, and then goes back to elm, overwintering in the egg stage in bark crevices. Woolly elm aphid (*E. americana*) has a similar life history, with ser-

Fig. III-39. Engraving of the wood made by the larvae of the smaller European elm bark beetle.

viceberry as the alternate host. Giant bark aphid (see aphids, under Sycamore), elm leaf aphid (*Tinocallis ulmifolii*), woolly elm bark aphid (*E. rielyi*), woolly hawthorn aphid (*E. crataegi*), and woolly pear aphid (*E. pyricola*) also feed on elm.

• *Control.* See under aphids, Chapter 11.

ELM COCKSCOMB GALL. *Colopha ulmicola.* Elongated galls that resemble the comb of a rooster are formed on the leaves as a result of feeding and irritation by wingless yellow-green aphids. Eggs are deposited in bark crevices in fall. Two other species, *C. ulmisaccula* and *Eriosoma langninosa,* behave similarly.

• *Control.* A malathion spray applied as the buds open in spring will destroy the so-called stem-mother stage.

ELM LACE BUG. *Corythucha ulmi.* This insect may do considerable damage to elms in eastern states. It first infests the tender foliage in spring, causing a characteristic spotting of the leaves, which later turn brown and die. Black specks of the excreta on the underside of the leaves are also characteristic.

• *Control.* See under lace bugs, chapter 11.

ELM LEAF MINER. *Fenusa ulmi.* Leaves of American, English, Scotch, and Camperdown elms are mined and blotched in May and June by white, legless larvae. Sometimes there may be as many as 20 in a single leaf, and defoliation may occur. Adult females are shining black sawflies, which deposit eggs in slits in the upper leaf surfaces.

• *Control.* See under leaf miners, Chapter 11.

SMALLER EUROPEAN ELM BARK BEETLE. *Scolytus multistriatus.* Trees in weakened condition are most subject to infestation by the smaller European elm bark beetle. The adult female, a reddish black beetle $^1/_{10}$ inch long, deposits eggs along a gallery (Fig. III-39) in the sapwood. The small white larvae that hatch from the eggs tunnel out at right angles to the main gallery. Tiny holes are visible in the bark when the adult beetles finally emerge. The beetle is of interest because it is one of the principal vectors of the Dutch elm disease fungus. Adult beetles will feed to a slight extent on buds and bark of twigs during summer. Native bark beetle, *Hylurgopinus rufipes,* which has similar life history, is also a Dutch elm disease vector.

• *Control.* Remove and destroy severely infested branches or trees, and fertilize and water weakened trees. Dursban, Lindane, or Methoxychlor should be applied to the trunks in April to control this pest and thus prevent it from spreading the Dutch elm disease fungus.

ASIAN LONGHORN BEETLE. See under Maple.

DOGWOOD TWIG BORER. *Oberea tripunc-* *tata.* This girdler causes small twigs of the elm to drop in May and June. The female partially girdles branches up to $^1/_2$ inch in diameter. When the eggs hatch, the grubs bore 4 to 5 inches down the center of the twig, which may have broken off at the girdled point. The grubs are dull yellow, $^3/_4$ inch long; they overwinter in the twigs. The adult beetles appear in the spring.

• *Control.* To destroy the larvae, gather and discard all fallen twigs as soon as they are noticed; to destroy the adults, spray the trees with Sevin or Methoxychlor in June.

ELM BORER. *Eutetrapha tridentata.* Weakened trees are also subject to attack by the elm borer, a white grub 1 inch long which burrows into the bark and sapwood and pushes sawdust out through the bark crevices. The adult is a grayish brown beetle $^1/_2$ inch long, with brick red bands and black spots. Eggs are deposited on the bark in June. The larvae overwinter in tunnels beneath the bark. The red-headed ash borer (see under Ash) also attacks elm.

• *Control.* See under borers, Chapter 11.

DOGWOOD BORER. See under Dogwood.

TWIG GIRDLER. See under Hackberry.

LEOPARD MOTH BORER. See under Maple.

SCALES. Many species of scales infest elms: brown elm, calico, camphor, citricola, cottony maple, elm scurfy, European fruit lecanium, walnut, gloomy, hickory lecanium, San Jose, obscure, European peach, frosted, Comstock mealybug, oystershell, Putnam and scurfy. The European elm scale, *Gossyparia spuria* (Fig. III-40), is a soft scale not protected by a waxy covering.

Fig. III-40. European elm scale, *Gossyparia spuria.*

Fig. III-41. Witches' broom on false cypress.

- *Control.* See under scales, Chapter 11.

MITES. The four-spotted mite, *Tetranychus canadensis*; Schoene spider mite, *Tetranychus schoene*; and elm spider mite, *Eotetranychus matthyssei*, infest elms, causing the leaves to turn yellow prematurely. An eriophyid mite, *Eriophyes ulmi*, causes bladderlike galls on elm leaves.

- *Control.* See under spider mites, Chapter 11.

OTHER INSECTS. Elm is attacked by elm calligrapha, elm sawfly, Japanese beetle, and linden looper, each discussed under Linden. Carpenterworm, fall webworm (see under Ash), redhumped caterpillar (see under Poplar), fruit tree leafroller (see under Crabapple), and forest tent caterpillar (see under Maple) infest elm.

EMPRESS-TREE (*Paulownia*)

The empress-tree, a native of China, is prized most for its showy clusters of violet flowers in early spring. It is relatively free of pests, but it is sensitive to extremes in the weather. No important insects are known to attack the empress-tree.

Diseases

LEAF SPOTS. *Ascochyta paulowniae* and *Phyllosticta paulowniae*. Two species of fungi cause leaf spots on this host in very rainy seasons.

- *Control.* See under leaf spots, Chapter 12.

MILDEW. *Phyllactinia guttata* and *Uncinula clintonii*. Two species of powdery mildew fungi occasionally infect empress-tree leaves.

- *Control.* See Chapter 12.

WOOD DECAY. *Polyporus spraguei* and *Trametes versicolor*. These fungi are constantly associated with a wood decay of this host.

- *Control.* See under wood decay, Chapter 12.

TWIG CANKER. *Phomopsis imperialis*. Twigs and small branches occasionally are affected by this disease.

- *Control.* Prune and destroy infected branches.

EUCALYPTUS. See Gum-Tree.

FALSE CYPRESS (*Chamaecyparis*)

Some, such as Port Orford cedar (*C. lawsoniana*), also called Lawson cypress, a handsome ornamental, and Alaska cedar (*C. nootkatensis*) are native to the Pacific Northwest while others (*C. thyoides*) are East Coast or Japanese natives. False cypress is also known as white cedar and is best grown in moist circumstances.

Diseases

BLIGHT. *Phomopsis juniperovora.* The leaves of false cypress may be attacked by this fungus. (See Phomopsis twig blight under Juniper.) The following cultivars, all of *Chamaecyparis pisifera*, are reported to be resistant to blight: Filifera Aureovariegata, Plumosa Aurea, Plumossa Argentea, Plumosa Lutescens, and Squarrosa Sulfurea. *Didymascella chamaecyparissi* also causes a destructive leaf and tip blight.

WITCHES' BROOM. *Gymnosporangium ellissii.* This fungus enters the leaves and travels down into the living bark of the twigs. The presence of the fungus stimulates the formation of a large number of buds that develop to form characteristic witches' brooms, shown in Fig. III-41. Eventually the branch with its broom dies. During the early spring (in April) brown, threadlike telial horns, about 1/4 inch long, grow out from infected branches. Spores from these horns are carried by the wind to infect the bayberry, *Myrica*, on which is produced a light orange-colored rust on the leaves that does considerable damage. The rust occasionally affects sweet-fern, *Comptonia*. When the seedlings of false cypress are attacked, the trees become dwarfed; trees 15 to 20 years old, if they live, may not be over a foot or two high. When young trees are infected at the growing point, the main trunk is prevented from developing normally and the tree is stunted. The fungus is deep-seated and may even be found in the pith region. Heavily broomed trees may die.

• *Control.* No effective control has been proposed other than separating the two hosts. The fungus acts slowly, so removal of the

Fig. III-42. Bags of the bagworm on false cypress twigs.

brooms can prevent the spread of the parasite to the bayberry.

SPINDLE BURL GALL. *Gymnosporangium biseptatum.* This rust fungus is more or less local in its infection, probably first infecting the leaves and then penetrating into the young branches. It stimulates an excessive growth of wood into long burls, which may be several inches in diameter. The branches that bear these burls eventually die. When infection occurs at the base of a young tree, the burl may continue to grow with the tree without doing particular damage. Burls 12 inches in diameter have been seen at the bases of trees of about the same width. Occasionally two species of rust attack the trees at the same point, and a combination of burl and witches' broom symptoms results. The alternate host of the rust that causes spindle burl gall is the serviceberry (*Amelanchier*). Ornamental cedars grown individually are rarely infected.

Another species of rust, *Gymnosporangium fraternum,* attacks the leaves of false cypress. It is of little consequence on this host but does some damage to chokeberry (*Aronia*), its alternate host.

• *Control.* Control measures are unavailable.

ROOT ROT. *Phytophthora lateralis.* This highly destructive disease affects native species in the Pacific Northwest. It is especially serious on Lawson cypress. The fungus infects the leaves, stems, and trunk in addition to the roots. Another species, *P. cinnamomi,* causes root rot of Lawson cypress seedlings in Louisiana.

• *Control.* No satisfactory controls have been developed.

Insects

Among the pests that attack false cypress are the arborvitae weevil (see under Arborvitae), spruce spider mite (see under Spruce), larvae of the imperial moth (*Eacles imperialis*), bagworm (Fig. III-42; see under Juniper),and the juniper scale (*Diaspis carueli*). Details on control of the bagworm are given under Juniper.

FIG (*Ficus*)

Of the many species of fig, some—such as Moreton Bay and weeping figs—are being used as street and landscape trees in California and Florida. Figs are subject to a number of fungal leaf spots, crown gall, twig and branch cankers, root rots, and nematode damage.

TWIG BLIGHT. *Diaporthe cinerescens.* We have observed that Benjamin figs grown indoors under low light intensities are subject to a twig blight. The causal fungus has a *Phomopsis* conidial stage. Moving the affected plants nearer to windows alleviates the problem.

GIANT WHITEFLY. *Aleurodicus dugesii.* This insect is active in coastal southern California (see under Citrus).

CUBAN LAUREL THRIPS. *Gynaikothrips ficorum.* Thrips cause distortion and rolling of young terminal foliage.

FIR (*Abies*)

Firs thrive best in light, porous, acid soils that are well drained yet are continually moist. They also require full sunlight but appear to grow more vigorously in valleys or protected places than near hilltops or in exposed situations. They do not do well in hot, dry climates.

Several parasitic microorganisms and insects attack firs but rarely produce extensive damage in ornamental plantings. Lack of vigor in many trees is more likely caused by an unfavorable environment. Firs are sensitive to air pollution injury.

Diseases

NEEDLE AND TWIG BLIGHT. In the northeastern United States, needle and twig blight occurs commonly on balsam fir and to a lesser extent on Colorado and Alpine firs. Noble and Fraser firs are also susceptible but are attacked infrequently.

• *Symptoms.* The needles of the current season's growth turn red and shrivel, and the new twigs are blackened and stunted. Severely infected trees appear to be scorched by fire or damaged by frost. The lower branches are most heavily infected. Needles infected in previous years remain attached to the twigs.

• *Cause.* The fungus *Rehmiellopsis balsameae*

causes needle and twig blight. It overwinters on diseased needles and twigs. In spring, fruiting bodies mature on these parts and release spores, which infect the newly developing needles. The fungus *Cenangium abietis* occasionally attacks firs but is more common on pine.

• *Control.* In ornamental plantings, control is possible by pruning and destroying infected twigs and by applying a copper fungicide three times at 12-day intervals, starting when the new growth begins to emerge from the buds.

LEAF CAST. *Bifusella abietis, B. faullii, Hypodermella mirabilis, H. nervata, Lophodermium autumnale,* and *L. lacerum.* When attacked by any one of these fungi, needles turn yellow, then brown, and drop prematurely. Elongated black bodies appear along the middle vein of the lower leaf surface. Spores shot from black fruiting bodies in summer land on young leaves where they germinate and penetrate the new growth.

• *Control.* Copper sprays applied as for needle and twig blight will control leaf cast.

CANKERS. *Cylindocarpon* sp., *Cytospora pini* and *C. abietis, Cryptosporium macrospermum, Scoleconectria balsamea,* and *S. scolecospora.* Occasionally, sunken dead areas on the trunk and branches of firs in ornamental plantings result from infection by one of the several fungi listed. On balsam fir, the fungus *Aleurodiscus amorphus* forms narrowly elliptical cankers with a raised border, which center around a dead branch on the main trunk of young trees.

• *Control.* Sanitation, avoidance of bark injuries, and fertilization to maintain the trees in good vigor are suggested.

ROOT ROT. *Phytophthora cinnamomi.* This fungus has been destructive to Fraser fir seedlings growing in southern Appalachian nurseries. See under root rots, Chapter 12.

ARMILLARIA ROOT ROT AND WOOD DECAY. These problems are caused, respectively, by the fungus *Armillaria mellea* and by fungi belonging to the genera *Phellinus, Odontia, Coniophora, Polyporus, Lenzites,* and *Haematostereum.*

• *Control.* See Chapter 12.

RUSTS. *Milesia fructuosa, Hyalospora aspidiotus, Uredinopsis mirabilis, U. osmundae, U. phe-*
gopteridis, Pucciniastrum pustulatum, P. goeppertianum, Melampsora abieticapraearum, Caeoma faulliana, Peridermium ornamentale,* and *Melampsora cerastii.* Most of these rust fungi attack forest firs but are seldom found on ornamental specimens. *Melampsorella caryophyllacearum* infection causes a common yellow witches' broom of balsam fir. Broomed shoots are upright and dwarfed with short yellow needles that drop in less than a year, leaving bare shoots. The rust is perennial in the broomed shoots; it also causes trunk and branch swellings.

• *Control.* Measures to control rusts are rarely adopted because the fungi do not cause much damage. Most of the rusts listed above have alternate hosts that are needed to complete the life cycles of the fungi. The elimination of the alternate host will result in nearly complete disappearance of the fungus. A rust expert must be consulted for the name of the alternate host before any eradicatory steps are taken. Periodic applications of sulfur sprays during summer will also control rusts, but these are not practical for large trees.

SOOTY MOLD. Needles of firs are covered more frequently than those of most other species of evergreens by black sootlike material. This substance consists of fungal tissues that exist on the secretions of aphids and other insects. The black mold will usually disappear if the insects are kept under control.

Insects

BALSAM TWIG APHID. *Mindarus abietinus.* The leaves and shoots of white and balsam firs as well as spruce are attacked by this green aphid, which is covered with white waxy secretions. Affected shoots are roughened and curled.

ROOT APHID. *Prociphilus americanis.* This insect feeds on roots of conifers.

BALSAM WOOLLY APHID. *Adelges piceae.* This introduced adelgid is becoming increasingly prevalent in the Northeast. It attacks twigs and buds and causes dieback of twigs and treetops.

• *Control.* See under aphids, Chapter 11.

BALSAM GALL MIDGE. *Paradiplosis fumifex.* Small, subglobular swellings at the base of the leaves are caused by this midge.

- *Control.* Spray the newly developing leaves in late April with malathion.

BARK BEETLE. *Pityokteines sparsus.* Seepage of balsam from the trunk, reddening of the needles, and death of the upper parts of the tree result from infestations of the balsam bark beetle, a pest $^1/_{10}$ inch long. Vigorous trees are attacked.

- *Control.* Prune and destroy infested parts.

CATERPILLARS. Several caterpillars feed on the needles of firs. These are the larval stage of the hemlock looper moth, the spotted tussock moth, the balsam fir sawfly, the Zimmerman pine moth, and the pine butterfly.

- *Control.* See under caterpillars, Chapter 11.

SPRUCE SPIDER MITE. See spider mite, under Spruce.

SCALE. Several kinds of scale insects, including the cottony-cushion, oystershell, and pine needle scale, infest firs.

- *Control.* See under scales, Chapter 11.

BAGWORM. See under Juniper.

SPRUCE BUDWORM. See under Spruce.

CEDAR TREE BORER. See under Redwood.

CAMBIUM MINER. See under Holly.

PALES WEEVIL. See under Pine.

Other Pests

DWARF MISTLETOE. *Arceuthobium campylopodum.* In California this mistletoe is a widespread serious pest of red fir (*Abies magnifica*) and white fir (*A. concolor*).

- *Control.* In valuable trees, remove dwarf mistletoe by pruning.

FIREWHEEL TREE (*Stenocarpus*)

This Australian native grows in mild climates, preferring rich, slightly acid soil. It is relatively free of diseases and pests.

FLAME-TREE. See Poinciana.

FRANKLIN-TREE (*Franklinia alatamaha*)

Franklinia, native to Georgia, is a small to medium-sized tree that blooms in the fall. It prefers partial shade, wind protection, and rich, acid soil.

Diseases

LEAF SPOT. *Phyllosticta gordoniae.* Spots on the leaves of this host occasionally occur in rainy seasons.

- *Control.* See under leaf spots, Chapter 12.

BLACK MILDEW. *Meliola cryptocarpa.* Leaves of the Franklin-tree may be covered by this black fungus.

ROOT ROT. *Phymatotrichum omnivorum.*

- *Control.* Measures are rarely adopted to control black mildew or root rot.

WILT. *Phytophthora* sp. This is a highly destructive wilt disease in container-grown Franklin-trees. Affected trees wilt suddenly during hot weather and then drop their leaves. Infected roots are brownish black in color, and black cankers occur along the stems.

- *Control.* See under root rots, Chapter 12.

Insects

SCALES. Three species of scale insects: red bay, *Chrysomphalus perseae*; walnut, *Aspidiotus juglansregiae*; and *Lecanium* sp. are known to attack the Franklin-tree.

- *Control.* See under scales, Chapter 11.

GIANT HORNET WASP. *Vespa crabro germana.* This wasp tears the bark not only from the Franklin-tree but also from birch, boxwood, poplar, willow, and other trees and shrubs. The wasp is dark reddish brown with orange markings on the abdomen. It is the largest wasp in the United States—1 inch long. The pest nests in trees, buildings, or underground.

- *Control.* Apply a Sevin spray to the trunk when the wasp begins to tear the bark in July. Blow Diazinon dust into wasp nests.

FRINGE-TREE (*Chionanthus*)

Native to the southeastern United States, fringe tree is small and usually multi-trunked.

Diseases

LEAF SPOTS. *Cercospora chionanthi, Phyllosticta chionanthi, Septoria chionanthi,* and *S. eleospora.* These four species of fungi are known to cause leaf spotting on this host.

- *Control.* See under leaf spots, Chapter 12.

Fig. III-43. Leaf spot on *Ginkgo biloba*, the cause of which is unknown.

POWDERY MILDEW. *Phyllactinia guttata.* The leaves of fringe-tree are occasionally affected by this disease.

- *Control.* See under powdery mildew, Chapter 12.

OTHER DISEASES. Fringe-tree is occasionally subject to several canker diseases caused by *Botryosphaeria dothidea, Phomopsis diatrypea,* and *Valsa chionanthi.*

- *Control.* Prune diseased branches.

Insects

SCALES. The rose scale *Aulacaspis rosea* and the white peach scale (see under Cherry) infest fringe-tree.

- *Control.* See under scales, Chapter 11.

GINKGO (*Ginkgo*)

Ginkgo is successfully being grown as a shade tree throughout the United States, except in the coldest northern states. It withstands urban conditions and is easy to transplant.

There are several cultivars available, and planting of male ginkgo is usually recommended because a disagreeable odor is associated with female fruit hulls.

Ginkgo, also known as maidenhair-tree, is unusually resistant to fungus and insect attack. A fungitoxic substance, α-hexenal, is believed by some to be responsible for its resistance to fungus diseases, although a waxy cuticle has also been implicated. Leaf spots have been attributed to three fungi: *Glomerella cingulata, Phyllosticta ginkgo,* and *Epicoccum purpurascens.* The damage by these fungi is negligible. The cause of another leaf spot (Fig. III-43) is not known with certainty; in Czechoslovakia a similar disease has been attributed to a virus.

Several wood-decaying fungi, including *Irpex lacteus, Trametes versicolor,* and *Fomes meliae,* have also been reported, but these occur rarely.

- *Control.* No effective control is known.

Insects and Other Pests

Few insects attack this tree. Among those occasionally found are the omnivorous looper, *Sabulodes caberata*; the grape mealybug, *Pseudococcus maritimus*; the white-marked tussock moth, *Orgyia leucostigma*; the American plum borer (see under Planetree); and the fruit tree leafroller, *Archips argyrospilus*.

The root-knot nematode *Meloidogyne incognita* has been reported on ginkgo in Mississippi.

• *Control.* Control measures are rarely applied.

GOLDEN-CHAIN (*Laburnum*)

This medium-sized tree does best in moist limestone soils and is hardy in the South.

Diseases

LEAF SPOT. *Phyllosticta cytisii.* Leaves are subject to spotting when infected by this fungus. The spot is at first light gray, later turning brown, and has no definite margin. The black fruiting bodies of this fungus dot the central part of the spot. Another fungus, *Cercospora laburni,* also causes leaf spots.

• *Control.* See under leaf spots, Chapter 12.

TWIG BLIGHT. *Fusarium lateritium.* Brown lesions on the twigs followed by blighting of the leaves above the affected area in very wet springs is characteristic of this disease. The sexual stage of this fungus is *Gibberella baccata.*

• *Control.* Prune and destroy infected twigs and spray as for leaf spot.

LABURNUM VEIN MOSAIC. Conspicuous veinbanding of *Laburnum alpinum* may be due to a virus.

• *Control.* Remove and destroy infected plants.

Insects and Related Pests

APHIDS. *Aphis craccivora.* The cowpea aphid, black with white legs, clusters at the tips of the branches. The bean aphid, *A. fabae,* also infests golden-chain.

• *Control.* See under aphids, Chapter 11.

GRAPE MEALYBUG. *Pseudococcus maritimus.*

The grape mealybug occasionally infests golden-chain both above and below ground.

• *Control.* Mealybugs infesting the branches and twigs can be controlled with malathion sprays. Those infesting the roots can be curbed by wetting the soil with Diazinon.

NEMATODE. *Meloidogyne hapla.*

• *Control.* Control measures are rarely practiced for this pest.

GOLDENRAIN-TREE (*Koelreuteria*)

This native of Asia tolerates a wide range of growing conditions.

Diseases

CORAL SPOT CANKER. *Nectria cinnabarina.* Small, depressed, dead areas in the bark near wounds or branch stubs are caused by this fungus. Tiny coral-pink bodies are formed on the dead bark.

• *Control.* Prune infected branches back to sound wood. Fertilize and water to maintain vigor.

OTHER FUNGAL DISEASES. Verticillium wilt can kill goldenrain-trees. The only other fungal diseases reported as affecting goldenrain-tree are a leaf spot caused by a species of *Cercospora* and root rot by *Phymatotrichum omnivorum.*

• *Control.* For wilt, see Chapter 12. The other diseases are rarely destructive enough to warrant control measures.

Insects

SCALES. Three species of scales infest goldenrain-tree: lesser snow (see under Poinciana); mining, *Howardia biclavis*; and white peach (see under Cherry).

• *Control.* See under scales, Chapter 11.

GUAVA (*Psidium guajava*)

Diseases

In Florida this host is subject to a leaf and fruit spot caused by the fungus *Glomerella cingulata*; a thread blight by *Ceratobasidium stevensii*; a leaf spot by *Cercospora psidii*; and a root rot by *Clitocybe tabescens.*

• *Control.* Effective control measures are unavailable.

Insects and Other Animal Pests

SCALES. Nine species of scale insects infest guavas: barnacle, black, chaff, Florida red, Florida wax, greedy, green shield, hemispherical, and soft.
• *Control.* See under scales, Chapter 11.

OTHER INSECTS. Occasionally guava is attacked by the Mexican fruit fly, *Anastrepha ludens*; the long-tailed mealybug, *Pseudococcus adonidum*; and the eastern subterranean termite, *Reticulitermes flaviceps*.
• *Control.* Control measures have not been developed.

SOUTHERN ROOT-KNOT NEMATODE. *Meloidogyne incognita.* This nematode infests guava roots in Florida.
• *Control.* This pest is difficult to control on plants growing outdoors.

GUMBO-LIMBO (*Bursera*)

Diseases

SOOTY MOLD. *Fumago vagans.* A heavy growth of this sooty mold develops on the plant as a result of infestation by the brown soft scale (see below). The results are sometimes very serious.
• *Control.* Spray with malathion or Sevin to control the scale insect that secretes the substance on which the sooty mold fungus lives.

OTHER DISEASE. Branches of this host may be killed by the fungus *Physalospora fusca.*
• *Control.* Prune infected branches.

Insects

BROWN SOFT SCALE. *Coccus hesperidum.* This scale insect does great damage to greenhouse plants but often remains inconspicuous because of its tendency to assume the color of the twigs or leaves upon which it is feeding. The lower sides of the leaves are sometimes heavily infested with light green scales, whose mature stages are often found on petioles, twigs, and the upper parts of the trunk. The scales are arranged longitudinally along the smaller branches and main stem. They are very flat and thin, more or less transparent. Though the young are usually sluggish, not migrating far from the mother scale, they do travel to the upper leaves. This scale is said to be a general feeder, attacking many plants in the greenhouse and also tropical fruit outdoors. Eight other species of scales may infest gumbo-limbo.
• *Control.* See under scales, Chapter 11.

GUM-TREE (*Eucalyptus*)

Gum-trees, also called eucalyptus, are largely natives of Australia. They are widely grown in California and some parts of the southwestern United States. They are fast growing and require a mild climate. When killed by frost in parts of northern California, gum-trees become a fire hazard the following season.

Diseases

LEAF SPOTS. *Hendersonia sp., Monochaetia monochaeta, Mycosphaerella moelleriana,* and *Phyllosticta extensa.* Several species of fungi cause leaf spots of gum-tree. They are rarely serious enough to justify control measures.

CROWN GALL. The bacterium *Agrobacterium tumefaciens* and many wood-decaying fungi also occur on this host.

OTHER FUNGAL DISEASES. Among other diseases of this host are canker and twig blight caused by *Botryosphaeria dothidea*; basal canker by *Cryphonectria cubensis*; root rot by *Armillaria mellea* in California; root rot by *Clitocybe tabescens* in Florida and by *Phymatotrichum omnivorum* in Texas.
• *Control.* Control measures for crown gall and Armillaria root rot are given in Chapter 12. The other diseases are not serious enough to warrant control practices.

EDEMA. Several species of *Eucalyptus* grown in greenhouses as ornamentals are subject to a physiological disease which is manifested by intumescences or blisterlike galls on the leaves. Sections of these galls show several layers of cells formed one above the other. These growths usually crack open and become rust-colored. The disease is difficult to diagnose because it looks so much like the work of a blister mite or a rust fungus. It is not caused by a parasite but results from the accumulation of too much water through poor ventilation of the greenhouse or through overwatering of the plants.

Insects and Related Pests

A large number of insects attack gum-trees. Among the more common are the cowpea aphid; three species of borers—California prionus, nautical, and Pacific flatheaded; the lygus bug; three kinds of caterpillars—California oakworm, omnivorous looper, and orange-tortrix; long-tailed mealybug; greenhouse thrips; and three species of mites—avocado red, platanus, and southern red.

 • *Control.* See under aphids, borers, caterpillars, and mites, respectively; Chapter 11.

BLUE GUM PSYLLID. *Ctenarytaina eucalypti.* This psyllid produces white, waxy material on blue gum and baby blue gum foliage in California. A parasitic wasp, imported from the gumtree's native Australia, now effectively controls this pest.

SCALES. Many species, including black, Cali-

fornia red, greedy, dictyospermum, and ivy scales, infest gum-trees.

 • *Control.* See under scales, Chapter 11.

EUCALYPTUS LONGHORN BORER. *Phoracantha semipunctata.* This potentially destructive pest was discovered in southern California in 1984. Trees die when the larvae tunnel into them and riddle the inner bark and cambium with frass-filled galleries. Adult beetles lay eggs under loose bark of moisture-stressed trees, generally avoiding healthy trees. Larvae attempting to penetrate healthy trees are smothered by copious quantities of gum produced by the tree. Stressed trees lack this defensive ability.

 • *Control.* Prevent drought stress. Burn or destroy infested firewood.

HACKBERRY (*Celtis*)

Common hackberry, despite witches' broom and nipple gall problems, is a tough tree, adapted to harsh growing conditions, as in the Great Plains. Sugar hackberry, grown in the South, and European hackberry are also tolerant of a wide range of soil conditions.

Fig. III-44. Hackberry nipple galls.

Diseases

LEAF SPOTS. *Cercospora spegazzini, Cercosporella celtidis, Cylindrosporium defoliatum, Phleospora celtidis, Phyllosticta celtidis, Pseudoperonospora celtidis,* and *Septogloeum celtidis.* Many fungi cause leaf spots on hackberry in rainy seasons or late in the season on senescing leaves.

• *Control.* Leaf spots are rarely serious enough to warrant control.

POWDERY MILDEW. *Pleochaeta polychaeta.* This fungus attacks both sides of the leaves, and the mildew is visible either as a thin layer over the entire surface or in irregular patches. The small black fruiting bodies, ascocarps, develop mostly on the side opposite the mildew.

• *Control.* See under powdery mildew, Chapter 12.

WITCHES' BROOM. In the eastern and central states the American hackberry is extremely susceptible to the witches' broom disease.

• *Symptoms.* The early symptoms are visible on the buds during the winter. Affected buds are larger, more open, and hairier than normal ones. Branches that develop from such buds are bunched together, producing a broom-like effect that is most evident in late fall or winter. The witches' brooms are more unsightly than harmful. Their presence, however, causes branches to break off more readily during windstorms; the exposed wood is then subject to decay.

• *Cause.* The exact cause of the witches' broom disease is not definitely known. An eriophyid mite, *Eriophyes celtis,* and the powdery mildew fungus *Sphaerotheca phytoptophila* are almost constantly associated with the trouble and are believed to be responsible for the deformation of the buds that results in the bunching of the twigs.

• *Control.* No effective control measures are known. Prune back all infected twigs to sound wood. Spraying with 1 part of lime sulfur in 10 parts of water in early spring might help. This should be followed with two applications of Kelthane at 2-week intervals starting in mid-May. The Chinese hackberry is less susceptible than the common hackberry and should be substituted in areas where the disease is prevalent. The southern hackberry (*C. mississippiensis*) is also less subject to witches' broom.

GANODERMA ROT. *Ganoderma lucidum.* This fungus is capable of attacking living trees, causing extensive decay of the roots and trunk bases. See under Maple.

WOOD DECAY. Hackberries are subject to decay by several different heart rot fungi. See Chapter 12.

Insects and Related Pests

HACKBERRY NIPPLE-GALL MAKER. *Pachypsylla celtidis-mamma.* Small round galls opening on the lower leaf surfaces and resembling nipples (Fig. III-44) are caused by a small jumping louse or psyllid. Another species of psyllid, *Pachypsylla celtidisvesicula,* produces blister galls.

• *Control.* Spray with Azatin, Carbaryl, Dursban, horticultural oil, Sevin, or Tempo in May when the leaves are one-quarter grown.

SCALES. Several species of scales including camphor, citricola, cottony-cushion, cottony maple, gloomy, hickory lecanium, obscure, oystershell, Putnam, San Jose, and walnut occasionally infest this host.

• *Control.* See under scales, Chapter 11.

MOURNING-CLOAK BUTTERFLY. See under Elm.

PAINTED HICKORY BORER. See under Hickory.

TWIG GIRDLER. *Oncideres cingulata.* This beetle chews a continuous notch around the twig near the point of egg deposition. The girdled twigs die and break off, and the larva completes its development there.

• *Control.* No chemical control is available. Pick up and destroy fallen branches.

RED-HEADED ASH BORER. See under Ash.

HARDY RUBBER TREE (*Eucommia*)

This tree, adapted to temperate regions, is unusually free of insects and diseases.

HAWTHORN (*Crataegus*)

Hawthorns, sometimes called thorns, are members of the same family as apples and pears and are frequently attacked by the same fungi, bacteria, and insects. There are many

Fig. III-45. Leaf blight of hawthorn caused by the fungus *Diplocarpon mespili.*

species and cultivars of hawthorn, and each has different susceptibilities to pests.

Diseases

LEAF BLIGHT. In late summer English hawthorn (*Crataegus oxyacantha*) and Paul's scarlet thorn (*C. oxyacantha pauli*) may be completely defoliated by the leaf blight disease. Cockspur thorn and Washington thorn appear to be much more resistant to the trouble.

• *Symptoms.* Early in spring small, angular, reddish brown spots appear on the upper leaf surface (Fig. III-45). As the season advances, the spots enlarge, then coalesce, and the leaves finally drop.

• *Cause.* The fungus *Diplocarpon mespili* (*Entomosporium mespili*) causes leaf blight. It is primarily a parasite of hawthorns, although some investigators believe it is also responsible for a similar disease of pear and quince. Numerous black, flattened, orbicular bodies are visible, with a hand lens, in the discolored spots on both leaf surfaces. These bodies contain spores that initiate numerous infections during rainy springs.

LEAF SPOTS. A large number of other fungi are known to cause leaf spots on hawthorns. Among the more common ones are *Cercospora confluens, C. apiifoliae, Cercosporella mirabilis, Cylindrosporium brevispina, C. crataegi, Gloeosporium crataegi, Hendersonia crataegicola, Septoria crataegi,* and *Monilinia johnsonii.* The latter

Fig. III-46. Cedar-quince rust of hawthorn caused by the fungus, *Gymnosporangium clavipes*.

also causes spots on the fruits, as does the scab fungus *Venturia inaequalis.*

• *Control.* See under leaf spots, Chapter 12.

FIRE BLIGHT. Hawthorns, especially the English hawthorn, are subject to the bacterial disease fire blight, discussed in Chapter 12.

RUSTS. At least twelve rust fungi attack hawthorns. Of these the cedar quince and cedar hawthorn rusts, *Gymnosporangium clavipes* and *G. globosum,* are the most common on ornamental species. The former (Fig. III-46) attacks both the stems and the fruits of hawthorn, causing deformed fruit. The twigs are also attacked and develop deformed, antlerlike branches. The fungus breaks out in little cuplike structures called cluster-cups from which quantities of bright orange spores are shed. These spores are borne by the wind to nearby red cedars and infect their leaves and young twigs. The fungus is perennial in the cedars but annual in the hawthorns. *G. globosum* produces light gray to brown spots on the leaves (Fig. III-47); the cluster-cups are long, slender, and tubelike. This fungus also attacks apple trees and seldom does

Fig. III-47. Cedar-hawthorn rust symptoms on hawthorn leaves.

much damage to hawthorn. Another stage of the fungus lives for two or three years on the red cedar before the small branches it has infected die. See cedar rusts, under Juniper.

• *Control.* See under rust diseases, Chapter 12.

POWDERY MILDEW. Two species of powdery mildew fungi, *Phyllactinia guttata* and *Podosphaera clandestina*, attack hawthorns.

• *Control.* See under powdery mildew, Chapter 12.

Insects and Other Pests

APHIDS. The four-spotted hawthorn aphid *Macrosiphum crataegi* is light green with four black spots and feeds mainly on hawthorn. Many other species of aphids infest hawthorns: apple, apple grain, rosy apple, woolly apple, and woolly hawthorn.

• *Control.* See under aphids, Chapter 11.

HAWTHORN LEAF-MINING SAWFLY. *Profenusa canadensis.* This wasplike sawfly leaf miner can be very destructive in hawthorn nurseries and ornamental plantings, especially to *Crataegus crus-galli* and certain other species and cultivars. The principal damage is done during May or June. The first symptom is a small channel in the leaf, which widens to a blisterlike area light brown in color. The miner begins work at one edge of the leaf near the stalk and continues on that side toward the point of the leaf. The inner parts of the leaf are usually completely consumed, with only the epidermis and veins remaining. Only the leaves that are unfolding are attacked. Much defoliation follows the work of these insects. The same species attacks many fruit trees, including apple, cherry, plum, quince, and sweet-scented crabapple.

• *Control.* See under leaf miners, Chapter 11.

APPLE AND THORN SKELETONIZER. *Choreutis pariana.* This pest feeds on leaves of apple, pear, and hawthorn in the northeastern United States. Infested leaves are often folded in two by the insect and fed upon from within. The adult moth is dark gray to reddish brown with a $^1/_2$-inch wing span. The fully grown larva is $^1/_2$ inch long, with a yellowish green body and a pale brown head. Hawthorn leaves may also be skele-

tonized by the pear slug, a sawfly larva (see under Cherry).

• *Control.* Spray with Sevin when the larvae begin to feed.

WESTERN TENT CATERPILLAR. *Malacosoma pluviale.* Tawny or brown caterpillars with a dorsal row of blue spots feed primarily on hawthorns as well as wild cherry and alder, making tents at the same time as the fall webworm. The moths are smaller and somewhat lighter than the eastern tent caterpillar adults.

Other caterpillars that may feed on hawthorns include the eastern tent, the forest tent, the redhumped, the variable oak leaf, and the walnut caterpillar. The caterpillar stages of the gypsy moth, the cecropia moth, and the western tussock moth also occur on this host.

• *Control.* See under caterpillars, Chapter 11.

BORERS. Four borers infest hawthorns—flatheaded apple tree, roundheaded (see under Mountain-Ash), pear, and shot-hole.

• *Control.* See under borers, Chapter 11.

LACE BUGS. *Corythucha cydoniae* and *C. bellula.* Lace bugs are not common on hawthorn but may occasionally be a problem. Once an infestation is established, the insects breed during summer and become numerous. Adults are about $^1/_8$ inch long and have lacy wings. Nymphs may be spiny. Lace bugs have sucking mouthparts; they feed on the undersides of the leaves, depositing small, brown, sticky spots of excreta.

• *Control.* See under lace bugs, Chapter 11.

PLANTHOPPERS. See under Cherry. Rose leafhopper, *Edwardsiana roseae,* also feeds on hawthorn.

• *Control.* Spray the trees forcefully with pyrethrum or rotenone sprays or a combination of both. The insects must be thoroughly wetted by the sprays.

TWO CIRCULI MEALYBUG. *Phenacoccus dearnessi.* Whitish nymphs lined up on hawthorn twigs and branches in fall appear as frost. This pest has many natural enemies.

SCALES. The following kinds of scales infest hawthorn: azalea bark, barnacle, cottony maple, European fruit lecanium, Florida wax, frosted, lecanium, Putnam, oystershell, scurfy, soft, San Jose, and walnut.

- *Control.* See under scales, Chapter 11.

TWO-SPOTTED MITE. *Tetranychus urticae.* During the summer months, especially in dry weather, these mites often become sufficiently numerous to injure foliage. Infested leaves take on a gray or yellow cast and may be covered with fine silky threads (see Chapter 11). Eriophyid mites, such as leaf blister mites, also attack hawthorns.

- *Control.* See under spider mites, Chapter 11.

HEMLOCK (*Tsuga*)

Hemlocks are native to the United States and are among our most graceful and highly prized evergreens. They grow best in a fairly damp soil where their roots may be cool. The soil must be well drained, however, and moderately acid. Hemlocks do not tolerate heat and drought. They are among the few evergreens that thrive near the trunks of large deciduous trees. Like many other conifers, hemlocks do not adapt to city conditions and rarely prosper in small front yards of suburban homes.

Hemlocks are less susceptible to diseases and most insect pests than other conifers such as firs, pines, and spruces.

Diseases

BLISTER RUST. *Pucciniastrum vaccinii* and *P. hydrangeae.* Young hemlocks and the lower leaves of older trees have yellowish blisters or pustules, from which the spores sift out during June and July. Rhododendron is an alternate host of *P. vaccinii*, which causes rust-brown leaf spots injurious on nursery plants. Wild and cultivated hydrangeas are alternate hosts of *P. hydrangeae.*

- *Control.* The best control for these rusts is to exclude or remove the alternate hosts from the region.

CANKERS. At least six species of fungi, *Botryosphaeria tsugae, Cytospora* spp., *Leucocytospora kunzei, Dermatea balsamea* (Fig. III-48), *Hymenochaete agglutinans,* and *Phacidiopycnis pseudotsugae,* are known to cause cankers on hemlocks.

- *Control.* Prune affected branches and spray with a copper fungicide if affected trees are particularly valuable.

LEAF BLIGHT. *Fabrella tsugae.* In late summer, leaves of Eastern hemlock turn brown and

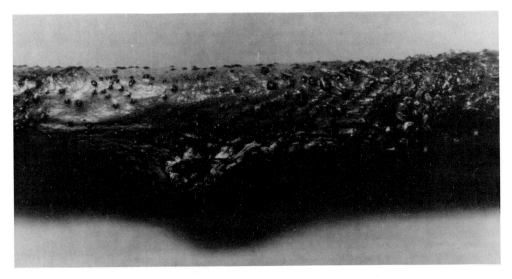

Fig. III-48. Fruiting bodies of *Dermatea* canker of hemlock.

drop prematurely when attacked by this fungus. Small black fruiting bodies of the fungus occur on the fallen leaves. These produce spores the following spring which initiate new infections.

• *Control.* Leaf blight rarely damages the trees enough to necessitate measures other than gathering and destroying fallen infected leaves in autumn.

SIROCOCCUS BLIGHT. *Sirococcus conigenus.* Western hemlock is susceptible to this shoot tip killing disease. Controls have not been developed.

NEEDLE RUST. *Melampsora farlowii* and *M. abietis-canadensis.* Eastern hemlock and, to a lesser extent, Carolina hemlock are attacked by these fungi. In late May or early June some of the new leaves turn yellow. Within 2 weeks the shoots to which these leaves are attached turn yellow, become flaccid, and droop. Most of the needles then drop from the affected shoots. Severely rusted trees appear as though their branch tips had been scorched by fire. Red, waxy, linear fungal bodies occur on the lower leaf surfaces, shoots, and cones.

• *Control.* See under rusts, Chapter 12.

HEARTWOOD ROT. *Ganoderma lucidum, Echinodontium tinctorum,* and *Coniophora puteana.* These fungi cause heart rot or decay of the tissues immediately beneath the bark at the base of the trunk, which results in the death of the tree.

• *Control.* See under wood decay, Chapter 12.

Abiotic Diseases

SUNSCORCH. Ornamental hemlocks are frequently subject to severe burning or scorching when the temperature reaches 95° F (35° C). The ends of the branches may be killed back several inches.

DROUGHT INJURY. Hemlocks are more sensitive to prolonged periods of drought than most other narrow-leaved evergreens. The damage is most severe on sites with southern exposures or on rocky slopes where the roots cannot penetrate deeply into the soil. Thousands of hemlocks died in the northeastern United States as a result of severe droughts in the years 1960 to 1966.

AIR POLLUTION. Hemlocks are sensitive to ozone. See Chapter 10.

Insects and Related Pests

HEMLOCK WOOLLY ADELGID. *Adelges tsugae.* This pest appears as white tufts on the bark and needles. It is capable of killing young ornamental hemlocks.

• *Control.* See under aphids, Chapter 11.

HEMLOCK BORER. *Melanophila fulvoguttata.* The adult, a flat, metallic-colored beetle with three circular reddish yellow spots on each wing cover, deposits eggs in bark crevices. The white, $1/2$-inch-long larvae bore wide, shallow galleries in the inner bark and sapwood.

• *Control.* See under borers, Chapter 11.

HEMLOCK LOOPER. *Lambdina fiscellaria.* Hemlocks may be completely defoliated by a pale yellow caterpillar with a double row of small black dots along its body, which is more than an inch long at maturity. The adult moth has tan to gray wings that expand to more than 1 inch. Two other species, *L. athasaria athasaria* and *L. fiscellaria lugubrosa,* also defoliate hemlocks.

• *Control.* See under caterpillars, Chapter 11.

HEMLOCK FIORINIA SCALE. *Fiorinia externa.* This scale may infest hemlock leaves, and occasionally those of spruce, causing them to turn yellow and drop prematurely. Both male and female scales are elongated. The females are pale yellow to brown and are almost completely covered with their own cast skins. There are two generations a year in the New England states.

HEMLOCK SCALE. *Abgrallaspis ithacae.* The adult female, circular and nearly black, infests the lower surfaces of hemlock leaves, causing premature leaf fall. In heavy infestations the scales may move to the twigs and branches. *Tsugaspidiotus tsugae* closely resembles *A. ithacae.* It is also circular, somewhat darker brown-black, and nippled at the center. Another hemlock scale, *A. pini,* also circular and black, infests hemlock, Douglas-fir, and many species of pine.

GRAPE SCALE. *Aspidiotus uvae.* Hemlock hedges can be destroyed by infestations of this small, dingy-white scale, which has yellowish nipples or exuviae. Japanese wax scale also attacks hemlock.

• *Control.* See under scales, Chapter 11.

SPRUCE LEAF MINER. *Taniva albolineana.* This species, more common on spruce, occasionally mines the leaves of hemlock.

• *Control.* See under leaf miners, Chapter 11.

HEMLOCK RUST MITE. *Nalepella tsugifoliae.* This eriophyid mite severely affects hemlocks in the eastern United States, giving them an unthrifty look.

• *Control.* See under eriophyid mites, Chapter 11.

SPIDER MITES. Hemlocks in ornamental plantings are extremely susceptible to several other species of mites. The spruce spider mite (see under Spruce) is prevalent on hemlock. The two-spotted mite, *Tetranychus urticae,* feeds on the undersides of the needles, sucking the juice from the cells; the needles turn pale and become spotted. Eggs and mites are usually covered with delicate webs.

• *Control.* See under spider mites, Chapter 11.

OTHER PESTS. Hemlocks are also subject to bagworm, cypress moth, blackheaded budworm, Japanese weevil (see under Holly), black vine weevil (see under Yew), fir flatheaded borer, spruce budworm, gypsy moth, and hemlock sawfly.

• *Control.* Control measures are rarely used.

HICKORY (*Carya*)

Hickories are not commonly planted as shade trees but are often found on wooded lot homesites. Pignut hickory grows on dry upland sites in the eastern United States; most other hickories grow in moist sites having deep soils. Most hickories have naturally deep taproots, which may make transplanting difficult, thus limiting their use as landscape trees.

Diseases

CANKER. Several canker diseases occasionally occur on hickory. These are associated with the fungi *Strumella coryneoidea, Nectria galligena,* and *Rosellinia caryae. Poria spiculosa* causes a serious canker rot of hickory stems. These cankers, with thick deep callus folds, appear as rough, circular trunk swellings having depressed centers.

• *Control.* Prune dead or weak branches. Avoid bark injuries and keep the trees in good vigor by fertilizing and by watering during dry spells. Keep borer and other insect infestations under control.

CROWN GALL. *Agrobacterium tumefaciens.* This bacterial disease occurs occasionally on hickory. Heavily galled limbs and branches resembling crown gall are actually caused by *Phomopsis* sp. infections.

• *Control.* Prune and destroy infected twigs or branches.

LEAF SPOTS. Several leaf-spotting fungi occur on hickory. Of these, *Gnomonia caryae* is the most destructive. It produces large, irregularly circular spots that are reddish brown on the upper leaf surface and brown on the lower. The margins of the spots are not as sharply defined as those of many other leaf spots. The minute brown pustules on the lower surface are the summer spore-producing bodies. Another spore stage develops on dead leaves and releases spores the following spring to initiate new infections. The fungus *Monochaetia monochaeta* occasionally produces a leaf spot on hickory but is more prevalent on oaks. The fungus *Marssoniella juglandis* also attacks hickory but is more destructive to black walnut. Two species of *Septoria, S. caryae* and *S. hicoriae,* also cause spots on hickory.

• *Control.* See under leaf spots, Chapter 12.

POWDERY MILDEWS. *Phyllactinia guttata* and *Microsphaera caryae.* These fungi cause mildewing of leaves.

• *Control.* Control measures are rarely adopted.

WITCHES' BROOM. *Microstroma juglandis.* This fungus, which causes a leaf spot of butternut and black walnut, is capable of causing a witches' broom disease on shagbark hickory. The brooms, best seen when the trees are dormant, are composed of a compact cluster of branches. Early in the growing season the leaves on these branches are undersized and curled with white, moldy growth on the lower leaf surface; later they turn black and drop off. Another witches' broom disease of hickory and pecan, called bunch disease, is caused by a phytoplasma.

• *Control.* No effective control measures have been developed.

TRUNK ROTS. *Ganoderma applanatum, Oxyporus populinus, Phellinus igniarius, Climacodon septentrionalis,* and *Inonotus andersonnii.* These fungi cause hickory trunk rots.

Insects

HICKORY LEAF STEM GALL ADELGID. *Phylloxera caryaecaulis.* In June, the sucking of this small insect produces hollow green galls on leaves, stems, and small twigs. The insides of the galls are lined with minute shiny aphidlike adelgids of varying sizes. By July, the galls turn black. Galls range from the size of a small pea to more than $1/2$ inch in diameter.

• *Control.* See under aphids, Chapter 11.

APHIDS. Black pecan and giant bark aphids attack hickory.

HICKORY BARK BEETLE. *Scolytus quadrispinosus.* Young twigs wilt as a result of boring by the bark beetle, a dark brown insect $1/5$ inch in length. The bark and sapwood are mined, and the tree may be girdled, by the fleshy, legless $1/4$-inch-long larvae that overwinter under the bark.

• *Control.* Spray the foliage with Sevin when the beetles appear in July. Remove and destroy severely infested trees, and peel the bark from the stump. Increase the vigor of weak trees by fertilization and watering.

CATERPILLARS. The leaves of park trees may be chewed by one of the following caterpillars: elm spanworm, hickory-horned devil (royal walnut moth), redhumped, walnut, yellow-necked, white-marked tussock moth, fall cankerworm, fall webworm, fruit tree leafroller, green mapleworm, and hickory tussock moth. Larvae of the last named are covered with stiff white hairs that may cause a rash when handled.

• *Control.* See under caterpillars, Chapter 11.

JUNE BUGS. *Phyllophaga* sp. The leaves may be chewed at night by light to dark brown beetles, which vary from $1/2$ to $7/8$ inch in length. The beetles rest in nearby fields during the day. The larva is $3/4$ to 1 inch long, white, and soft-bodied with a brown head. Three or more years are required for completion of the life cycle of most species.

• *Control.* See under foliage-feeding beetles, Chapter 11. The larval stage, which feeds on grasses in lawns and golf courses, can be controlled by treating the lawns with Diazinon or Dursban.

PECAN CIGAR CASEBEARER. *Coleophora caryaefoliella.* The leaves are mined, turn brown, and drop when infested by the pecan cigar casebearer, a larva $1/5$ inch long with a black head. The adult female is a moth with brown wings that have fringed hairs along the edge and a spread of $2/5$ inch. The larvae overwinter on twigs and branches in cigar-shaped cases $1/8$ inch long.

• *Control.* See under caterpillars, Chapter 11.

PAINTED HICKORY BORER. *Megacyllene caryae.* The sapwood of recently killed trees is soon riddled by painted hickory borers, creamy white larvae that attain a length of $1/4$ inch. The $3/4$-inch-long adult beetle, which is dark brown with zigzag lines on the back, lays its eggs in late May or early June. Dogwood and red-headed ash borers also attack hickory (see under Dogwood and Ash).

• *Control.* See under borers, Chapter 11.

OAK TWIG PRUNER. *Elaphinoides villosus.* Twigs weakened by the tunneling activity of the $1/2$-inch-long larva are broken off by the wind and fall to the ground. The larvae overwinter inside these twigs. The adult is a reddish brown beetle $1/4$ inch long. Another insect, the twig girdler, also attacks hickory (see under Hackberry).

• *Control.* Gather and destroy severed branches and twigs in autumn or early spring.

SCALES. Several species of scales—grape, obscure, walnut, and Putnam—occasionally infest hickories.

• *Control.* See under scales, Chapter 11.

MITE. *Eotetranychus hicoriae.* This pest occasionally infests hickory leaves, causing them to turn yellow and drop prematurely. An eriophyid gall mite also attacks hickory.

• *Control.* See under spider mites, Chapter 11.

OTHER INSECTS. Butternut woolly worm (see under Walnut), Asiatic oak weevil (see under Oak), maple leafhopper (see leafhoppers, under Maple), and planthoppers (see under Cherry) also feed on hickory.

Fig. III-49. Blight of American holly caused by the bacterium *Corynebacterium ilicis.*

HOLLY (*Ilex*)

The American holly, *Ilex opaca*, native to the eastern United States, is not subject to large numbers of pests. Most of the problems experienced with trees in ornamental plantings result from improper transplanting practices or unfavorable soil conditions. In localities where it can withstand the winters, holly will thrive in almost any type of soil that is well drained and contains considerable amounts of humus. Incorporating several bushels of well-rotted oak leaf mold into the soil at transplanting time will help to provide the conditions favored by this tree. When the tree is established in its new site, cottonseed meal should be occasionally worked into the soil to supply nitrogen.

Many other species of holly, both of native and Asiatic origin, are also grown. Unless specifically noted, the diseases and insects described here affect primarily the American holly. American holly leaves remain attached for 3 years and are shed in the spring, thus giving foliar pests plenty of time to attack.

Diseases

BACTERIAL BLIGHT. *Corynebacterium ilicis.* Leaves and shoots of the primary growth appear scorched in June and July. Diseased shoots wilt, droop, and dry out but persist. The infection progresses into the woody shoots of the previous year's growth where the leaves turn black (Fig. III-49).

• *Control.* Copper fungicides will probably control this disease. Excessive use of nitrogenous fertilizers and cultivation of soil beneath the trees increase their susceptibility to bacterial blight.

CANKER. *Botryosphaeria dothidea, Diaporthe eres, Nectria coccinea, Physalospora ilicis, Phomopsis (Diaporthe) crustosa,* and *Diplodia* sp. Sunken areas on the twigs and stems may be caused by these fungi.

• *Control.* Prune diseased branches and spray with copper fungicides several times in late spring.

ANTHRACNOSE. *Gloeosporium ilicis.* Leaves of native holly (*Ilex opaca*) develop dead blotches that look like winter scorch symptoms but with a prominent black line margin.

• *Control.* See under Anthracnose, Chapter 12.

LEAF ROT, DROP. *Pellicularia filamentosa.* As a result of invasion by the *Rhizoctonia* stage of the causal fungus, the leaves of American holly cuttings may decay and drop about 2 weeks after the cuttings are inserted into rooting medium. The disease first appears as a cobwebby coating, a combination of the fungus threads and grains of sand adhering to the undersides of the leaves which touch the sand.

• *Control.* Insert cuttings in clean fresh sand or in pasteurized old sand. Do not use cuttings taken from holly branches that touch the ground.

LEAF SPOTS. *Cercospora ilicis, C. ilicicola, C. pulvinula, Cylindrocladium* sp., *Englerulaster orbicularis, Gloeosporium aquifolii, Macrophoma phacidiella, Microthyriella cuticulosa, Phyllosticta concomitans, P. terminalis, Rhytisma ilicinicola, R. velatum, Sclerophoma* sp., and *Septoria ilicifolia.* Many fungi cause brown spots of the leaves.

• *Control.* See under leaf spots, Chapter 12.

POWDERY MILDEWS. *Microsphaera nemopanthis* and *Phyllactinia guttata.* In the South, holly leaves may be affected by these mildew fungi.

• *Control.* See under powdery mildew, Chapter 12.

SPOT ANTHRACNOSE. *Elsinoë ilicis.* Leaves of Chinese holly (*I. cornuta*) in the South are occasionally affected by this disease. Two types of lesions occur on the leaves: numerous tiny black spots and a large leaf-distorting spot more than an inch in length, which is confined to half of the leaf blade. Lesions may also occur on the shoots and berries.

• *Control.* Periodic applications of copper fungicides will provide control.

TAR SPOT. Native holly and English holly (*I. aquifolium*) are subject to the tar spot disease.

• *Symptoms.* Yellow spots appear on the leaves during late May. These turn reddish brown and finally black by fall. A narrow border of yellow tissue remains around the darkened spots. Premature defoliation seldom occurs.

• *Cause.* Tar spot is caused by the fungus *Phacidium curtisii.* Spores produced in the blackened areas initiate new infections in early spring.

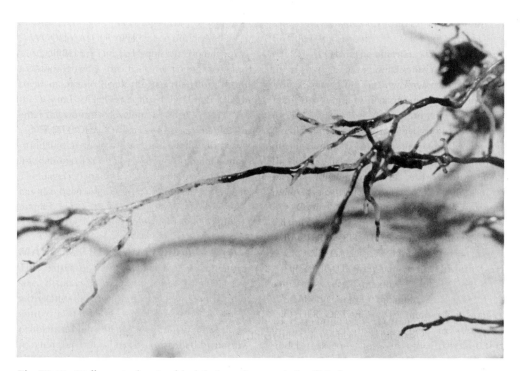

Fig. III-50. Holly roots showing black lesions characteristic of black root rot.

• *Control.* Gather and destroy badly spotted leaves. Spray with a copper fungicide several times at 2-week intervals starting in late spring. Such sprays may cause slight injury if the season is cool.

BLIGHT. *Botrytis cinerea.* Blossoms and new growth are blighted in wet seasons.

TWIG DIEBACK. Black stem cankers and black spots on the leaves of English holly grown in the Pacific Northwest are typical of this fungal disease. During cool rainy weather, complete defoliation and severe twig blighting of the lowermost branches are common.

• *Cause.* The fungus *Phytophthora ilicis* causes this disease.

• *Control.* Although no control measures have been developed, it is highly probable that periodic applications of copper fungicides will provide control.

BLACK ROOT ROT. *Thielaviopsis basicola* (also referred to as *Chalara elegans*). This fungus, the cause of Japanese and blue holly decline and death, also infects American holly roots (Fig. III-50). See under root rots, Chapter 12.

OTHER DISEASES. A disease called yellow leaf spot is caused by a virus; a red leaf spot is caused by a species of the fungus *Sclerophoma*; and several species of nematodes cause root-knot.

• *Control.* Control measures are usually unnecessary.

Nonparasitic Diseases

LEAF SCORCH. A browning or scorching of the leaves, common on holly in late winter or early spring, is of nonparasitic origin. Occasionally it is caused by the presence of water or ice on the leaves at the time the sun is shining brightly. This causes a scalding, followed by invasion by secondary organisms and finally by scorching.

Hollies planted in wind-swept areas are also more susceptible to so-called winter drying. The leaves in late winter or early spring lose water faster than it can be replaced through the roots. As a result the leaf edges wilt and turn brown. In exposed situations, newly transplanted holly should be protected with some sort of windbreak or sprayed with an antidesiccant such as Wilt-Pruf NCF.

Fig. III-51. Blotch on holly leaf caused by the leaf miner *Phytomyza ilicis.*

Insects

HOLLY LEAF MINER. *Phytomyza ilicis.* Yellow or brown serpentine mines or blotches in leaves are produced by the leaf miner, a small yellowish white maggot, $1/6$ inch long, that feeds between the leaf surfaces (Fig. III-51). The adult is a small black fly that emerges about May 1 and makes slits in the lower leaf surfaces, where it deposits eggs. Prior to egg laying, females may puncture leaves and feed on the sap oozing from the wounds.

Another species, the native leaf miner, *P. ilicicola*, produces very slender mines and may occur on the same tree as *P. ilicis.* Holly leaf miners are the most damaging pest of landscape hollies.

• *Control.* See under leaf miners, Chapter 11.

BEETLES. The black blister beetle, the Japanese beetle, and the potato flea beetle occasionally infest hollies.

• *Control.* See under foliage feeding beetles, Chapter 11.

BERRY MIDGE. *Asphondylia ilicicola.* The larvae of this pest infest holly berries and prevent them from turning red in fall.

• *Control.* Where only a few trees are involved, hand picking and destruction of infested berries should keep this pest under control.

Diazinon spray applied in early June will control the midge where many trees are being grown.

CAMBIUM MINER. *Marmara* and other spp. Thin-barked trees such as young holly, fir, pine, birch, and cherry show serpentine mines in the bark. These larvae feed primarily on phloem and generally do little damage. Controls are not normally used.

BUD MOTH. *Rhopobota naevana ilicifoliana.* Holly in the Pacific Northwest is subject to this pest, the larval stage of which feeds on the buds and terminal growth inside a web.

• *Control.* See under caterpillars, Chapter 11.

HOLLY SCALE. *Dynaspidiotus britannicus.* Circular, flat, $^{1}/_{16}$-inch scales infest the berries, leaves, and twigs of holly in the West.

PIT-MAKING SCALE. *Asterolecanium puteanum.* This scale has become prevalent on American holly and Yaupon (*I. vomitoria*) in the southern states. Oval, $^{1}/_{16}$ inch in diameter when mature, and pale yellow in color, this insect embeds itself in the bark and causes a pitted and swollen condition of the stems similar to that produced by the golden oak scale. Branches are distorted, the leaves take on an abnormal color, and at times there is considerable dieback from the branch tips.

OTHER SCALES. Many other species of scale insects attack hollies: black, cottony maple leaf, cottony taxus, Japanese wax, nigra, latania, purple, dictyospermum, gloomy, walnut, euonymus, California red, greedy, lecanium, oleander, oystershell, peach, soft, and tea. Comstock mealybug also feeds on holly.

• *Control.* See under scales, Chapter 11.

JAPANESE WEEVIL. *Callirhopalus bifasciatus.* The leaves of ash, elm, dogwood, hemlock, and oak, as well as of holly, are occasionally attacked by this pest. The beetles are about $^{1}/_{4}$ inch in length, varying from light to dark brown, with

striations on the wing covers. Weevil feeding results in notches cut into the leaf margin.

• *Control.* See under weevils, Chapter 11.

WHITEFLY. See under Dogwood.

APHIDS. *Toxoptera aurantii.* The citrus aphid feeds on holly.

PSYLLID. *Metaphalaria ilicis.* The yaupon psyllid forms a reddish gall on yaupon holly.

Other Pests

SOUTHERN RED MITE. *Oligonychus ilicis.* This mite, most active in spring and fall, has become a serious pest of holly.

• *Control.* See under spider mites, Chapter 11.

HONEYLOCUST. See Locust, Honey

HOP-HORNBEAM (*Ostrya*)

Hop-hornbeam, also known as ironwood, is a slow-growing tree with a rounded crown and slender, pendulous, often contorted branches. This American native is tolerant of a wide range of soils.

Diseases

Hop-hornbeam is susceptible to cankers caused *by Aleurodiscus sp., Nectria* sp., and *Strumella coryneoidea*; to leaf spots by *Cylindrosporium dearnessi* and *Septoria ostryae*; to powdery mildews by *Microsphaera ellisii, Phyllactinia guttata,* and *Uncinula macrospora*; to root rots by *Armillaria mellea* and *Clitocybe tabescens*; to a leaf blister by *Taphrina virginica*; and to a rust caused by *Melampsoridium carpini. Gnomoniella carpinea* (imperfect, *Monostichella robergei*) causes twig cankers and leaf spots of ironwood.

• *Control.* The controls for these diseases are discussed in Chapter 12.

Insects

Among the insects that attack hop-hornbeam are birch lace bug, melon aphid, pitted ambrosia beetle, two-lined chestnut borer, and two species of scales, cottony-cushion and latania.

• *Control.* These insects are rarely serious enough to require control measures.

HOP-TREE (*Ptelea*)

Diseases

LEAF SPOTS. *Cercospora afflata, C. pteleae, Phleospora pteleae, Phyllosticta pteleicola,* and *Septoria pteleae.* These species of fungi cause leaf spots on hop-tree.

• *Control.* Pick off and destroy spotted leaves. Other control measures rarely become necessary.

RUST. *Puccinia windsoriae.* This fungus occasionally occurs on hop-tree. The alternate stage of the fungus is found on grasses.

• *Control.* No controls are required.

ROOT ROT. *Phymatotrichum omnivorum.*

• *Control.* Control measures have not been developed.

Insects

TWO-MARKED TREEHOPPER. *Enchenopa binotata.* These sucking insects, $^1/_4$ inch in length, with a long, proboscislike head portion, resemble miniature quail or partridges; they are dark brown in color with two white spots. When disturbed, the insects jump very rapidly from place to place. They secrete honeydew, and the egg masses are covered by a snow-white frothy substance (Fig. III-52)

Fig. III-52. Treehopper (*Enchenopa binotata*) egg masses on hop-tree.

like that secreted by spittle insects but much firmer in consistency. From a distance, infested branches resemble those infested with woolly aphids or cottony-cushion scale. Walnut, redbud, hickory, locust, and sycamore are also infested.

• *Control.* Spray with malathion or with a pyrethrum-rotenone compound when the insects are young.

SCALE. *Pseudaulacaspis pentagona.* This pest, also known as West Indian peach scale, infests a wide variety of trees in the warmer parts of the country.

• *Control.* See under scales, Chapter 11.

HORNBEAM (*Carpinus*)

European hornbeam has an attractive pyramidal shape. It is relatively tolerant of air pollution.

Diseases

LEAF SPOTS. *Clasterosporium cornigerum, Monostichella robergei, Gnomoniella fimbriella, Phyllosticta* sp., and *Septoria carpinea.* These fungi cause leaf spotting of hornbeam.

• *Control.* These leaf spots are rarely serious enough to warrant control measures. See under leaf spots, Chapter 12.

CANKERS. *Pezicula carpinea* and *Nectria galligena.* These two species of fungi frequently cause bark cankers, sometimes leading to severe dieback of branches.

• *Control.* Badly cankered trees cannot be saved. Prune out twigs and branches of mildly infected ones. The use of copper sprays may also be justified in some situations.

TWIG BLIGHT. *Fusarium lateritium.* In the South this fungus causes a twig blight on hornbeam. The sexual stage of this fungus is *Gibberella baccata.*

• *Control.* Prune and destroy infected twigs on valuable ornamental specimens.

OTHER PROBLEMS. The felt fungus, *Septobasidium curtisii,* occurs on a number of trees in the South. The mycelium is purplish black and covers the branch as well as the scale insect the fungus parasitizes.

• *Control.* Control measures are not necessary.

Insects

SOURGUM SCALE. *Phenacaspis nyssae.*

This scale is nearly triangular, flat, and snow-white.

• *Control.* See under scales, Chapter 11.

TWO-LINED CHESTNUT BORER. See under Oak. This pest attacks European hornbeam following severe winters, killing branches in the upper crown and sometimes the entire tree.

LEAFHOPPER. See under Locust, Honey.

MAPLE PHENACOCCUS. See under Maple.

HORSECHESTNUT (*Aesculus*)

Valued for its shape and palmate leaf, horsechestnut tends to be messy and troublesome because of leaf blotch and scorch. Other species in this genus are called buckeye, and some are good landscape trees. Buckeye and horsechestnut are subject to many of the same diseases, though buckeye generally is less damaged. Unlike buckeye, horsechestnut is not native to the United States.

Diseases

ANTHRACNOSE. *Glomerella cingulata.* Leaf petioles, midribs, and veins turn brown, distinguishing this disease from leaf blotch. In addition, terminal shoots become blighted down to several inches below the buds. Diseased tissue is shrunken, and the epidermis and young bark are ruptured; pustules are formed containing the pink spores of *Colletotrichum,* the imperfect stage of the anthracnose fungus.

• *Control.* See under anthracnose, Chapter 12.

CANKER. *Nectria cinnabarina.* This disease is said to attack the branches and to cause much defoliation of old trees; however, the fungus is apparently mostly saprophytic.

LEAF BLOTCH. *Guignardia aesculi.* This fungal disease is very serious in nurseries, where it often causes complete defoliation of the stock. The spots may be small but can also become so large that they include nearly all the leaf. At first they are merely discolored and water-soaked in appearance; later they turn a light reddish brown with a very bright yellow marginal zone (Figs. III-53 and 54). When the whole leaf is infected it becomes dry and brittle and usually falls. The small black specks seen in the center of the spot are the fruiting bodies of *Phyllosticta,* the imperfect stage of the fungus. The leaf stalks are also attacked. This leaf blotch is very similar to scorch, often seen on shade trees along streets and in city parks. The two diseases can be distinguished by the presence of small, black, pimplelike fruiting bodies on the leaf blotch caused by the fungus. The fungus overwinters on the

Fig. III-53. Early stage of the horsechestnut leaf blotch disease.

Fig. III-54. Advanced stage of the disease.

old leaves where it produces its perfect stage; the ascospores are the means by which the disease is spread in the spring. The first signs of infection may not appear until sometime in July. The 'Ruby Red' horsechestnut (*Aesculus* x *carnea*) is very susceptible to leaf blotch, whereas the yellow buckeye (*Aesculus flava*) is less susceptible.

• *Control.* See under leaf spots, Chapter 12.

LEAF SPOT. *Septoria hippocastani.* Small brown circular spots occasionally develop on the leaves of this host. The slender spores may be seen when the fruiting structures are observed under the microscope.

• *Control.* The sprays recommended for leaf blotch will also control this fungus.

POWDERY MILDEW. *Uncinula flexuosa.* This disease is prevalent in the Midwest, where the undersides of leaves frequently are covered with white mold. The fruiting bodies of the winter stage of the fungus appear as small black dots over the mold.

• *Control.* See under powdery mildew, chapter 12.

WOUND ROT. *Collybia velutipes.* The fungus enters through wounds and destroys the wood, later forming clusters of mushroomlike fruiting bodies. It is one of the fungi that may still be attached to the trees in winter. The fruiting bodies have dark brown, velvety stalks. Other fungi such as *Polyporus, Stereum, Hypoxylon,* and *Trametes* may also decay living horsechestnut.

• *Control.* Remove limbs that have been killed and cover the cut surface to prevent the entrance of fungal spores into the wound.

OTHER DISEASES. Horsechestnut is susceptible to three stem diseases: wilt caused by *Verticillium albo-atrum*; canker by *Diaporthe ambigua*; and bleeding canker by *Phytophthora cactorum.* The first is discussed in Chapter 12 and the third under crown canker in Dogwood.

Abiotic Scorch

Some horsechestnut trees are susceptible to nonparasitic leaf scorch. The scorching usually becomes evident in July or August. First the margins of the leaves become brown and curled. Within 2 or 3 weeks the scorch may extend over the entire leaf. Some observers report that scorch is more prevalent in hot, dry seasons, but serious injury also has been noted in wet seasons. Trees that are prone to scorch will show symptoms every year regardless of the kind of weather, whereas others nearby may show none.

• *Control.* Prune susceptible trees and provide them with good growing conditions. Fertilize and water when necessary.

Insect Pests

PETIOLE BORER. *Proteoteras aesculana.* In springtime, look for drooping, wilted leaves and a hole in the petiole made by the larva of this moth.

• *Control.* No controls have been developed.

COMSTOCK MEALYBUG. See under Catalpa.

WHITE-MARKED TUSSOCK MOTH. See under Elm.

JAPANESE BEETLE. See under Linden.

ELM SPANWORM. See under Elm.

APHIDS. *Drepanaphis* and *Periphyllus* spp. These aphids infest horsechestnut.

LEAFHOPPERS. *Empoasca* sp. This pest causes premature defoliation of California buckeye.

WALNUT SCALE. See under Walnut.

BAGWORM. See under Juniper.

FLATHEADED BORER. See under Maple.

ASIAN LONGHORN BEETLE. See under Maple.

JACARANDA (*Jacaranda*)

This native of Brazil produces showy lavender flowers in spring. Leaves are double compound, giving the tree a feathery look. Jacaranda, planted as a street and landscape tree in southern California, tends to be weak wooded with narrow branch crotch angles. Pest problems are minimal.

Insects

BEAN APHID. *Aphis fabae.* This aphid may form dense colonies.

JAPANESE LILAC TREE (*Syringa*)

This is a hardy, small tree that is drought resistant.

JAPANESE PAGODA-TREE
See Pagoda-Tree, Japanese.

JAPANESE SNOWBELL (*Styrax*)

This small tree has pendant flowers and grows best in well-drained, rich soil. It suffers few problems.

JUNEBERRY. See Serviceberry.

JUNIPER (*Juniperus*)

Because of their evergreen habit, diversity of form, and other desirable qualities, the species and varieties in the genus *Juniperus,* known commonly as cedars and junipers, are extensively used in ornamental plantings.

Diseases

CEDAR RUSTS. The rust fungi that attack *Juniperus* occasionally produce material damage to junipers, but are a greater threat to apples, crabapples, hawthorns and related trees growing nearby. Indeed, because this disease requires the two different host plants, in some regions where apples are an important crop, cedars and junipers are routinely eradicated to suppress rust disease. On juniper, the fungus often puts on a bright orange show (Fig. III-55) in moist spring weather. This disease and control suggestions are discussed in Chapter 12.

PHOMOPSIS TWIG BLIGHT. Also called tip blight, this disease occurs on seedlings and nursery stock as well as on 8- to 10-foot trees in ornamental plantings and on larger native red cedars in some parts of the country. The disease becomes progressively less serious, however, as the trees become older, and little damage occurs on trees over 5 years old.

Though primarily a disease of the eastern red cedar (*J. virginiana*) and its horticultural varieties, twig blight also affects more than a dozen other groups, among which are arborvitae, cypresses, retinosporas, white cedar, and other species of *Juniperus.*

• *Symptoms.* The tips of branches turn brown in early summer (Fig. III-56), followed by progressive dieback until an entire branch or even an entire young tree is killed. This disease is easily confused with Kabatina twig blight.

• *Cause.* Twig blight is caused by the fungus *Phomopsis juniperovora.* Older trees in the vicinity of an evergreen nursery may harbor the fungus on diseased twigs. The fungus may also be carried on small, diseased branch tips that accompany purchased seed. In spring and summer, large numbers of spores are produced in

Fig. III-55. Orange galls of the cedar-apple rust fungus *Gymnosporangium juniperi-virginianae* showing gelatinous "horns."

Fig. III-56. Phomopsis twig blight of juniper.

gray to black fruiting bodies on infected twigs and branches. During rainy periods the spores ooze out and are spread to other plants by wind, rain, and laborers.

• *Control.* Where practicable, prune out and destroy infected branch tips. Spray with Benomyl, Cleary's 3336, Banner MAXX, or Zyban plus a spreader-sticker once in the fall and three or four times at 2-week intervals in spring, starting when warm weather begins.

Resistant cultivars and varieties are: Chinese juniper—'Femina', 'Iowa', 'Keteleeri', 'Pfitzeriana Aurea', 'Robusta', *sargentii*, *sargentii* 'Glauca', and 'Shoosmith'; common juniper—'Ashfordii', 'Aureospica', 'Depressa', 'Hulkjaerhus', 'Prostrata Aurea', 'Repanda', and 'Suecica'; creeping juniper—'Depressa' and 'Procumbens'; Savin juniper—'Broadmoor', 'Knap Hill', and 'Skandia'; western red cedar—'Silver King', 'Campbellii', 'Fargesii Prostrata', and 'Pumila'; red cedar—'Tripartita'.

KABATINA TWIG BLIGHT. *Kabatina juniperi.* This fungus causes a serious twig blight of juniper with symptoms similar to those of Phomopsis tip blight. Tips turn brown and die in spring as new growth begins, suggesting that

infections may occur during the previous season. Since the fungus enters through wounds, infection may be associated with insect feeding.

• *Control.* We have found no fungicides that control the disease well. The following juniper cultivars have been found to be resistant to Kabatina blight: Chinese juniper—'Aurea Gold Coast', 'Hetzii Glauca', 'Hetzii', 'Keteleeri', 'Mint Julep', 'Pfitzeriana Aurea', 'Pfitzeriana', *sargenti* 'Glauca', *sargentii* 'Viridis', and 'Spartan'; common juniper—'Hibernica' and 'Hornibrooki'; Dahurian juniper—'Expansa'; creeping juniper—'Andorra' and 'Marcella'; japgarden juniper—'Nana' and 'Variegata'; Savin juniper—*tamariscifolia*; eastern red cedar—'Prostrata Glauca'. It is important to remember that when junipers are evaluated for disease reactions, resistance does not equal immunity—some disease may appear on plants considered resistant.

BOTRYOSPHAERIA DIEBACK. *Botryosphaeria stevensii.* This fungus causes cankers and branch dieback of junipers weakened by heat and drought. Rocky Mountain juniper is more susceptible to this disease than other junipers.

CERCOSPORA BLIGHT. *Pseudocercospora juniperi (Cercospora sequoiae* var. *juniperi).*

Fig. III-57. Juniper bagworms.

Cercospora blight can be distinguished from Phomopsis and Kabatina blights because Cercospora blight leaves trees with healthy tips while blighted foliage occurs away from the tip. Bordeaux mixture applied in late spring and early summer will prevent Cercospora blight.

ROOT ROT. *Phytophthora cinnamomi.* Although many junipers are considered to be resistant to this fungus, some cultivars are susceptible.

• *Control.* See under root rots, Chapter 12.

WOOD DECAY. Several species of wood-decay fungi, such as *Heterobasidion annosus* and *Antrodia juniperina,* are occasionally found on junipers though they rarely become destructive enough to warrant special treatments. See under wood decay, Chapter 12.

Abiotic Diseases

Hybrid junipers may be seriously injured and even killed by coatings of ice which last for several days. Individual branches or whole trees may die from the aftereffects.

Insects and Related Pests

ROCKY MOUNTAIN JUNIPER APHID. *Cinara sabinae.* Twig growth may be checked and the entire tree weakened by heavy infestations of the Rocky Mountain aphid, a reddish brown insect $1/8$ inch long. The honeydew that it secretes is a good medium for the sooty mold fungus which can completely coat the leaves and further weaken the tree.

• *Control.* See under aphids, Chapter 11.

RED CEDAR BARK BEETLE. *Phloeosinus dentatus.* The adult beetle is about $1/16$ inch long. It lays its eggs in narrow excavations about 1 or 2 inches long. As the young grubs hatch, they bore out sideways, making galleries of a characteristic pattern that resembles the markings made in elms by the elm bark beetle.

• *Control.* The cedar bark beetle is more likely to attack recently transplanted trees or those that are suffering from lack of water. Spraying with methoxychlor may help.

BAGWORM. *Thyridopteryx ephemeraeformis.* Red cedars are among the most susceptible of ornamentals to the attacks of bagworms. The caterpillar builds around itself an elongated sack (Fig. III-57), which increases to 2 to 3 inches in length as the insect grows. The presence of these feeding insects is likely to be overlooked since the protecting bags are made of green leaves. The adult stage is a moth, and the pest overwinters as eggs in the old female bags. Bagworms have a wide host range, which includes needled and broad-leaved trees.

• *Control.* Pick off the bags by hand or cut them off with a pruning pole in late summer, fall, or winter and destroy them. If necessary, spray with Carbaryl, Diazinon, Dursban, Dycarb, Dylox, Malathion, Mavrik, Pounce, Orthene, Sevin, or Talstar in late spring when the young caterpillars begin to feed.

JUNIPER MIDGE. *Contarinia juniperina.* Blisters at the base of the needles and death of leaf tips are produced by small yellow maggots. The adult stage is a small fly.

• *Control.* Spray the leaves in mid- to late April with Dimethoate or Cygon.

JUNIPER SCALE. *Carulaspis juniperi.* The needles, particularly of the Pfitzer juniper, turn yellow as a result of sucking by tiny circular

Fig. III-58. Juniper scale, *Carulaspis carueli.*

scales that are at first snow-white then turn gray or black (Fig. III-58). The pest overwinters as the female adult. Greedy and black scales also attack juniper.

• *Control.* See under scales, Chapter 11.

JUNIPER WEBWORM. *Dichomeris marginella.* The twigs and needles are webbed together, and some turn brown and die when infested by the juniper webworm, a $^1/_2$-inch-long brown larva with longitudinal reddish brown stripes. The adult female, a moth with a wingspread of $^3/_5$ inch, appears in June and deposits eggs that hatch in 2 weeks. The insect overwinters in the immature larval stage. This pest also attacks the creeping juniper (*J. horizontalis*).

• *Control.* See under caterpillars, Chapter 11.

CYPRESS TIP MOTH. *Argyresthia cupressella.* This moth attacks foliage tips.

JUNIPER MEALYBUG. *Pseudococcus juniperi.* This dark red mealybug infests junipers in the Midwest.

• *Control.* See under scales, Chapter 11.

TAXUS MEALYBUG. See under Yew.

TWO-SPOTTED MITE. See under Hawthorn.

SPRUCE SPIDER MITE. See under Spruce.

OTHER INSECTS. *Periploca nigra,* juniper twig girdler, is a pest in California. *Clastoptera juniperina* is a western spittlebug of juniper. Arborvitae weevil and leaf miner also feed on juniper (see under Arborvitae). A borer, *Isophrietis* sp., girdles juniper seedling stems.

KATSURA-TREE (*Cercidiphyllum*)

Katsura is tolerant of shade and moist soil and grows well in the eastern and northern United States. The only diseases recorded on this host are cankers caused by species of *Phomopsis* and *Dothiorella*; Verticillium wilt; and root rot caused by *Armillaria*. Pruning cankered branches below the infected area should keep the canker disease under Control.

See Chapter 12 for the control of Armillaria root rot.

KENTUCKY COFFEE TREE (*Gymnocladus*)

Kentucky coffee tree has bipinnate compound leaves and produces large, coarse, heavy pods.

It grows best in rich, light soils. This species is remarkably free of fungal parasites and insect pests.

Diseases

LEAF SPOT. Three fungi, *Cercospora gymnocladi, Phyllosticta gymnocladi,* and a species of *Marssonia,* have been reported on this host.

• *Control.* Special control measures are rarely required. See Chapter 12.

OTHER DISEASES. A root rot caused by the fungus *Phymatotrichum omnivorum* and a root and butt rot by *Ganoderma lucidum* are the only other fungal diseases known on *Gymnocladus.*

• *Control.* Control measures have not been developed.

Insects

OLIVE SCALE. *Parlatoria oleae.* This insect is purplish brown and its female shell is ovate, circular, dirty gray, and very small. The female begins laying eggs in spring. Walnut scale also attacks Kentucky coffee tree.

• *Control.* See under scales, Chapter 11.

LARCH (*Larix*)

Larches grow in almost any type of soil, including clay and limestone. They do best, however, in moist, loamy soils and in full sunshine. They do not tolerate dry soils or sandy hillsides in climates where the summers are hot. The varieties most commonly used in ornamental plantings are the American larch or tamarack, the European larch, the Japanese larch, and the western larch.

Diseases

EUROPEAN LARCH CANKER. This disease has been particularly destructive on larches in Europe for a long time. It also infects larches in the coastal regions of New England and eastern Canada. European and American larches are known to be very susceptible to canker, whereas the Japanese larch is relatively resistant.

• *Symptoms.* Stems of young trees and branches of small diameter die suddenly as the result of girdling by cankers. Cankers that form on trunks and large branches slowly increase in

size each year, resulting in considerable distortion and swelling. A heavy flow of resin is evident from the cankers. White, hairy, cup-shaped fruiting bodies, about $1/4$ inch in diameter, develop in the cankered tissues.

• *Cause.* The fungus *Lachnellula willkommii* causes larch canker. It enters through wounds and destroys the inner bark and cambium. Frost injury may also be associated with canker.

• *Control.* European investigators have found that selecting favorable planting sites and maintaining trees in good vigor discourages severe outbreaks of the disease.

OTHER CANKERS. Four other species of fungi cause cankers: *Dasyscypha ellisiana, Aleurodiscus amorphus, Leucostoma (Valsa) kunzei,* and *Phomopsis* species, some of which are associated with senescing tissues.

• *Control.* Canker diseases are difficult to control.

Keeping trees in good vigor by fertilizing and watering when needed will help to reduce the severity of infection.

LEAF CAST. *Hypodermella laricis.* The needles of the American and western larches and the spur shoots that bear them may be killed by this disease. The earliest symptom is yellowing, which is followed by browning of the needles. Very small, elliptical, black fruiting bodies of the fungus appear on the dead leaves during the winter.

Several other fungi, including *Cladosporium* sp., *Lophodermium laricis, L. laricinum,* and *Meria laricis,* also cause leaf cast or leaf blights of larch. The leaf cast diseases are most common on ornamental larches in the western United States.

• *Control.* See under leaf spots, Chapter 12.

NEEDLE RUSTS. Three rust fungi attack larch needles. They develop principally on the needles nearest the branch tips. Affected needles turn yellow and have pale yellow fungal pustules on the lower surfaces.

The fungus *Melampsora paradoxa* occurs on American, European, western, and Alpine larches. Its alternate hosts are several species of willows. Larches are infected only by spores from willows, but the spores from willows can reinfect willows.

The fungus *Melampsora medusae* attacks American larch and its alternate host, poplar. As with *M. paradoxa,* only spores from poplar can infect larch.

Melampsoridium betulinum affects American larch and several species of birches. The spore stage on birch can infect birch as well as larch.

• *Control.* Where larches are the more valuable ornamental plantings, the removal of the alternate host will prevent infection. Infected needles should be submitted to a rust specialist for determination of the exact species involved before attempts to eradicate the alternate hosts are made.

WOOD DECAY. Several species of fungi are constantly associated with the various types of wood decay of larch. Those most common in the eastern United States are *Heterobasidion annosum, Phellinus pini,* and *Phaeolus schweinitzii.* They are found mainly on older, neglected trees.

• *Control.* See under wood decay, Chapter 12.

Abiotic Disease

AIR POLLUTION. Larch is sensitive to sulfur dioxide and ozone. See Chapter 10.

Insects

LARCH CASEBEARER. *Coleophora laricella.* In May and June, larches infested with this caterpillar suffer from an extensive browning of the leaves. The leaves are mined by a small caterpillar that uses pieces of the needles to form a cigar-shaped case $^1/4$ inch long. The black-headed caterpillar eats a hole in the leaf either at the end or in the middle and feeds as a miner in both directions as far as possible without leaving the case. The miners winter in the cases, which are attached to the twigs. The small, silvery-gray moths emerge in late June or July.

• *Control.* See under caterpillars, Chapter 11.

LARCH SAWFLY. *Pristiphora erichsonii.* The needles are chewed by $^3/4$- to 1-inch-long, olive green larvae covered with small brown spines. The adult is a wasplike fly with a wingspread of $^4/5$ inch. Eggs are deposited in incisions on twigs in late May and June. The larvae overwinter in brown cocoons on the ground.

• *Control.* See under sawflies, Chapter 11.

WOOLLY LARCH ADELGIDS. *Adelges laricis* and *A. lariciatus.* White woolly patches adhering to the needles are typical signs of this pest. The adults are actually adelgids and are hidden beneath the woolly masses. The winged adults migrate to pines, and another generation returns to the larch the following season. Eggs are deposited at the bases of the needles in the spring. Young adelgids overwinter in bark crevices.

• *Control.* See under aphids, Chapter 11.

OTHER INSECTS. The larvae of the gypsy moth and the white-marked tussock moth (see under Elm), the Japanese beetle (see under Linden), and *Orgyia pseudotsugata,* the Douglas-fir tussock moth, also chew larch leaves.

• *Control.* See Chapter 11.

LINDEN (*Tilia*)

Linden is also called basswood. The larger-leaved American basswood and the smaller leaved European linden are both used in landscapes. They grow best in fertile, moist soils, but they can tolerate variable soils, heat, and drought associated with urban circumstances.

Diseases

CANKER. *Nectria cinnabarina.* Twigs and larger branches bear cinnabar-colored fruiting bodies of the fungus, each body about the size of a pinhead. These ascocarps break through the bark and can be seen without a hand lens. The same or similar fungi attack apples, oaks, and other trees.

Other cankers on linden are caused by *Aleurodiscus griseo-cana* and *Strumella coryneoidea.*

• *Control.* Cut out and destroy all cankered branches and remove and destroy twigs and branches that have fallen to the ground.

ANTHRACNOSE. *Glomerella cingulata (Gnomonia tiliae).* Leaf spots and blotches appear along the veins; defoliation of blighted leaves may result.

Fig. III-59. Japanese beetles on linden and the characteristic appearance of the chewed leaves. **Fig. III-60.** Upper inset: Japanese beetle.

• *Control.* See under anthracnose, Chapter 12.

LEAF BLIGHT. *Cercospora microsora.* Circular brown spots with dark borders characterize this disease. The spots are very numerous, sometimes causing the entire leaf to turn brown and fall off. Young trees are most seriously affected. The sexual stage of the causal fungus is *Mycosphaerella microsora.*

• *Control.* The same as for leaf blotch, below.

LEAF BLOTCH. European lindens are occasionally affected.

• *Symptoms.* Splotchy brown spots with irregular margins occur in various parts of the leaf.

• *Cause.* Leaf blotch is caused by the fungus *Asteroma tiliae,* which is thought to overwinter on diseased fallen leaves.

• *Control.* Gather and destroy fallen leaves.

LEAF SPOTS. *Phlyctema tiliae* and *Phyllosticta*

tiliae. These leaf spots are relatively rare, so control measures are unnecessary.

POWDERY MILDEWS. *Microsphaera alni, Phyllactinia guttata,* and *Uncinula clintonii.* Lindens are susceptible to these powdery mildew fungi. They rarely cause enough damage to require control measures.

OTHER DISEASES. Other diseases reported on lindens include canker caused by *Botryosphaeria dothidea* and wilt by *Verticillium albo-atrum.*

• *Control.* See under Sycamore and Chapter 12.

Insects and Related Pests

LINDEN APHID. *Myzocallus tiliae.* Sap is sucked from the leaves by this yellow and black aphid.

• *Control.* See under aphids, Chapter 11.

JAPANESE BEETLE. *Popillia japonica.* This metallic green-bronze beetle skeletonizes the leaves of its host in midsummer. It is attracted to certain trees in a planting and feeds high up on a sunny side (Figs. III-59 and 60). The host range of Japanese beetle is very wide, but linden, sassafras, and horsechestnut are favored. The larval stage is a grub that feeds on roots of turf grasses in late summer.

• *Control.* See under foliage-feeding beetles, Chapter 11.

ELM CALLIGRAPHA. *Calligrapha scalaris.* Ragged holes appear in leaves chewed by creamy-white larvae with yellow heads. The adult beetle is ³/₈ inch long, oval, yellow, with green spots on the wing covers and a broad, irregular, coppery green stripe down the back. Lemon yellow eggs are deposited on the lower leaf surface in late June or early July. The beetles hibernate in the ground.

• *Control.* Spray the foliage with Sevin as the larvae appear.

LINDEN LOOPER. *Erannis tiliaria.* This pest, also known as the basswood looper, infests apple, birch, elm, hickory, and maple. The caterpillars are 1 ¹/₂ inches long at maturity, bright yellow, with ten longitudinal wavy black lines down the back (Fig. III-61). The moth, buff colored with a 1³/₄-inch wingspread, deposits its eggs from October to November.

CATERPILLARS. Among the many caterpillars that chew the leaves of this host are cankerworms, yellow-necked caterpillar, elm spanworm, variable oak leaf caterpillar, and the larvae of the gypsy, cynthia, cecropia, mourning-cloak, and white-marked and hickory tussock moths.

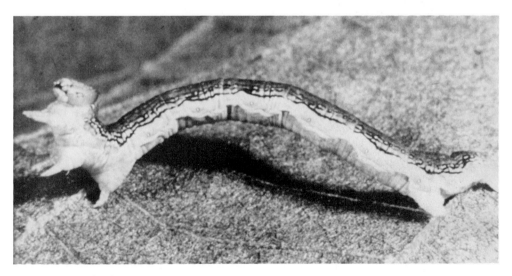

Fig. III-61. Linden looper.

- *Control.* See under caterpillars, Chapter 11.

BASSWOOD LEAF MINER. *Baliosus ruber.* The beetles feed on the underside of the leaves, eating out all tissues except the veins. The larvae also work on the underside, making large, blisterlike mines. The foliage turns brown, withers, and falls off.

- *Control.* See under leaf miners Chapter 11.

ELM SAWFLY. *Cimbex americana.* These smooth caterpillars, pale green with a black stripe down the middle of the back, are about 1 inch long. They curl up tightly when at rest. Elm, willow, maple, and poplar are other hosts frequented.

- *Control.* See under sawflies, Chapter 11.

LINDEN MITE. *Eotetranychus tiliarium.* In midsummer, especially in dry weather, the leaves become infested with this mite, causing them to turn brown and dry up.

- *Control.* See under spider mites, Chapter 11.

ERIOPHYID MITE. *Phytoptus tiliae.* In late spring, leaves infested with this eriophyid mite show light green feltlike galls on the lower leaf surface, which later appear as light brown woolly patches. An eriophyid mite causes elongated "nail galls" on upper leaf surfaces. Eriophyid mites have little effect on tree health.

- *Control.* See under eriophyid mites, Chapter 11.

EUROPEAN LINDEN BARK BORER. *Chrysoclista linneela.* This whitish larva with a light brown head bores into the bark of lindens. It does not affect the cambial area but honeycombs the bark to such an extent that decay-producing organisms have easy access.

- *Control.* No control measures have been developed.

LINDEN BORER. *Saperda vestita.* Broad tunnels beneath the bark near the trunk base or in roots are made by the linden borer, a slender white larva 1 inch long. The adult, a yellowish brown beetle $^3/_4$ inch long, with three dark spots on each wing cover, feeds on green bark. Eggs are deposited in small bark crevices made by the beetle.

Other borers that attack lindens are the flatheaded (see under Maple), the American plum (see under Planetree), red-headed ash (see under Ash), and the brown wood.

- *Control.* See under borers, Chapter 11.

SCALES. Nine species of scale insects infest this host: cottony maple, European fruit lecanium, oystershell, Putnam, San Jose, terrapin, tuliptree, walnut, and willow.

- *Control.* See under scales, Chapter 11.

WALNUT LACE BUG. See under Walnut.

TWIG GIRDLER. See under Hackberry.

LOCUST, BLACK (*Robinia*)

This tree, a native of the eastern United States, is a quick-growing, short-lived tree. Black locust tolerates heat, drought, and neglect. Globe-shaped thornless types are potentially useful landscape trees.

Diseases

CANKER. *Aglaospora anomala, Nectria galligena,* and *Diaporthe oncostoma.* Cankers on twigs and death of the distal portions may be due to any one of these fungi.

- *Control.* Prune and destroy infected twigs.

DAMPING-OFF. *Phytophthora parasitica.* In nurseries, serious damage to seedlings from 1 to 3 weeks old may be caused by this fungus. The young plants droop and their cotyledons curl. This is followed by wilting and the collapse of the entire seedling, which decays within a few days.

- *Control.* Use clean soil or steam-pasteurize old soil for setting out seedlings.

LEAF SPOTS. *Cladosporium epiphyllum, Cylindrosporium solitarium, Gloeosporium revolutum, Phleospora robiniae,* and *Phyllosticta robiniae.* Many fungi cause leaf spots of black locust.

- *Control.* Control measures are seldom practiced.

POWDERY MILDEW. *Erysiphe polygoni, Microsphaera diffusa,* and *Phyllactinia guttata.* White coating of the leaves by these fungi occurs only occasionally on black locust.

- *Control.* These mildews are never serious enough to warrant control measures.

WOOD DECAY. Nearly all the older black locusts growing along roadsides and in groves in the eastern United States harbor one of several wood-decay fungi. Many of these decays have followed infestation of the locust borer, *Megacyllene robiniae.*

Fig. III-62. Branch dieback of black locust caused by locust borer infestations.

The fungus *Phellinus robiniae* causes a spongy yellow rot of the heartwood. It infects the trunk through tunnels made by the locust borer or through dead older branches. After extensive decay of the woody tissues, the fungus grows toward the bark surface where it produces hard, woody, bracket- or hoof-shaped fruiting structures nearly 1 foot wide. The upper surface of the structure is brown or black and is cracked; the lower surface is reddish brown.

• *Control.* See under wood decay, Chapter 12.

WITCHES' BROOM. Black locust and honey locust are subject to this condition. The disease, though common on the sprouts, rarely occurs on the large trees. It is characterized by the production, in late summer, of dense clusters or bunches of twigs from an enlarged axis. The bunched portions ordinarily die during the following winter. Phytoplasmas have been associated with several witches broom diseases.

• *Control.* No control measures are known. Infected trees appear to recover naturally.

Insects

LOCUST BORER. *Megacyllene robiniae.* The young grubs first bore into the inner bark and sapwood. Galleries may extend in all directions into the wood, which becomes discolored or blackened. The trees become badly disfigured. Young plantings may be entirely destroyed (Fig. III-62). The adult is a black longhorned beetle about $^3/_4$ inch long, spotted with bright yellow, transverse, zigzag lines (Fig. III-63). Adults feed on pollen from goldenrod blossoms.

• *Control.* See under borers, Chapter 11.

LOCUST LEAF MINER. *Odontota dorsalis.* The beetles live through the winter and attack the young leaves in early May. They skeletonize the upper surface and lay their eggs on the under surface. The larvae enter the leaf and make irregular mines in the green tissue. Heavily mined foliage appears reddish brown from a distance. A second generation of beetles emerges in September.

• *Control.* See under leaf miners, Chapter 11.

LOCUST TWIG BORER. *Ecdytolopha insiticiana.* Elongated, gall-like swellings 1 to 3 inches long on the twigs result from feeding and irri-

Fig. III-63. Adult stage of the locust borer, *Megacyllene robiniae.*

tation by the pale yellow larvae of the locust twig borer. The adult female is a grayish brown moth with a wing expanse of $^3/_4$ inch. The pest overwinters in the pupal stage among fallen leaves.

• *Control.* Prune and destroy infested twigs in August and gather and destroy fallen leaves in autumn.

SCALES. Many species of scale insects infest black locust: black, cottony-cushion, cottony maple, frosted, greedy, oystershell, Putnam, San Jose, soft, and walnut.

• *Control.* See under scales, Chapter 11.

OTHER INSECTS. Carpenterworm (see under Ash), yellow-necked caterpillar, Asiatic oak weevil, honey locust plant bug, silver-spotted skipper caterpillar, greenhouse whitefly, bagworm, and treehoppers infest locust trees.

Other Pests

DODDER. *Cuscuta* sp. This well-known flowering vine, which grows as a parasite on various plants, causes considerable damage to seedlings. Dodder appears as a wiry, yellowish-orange annual vine. It lacks chlorophyll and takes its nutrition from other plants. It may also be a factor in transmission of certain viral diseases in border plantings.

• *Control.* Dacthal herbicide spray applied to the soil in early spring will prevent germination of dodder seed. Hand pulling or even killing of host plants may be needed once dodder becomes established.

LOCUST, HONEY (*Gleditsia*)

Honey locust is native to the eastern and central United States. It tolerates a wide range of growing conditions. Only the thornless types should be used in the landscape.

Diseases

CANKERS. *Thyronectria austro-americana* (imperfect, *Gyrostroma*), *T. denigrata*, *Tubercularia ulmea*, *Nectria cinnabarina* (imperfect, *Tubercularia vulgaris*), *Cytospora gleditschiae*, and *Dothiorella* sp. These fungi kill the bark, cambium, and outer sapwood of honey locust. Canker disease symptoms include sunken, dead areas of bark on the trunk, limbs, or branches; chlorotic, reduced foliage; premature fall col-

oration; early leaf drop; and twig and branch dieback. Signs of the fungi often appear in the cankers as yellow-brown, reddish brown, salmon, or black fungal fruiting structures on the bark. Years ago P. P. Pirone found that *Thyronectria* killed a number of honey locusts at the United Nations gardens in New York. (Fig. III-64). This fungus causes extensive areas of necrosis, cracking, and peeling of the trunk bark along with a brown discoloration.

• *Control.* Prevent wounds and promote tree vigor by providing mulch, a little fertilizer, and water. The cultivars 'Holka', 'Imperial', Shade-

Fig. III-64. The honey locust (*Gleditsia triacanthos*) at left in the United Nations gardens died as a result of infection by the fungus *Thyronectria austro-americana*.

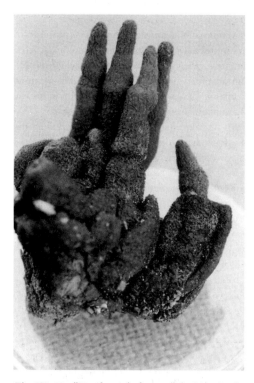

Fig. III-65. "Dead man's fingers," fruiting bodies of the Xylaria root rot fungus.

master', and 'Thornless' are disease tolerant, 'Moraine' and 'Skyline' are moderately tolerant; and 'Sunburst' is most susceptible.

POWDERY MILDEW. *Microsphaera ravenelii.* This mildew fungus is widespread on honey locusts.

• *Control.* Control measures are rarely used for this disease on this host.

TAR SPOT. *Plagiosphaeria gleditschiae.* This is a serious blight in the southern states. Numerous black asexual fruiting bodies (acervuli) of the fungus develop on the lower side of the leaves. The ascocarpic sexual stage develops throughout the summer and lives through the winter. In the Midwest other leaf spots are caused by *Cercospora condensata* and *C. olivacea.*

• *Control.* Gathering and destroying of all fallen leaves should provide practical control.

RUST. *Ravenelia opaca.* One rust disease is known to occur on honey locust.

• *Control.* No control measures are necessary,

WOOD DECAY. Like most trees, honey locusts are subject to wood-decaying fungi. Among the more prevalent ones are a species of *Fomes, Laetiporus sulphureus, Daedalea ambigua, D. elegans, Ganoderma lucidum,* and *Xylaria mali. G. lucidum* occurs on living trees, which suggests that it is a vigorous parasite, as it is on species of maple. Dead man's fingers (Fig. III-65), the *Xylaria* fruiting bodies, have been observed in lawns near declining honey locust trees. *Xylaria* also causes root rot. A collar rot of honey locust has been associated with overwatering, but no specific microbes have been implicated.

• *Control.* No effective controls are known. Avoid mechanical injuries around the base of the tree, provide good growing conditions; and mulch, fertilize, and water the tree to maintain good vigor. Avoid overwatering. See under wood decay, Chapter 12.

Abiotic Disease

AIR POLLUTION. Honey locust is very sensitive to ozone. See Chapter 10.

Insects and Related Pests

HONEY LOCUST BORER. *Agrilus difficilis.* A flatheaded borer burrows beneath the bark of honey locust and eventually may girdle the tree. Large quantities of gum exude from the bark near the infested nodes. The adult beetles, which emerge in June, are elongate, $1/2$ inch long, and black with a metallic luster. The borer preferentially attacks stressed trees.

• *Control.* See under borers, Chapter 11.

POD GALL MIDGE. *Dasineura gleditschiae.* This midge causes globular galls $1/8$ inch in diameter at the growing tips (Fig. III-66). The pest seems to prefer some of the thornless varieties such as 'Moraine' and 'Sunburst' to the ordinary honey locust. The adult midge appears in April when the leaves begin to emerge. It deposits eggs singly or in clusters among the young leaflets. The larvae hatch within a few days and begin to feed on the inner surface of the leaflets.

• *Control.* Carbaryl, Precision, or Sevin sprays applied in May control this pest. Because

Fig. III-66. Pod gall of honey locust caused by the insect *Dasineura gleditschiae.* Uninfested leaves are seen on the lower right side.

the pest has many natural enemies, outbreaks may be brief.

WEBWORM. Damage (Fig. III-67) is similar to that observed on silk-tree (see under Silk-Tree). Most tolerant to least tolerant honey locust cultivars are 'Moraine', 'Skyline', 'Shademaster', 'Imperial', and 'Sunburst'. Winter survival of pupae in the North is thought to be enhanced when relatively warm overwintering sites such as buildings are located within a few yards of infested honey locusts.

• *Control.* See under caterpillars, Chapter 11.

HONEY LOCUST PLANT BUG. *Diaphnicoris chlorionis.* Discoloration of leaves and stunting of new growth is caused by this widely distributed pest (Figs. III-68 and 69). Complete defoliation may occur during heavy infestations. Nymphs emerge from eggs and begin feeding in spring as new growth begins. Adults are $3/16$ inch long and pale green. Both adults and nymphs are difficult to detect because they blend with foliage and growing tips on which they feed. Yellow-leaved cultivars of honey locust, such as 'Sunburst,' are more susceptible to this bug than are green-leaved ones like 'Shademaster.'

• *Control.* See under plant bugs, Chapter 11.

Fig. III-67. Mimosa webworm damage on honey locust.

Fig. III-69. Adult of honey locust plant bug.

Fig. III-68. Honey locust plant bug injury and discoloration of foliage.

LEAFHOPPER. *Macropsis fumipennis.* These active, pale, ¹/₈-inch-long, wedge-shaped sucking insects cause stippled, pale leaves.

• *Control.* Control is normally not needed. Spray with Diazinon, Malathion, Merit, Orthene, Pounce, or Sevin, when leafhoppers are first noticed in June, July, or Augus.,

POTATO LEAFHOPPER. See under Maple. The toxic saliva of this insect causes leaves to be stunted, browned, and curled.

SPIDER MITE. *Eotetranychus multidigituli.* This mite causes yellow stippling of the leaves, which drop prematurely. Defoliated trees usually leaf out again in late summer but are considerably weakened.

• *Control.* Some insecticides applied for control of other honey locust insects may kill mite predators, thus worsening mite infestations. See under spider mites, Chapter 11.

BAGWORM. See under Juniper.

SPRING AND FALL CANKERWORMS. See under Elm.

WALNUT CATERPILLAR. See under Walnut.

TWIG GIRDLER. See under Hackberry.

SCALES. Cottony maple (see under Maple), San Jose (see under Cherry), walnut (see under Walnut), black, and hickory lecanium scales feed on honey locusts.

OTHER INSECTS. Cowpea aphid (see Chapter 11) and whitefly attack honey locust.

MADRONE. See Strawberry-Tree.

MAGNOLIA (*Magnolia*)

Many species of magnolia are native to the southeastern and eastern United States. The three major groups include evergreen, late-blooming deciduous, and early-blooming deciduous types. Early-blooming magnolias should not be planted in protected, sunny locations, but rather in areas where bloom will be delayed and thus frost damage to flowers avoided.

Magnolias are subject to a number of

fungal diseases and insect pests. Only the more important are treated here.

Diseases

BLACK MILDEWS. *Irene araliae, Meliola amphitrichia, M. magnoliae,* and *Trichodothis comata.* A black, mildewy growth covers the leaves of magnolias in the Deep South.

• *Control.* Magnolias may be sprayed with wettable sulfur.

LEAF BLIGHT. *Ceratobasidium stevensii.* The leaves of *Magnolia grandiflora* may be blighted by this fungus, which also affects apple, citrus, dogwood, Japanese persimmon, pecan, quince, and many shrubs in the South.

• *Control.* Valuable specimens can be protected with two or three applications of copper fungicides.

LEAF SPOTS. *Alternaria tenuis, Cladosporium fasciculatum, Mycosphaerella* spp., *Colletotrichum* sp., *Coniothyrium fuckelii, Epicoccum nigrum, Exophoma magnoliae, Glomerella cingulata, Hendersonia magnoliae, Micropeltis alabamensis, Phyllosticta cookei, P. glauca, P. magnoliae, Septoria magnoliae,* and *S. niphostoma.* These species of fungi cause leaf spots on magnolias.

• *Control.* See under leaf spots, Chapter 12.

LEAF SCAB. *Elsinoë magnoliae* (imperfect *Sphaceloma*). In the Deep South *Magnolia grandiflora* leaves may be spotted by this fungus.

• *Control.* The disease is not serious enough to warrant control measures.

DIEBACK. *Phomopsis* sp. Cankers with longitudinal cracks in the bark are formed on the larger limbs and trunks. The wood is discolored to blue-gray. The bark is dark brown over the affected areas. Apparently healthy branches also may be discolored.

• *Control.* No suggestion for control has been made.

NECTRIA CANKER. *Nectria magnoliae.* This fungus produces symptoms similar to those of *N. galligena* (see canker, under Walnut), but it infects only magnolias and tuliptrees.

• *Control.* Prune and destroy cankered branches. Keep trees in good vigor by watering, spraying, and fertilizing when necessary.

WOOD DECAY. A heart rot associated with the fungi *Fomes geotropus* and *F. fasciatus* has been reported on magnolia. Affected trees show sparse foliage and dieback of the branches. In early stages the rot is grayish black, with conspicuous black zone lines near the advancing edge of the decayed area. The mature rot is brown. The causal fungi gain entrance through wounds.

• *Control.* See under wood decay, Chapter 12.

ALGAL SPOT. *Cephaleuros virescens.* Leaves and twigs infested by this alga have velvety, reddish brown patches.

• *Control.* Control measures are usually unnecessary.

OTHER DISEASES. Bacterial blight (see under Cherry) causes magnolia leaf spots. Magnolia is also susceptible to wilt caused by *Verticillium albo-atrum.*

Insects

MAGNOLIA SCALE. *Neolecanium cornuparvum.* Underdeveloped leaves and generally weak trees may result from heavy infestations of the magnolia scale, a brown, varnishlike hemispherical scale, $1/2$ inch in diameter, with a white, waxy covering (Fig. III-70). The young scales appear in August and overwinter in that stage. Feeding females produce large quantities of honeydew.

TULIPTREE SCALE. *Toumeyella liriodendri.* This scale can also infest magnolias and lindens.

• *Control.* See under scales, Chapter 11.

OTHER PESTS. The following scales also occur on magnolias: black, California, chaff, cottony-cushion, European fruit lecanium, Florida wax, Glover, greedy, oleander, purple, and soft. Spray the young, crawling stages of these scales in May and June. See Chapter 11.

The Comstock mealybug, the omnivorous looper caterpillar, and the citrus whitefly also infest magnolias. Malathion sprays, applied when the pests appear, give good control.

A species of eriophyid mite infests *Magnolia grandiflora* in the South. For control, see under eriophyid mites, Chapter 11.

The sassafras weevil (see under Sassafras) often infests magnolias (Fig. III-71).

Planthoppers (see under Cherry) and

Fig. III-70. Magnolia scale, *Neolecanium cornuparvum.*

Fig. III-71. Sassafras weevil damage to magnolia.

greenhouse thrips (see under Maple) also attack magnolia.

MAIDENHAIR. See Ginkgo.

MANGO (*Mangifera indica*)

Mango produces an important fruit in the tropics. This tree with a wide-spreading crown is grown in some landscapes in southern Florida and California.

Diseases

ANTHRACNOSE. *Glomerella cingulata.* This is perhaps the most prevalent disease of mango in the South. Leaves are spotted, flowers and twigs blighted, and fruits rotted by this fungus.

• *Control.* See under anthracnose, Chapter 12.

POWDERY MILDEW. *Oidium mangiferae.*

Flower panicles and foliage may be distorted when infected by this powdery mildew fungus.

• *Control.* See under powdery mildew, Chapter 12.

OTHER FUNGAL DISEASES. In Florida mango is subject to twig blight caused by a species of *Phomopsis*; leaf spots by *Pestalotiopsis mangiferae, Phyllosticta mortoni,* and *Septoria* sp.; scab by *Elsinoë mangiferae*; and sooty mold by species of *Capnodium* and *Meliola.*

• *Control.* Control measures are rarely necessary.

Insects

SCALES. Many species of scale insects including California red, tea, and armored (*Lindingaspis*) infest mango.

• *Control.* Spray with malathion or Sevin.

OTHER INSECTS. The long-tailed mealybug and greenhouse thrips also infest mango.

• *Control.* Same as for scales.

MAPLE (*Acer*)

The many species of maple in the United States range from shrub-sized patio trees to forest giants. Their leaves are palmately lobed, easily recognized (as on the Canadian flag), and set in opposite pairs. Leaf size ranges from a few inches to a foot; color is variable in summer and often spectacular in fall. Distinctive winged seeds (samaras) consist of paired seeds set in flat membranous wings that facilitate dispersal. Some maples are desired landscape trees; others are prohibited as street trees. Some, like boxelder, are adapted to harsh growing conditions and may be desirable in the Great Plains while being undesirable in the East. Norway maple is said to be adapted to urban environments.

Fungal and Bacterial Diseases

BASAL CANKER. An early symptom is a thin crown resulting from a decrease in the number and size of the leaves. Trees die within a year or two following this period of weak vegetative growth. A more striking symptom is the presence of cankers at the base of the trunk near the soil line. The inner bark, the cambium, and in many instances the sapwood are reddish brown in the cankered area (Fig. III-72). Death occurs when the entire root system decays or when the cankers completely girdle the trunk.

• *Cause.* Basal canker is caused by the fungus *Phytophthora.* A similar disease was found to be caused by *P. cactorum.* The fungus appears to be most destructive on trees growing in poorly drained or shallow soils. Phloem and young xylem are killed by this disease.

• *Control.* See under root rots, Chapter 12.

BLEEDING CANKER. This highly destructive disease affects Norway, red, sycamore, and sugar maples as well as oaks, elms, and American beech. The severity of the disease is associated in some way with hurricane damage.

• *Symptoms.* The disease is named for its most characteristic primary symptom, the oozing of sap from fissures overlying cankers in the bark. Infected inner bark, cambium, and sapwood develop a reddish brown necrotic lesion that commonly exhibits an olive green margin. These symptoms differ markedly from those produced in trees affected with basal canker. A secondary symptom, the wilting of the leaves and dying back of the branches, may be due to a toxic material secreted by the causal fungus.

Fig. III-72. Bark cut away to show the reddish brown discoloration of the sapwood produced by the fungus *Phytophthora.*

• *Cause.* Bleeding canker is caused by the fungus *Phytophthora cactorum. P. cactorum* is also responsible for the crown canker disease of dogwoods. The same organism kills the growing tips of rhododendrons during rainy seasons and is known to attack a large number of other trees, including apple, apricot, cherry, peach, and plum.

• *Control.* No effective control measures are known.

ANTHRACNOSE. *Kabatiella apocrypta.* In rainy seasons this disease may be serious on sugar and silver maples, on boxelder, and to a lesser extent on other maples. The spots are light brown and irregular in shape and may enlarge and run together, causing the death of the entire leaves. Partially killed leaves appear as if scorched. Another anthracnose caused by *Discula* also attacks maples.

• *Control.* See under anthracnose, Chapter 12.

LEAF SPOT (PURPLE EYE). *Phyllosticta minima.* The spots are $1/4$ inch or more in diameter, more or less irregular, with brownish centers and purple-brown margins. The black pycnidia of the fungus develop in the center of the spots. This disease is most severe on red, sugar, and silver maples but also occurs on Japanese, Norway, and sycamore maples.

• *Control.* See under leaf spots, Chapter 12.

BACTERIAL LEAF SPOT. *Pseudomonas aceris.* The leaves of Oregon maple (*A. macrophyllum*) may be spotted by a bacterium. Spots vary from pinpoint dots to areas $1/4$ inch in diameter. They first appear to be watersoaked and are surrounded by a yellow zone; later they turn brown and black.

• *Control.* Preventive sprays containing a copper fungicide applied in early spring should control this organism.

LEAF BLISTER. *Taphrina sacchari.* The lesions produced by this fungus are circular or irregular in shape (Fig. III-73), pinkish or buff on the underside and ochre or buff on top. Sugar and black maple are most susceptible. Blistering, curling, and blighting of other maples are produced by different species: *Taphrina lethifer* affects mountain maple; *T. aceris,* Rocky Mountain hard maple; *T. dearnessii,* red maple; and *T. carveri,* silver maple. *Taphrina* spores are released from diseased leaves in summer and lodge in bud scales, where they lie dormant until new leaves emerge the following spring, at which time infection occurs.

• *Control.* Leaf blisters on valuable specimens can be prevented by spraying dormant lime sulfur or Ferbam just before growth starts in spring.

BULL'S-EYE SPOT. This name describes a spot that occurs in the eastern United States on red, silver, sugar, and sycamore maples located in especially shaded spots. The spots show a distinct target pattern with layers of concentric rings (Fig. III-74). The causal organism is *Cristulariella depraedeus* or *C. moricola.*

Fig. III-73. Maple leaf blister.

Fig. III-74. Bull's-eye spot of sycamore maple.

Fig. III-75. P. P. Pirone examining a dying Norway maple infected with the fungus *Ganoderma lucidum*. He is holding a live branch cut from the healthy side of this tree. The fungus bodies are at the base of the tree.

• *Control.* Control measures have not been developed.

TAR SPOT. *Rhytisma acerinum.* Street maples are seldom infected with this fungus, but red maples in forests may be prematurely defoliated. The spots are irregular, shining black tarlike discolorations up to ¹/₂ inch in diameter, which develop on the upper sides of the leaves. The dark color is due to the masses of brown or black mycelium.

• *Control.* See under leaf spots, Chapter 12.

OTHER FUNGAL LEAF SPOTS. *Rhytisma punctatum* causes a minute black spotting on many species of maples. It is rare in the East but prevalent on the Pacific Coast. A number of other leaf spots that occur occasionally are caused by fungi belonging to the genera *Cercospora, Venturia, Didymosporina, Septoria, Cylindrosporium,* and *Monochaetia.*

POWDERY MILDEWS. *Uncinula circinata, Microsphaera aceris,* and *Phyllactinia guttata.* These mildews are rarely serious. See under powdery mildew, Chapter 12.

NECTRIA CANKER. *Nectria cinnabarina.* Cankers appear on twigs and branches and occasionally develop on the trunks to such an extent

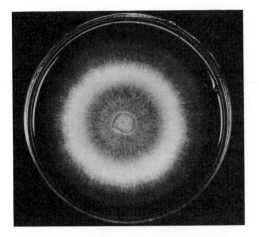

Fig. III-76. A culture of the *Ganoderma* fungus growing on a synthetic medium in the laboratory.

as to cause the death of weak trees. Reddish fungal fruiting bodies develop in large numbers. Although the fungus is most common on maples and lindens, it attacks a wide variety of other hardwood trees. Another fungus, *Nectria galligena*, causes the development of target cankers accompanied by the formation of thick calluses that later become diseased and leave an open wound. Cankers should be cut out well beyond the diseased areas. The vigor of the trees should be increased by mulching, a little fertilization, and watering as needed.

EUTYPELLA CANKER. *Eutypella parasitica.* Boxelder and Norway, red, and sugar maples occasionally show another type of canker, which differs strikingly from that produced by either of the Nectria fungi. The cankers are irregularly circular and contain broad, slightly raised concentric rings of callus tissue. The cankered tissue is firmly attached to the wood with heavy, white-to-buff, fan-shaped wefts of fungal tissue under the bark near the margins. Tiny black fungal bodies are present in the centers of the old cankers.

Other cankers may be caused by the following fungi: *Botryosphaeria dothidea, Diaporthe* spp., *Cytospora* sp., *Fusarium solani, Nectria* sp., *Botryosphaeria obtusa, Septobasidium fumigatum, Valsa sordida, V. ambiens* subsp. *leucostomoides, Steganosporium ovatum,* and *Cryptosporiopsis* sp.

• *Control.* Cankers on trunks can rarely be eradicated by surgical methods once they have become very extensive. Some cankers weaken the tree enough that trunk breakage can occur during strong winds. Those on branches can be destroyed by removing and discarding the affected members. Removal of dead branches, avoidance of unnecessary injuries, and maintenance of trees in vigorous condition by mulching, fertilization, and watering are probably the best means of preventing canker formation.

PHOMOPSIS BLIGHT. *Phomopsis acerina.* The death of Norway maples along city streets has been attributed to infection by this fungus. Galls and burls of red and sugar maple stems and branches have been attributed to *Phomopsis* species.

• *Control.* No controls have been developed.

GANODERMA ROT. *Ganoderma lucidum.* Rapid decline and death of many trees growing along city streets were found to be caused by this fungus. It forms large, reddish fruit bodies with a varnishlike coating at the base of the infected tree or on its surface roots (Figs. III-75, 76, and 77). Red and Norway maples appear to be most susceptible.

• *Control.* See under wood decay, Chapter 12.

TRUNK DECAY. In New England, sugar maples tapped for their sap are subject to a serious trunk decay caused by the fungus *Valsa leucostomoides.* In longitudinal section the affected areas appear as truncated cones with pale yellow centers bordered by deep olive or greenish black streaks. Adherence to proper tree tapping patterns and hole treatments have been developed for this specialized industry. Wood decay (Chapter 12) is especially damaging to red maple. Fungi such as *Ganoderma applanatum, Climacodon septentrionale, Cerrena unicolor,* and *Oxyporus populinus* may be involved.

ARMILLARIA ROOT ROT. See Chapter 12.

XYLARIA ROOT ROT. *Xylaria mali.* The infected root appears black with a carbonaceous crust. The fruiting body of *Xylaria,* the dead man's finger fungus, is a dark, upright mass that protrudes through the soil. There is no control for Xylaria root rot.

BACTERIAL WETWOOD. See Chapter 12.

Fig. III-77. Close-up of the *Ganoderma* fungus at the base of a dying maple tree.

VERTICILLIUM WILT. Already discussed in detail in Chapter 12, wilt is perhaps the most important vascular disease of maples. Positive diagnosis of this disease is possible only by collecting adequate specimens of discolored sapwood beneath the bark of the trunk near the soil line and culturing the material in the laboratory. The *Verticillium* fungus is not the only fungus capable of discoloring the sapwood of maples.

SAPSTREAK. *Ceratocystis coerulescens.* Typical symptoms include thinning crowns with undersized chlorotic foliage, followed by death of the tree. Trunk xylem tissues are discolored. Sapstreak also affects tuliptree in woodlands.

• *Control.* No effective control has been developed.

MOSAIC. Viral diseases of maple have been reported.

Abiotic Disease

MAPLE DECLINE. Weakening and death of sugar maples along heavily traveled highways in the northeastern United States has been attributed to excessive use of de-icing salts (sodium and calcium chlorides) during the winter months. See Chapter 10. Other stress-inducing factors such as drought, restricted root space, or

injuries can initiate decline. Canker and decay fungi and insect borers may attack weakened trees.

LEAF SCORCH. See Chapter 10.

GIRDLING ROOTS. See Chapter 10.

Insects and Mites That Attack Leaves

BOXELDER BUG. *Boisea trivittata.* The leaves of boxelder are injured by the sucking of the bright red nymphs as well as the adults, which are stout, $1/2$ inch-long, grayish black bugs with three red lines on the back (Fig. III-78). All stages of this pest are clustered on the bark and branches in early fall. The four-lined plant bug, *Poecilocapsus lineatus,* also feeds on maple, causing circular brown leaf spots.

• *Control.* Spray the bark with Carbaryl, Dursban, Orthene, or Sevin when clusters of bugs appear in late May or early June. Because this bug feeds on seeds of pistillate (female) trees, one should plant only staminate (male) trees.

FOREST TENT CATERPILLAR. *Malacosoma distria.* These caterpillars are bluish with a row of diamond-shaped white spots along the back (Fig. III-79). They feed individually on leaves of maple, birch, oak, and poplar, but do not form a

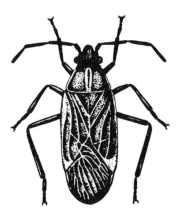

Fig. III-78. Boxelder bug.

tent in the forks of smaller branches as do the eastern tent caterpillars (*M. americana*), which have a white stripe down the back. The egg masses of the two species are similar: about $1/2$ inch long, they completely surround the twig and have a brown, varnished appearance. Fall webworm (see under Ash) also attacks maple.

• *Control.* See under caterpillars, Chapter 11.

GREENSTRIPED MAPLEWORM. *Dryocampa rubicunda*. These caterpillars are pale yellowish green, $1^{1}/2$ inches long, with large black heads; the stripes along the back are alternately pale yellowish green and dark green. The insects pre-

fer red and silver maples. A close relative, the orange striped oakworm (see under Oak), also feeds on maple.

ORANGEHUMPED MAPLEWORM. *Symmerista leucitys*. This caterpillar feeds on maple foliage, skeletonizing leaves.

• *Control.* See under caterpillars, Chapter 11.

BUFFALO TREEHOPPER. *Stictocephala bubalus*. This insect damages maples by scarring their twigs during egg laying.

LEAFHOPPERS. Japanese leafhopper (see under Mountain-Ash) and rose leafhopper, *Edwardsiana roseae*, attack maple. The Norway maple may become seriously infested with yellowish leafhoppers (*Alebra albostriella*); they also cause swellings on the twigs by depositing eggs under the young bark. Leaf symptoms appear as a chlorotic stippling or flecking, resembling ozone injury.

POTATO LEAFHOPPER. *Empoasca fabae*. These insects, migrating from warmer climates, feed on foliage in late spring, causing stunted shoot growth with brown, curled, or distorted leaves.

• *Control.* Early detection of the pest in spring followed by treatment with insecticides such as Carbaryl, Dursban, Diazinon, Malathion, Merit, Pounce, or Sevin is effective.

LEAF STALK BORER. *Nepticula sericopeza*. Though these small borers more commonly

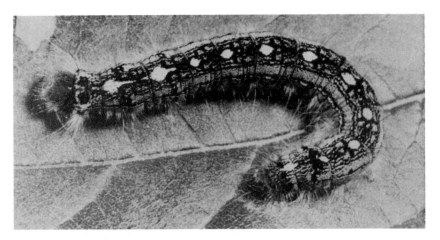

Fig. III-79. Forest tent caterpillar.

infest the fruit of Norway maples, they may attack the leaf stalks in June. The borers tunnel in the stalks, causing a black discoloration about $^1/_2$ inch from the base. The lower end of the stalk is somewhat enlarged. These borers occasionally cause severe defoliation. The adult is a minute moth with spinelike hairs on the surface of its wings.

• *Control.* Spray with a horticultural oil in early spring to destroy the cocoons.

PETIOLE BORER. *Caulocampus acericaulis.* This smooth yellowish sawfly larva causes leaf fall of sugar maple. It tunnels in the upper end of the leaf stalk about $^1/_2$ inch from the blade. The leaf blades and sometimes the leaf stalk itself fall off in May and June. Only the leaves of the lower branches are usually seriously infested.

• *Control.* This pest is rarely serious enough to warrant control measures.

MAPLE LEAF CUTTER. *Paraclemensia acerifoliella.* Much defoliation of sugar maple and beech trees results from the work of this $^1/_4$-inch-long caterpillar. It cuts out small sections of the leaves and forms a case in which it hides while it feeds. It skeletonizes in a ringlike pattern, the center of which may fall out.

• *Control.* Rake up and discard all leaf litter in fall to kill the hibernating pupae.

MAPLE TRUMPET SKELETONIZER. *Epinotia aceriella.* Red and sugar maple leaves are folded loosely by small green larvae which develop inside a long trumpetlike tube. Fruit tree leaf roller (see under Crabapple) also feeds on maple.

• *Control.* These pests are rarely serious enough to warrant control.

NORWAY MAPLE APHID. *Periphyllus lyropictus.* The leaves of Norway maple are subject to heavy infestation by these hairy aphids, which are greenish with brown markings. Like other aphids they secrete large amounts of honeydew. The attacked leaves become badly wrinkled, discolored, and reduced in size. Defoliation may follow. Several other species, including *Drepanaphis acerifoli*, *Periphyllus aceris*, and *P. negundinis*, also occur on maples.

• *Control.* See under aphids, Chapter 11.

GREENHOUSE THRIPS. *Heliothrips haemorrhoidalis.* This insect feeds on maple leaves with its piercing-sucking mouthparts, causing stippling, glazing, and browning of leaves. It produces large quantities of small, dry droplets of excrement.

• *Control.* Spray in early summer with Diazinon, Malathion, Orthene, or Sevin, being careful to cover the undersides of the leaves.

OCELLATE LEAF GALL. *Acericecis ocellaris.* These galls are less than $^1/_2$ inch in diameter and are located within a spot with cherry red margins that is easily confused with purple eye leaf spot. They occur on red maple but are usually not injurious. The small maggots, which may be seen at the center of the spot, soon drop off.

OTHER LEAF-FEEDING INSECTS. Other insects that attack maple leaves include the elm sawfly, *Cimbex americana*; brown-tail moth, *Nygmia phaeorrhoea*; gypsy moth (see under Elm) in certain eastern states; white-marked tussock moth (see under Elm); American dagger moth, *Acronicta americana*; saddled prominent caterpillar, *Heterocampa guttivitta*; and the oriental moth, *Cnidocampa flavescens*. The spring and fall cankerworms, *Paleacrita vernata* and *Alsophila pometaria* (also known as inchworms and measuring-worms); bagworms (see under Juniper); linden looper (see under Linden); yellow-necked caterpillar, io and cecropia moth larvae; green mapleworm; and elm spanworm all feed on maple foliage. See under caterpillars, Chapter 11.

BLADDER-GALL AND OTHER ERIOPHYID MITES. *Vasates quadripes.* The upper surfaces of maple leaves are often covered with small, green, wartlike galls, which later turn blood red. These are caused by the feeding of this mite. If the galls are very numerous, the leaves become deformed. Other maple galls are caused by the mites *Vasates aceris-crumena* and *Eriophyes elongatus*. The galls are about $^1/_5$ inch long, tapering at both ends (fusiform). They develop on the upper side of the leaf and, when numerous, render the foliage unsightly. Several other species of *Eriophyes* produce large blotches along the veins or between them. Brilliant purple, red, or pink minute blisterlike or pilelike growths are caused by the mites.

• *Control.* See under eriophyid mites, Chapter 11.

SPIDER MITES. The mite *Paratetranychus*

Fig. III-80. Maple phenacoccus (*Phenacoccus acericola*) on sugar maple leaf.

aceris occasionally infests maple leaves, causing mottling and premature yellowing. See under spider mites, Chapter 11.

Scale Insects

MAPLE PHENACOCCUS. *Phenacoccus acericola.* The leaves of sugar maple may be covered on the underside by cottony masses that envelop the females in July (Fig. III-80). The males collect in the crevices of the bark and give it a white, chalky appearance. There are several generations a year. Predacious insects feed on the eggs of this scale.

COTTONY MAPLE SCALE. *Pulvinaria innumerabilis.* Silver maples are especially susceptible to infestation by this scale. Smaller branches are often covered with large cottony cushionlike masses about $^1/_2$ inch in diameter that contain as many as 500 eggs in June. The scale itself is brown, $^1/_8$ to $^1/_4$ inch in diameter. The young move out and infest the leaves; later they migrate to the branches, especially the undersides.

Cottony maple leaf scale, *P. acericola*, attacks maples and other trees throughout the eastern United States. Cottony-cushion scale (see under Acacia), with its large cottony white and fluted egg sac, and cottony taxus scale, another soft scale, also attack maple.

OTHER SCALES. Several other scales attack maples. Among them are the terrapin, gloomy, Japanese, black, obscure, walnut, calico, ivy, and banded oystershell scales. Comstock mealybug also attacks maple.

- *Control.* See under scales, Chapter 11.

Wood Borers

Limbs infested with borers are the first to break after a heavy snowstorm or after they have been coated with ice. Five or six species of wood-boring insects infest maples.

FLATHEADED BORER. *Chrysobothris femorata.* Trees in poor vigor are attacked by this light yellow larva, 1 inch long, which builds flattened galleries below the bark, often girdling the tree. The adult is a dark coppery brown beetle, $^1/_2$ inch in length, which deposits eggs in bark crevices in June and July. The pest overwinters in the wood in the pupal stage.

This borer, sometimes referred to as flatheaded apple tree borer, may attack newly transplanted maples in the nursery. A similar pest, *C. mali*, the Pacific flatheaded borer, is also a threat to newly transplanted trees.

LEOPARD MOTH BORER. *Zeuzera pyrina.* This brown-headed borer is white or slightly pinkish with blackish spots on the body. Full-grown borers may be 2 to 3 inches long and nearly $^1/_2$ inch in diameter. They produce cylindrical burrows, in which they overwinter, and also make wide cavities that so weaken the limbs that they break off very readily. Soft maples are especially susceptible. Adult female moths are white with dark spots and are weak fliers.

METALLIC BORER. *Dicerca divaricata.* The larvae of these brass- or copper-colored beetles frequently invade the limbs of peach, cherry, beech, maple, and other deciduous trees. The beetles live 2 or 3 years as borers in the trees and lay their eggs in August and September. The adults have been known to cause much defoliation.

PIGEON TREMEX. *Tremex columba.* These insects, also called horntails, bore round exit holes in the bark about $^1/_4$ inch in diameter. The larvae usually attack trees that are dying from the attacks of fungi or other species of borers.

SUGAR MAPLE BORER. *Glycobius speciosus.* The appearance of dead limbs among leafy branches or dead areas on branches and trunks is evidence of this insect's presence.

ASIAN LONGHORN BEETLE. *Anoplophora glabripennis.* This black beetle with white spots and long black and white striped antennae is native to the Orient. The insect is a quarantined pest but has been found in the East and in the Chicago area, apparently entering the United States on infested wooden packing material. Efforts are under way to eradicate it. This borer perforates the wood with large galleries; exit holes are $3/8$ to $1/2$ inch in diameter. This pest has a wide host range including ash, birch, maple, willow, and plane.

• *Control.* See under borers, Chapter 11.
CARPENTERWORM. See under Ash.

Other Pests

BIRCH LACE BUG. See under Birch.
CAMBIUM MINER. See under Holly.
FRUIT TREE LEAF ROLLER. See under Crabapple.
JAPANESE BEETLE. See under Linden.
LEAF MINER. See Chapter 11.
TWIG PRUNER. See oak twig pruner, under Hickory.
WHITEFLY. See under Dogwood.
WOOLLY ALDER APHID. See under Alder.
SQUIRRELS. The red squirrel, *Tamiasciurus hudsonicus,* bites the bark of young red and sugar maples to drink the sap that flows from the wounds. Cankers, caused by fungi which enter the wounds, develop.
NEMATODES. Dieback of sugar maples in woodland areas has been associated with heavy infestations of several species of root nematodes. The dagger nematode, *Xiphinema americanum,* has also been associated with a decline of sugar maples.

• *Control.* Controls have not been developed.

MAYTEN TREE (*Maytenus*)

This native of Chile, which reaches medium size, displays a weeping habit. Mayten tree withstands hot weather in regions with mild winters and tolerates salty soils. It has few pest problems. Nigra scale, *Saissetia nigra,* attacks mayten tree but is largely controlled by natural parasites.

MELALEUCA. See Cajeput.

MESQUITE (*Prosopis*)

A tree for hot, dry regions, mesquite tolerates drought and alkaline soil.

MIMOSA. See Silk-Tree.

MOUNTAIN-ASH (*Sorbus*)

Because mountain-ashes belong to the rose family, they are subject to most of the same diseases and insects common in this family. Cultivars of European mountain-ash are generally more readily available than American mountain-ash cultivars.

Bacterial and Fungal Diseases

CANKER. *Cytospora chrysosperma, C. massariana, C. microspora,* and *Fusicoccum* sp. These fungi occur occasionally on mountain-ash, especially on weakened trees.

• *Control.* Prune severely affected branches and fertilize. The use of highly nitrogenous fertilizers may increase the tree's susceptibility to fire blight, however.

LEAF SPOTS. The alternate stages of the rust fungi *Gymnosporangium aurantiacum, G. cornutum, G. globosum, G. tremelloides, G. nelsoni, G. nootkatense,* and *G. libocedri* occur on juniper or incense cedar species. On mountain-ash leaves, circular, light yellow, thickened spots first appear during summer. Later, orange cups develop on the lower surfaces of these spots. *Diplocarpon mespili, Phyllosticta sorbi,* and *Venturia inaequalis* also cause leaf spots.

• *Control.* See under rusts and leaf spots, Chapter 12.

BACTERIAL BLIGHT. See under Cherry.
CROWN GALL. See Chapter 12.
FIRE BLIGHT. See Chapter 12. This disease is sometimes very damaging.
SCAB. See under Crabapple.
OTHER DISEASES. Mountain-ash is subject to many other diseases. These are discussed under rosaceous hosts in other parts of this book.

Abiotic Disease

AIR POLLUTION. European mountain-ash is sensitive to ozone. See Chapter 10.

Insects

APHIDS. The rosy apple and woolly apple aphids frequently infest this host.

- *Control.* See under aphids, Chapter 11.

JAPANESE LEAFHOPPER. *Orientus ishidae.* This insect causes a characteristic brown blotching bordered by a bright yellow margin, the yellow zone merging into the color of the leaf.

- *Control.* Spray with Diazinon, Dursban, Merit, Sevin, or Malathion in late May or early June.

MOUNTAIN-ASH SAWFLY. *Pristiphora geniculata.* Green larvae with black dots feed on mountain-ash leaves from early June to mid-July, leaving only the larger veins and midribs. The adults, yellow with black spots, deposit eggs on the leaves in late May.

- *Control.* See under sawflies, Chapter 11.

PEAR SLUG. See under Cherry.

PEAR LEAF BLISTER MITE. *Phytoptus pyri.* Tiny brownish blisters on the lower leaf surface and premature defoliation result from infestations of the pear leaf blister mite, a tiny, elongated, eight-legged pest, $1/125$ inch long. The mites overwinter beneath the outer bud scales. Eggs are deposited in spring in leaf galls, which develop as a result of feeding and irritation by the adult.

- *Control.* See under eriophyid mites, Chapter 11.

MOUNTAIN-ASH SPIDER MITE. See Chapter 11.

ROUNDHEADED BORER. *Saperda candida.* Trees are weakened and may be killed by the roundheaded borer, a light yellow, black-headed, legless larva 1 inch long. Galleries in the trunk near the soil level, frass at the base of the tree, and round holes the diameter of a lead pencil in the bark are typical signs of this pest. The adult is a beetle $3/4$ inch long, brown, with two white longitudinal stripes on the back. It emerges in April and deposits eggs on the bark.

- *Control.* See under borers, Chapter 11.

DOGWOOD BORER. See borers, under Dogwood.

AMERICAN PLUM BORER. See under Planetree.

FLATHEADED BORER. See under Maple.

SCURFY SCALE. *Chionaspis furfura.* A grayish scurfy covering on the bark indicates the presence of this pest. The scale covering the adult female is pear-shaped, gray, and about $1/10$ inch long. The pest overwinters as purple eggs under the female scale.

SCALES. Several other species of scale insects infest mountain-ash: black, globose, frosted, walnut, cottony maple, oystershell, and San Jose.

- *Control.* See under scales, Chapter 11.

BIRCH LACE BUG. See under Birch.

MULBERRY (*Morus*)

Red mulberry, native to North America, has few serious diseases and pests.

Diseases

BACTERIAL BLIGHT. *Pseudomonas mori.* Watersoaked spots appear on leaves and shoots; they later become sunken and black. The leaves are distorted, and black stripes develop on the shoots. The leaves at the tips of twigs wilt and dry up.

- *Control.* Some control may be obtained on young trees by pruning dead shoots in autumn and spraying with bordeaux mixture the following spring. Streptomycin sprays may also be effective.

BACTERIAL WETWOOD. This can be a serious problem on mulberry. See Chapter 12.

LEAF SCORCH. Caused by xylem-inhabiting bacteria, leaf scorch occurs on red mulberry in the East. See Chapter 12.

LEAF SPOTS. *Cercospora moricola, C. missouriensis,* and *Cercosporella mori.* The leaves of mulberry are spotted by these fungi in very rainy seasons. Sometimes, defoliation may result.

- *Control.* See under leaf spots, Chapter 12.

POPCORN DISEASE. *Ciboria carunculoides.* This disease, known only in the southern states, is largely confined to the carpels of the fruit. It causes them to swell and remain greenish, and it interferes with ripening.

• *Control.* The disease is of little importance. It does not lessen the value of the tree as an ornamental.

FALSE MILDEW. *Mycosphaerella mori.* The foliage of mulberries grown in the southern states may suffer severely from attacks of this fungus. It appears in July as whitish, indefinite patches on the undersides of the leaves. Yellowish areas then develop on the upper sides. The fungal threads that will produce spores emerge from the stomata on the underside and spread out to form a white, cobweblike coating; the general appearance is that of a powdery mildew. The asexual spores are colorless, each composed of several cells. The infected leaves fall to the ground, and the overwintering or ascocarpic stage matures in spring on these leaves.

• *Control.* See under leaf spots, Chapter 12.

CANKERS. *Cytospora* sp., *Dothiorella* sp., *D. mori*, *Gibberella baccata* f. *moricola*, *Nectria* sp. and *Stemphyllium* sp. These species of fungi may cause cankers on twigs and branches and dieback of twigs. They can be distinguished only by microscopic examination or laboratory tests.

• *Control.* Prune and destroy dead branches. Keep trees in good vigor by mulching, watering and fertilizing.

POWDERY MILDEWS. *Phyllactinia guttata* and *Uncinula geniculata.* The lower leaf surface is covered by a white, powdery coating of these fungi.

• *Control.* See under powdery mildew, Chapter 12.

Abiotic Disease

AIR POLLUTION. Mulberry is sensitive to sulfur dioxide. See Chapter 10.

Insects and Related Pests

CITRUS FLATID PLANTHOPPER. *Metcalfa pruinosa.* This very active insect, also called mealyflata, is $1/4$ inch long with purple-brown wings and is covered with white woolly matter.

• *Control.* Spray with pyrethrum or rotenone or a combination of the two.

CERAMBYCID BORER. *Dorcaschema wildii.* In the South this borer mines large areas of cambium and tunnels into the wood of mulberry trees. Branches or even entire trees are girdled and killed.

• *Control.* See under borers, Chapter 11.

SCALES. The following scales infest mulberry: California red, cottony maple, hickory lecanium, ivy, Glover, dictyospermum, Florida wax, greedy, olive parlatoria, peach, San Jose, and soft.

• *Control.* See under scales, Chapter 11.

OTHER PESTS. The Comstock mealybug and the mulberry whitefly also infest this host.

In dry seasons the two-spotted mite, *Tetranychus urticae,* may be abundant and injurious, causing leaves to become mottled and yellow.

AMERICAN PLUM BORER. See under Planetree.

PLANTHOPPER. See under Cherry.

NORFOLK ISLAND PINE (*Araucaria*)

This beautiful evergreen tree does well outdoors in the warmer parts of the United States. It is also grown indoors as a potted plant.

Diseases

BLIGHT. *Cryptospora longispora.* The lower branches are attacked first, and the disease gradually spreads upward. As the entire branch becomes infected, the tip end becomes bent. The limbs die and then break off at the tip ends. Plants 5 or 6 years old are soon killed by this disease.

• *Control.* The infected branches should be pruned off and destroyed as soon as discovered. Seeds of *Araucaria excelsa* imported from Norfolk Island Territory, near New Zealand, frequently harbor the causal fungus. Hence they are dipped in sulfuric acid by plant quarantine inspectors before being released to nurserymen.

CROWN GALL. *Agrobacterium tumefaciens.* This host has been proved experimentally to be susceptible. A typical gall is smooth and up to 1 inch in diameter.

- *Control.* Prune infected branches.

BLEEDING CANKER. *Botryosphaeria rhodina* and *Dothiorella* sp. Two fungi are associated with this disease in Hawaii, where Norfolk Island pines are grown as Christmas trees and are used as windbreaks and as ornamentals.

- *Control.* Control measures have not been developed.

Insects

MEALYBUG. The golden mealybug (*Pseudococcus aurilanatus*), cypress mealybug (*P. ryani*), and citrus mealybug (*Planococcus citri*) are all serious pests of Norfolk Island pine in California.

- *Control.* See under scales, Chapter 11.

SCALES. Five species of scales are known to attack this host: araucaria (*Eriococcus araucariae*), pure white, with feltlike oval sacs enclosing bodies and eggs of females; black araucaria (*Lindingaspis rossi*), an almost black species resembling the Florida red scale; chaff (*Parlatoria pergandii*), brownish gray, circular to elongate, smooth, semitransparent; Florida red (see under Camellia), an armored, $^1/_{12}$-inch, circular, reddish brown to nearly black scale; and soft (*Coccus hesperidum*), a flat, soft, oval, yellowish brown, $^1/_8$-inch-long species.

- *Control.* See under scales, Chapter 11.

OAK (*Quercus*)

There are many native oaks throughout North America. Many species are desirable trees for large spaces. Oaks are generally durable and long-lasting. The open-branched habit of species like northern red oak, bur oak, and white oak facilitates passage of utility wires so these species suffer less topping than the pin oak. Some oaks become chlorotic due to alkaline soil conditions.

Diseases

OAK WILT. Arborists in many regions of the country must deal with this highly publicized disease of oaks. The disease has been found from Kansas and Nebraska eastward to western New York state, and from Minnesota southward to Texas. It is most damaging to live oaks in the South and to red oaks in the upper Midwest, but white oaks are also susceptible.

- *Symptoms.* A dull, pale, water-soaked appearance at the edges of leaves progresses toward the midrib and base. The edges may turn brown or bronze, and the leaves quickly drop from the tree. Live oak leaves may develop brown midribs and veins. Spring infections may result in a lesser amount of leaf drop, with the browned leaves remaining attached. Fungal mats are found under the bark of infected trees in the spring.

- *Cause.* The fungus *Ceratocystis fagacearum*, once known as *Chalara quercina*, causes wilt. The fungus causes infections primarily through wounds made by humans or other agents. It is readily transmitted by tools used by arborists and foresters. The most susceptible period appears to be during spring wood development in the tree.

The causal fungus is spread by root grafts and by several insects, including fruit flies, Nitidulid beetles, and the flatheaded borer *Chrysobothris femorata,* as well as the mite *Garmania bulbicola.* This fungus has also been recovered from a species of bark beetle and from the two-lined chestnut borer. Two species of oak bark beetles, *Pseudopityophthorus minutissimus* and *P. pruinosus,* have been found capable of carrying the oak wilt fungus. Tools and climbing spurs used by lumberjacks, foresters, or arborists are other ways by which the fungus is spread. Squirrels are also believed to be vectors. Tree species belonging to the red oak group appear to be the most susceptible, whereas the white oak is markedly resistant. In nature the fungus has been found on Chinese chestnuts and related genera, and it has been transmitted experimentally to a wide variety of trees, including apple.

- *Control.* No effective control is known. Eradicate and destroy infected trees and firewood. Infection centers develop where diseased trees are not removed and destroyed. Severing root grafts mechanically or chemically can halt further spread. Because insect transmission occurs in spring and because the oak wilt fungus appears to be most infectious early in the growing season when the new spring wood vessels are developing, oaks should be pruned only in fall and winter. Injections with the fungicide Alamo have proven successful in some circumstances.

BACTERIAL LEAF SCORCH. *Xylella fastidiosa.* This disease, discussed in more detail in Chapter 12, is widespread in the southeastern and mid-Atlantic states. It occurs on almost a dozen oak species, especially pin and turkey oaks. Symptoms appear annually in the late summer as marginal leaf scorch and premature defoliation. Affected branches are slightly delayed in leafing out in the spring. Over a period of years branches begin to die back, and trees gradually decline and die. The bacterium is thought to be vectored by xylem-feeding leafhoppers.

• *Control.* Use tree maintenance practices that promote good growing conditions so that the disease effects are not compounded by stress. Plant replacement trees that can become established as the infected trees gradually die.

ANTHRACNOSE. *Apiognomonia quercina.* This fungus, whose asexual stage is *Discula quercina,* is common in the northern states especially on white oak, but also on red oaks, American elm, and black walnut. Spots on the leaves run together, causing the appearance of a leaf blotch or blight. The blotches are light brown; dead areas follow the veins or are bounded by larger veins. Infection may occur in midsummer. Defoliation, which may result in the death of weak trees, may occur during rainy weather, which favors the disease.

• *Control.* See under anthracnose, Chapter 12.

LEAF BLISTER. During cool, wet springs almost all species of oak are subject to the leaf blister disease. Circular, raised areas ranging up to $^1/_2$ inch in diameter are scattered over the upper leaf surface, causing a depression of the same size on the lower surface (Fig. III-81). The upper surface of the bulge is yellowish white and the lower yellowish brown. The leaves remain attached to the tree, and there is rarely any noticeable impairment in their functions.

• *Cause.* Leaf blister of oaks is caused by the fungus *Taphrina coerulescens.* New infections in

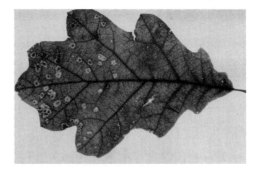

Fig. III-81. Leaf blister of oak.

spring are caused by spores from overwintered leaves and possibly by spores lodged in bud scales and on twigs.

• *Control.* Make a single application of Captan, Ferbam, Fore, or Mancozeb before bud swelling time in spring, using a power sprayer to coat buds and twigs thoroughly.

ACTINOPELTE LEAF SPOT. *Tubakia dryina.* This disease causes brown blotchy leaf spots of pin and red oaks, often late in the season. Where it occurs earlier, it is an indicator of tree stress. We have found that oaks suffering from being planted too deeply or from lime-induced iron chlorosis are especially prone to this disease.

LEAF SPOTS. Like most other trees, oaks are subject to a number of leaf spot diseases. These rarely cause much damage to the trees, since they only develop extensively late in the growing season. The following fungi are known to cause leaf spots: *Cylindrosporium microspilum, Dothiorella phomiformis, Elsinoe quercus-falcatae, Gloeosporium septorioides, G. umbrinellum, Leptothyrium californicum, Marssonina martini, M. quercus, Microstroma album, Monochaetia monochaeta, Phyllosticta tumericola, P. livida, Septogloeum querceum, Septoria quercus,* and *S. quercicola.*

• *Control.* The gathering and destruction of all fallen leaves is usually sufficient to keep down outbreaks in seasons of normal rainfall. Valuable oaks may be protected by several applications of Benlate, copper, or zineb sprays at 2 week intervals, starting in early spring when the leaves unfold.

POWDERY MILDEW. In the southern and

western states *Sphaerotheca lanestris* is the most troublesome mildew producer. It forms a white mealy growth, which later turns brown, on the undersides of the leaves. The entire surface of the leaf, as well as the tip ends of the twigs, may be covered with the brown feltlike mycelium. On California live oak the fungus produces white mycelium and may induce witches' brooms.

Other powdery mildew fungi affecting oaks include *Brasiliomyces trina, Microsphaera* spp. and *Phyllactinia guttata*. Powdery mildew normally attacks oaks late in the growing season, doing little harm; however, in nurseries, highly susceptible English oaks often require season-long treatment.

• *Control.* See under powdery mildew, Chapter 12.

BASAL CANKER. *Phaeobulgaria inquinans.* This fungus frequently develops in crevices of the sunken bark that overlies basal cankers. It enters through open wounds, then invades the surrounding bark and sapwood, eventually girdling and killing the tree. The mature fruiting bodies are cup- or saucer-shaped and grow from short stems that extend into the bark. When moist, the fruiting bodies look and feel like rubber.

STRUMELLA CANKER. *Urnula craterium* (imperfect, *Conoplea*). Although primarily a disease of forest oaks, this canker occasionally affects red and scarlet oaks in ornamental plantings. American beech, chestnut, red maple, tupelo, and pignut and shagbark hickories are also susceptible. Several types of cankers are produced on the trunks, depending on the age of the tree and the growth rate of the causal fungus. Smooth-surfaced, diffuse, slightly sunken cankers are common on young trees 3 to 4 inches in diameter. Cankers on older trees have a rough surface ridged with callus tissue. Open wounds may be present in the center of large cankers as a result of secondary decay and shedding of the bark.

HYPOXYLON CANKER. Recognizable by the gray or tan crusty fungus tissue (stroma) over the cankered area, canker caused by *Hypoxylon atropunctatum* appears on trees under stress from drought, construction damage, or other injuries. Trees weakened by *Hypoxylon* can be a safety hazard in urban areas.

OTHER CANKERS. *Nectria galligena,* cause of nectria canker (see canker, under Walnut), also attacks oak. Cankers on red oaks (*Quercus borealis maxima*) were found to be caused by the fungus *Fusarium solani*; those on stressed oaks growing at high temperatures by *Lasiodiplodia theobromae*; those on scarlet oak and live oak (*Q. virginiana*) by *Endothia parasitica* (the chestnut blight fungus); and those on pin oak (*Q. palustris*) by *E. gyrosa.*

Canker-rots of oak are caused by *Inonotus hispidus, Phellinus spiculosus,* and *Irpex mollis.* Canker-rots can cause affected trees to be a hazard in urban areas. Smooth-patch canker disease is caused by *Aleurodiscus oakesii.*

• *Control.* Prune cankered, dying, and dead branches to eliminate an important possible source of inoculum. Remove small cankers on the trunk by surgical means; mulch, fertilize, and water the trees to improve their vigor.

RUST. *Cronartium quercuum.* The alternate hosts of this fungus are species of pine, such as *Pinus rigida* and *P. banksiana,* to which some damage is done by the development of gall-like growths. On oak leaves small yellowish spots first appear on the undersides, and later brown, bristlelike horns of spores develop.

• *Control.* Little damage is done to the oak; hence no control has been found necessary.

TWIG BLIGHTS. Chestnut oak and, at times, red and white oaks in ornamental plantings are affected by twig blights.

• *Symptoms.* A sudden blighting of the leaves on twigs and branches scattered over the tree is the most striking symptom. The dead twigs and small branches, with their light brown, dead leaves still attached, are readily visible at some distance from the tree because of the sharp contrast with unaffected leaves near them. Diseased bark becomes sunken and wrinkled, and the sapwood beneath is discolored.

• *Cause.* Several species of fungi are cause twig blight: *Botryosphaeria quercuum* (imperfect, *Dothiorella quercina*), *Coryneum kunzei, Diplodia longispora, Pseudovalsa longipes,* and *Botryosphaeria obtusa.* The anthracnose pathogen also causes twig blight. Most of the fungi overwinter in dead twigs and in cankers on larger branches. During rainy weather the following spring spores ooze in large numbers from the dead areas and splash onto young shoots and into injured bark, where they germinate and cause new infections.

• *Control.* During summer, prune infected twigs and branches to sound wood. As a rule, cutting to about 6 inches below the visibly infected area will ensure removal of all fungus-infected tissue in the wood. In severely weakened trees, considerably more tissue must be removed, as the fungus may penetrate down the branch for a distance of 2 or more feet.

Valuable trees should be mulched, fertilized and watered.

ARMILLARIA ROOT ROT. *Armillaria mellea.* Oaks are a favored host of this disease, which is discussed in Chapter 12.

ROOT AND BUTT ROT. *Ganoderma lucidum (Polyporus lucidus).* This fungus causes internal decay of the base of the tree without necessarily killing it. It infects through wounded, dead, and scarred roots and butts and sometimes produces large fruiting bodies at the tree base. *Hericum erinaceus, Inonotus dryadeus, Polyporus fissilis, Laetiporus sulphureus,* and *Pleurotus sapidus* also cause butt rot. *Corticium galactinum* may cause root rot of urban trees; *Phymatotrichum omnivorum,* cause of Texas root rot, attacks seedlings.

WOOD DECAY. Oaks are subject to attacks by a number of fungi that cause wood decays. Fungi such as *Globifomes graveolens, Phellinus everhartii, Stereum gausapatatum, S. subpileatum,* and *Inonotus andersonii* attack living trees. The butt rot fungi *Polyporus fissilis, Laetiporus sulphureus, Hericum erinaceus,* and *Pleurotus sapidus* also decay wood. The same tree may be infected by more than one of these fungi, although in most cases only one is involved.

• *Control.* See under wood decay, Chapter 12.

RINGSPOT. Plant pathologists have found viruslike particles in black and blackjack oak leaves exhibiting ringspot symptoms.

The symptoms were found in trees with dead crowns as well as in vigorous saplings.

• *Control.* Control measures have not been developed.

DECLINE. Oak decline following periods of stress such as construction injury, drought, and insect defoliation has been observed on landscape trees.

Abiotic Disease

CHLOROSIS. Leaf yellowing between the veins of oaks is often the result of iron deficiency. In severe cases, tree growth may be stunted, dead areas may develop on leaves, and trees may die. The problem is often referred to as iron chlorosis or lime-induced chlorosis. Chlorosis occurs on trees growing in neutral or alkaline soils (pH 7.0 or above). Pin oaks grown in many midwestern locations are subject to this disorder.

• *Control.* Iron can be added to the soil in the form of iron chelates, but high pH soils may still tie up the iron. Iron as iron sulfate, iron chelate, or iron citrate may be sprayed on the foliage, implanted into the trunk (Medicaps), or injected via feeder tubes (Mauget). Foliage will turn green in response to these treatments, and the effects may last several months to a few years.

Sulfur should be applied to the root zone to decrease the soil pH. Depending on soil tests, 60–100 pounds/1000 square feet can be applied, and once the soil pH has been changed, the effects should be long-lasting. Elemental sulfur used for soil pH change is more effective if it is worked into the soil; thus the pH of planting site soil should be tested before planting so that sulfur can be incorporated if necessary.

Avoid using alkaline "hard" water for irrigation. Remove masonry materials from soil in oak planting sites and avoid planting sites, located near masonry walls or other lime-containing construction.

AIR POLLUTION. Pin, scarlet, and white oaks are sensitive to ozone. See Chapter 10.

Leaf-Eating Insects

There are so many species of insect larvae that feed on oak leaves that it is impossible to list them all. The following are some of the more important.

GYPSY MOTH. *Lymantria dispar.* Oak is a

favored host for gypsy moth. For a description and remedies, see under Elm.

YELLOW-NECKED CATERPILLAR. *Datana ministra.* The leaves are chewed by this caterpillar, a black and yellowish white-striped larva 2 inches long with a bright orange-yellow segment behind its head. The adult female moth is cinnamon brown with dark lines across the wings, which have a spread of 1¹/₂ inches. Eggs are deposited in batches of 25 to 100 on the lower leaf surface. The insect hibernates in the pupal stage in the soil. The walnut caterpillar (see under Walnut) also feeds on oak.

• *Control.* See under caterpillars, Chapter 11.

SADDLEBACK CATERPILLAR. *Sibine stimulea.* This 1-inch-long, broad, spine-bearing, red caterpillar with a large green patch in the middle of its back occasionally chews the leaves of ornamental and streetside oaks. The adult is a small moth with a wingspread of 1¹/₂ inches. The upper wings are dark reddish brown; the lower, a light grayish brown.

• *Control.* The same as for the yellow-necked caterpillar.

OAK SKELETONIZER. *Bucculatrix ainsliella.* The leaves of red, black, and white oaks may be skeletonized by a yellowish green larva, ¹/₄ inch long when fully grown. The adult, a moth with a wing expanse of ⁵/₁₆ inch, is creamy white, more or less obscured by dark brown scales. A related insect, the oak-ribbed casemaker, *B. albertiella*, feeds on oaks and makes distinctive elongate white cocoons with longitudinal ribs.

• *Control.* See under caterpillars, Chapter 11.

OAK LEAF TIER. *Croesia purpurana.* Buds and leaves are chewed by the larval stage of this pest.

• *Control.* Spray with Sevin, Malathion, or Diazinon before the buds begin to break.

OAK WEBWORM. *Archips fervidanns.* The larva spins a dense webbing around the feeding site.

CALIFORNIA OAKWORM. *Phryganidia californica.* The larvae of this tan to grayish moth feed on and defoliate many kinds of oak trees in California.

WESTERN TUSSOCK MOTH. *Orgyia vetusta.* This insect feeds on oak.

MAPLE TRUMPET SKELETONIZER. See under Maple.

OAK LEAFROLLER. *Archips semiferanus.* The ⁴/₅-inch black-headed green larvae fold or roll individual leaves together, forming a protected enclosure. They feed on leaves and cause defoliation. Chemical controls may be needed to protect high-value trees.

CALIFORNIA TENT CATERPILLAR. See under Strawberry-Tree.

FOREST TENT CATERPILLAR. See under Maple.

EASTERN TENT CATERPILLAR. See under Willow.

ELM SPANWORM. This destructive pest attacks red and white oaks. See under Elm.

FALL CANKERWORM. See under Elm.

ORANGESTRIPED OAKWORM. *Anisota senatoria.* The black 2-inch larva with eight narrow yellow longitudinal stripes feeds in late summer, consuming leaf blades. The adult is a yellowish red moth.

REDHUMPED OAKWORM. *Symmerista canicosta.* This caterpillar is capable of skeletonizing oak leaves.

VARIABLE OAKLEAF CATERPILLAR. *Heterocampa manteo.* The yellowish green 1¹/₂-inch larvae feed on leaves in early summer and again in early fall. The adult is a gray moth with dark forewing lines.

• *Control.* See under caterpillars, Chapter 11.

OTHER FOLIAGE FEEDING CATERPILLARS. Among other insects that chew oak leaves are the caterpillars of the io moth, *Automeris io*; the cecropia moth, *Hyalophora cecropia*; the luna moth, *Actias luna*; the American dagger moth, *Acronicta americana*; and the satin moth, *Stilpnotia salicis.* The spring cankerworm, *Paleacrita vernata*; green fruitworm, *Lithophane antennata*; linden looper, *Erannus tiliaria*; fruit tree leaf roller, *Archips argyrospila*; and pinkstriped and spiny oakworms, *Anisota virginiensis* and *A. stigma*, also feed on oaks.

• *Control.* See Chapter 11.

OAK BLOTCH LEAF MINERS. *Cameraria hamadryadella* and *C. cincinnatiella.* The solitary leaf miner makes pale blotches on many kinds of oak leaves. One leaf miner larva is found in each blotch. The gregarious leaf miner makes similar blotches on leaves of white oak, but each blotch contains 10 or more larvae.

- *Control.* See under leaf miners, Chapter 11.

OAK SHOT-HOLE LEAF MINER. *Agromyza viridula.* The adult female fly makes punctures in expanding leaves in early spring, which appear as groups of small leaf holes in summer. This is a minor problem.

PIN OAK SAWFLY. *Caliroa lineata.* The lower surfaces of pin oak leaves in the upper part of the tree are chewed by $^3/_8$- to $^1/_2$-inch-long greenish larvae or slugs, which leave only the upper epidermal layer of cells and a fine network of veins. Injured leaves turn a golden brown, and when the feeding by the larvae is extensive, the injury can be readily distinguished at a considerable distance. The adult stage is a small, shiny-black, four-winged insect $^1/_4$ inch long.

SCARLET OAK SAWFLY. *Caliroa quercuscoccineae.* Dark green to black larvae, or slugs, of this insect pest damage black, pin, and scarlet oak leaves in the same way as pin oak sawflies do.

- *Control.* See under sawflies, Chapter 11.

OAK SAWFLY. *Periclista* sp. This insect occasionally feeds on oak.

ASIATIC OAK WEEVIL. *Cyrtepistomus castaneus.* Severe damage to the leaves of oaks and chestnuts can be caused by a deep red or blackish weevil, $^1/_4$ inch long, with scattered metallic green scales. Its long antennae are characteristic of only one other species in this country, the longhorned weevil, *Calomycteris setarius.* Like the elm leaf beetle and the boxelder bug, they move into homes in the fall to hibernate. The Japanese weevil, *Callirhopalus bifasciatus,* also feeds on oak.

- *Control.* See under weevils, Chapter 11.

WALKINGSTICK. *Diapheromera femorata.* The walkingstick defoliates oaks in the East. Nymphs and adults are slender with long thin legs and antennae. They are $2^1/_2$ to 3 inches long and resemble twigs of their host when still. Chemical control is occasionally needed, since their presence can be a nuisance in high-use areas.

Borers

TWO-LINED CHESTNUT BORER. *Agrilus bilineatus.* Large branches are killed on some trees as a result of the formation of tortuous galleries underneath the bark by the two-lined chestnut borer, a white flatheaded larva $^1/_2$ inch long. The adult, a slender greenish black beetle $^3/_8$ inch long, appears in late June. The larvae pass the winter beneath the bark. Weakened European hornbeam, beech, and American chestnut are also attacked by this pest. A slightly larger relative, *A. acutipennis,* attacks members of the white oak group.

FLATHEADED BORER. See under Maple.

TWIG BORERS. *Agrilus angelicus.* The oak twig girdler, a flatheaded borer, is responsible for patches of dead foliage of California live oak. The oak twig pruner (see under Hickory) and the twig girdler (see under Hackberry) also infest oak.

CARPENTERWORM. See under Ash. This insect produces unsightly scars on oaks.

OAK CLEARWING BORER. *Paranthrene simulans.* This gray, 1-inch-long, black-headed borer attacks the lower trunk of red and white oaks in the East. Sap spots and clumped frass are emitted from galleries $^1/_3$ inch in diameter. Adults are black and orange-banded beelike clear-winged moths.

RED OAK BORER. *Enaphalodes rufulus.* The shiny white larva with tiny legs mines under the bark the first year and tunnels into the wood the second year, producing frass. Adults are 1-inch-long gray beetles with long antennae.

WHITE OAK BORER. *Goes tigrinus.* The 1- to $1^1/_2$-inch legless whitish larva feeds on wood of young trees. The adult beetle is white and brown mottled, almost 1 inch long.

OAK TIMBERWORM. *Arrhenodes minutus.* The whitish larvae of this weevil invade wounds of mature trees.

COLUMBIAN TIMBER BEETLE. *Corthylus columbianus.* These black to reddish brown beetles are less than $^1/_4$ inch long and construct galleries $^1/_{10}$ inch in diameter. Even vigorous trees may be attacked.

PIN-HOLE BORERS. *Platypus* and *Xyleborus* spp. These brown ambrosia beetles are $^1/_8$ inch long and attack injured, weakened, stressed, and dying trees.

ROOT BORERS. *Prionus imbricornis* and *P. laticollis.* Several years of feeding by the creamy white larvae on oak roots causes tree decline. Larvae grow to 3 inches, and the dark brown adult beetles are $1\frac{1}{2}$ inches long. Stressed trees are most likely to be attacked.

OAK BARK BEETLE. *Pseudopityophthorus pubipennis.* These beetles and their larvae tunnel the inner bark of oaks already under stress.

OTHER BORERS. Oaks may be invaded by other borers, including the carpenterworm (see under Ash); the leopard moth borer (see under Maple); dogwood borer (see borers, under Dogwood); red-headed ash borer (see under Ash); beech borer, *Goes pulverulentis*; oak branch borer, *G. debilis*; and oak stem borer, *Aneflormopha subpubescens.*

- *Control.* See under borers, Chapter 11.

Scale Insects

GOLDEN OAK SCALE. *Asterolecanium variolosum.* Shallow pits are formed in the bark by these circular, greenish gold scales. Sometimes referred to as pit scales, they attain a diameter of only $\frac{1}{16}$ inch. Infested trees have a ragged, untidy appearance. Young trees may be killed outright, the lower branches dying first.

LECANIUM SCALES. *Lecanium corni* and *L. quercifex.* Branches and twigs infested with these pests are covered with brown, down-enveloped, hemispherical scales. The adults are $\frac{1}{6}$ inch in diameter. The winter is passed in the partly grown stage. *L. quercitronis* infests several species of oaks in the eastern states and live oak in California.

OAK GALL SCALE. *Kermes pubescens.* These large, globular, red and brown mottled scale insects about $\frac{1}{8}$ inch in diameter (Figure III-82) infest the twigs, leaf stalks, and midribs and tend to gather on the young buds. The leaves become distorted and many of the twigs are killed, but the tree as a whole is not usually seriously injured.

OBSCURE SCALE. *Melanaspis obscurus.* Tiny, circular, dark gray scales, $\frac{1}{10}$ inch in diameter, may occasionally cover the bark of twigs and branches. The pest overwinters in the partly

Fig. III-82. Scale (*Kermes pubescens*) on bur oak. The scale insects closely resemble the oak buds in the axils of the leaves and at the tips of the twigs.

grown adult stage. Heavy infestations kill branches and weaken trees.

PURPLE SCALE. *Lepidosaphes beckii.* The female of this species has an elongated, oyster-shaped, slightly curved brown or purplish body. This species infests citrus trees primarily, but it also occurs on many other species in California. Cottony-cushion scale (see under Acacia) attacks oak, as does cottony maple (see under Maple) and California red scale.

- *Control.* See under scales, Chapter 11.

Other Insects

GALLS. There are hundreds of kinds of galls on oak leaves, twigs, and flowers, and a special manual is needed to identify them. Galls are caused by many species of flies, nonstinging wasps, and mites; some of these species have larvae that feed and grow completely protected inside the galls. The life cycles of gall insects and mites vary according to the species involved. The

Fig. III-83. Wool sower gall caused by the gall wasp *Callirytis seminator.*

pests overwinter either on the trees or on the ground. The adults emerge in spring and travel to the leaves and twigs, in which they deposit eggs. The young that hatch from the eggs then mature in the galls that form around them.

One of the most beautiful galls is the wool sower (Fig. III-83), produced by the gall wasp *Callirytis seminator.* It is found on white, chestnut, and basket oaks. On leaves, spined turban oak galls, which are pink, and oak cone galls, which are red, are highly visible but relatively benign galls caused by two wasp species, *Antron douglasii* and *Andricus kingi.* Similarly, the jumping oak gall wasp, *Neuroteres salatorius* causes spherical galls on the undersides of leaves in the white oak family. Most kinds of galls on oaks rarely affect the health of the trees. However, some such as the gouty oak gall (Fig. III-84), the horned oak gall, and the oak-potato gall (Figs. III-85 and 86), affect the health as well as the appearance of the tree by killing branches.

Horned oak galls are caused by *Callirytis cornigera.* Mature galls are 3 years old with numerous cone-shaped horns, giving galls their name. Gall horns contain larvae which have grown up from eggs laid on twigs 3 years previously. This insect also has a leaf gall phase. Horned oak galls may provide residence for many other insects, including other oak pests or even pests of other trees. Another common gall is the oak-apple gall (Fig. III-87).

• *Control.* Normally, controls are not needed. Spraying trees either with dormant lime sulfur or with a dormant miscible oil before growth starts in spring will destroy some of the pests overwintering on the branches. Valuable trees should also be given a Carbaryl or Sevin spray in mid-May when the new leaves and shoots are expanding. Heavily infested branches should be pruned and destroyed before the adults emerge.

OAK PHYLLOXERAN. *Phylloxera* sp. Phylloxerans are very small aphidlike insects that attack oak buds and developing leaves, causing the leaves to curl and twist, thus marring their normal appearance. They are usually found in clusters.

OAK LEAF APHID. *Myzocallis* sp. See Chapter 11.

GIANT BARK APHID. See aphids, under Sycamore,

WOOLLY APHIDS. *Stegophylla* sp. These can be found on several species of oak.

MAPLE LEAFHOPPER. See leafhoppers, under Maple.

OAK LACE BUG. *Corythucha arcuata.* The leaves of many oaks, particularly the white oak species, may turn whitish gray and become defoliated when heavily infested with this lace bug. The upper surface of the leaf shows chlorotic flecks or stippling, and the under surface may be covered with tiny black spiny nymphs and adults with lacy, transparent wing covers. Tiny black tarlike splotches, insect fecal wastes, are also present.

• *Control.* Spray with malathion, Orthene, Sevin, methoxychlor, or Guthion in May when bugs appear, and again in July if needed.

STANFORD WHITEFLY. *Tetraleurodes stanfordi.* This pest infests coastal oaks in California.

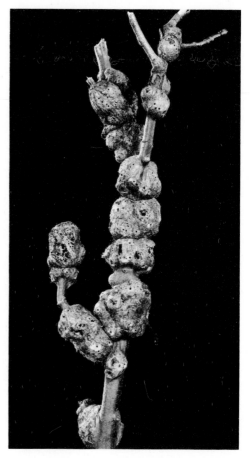

Fig. III-84. Gouty oak gall caused by the gall wasp *Andricus punctatus.*

Crown whitefly, *Aleuroplatus coronata* also infests oaks; high whitefly populations can be unsightly but do little harm to the tree.

PERIODICAL CICADA. See under Crabapple. Oaks can be attacked.

Other Pests

OAK MITE. *Oligonychus* and *Eotetranychus* spp. Mottled yellow foliage results from the sucking of the leaf juices by the oak mite, a yellow, brown, or red eight-legged pest. The mite usually overwinters in the egg stage.

OAK RED MITE. *Oligonychus bicolor.* This pest also attacks oak.

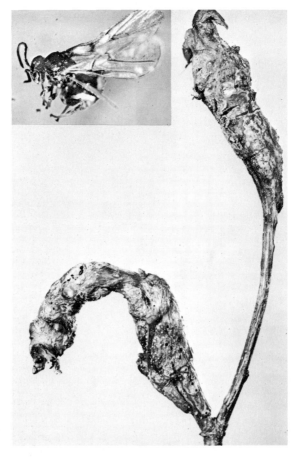

Fig. III-85. Oak-potato gall. **Fig. III-86.** Inset: Adult of *Neuroterus batatus*, the cause of oak-potato gall.

Fig. III-87. Oak-apple gall caused by a species of gall wasps.

• *Control.* See under spider mites, Chapter 11.

LIVE OAK ERINEUM MITE. *Eriophyes mackiei.* This pest causes raised blisters on live oaks in California.

OLIVE (*Olea*)

This Mediterranean native is planted in mild regions such as California. It transplants easily, and fruitless types such as the cultivar 'Swan Hill' are available.

Diseases

LEAF SPOTS. *Cycloconium oleaginum, Gloeosporium olivarum, Cercospora cladosporioides,* and *Asterine oleina.* These fungi cause leaf spots of olive.

OLIVE KNOT. *Pseudomonas syringae* pv. *savastanoi.* This bacterial disease results in swellings, galls, and cankers on olive and also on European ash.

ROOT ROTS. *Armillaria mellea* and *Phymatotrichum omnivorum.* These fungi cause root rots of olive.

WILT. *Verticillium albo-atrum.* This fungus causes a devastating wilt of olive. See Chapter 12.

Insects

SCALES. A California scale (*Parlatoria pittospori*), and black, greedy, ivy, latania, California red, dictyospermum, and euonymus scales (see Chapter 11) attack olive. Comstock mealybug also feeds on olive.

AMERICAN PLUM BORER. See under Plane-tree.

ORCHID-TREE (*Bauhinia*)

Several species of orchid-trees are grown as spectacular flowered lawn specimens and as framing for small houses in Florida and other warm parts of the country. Orchid-tree prefers warm, well-drained soil, and it does not tolerate drought and excessive heat.

Diseases

LEAF SPOT. *Colletotrichum* and *Phyllosticta* spp. The former has been reported from Texas and the latter from Florida.

• *Control.* Controls are usually unnecessary.

Insects

The Cuban may beetle, *Phyllophaga bruneri,* two species of mealybugs—citrus and long-tailed—and 19 species of scales attack orchid-trees.

• *Control.* See Chapter 11.

OSAGE-ORANGE (*Maclura*)

Osage-orange is typically a small, thorny tree adaptable to a wide range of sites. Occasional specimens, sometimes multitrunked, grow to 60 feet with an irregular crown.

Diseases

LEAF SPOTS. Leaves are occasionally spotted by four species of fungi: *Cercospora maclurae, Ovularia maclurae, Phyllosticta maclurae,* and *Septoria angustissima.*

• *Control.* Leaf spots are rarely serious enough to warrant control measures.

OTHER DISEASES. Osage-orange is susceptible to a leaf blight caused by *Sporodesmium maclurae,* rust by *Physopella fici,* Verticillium wilt, and *Phymatotrichum omnivorum* root rot.

The wood of osage-orange was found to contain about one percent of the chemical 2,3,4,5-tetrahydroxystilbene, which is toxic to a number of fungi. This may explain why this host is remarkably resistant to decay-producing fungi.

Insects

SCALES. Five species of scale insects infest osage-orange: cottony-cushion, cottony maple (see under Maple), European fruit lecanium, Putnam, and San Jose.

The citrus mealybug and the citrus whitefly also attack this tree.

• *Control.* See under scales, Chapter 11.

PAGODA-TREE, JAPANESE (*Sophora japonica*)

This member of the legume family produces clusters of white, showy flowers in summer. It should be more widely planted as a specimen tree. Japanese pagoda-trees do well in cities; the single-stemmed form 'Regent' is best suited for use along city streets. The falling flowers in summer will stain automobiles, and hence pagoda tree should not be planted in parking lots or similar areas.

Winter injury and branch breakage can be problems.

Diseases

CANKER. *Fusarium lateritium.* Oval, 1- to 2-inch cankers, with definite, slightly raised, dark red-brown margins and light tan centers occasionally appear on this host. When the cankers completely girdle the stem, the distal portion dies. The sexual stage of this fungus is *Gibberella baccata.* Another fungus, *Cytospora sophorae,* is also found on dead branches. Both fungi are usually associated with frost damage in late fall or early spring.

• *Control.* Prune diseased or dead branch tips to sound wood and apply a copper fungicide several times during the growing season.

DAMPING-OFF. *Pellicularia filamentosa.* This fungus causes a damping-off of seedlings.

• *Control.* Use steam-pasteurized soil or soil treated with PCNB (Terraclor).

TWIG BLIGHT. *Nectria cinnabarina* and *Diplodia sophorae.* Two fungi cause cankers and dieback of twigs of this host.

• *Control.* The same as for canker.

OTHER FUNGAL DISEASES. Japanese pagoda-tree is also subject to powdery mildew caused by *Microsphaera,* root rot by *Phymatotrichum omnivorum,* leaf spot by *Phyllosticta sophorae,* and rust by *Uromyces hyalinus.*

• *Control.* Controls are unnecessary.

VERTICILLIUM WILT. See Chapter 12.

Insects

SCALES. Cottony-cushion scale, *Icerya purchasi,* and long soft scale, *Coccus elongatus,* have been reported on Japanese pagoda-tree.

• *Control.* See under scales, Chapter 11.

OTHER PESTS. Nematodes, *Meloidogyne* sp., and mistletoe, *Phoradendron flavescens,* also attack Japanese pagoda-tree.

• *Control.* Control measures are not available.

PALMS (*Palmaceae*)

Diseases

LETHAL YELLOWING. This highly fatal disease, caused by a phytoplasma, was first observed in Jamaica nearly a century ago. In recent times it has been responsible for the death of hundreds of thousands of coconut palms (*Cocos nucifera*) in southern Florida. First observed in the Key West area in 1955, it has now spread throughout much of Florida and into south Texas. The vector for this organism is a planthopper.

An early symptom is the appearance of dead tips of the inflorescences when they emerge from the spathes. Leaf yellowing of the lower fronds follows. Another symptom is "shelling," the premature dropping of coconuts regardless of size. The disease progresses until the bud becomes necrotic and the tree dies.

In Florida lethal yellowing has also resulted in death of *Veitchia, Arikuryroba, Washingtonia,* and *Pritchardia* palms.

• *Control.* No practical control is presently available, although cessation of symptom development and resumption of normal growth has been obtained by injecting affected coconut palm trees with oxytetracycline HCl (Terramycin) antibiotic.

The rate of lethal yellowing spread reportedly decreases in trees injected with oxytetracycline at 4-month intervals, as compared with untreated trees. The treatment may be economically viable for coconut palms grown primarily for ornamental purposes.

One promising method of coping with lethal yellowing is to use resistant varieties. Seed nuts of resistant palms have been imported from Jamaica to be used as replacements. 'Malayan Dwarf', 'Fiji Dwarf', and 'Panama Tall' coconut palms are some of the more resistant kinds presently available. Other resistant or tolerant palms include areca, cabbage, MacArthur, panrotis, pindo, pygmy date, queen, royal, and solitaire.

BUTT ROT. *Ganoderma zonatum.* This is one of the most serious fungal diseases of palms in

Florida. The spores of the fungus enter through wounds at the base of the tree made by lawn mowers or other instruments. The lower leaves die and new leaves are stunted. Typical fruiting bodies of the fungus appear at the base of the infected tree near the soil line.

• *Control.* Remove diseased trees and avoid wounding healthy trees.

FALSE SMUT, LEAF SCAB. *Graphiola phoenicis.* Palms belonging to the genera *Arecastrum, Arenga, Howea, Phoenix, Roystonea,* and *Washingtonia* suffer from a parasite that causes a yellow spotting of the leaves and the formation of numerous small black scabs or warts. The outer parts of these fruiting bodies are dark, hard, and horny and many long, flexuous, sterile hyphae protrude from the inner parts. Powdery yellow or light brown masses of spores arise within the inner membrane of the structure. Severely infected leaves soon die.

This disease is most troublesome in areas with consistently high humidities.

• *Control.* Cut out and destroy infected leaves or leaf parts at the first sign of the disease and spray with a copper fungicide. On greenhouse palms, control insects with insecticides rather than by syringing, a practice that helps to spread false smut. Resistant or tolerant varieties may be available.

LEAF SPOT. *Stigmina palmivora.* This leaf spot is especially serious in greenhouses where insufficient light is provided. The spots are small, round, yellowish, and transparent. They often run together to form large irregular gray-brown blotches, which may result in death of the leaf. The spores are long, club-shaped, many-celled, and brown. They are formed on tufts of short basal cells.

LEAF SPOTS AND BLIGHTS. *Pestalotiopsis palmarum.* The fungus infects numerous palms in the axils of the leaves and where leaflets are attached to the leaf stalk. The fungus then penetrates into deeper tissues, causing a brown discoloration. Gray-brown spots develop on the leaf blades and run together to form large blotches. Brown septate spores with characteristic appendages develop in black masses on the upper parts of the leaves.

A serious leaf blight of *Washingtonia* is caused by *Calonectria* spp. (*Cylindrocladium*). Numerous small dark brown spots with light-colored margins disfigure the leaves.

Additional fungal leaf spots of palms include Pseudocercospora leaf spot (*Pseudocercospora rhapisicola*), anthracnose (*Colletotrichum gloeosporioides*), Annellophora leaf spot (*Annellophora phoenicis*), tar spot (*Catacauma sabal*), rachis blight (*Serenomyces californicum*), and spots caused by *Bipolaris* sp. and *Exserohilum* sp. A bacterium, *Pseudomonas avenae,* also causes leaf spot of palms.

• *Control.* See under leaf spots, Chapter 12.

PINK ROT. *Gliocladium vermoeseni.* Along the Pacific Coast in southern California this fungus causes diseases that result in great damage to ornamental palms growing under stressful conditions: leaf base rot of *Phoenix,* bud rot of *Washingtonia,* and trunk canker of *Cocos romanzoffiana* and trunk rot of reed palm. Among the effects are a successive decay of the leaf bases from the oldest to the youngest of the tightly folded bud leaves and the weakening and breaking of the trunk. The pathogen produces masses of dusty pink or orange spores on infected tissues.

• *Control.* Avoid wounds. Control measures have not been developed.

STEM AND ROOT ROTS AND WILTS. Species of several genera of fungi have been reported to cause root disease of palms in Florida and several western states. Fusarium wilt of palms in California is caused by *Fusarium oxysporum,* and a bacterium, *Xanthomonas* sp., has been associated with a wilt and trunk rot of coconut and Cuban royal palm. A root rot associated with *Pythium* is accompanied by yellowing and wilting of the leaves, one after another, until the bud falls from the top of the plant. *Fusarium* has been associated with stem and root rot of the royal palm (*Roystonea*). *Armillaria mellea* is thought to cause root rot of the date palm (*Phoenix*) in California.

In certain palm houses numerous fruiting bodies of *Xylaria schweinitzii* have been found growing from roots of *Howea* and other palms. Since another species of *Xylaria* is known to cause root rot of apple and landscape trees, the poor condition of some greenhouse palms may possibly be due to this fungus. No control has been suggested. Removal of the 2-inch-long black, club-shaped fruiting bodies may help reduce inoculum.

BUD ROT AND WILT. *Phytophthora palmivora.* At various times, wilting and bud rot of the coconut palm have been attributed to this fungus.

• *Control.* When this disease occurs in ornamental plantings, the infected plant must be removed.

BLACK SCORCH AND HEART ROT. *Ceratocystis paradoxa.* Another name for this disease is Thielaviopsis bud rot, and it is the imperfect, *Chalara paradoxa,* stage of the pathogen that is normally associated with the disease. This disease may be destructive to date palms of plantations and ornamental gardens in California. It is found in all structures of the plant except the roots and stems. Lesions are dark brown or black, hard and carbonaceous, and resemble a black scorch. The fungus gains entrance even in the absence of wounds. The disease is most serious as a terminal bud rot.

• *Control.* All infected fronds, leaf bases, and flower parts should be pruned out and destroyed and the cuts disinfested. Copper sprays give promise of control.

Insects and Related Pests

PALM APHID. *Cerataphis variabilis.* In its wingless form this aphid is often mistaken for a whitefly because it is dark and disc-like with a white fringe. The green peach aphid, *Myzus persicae,* also infests palms.

• *Control.* See under aphids, Chapter 11.
PALM LEAF SKELETONIZER. *Homaledra sabaliella.* In Florida this is a major pest, feeding on the leaves of many species of palms under a protective web of silk. Leaves are blotched, then shrivel and die.

• *Control.* Cut out and destroy infested fronds, or spray with Sevin.
MEALYBUGS. Four species of mealybugs infest palms: citrus, ground, long-tailed, and palm.

• *Control.* See under scales, Chapter 11.
SCALES. Palms are extremely susceptible to many species of scale insects. At least 23 species of scales have been recorded on this plant family, particularly those grown outdoors. See Chapter 11.

THRIPS. *Heliothrips haemorrhoidalis, H. dracaenae,* and *Hercinothrips femoralis.* Although these species have been reported on palms, they are not considered serious pests. They are not troublesome where scale insects and mealybugs are controlled.

MITES. Three species of mites-banksgrass, privet, and tumid spider-infest palms.

• *Control.* See under spider mites, Chapter 11.
FULLER ROSE BEETLE. See under Camellia.

PAPERBARK TREE
(*Melaleuca quinquenervia*)

This native of Australia is becoming a pest by rapidly taking over vast areas of the Everglades swamps in Florida. Insect pests that specifically feed on this tree are being introduced to manage this pest tree. It is often used as a street and landscape tree in southern California and is admired for its unusual bark.

Insects

Paperbark tree is host to the Melaleuca leaf weevil, *Oxyops vitiosa*; Melaleuca psyllid, *Boreioglycaspis melaleucae*; melaleuca sawfly, *Lophyrotoma zonalis*; and a tube moth, *Poliopaschia lithochlora.*

PAPER-MULBERRY
(*Broussonetia*)

Diseases

CANKER. *Fusarium solani.* Branch dieback results from cankers produced by this fungus, which also infects cottonwood, red oak, and sweetgum.

• *Control.* Prune infected branches. Keep trees in good vigor by fertilizing and by watering during dry spells.
ROOT ROT. *Phymatotrichum omnivorum.* This root rot, common on many plants in the

southern states, has been reported from Texas on the paper-mulberry tree.

• *Control.* Control measures have not been developed.

OTHER DISEASES. Among the other diseases reported are a dieback and canker caused by the fungus *Nectria cinnabarina*; a leaf spot by *Cercosporella mori*; a mistletoe disease by *Phoradendron flavescens*; and root knot by the nematode *Meloidogyne incognita*.

• *Control.* These diseases are seldom serious enough to warrant control measures.

PAULOWNIA
See Empress-Tree.

PEACH, FLOWERING. See Cherry.

PEAR, ORNAMENTAL (*Pyrus*)
The flowering pear (*Pyrus calleryana* 'Aristocrat', 'Bradford', 'Capitol', 'Redspire', 'Whitehouse', and others) is a nicely shaped medium-sized tree and is relatively disease-free.

Diseases

FIRE BLIGHT. Callery pears have been touted as resistant to fire blight, a serious disease of pears grown for fruit. In some seasons fire blight (Fig. III-88) has been observed on all callery pear cultivars planted in a particular region, at seriously damaging levels on some cultivars. See Chapter 12.

BACTERIAL BLIGHT. *Pseudomonas syringae* pv. *syringae.* This disease resembles fire blight. See under Cherry.

RUST. *Gymnosporangium globosum.* Cedar hawthorn rust causes brown leaf spots and petiole swellings when the aecial form of the fungus appears on ornamental pears. See chapter 12 for fungus life cycle details and control information.

CANKER. Trees with cankers on their trunks may be infected by a species of *Coniothyrium.*

Insects

PEAR LEAF BLISTER MITE. This mite causes reddish blisterlike swellings on the upper leaf

Fig. III-88. Callery pear fire blight.

surface, which can coalesce and cause leaf browning. See under Mountain-ash.

PEAR SAWFLY. See under Cherry (pear slug).

PECAN (*Carya illinoensis*)
A large nut tree of the South and some lower midwestern areas of the United States, pecan is seldom used in landscapes, perhaps because of its susceptibility to disease. Pecan disease and insect problems are similar to those of hickory. Pecans are subject to zinc deficiency in alkaline soil.

Diseases

BROWN LEAF SPOT. *Cercospora fusca.* This leaf spot is common throughout the pecan-growing areas in the South. Downy spot, caused by the fungus *Mycosphaerella caryigena*, also occurs in the same localities.

• *Control.* See under leaf spots, Chapter 12.

CROWN GALL. *Agrobacterium tumefaciens.* The roots of young pecan trees may be infected with this bacterial disease. Occasionally older trees in orchards are also infected.

• *Control.* See under crown gall Chapter 12.

SCAB. *Cladosporium effusum.* This is one of the most destructive diseases of pecans in the Southeast. The fungus attacks leaves, shoots, and nuts.

• *Control.* The local Cooperative Extension office can provide details on materials and timing of sprays for pecan scab.

OTHER DISEASES. Pecans are subject to a number of diseases that infect other species of *Carya.* They are not treated in detail here because the pecan is not an important ornamental tree. See Hickory.

Insects

BORERS. A large number of borers infest pecans. Following are the most important: dogwood, flatheaded apple tree, pecan, pecan carpenterworm, shot-hole, twig girdler, and twig pruner.

• *Control.* See under borers, Chapter 11.

PEPPERIDGE TREE. See Tupelo.

PEPPER-TREE (*Schinus molle*)

This native of South America grows in mild regions and is tolerant to some frost, poor, or dry soils, and wind. It is considered a messy tree because it sometimes sheds twigs and leaves.

Diseases

Heart rot caused by the fungi *Ganoderma applanatum, Polyporus dryophilus, P. farlowii, Laetiporus sulphureus,* and *Trametes versicolor* have been reported from California. These fungi are difficult to control.

Root rot caused by *Armillaria mellea* and wilt by *Verticillium albo-atrum* also affect pepper-tree. The controls for these fungal diseases are discussed in Chapter 12.

Insects

PEPPER-TREE PSYLLID. *Calophya rubra.* Nymphs of this pest can cause distortion and feeding pits in leaves.

OTHER INSECTS. The omnivorous looper caterpillar, citrophilus mealybug, citrus thrips, and thirteen species of scales including black, hemispherical, cottony-cushion, greedy, and ivy infest the pepper-tree.

• *Control.* See Chapter 11.

PERSIAN PARROTIA (*Parrotia*)

This native of Iran is relatively pest-free.

PERSIMMON (*Diospyros*)

American persimmon (*D. virginiana*), native to the eastern United States, tolerates a wide range of soils and climates. Persimmons in the wild and those grown for their fruits are subject to a great number of diseases and pests. Ornamentals, however, are relatively free of problems.

Diseases and Insects

LEAF SPOT. *Cercospora fuliginosa.* This fungus causes small black spots.

WILT. *Acremonium diospyri.* This has been a serious problem of persimmons in the Tennessee Valley.

TWIG BLIGHT. *Botryosphaeria diplodia.* Infection causes twig dieback.

SCALES. White peach, European fruit lecanium, and greedy scales, as well as Comstock and long-tailed mealybug, attack persimmon.

AMERICAN PLUM BORER. See under Plane-tree.

TWIG GIRDLER. See under Hackberry.

FULLER ROSE BEETLE. See under Camellia.

ASIATIC OAK WEEVIL. See under Oak.

PINE (*Pinus*)

Forty different pines are native to North America, and many non-native species are commonly grown. Their size, shape, and adaptability to landscape use vary.

Although pines fare better than hemlock, fir, or spruce in the environment of larger cities, they grow best in small towns and in the country. Among the species most commonly used in ornamental plantings are Austrian, mugo, Scots, and white pines. Each is

Fig. III-89. Single needle cluster of Austrian pine greatly magnified to show fruiting bodies of the causal fungus, *Sphaeropsis sapinea*.

subject to a number of fungal diseases and insect pests.

Many pines have a generally pyramidal shape that becomes flat-topped as they mature. Although some species will live 80, 100, or 200 years in good native sites, landscape pines often become aged, or flat-topped, in much less time owing to generally poor urban growing conditions. Scots and Austrian pines are a dubious choice in landscapes where pine tip blight and pine wilt diseases are prevalent.

Diseases

DIEBACK, TIP BLIGHT. Dieback or tip blight is a disease of major importance on several of the most desirable pines for ornamental plantings. Austrian pine is, without question, the most susceptible to the disease. Scots, red, mugo, ponderosa, and white pines are decreasingly susceptible in the order named. Douglas-fir and blue spruces are occasionally attacked. Tip blight tends to appear on trees growing under stressful conditions such as drought, shade, or competition with other trees in the landscape.

• *Symptoms.* The most prominent symptoms are stunting of the new growth and browning of the needles. The lower branches are most heavily infected, although in wet springs the branches over the entire tree may show browned tips.

Tip blight is often confused with troubles caused by low temperatures, drought, winter drying, and pine shoot moth injury. It can be differentiated from these by the presence of small, black, pin-point fungal bodies, which break through the epidermis at the base of the needle (Fig. III-89). These bodies also occur on affected twigs and cone scales but are less readily discernible than on needles, especially if the examination is made soon after the needles turn yellow and wilt. Placing such material in a moist atmosphere for a day or so will induce the black bodies to break through the epidermis, thus becoming visible.

• *Cause.* The fungus *Sphaeropsis sapinea*, formerly called *Diplodia pinea*, causes tip blight. The generic name is used to distinguish this disease from twig blights caused by other microorganisms. The fungus overwinters in infected needles, twigs, and cones. In spring, the small fruiting bodies release egg-shaped, light brown spores, which are splashed by rain and wind to the newly developing needles. The fungus grows down through the needles and into the twigs, where it destroys tissues as far as the first node.

P. P. Pirone demonstrated that *Sphaeropsis sapinea* could attack Austrian, mugo, Scots, and white pines, as well as Douglas-fir and Norway spruce. It has also been associated with a root and root-collar rot of red pine in nurseries, where such trees were completely destroyed by the disease.

The fungus *Sirococcus conigenus* also causes a shoot blight of red pine (*Pinus resinosa*) *in* the Great Lake states.

• *Control.* Control is difficult for the tip blight phase of the disease on older trees. As soon as the blight is noticed, the infected needles, twigs, and cones should be pruned back to sound tissues and destroyed. Pruning should be done when the branches are dry, because there is less danger of spreading the spores by contact with the operator and with tools. Where infection has been particularly severe, preventive fungicides are also recommended. Banner Maxx, Bayleton, Cleary's 3336, Daconil 2787, or fixed copper sprays should be applied very early in spring, starting when the buds swell and then twice more at 10-day intervals. In rainy springs, a fourth application about 7 days after the third may be necessary. As a supplementary treatment, watering of the trees during dry seasons is suggested.

No cure is possible for plants that are infected at the base. The use of steam-pasteurized soil is suggested for avoiding root and stem base infections of seedlings in nurseries.

CANKERS. A large number of canker diseases occur on all pines; they are more common on trees in forests than in ornamental plantings.

The fungus *Ascocalyx abietina* (formerly called *Scleroderris* and *Gremeniella*) has caused serious losses of red and Scots pines. The disease, called Scleroderris canker, causes needles of infected shoots to discolor at the base and drop.

Infection usually starts in the lower branches and then moves upward. Cankers develop along the main stem, eventually girdling the stem and causing death of the entire tree. Scolytid beetles are suspected of being vectors of the fungus. Control measures suggested for the dieback, or tip blight, of pines (see below) might be helpful in combating this disease.

The fungus *Tympanis pinastri* attacks red pine and, to a lesser extent, white pine. It produces elongated stem cankers with or without definite margins and with depressed centers that become roughened and open after 2 or 3 years. Each canker centers at a node, indicating that the fungus enters the branch at the base of the lateral branch. Trees in a weakened condition are most susceptible.

Other cankers on pines are associated with the fungi *Dasyscypha pini, Atropellis pinicola, A. piniphila, A. tingens, Caliciopsis pinea, Fusarium subglutinans* f. sp. *pini* (the southern pitch canker fungus), *Sphaeropsis* sp., *Phomopsis* sp., *Valsa* spp. (imperfect *Cytospora*), and *Leucostoma kunzei*.

• *Control.* Removal of dead and weak branches, avoidance of bark injuries, mulching, and fertilization to increase the vigor of the trees are suggested. Surgical treatment of cankers should be attempted only on valuable specimens. Sprays of Daconil 2787 can be applied in spring.

CENANGIUM TWIG BLIGHT. *Cenangium ferruginosum.* The twigs of exotic species of pine and white pine may be killed by this fungus. Early symptoms are the dying of terminal buds and reddening of the needles. Infection rarely spreads beyond the current season's growth.

• *Control.* Affected twigs should be pruned to sound wood and destroyed. Because twig blight is usually severe only on weakened trees, mulching, fertilization and watering are suggested for valuable pines.

NEEDLE CASTS. *Bifusella linearis, B. striiformis, Elytroderma deformans, Ploioderma (Hypoderma) desmazerii, P. hedgecockii, P. lethale, P. pedatum, P. pini, Hypodermella* sp., *Lophodermium seditiosum, Lophodermella* spp., *Cyclaneusma (Naemacyclus) minus,* and *Mycosphaerella dearnessii (Scirrhia acicola)* (Fig. III-90). Many species of fungi cause leaf yellowing and, at times, premature defoliation of pines primarily in nurseries and in forest and Christmas tree plantings. Often, browning and defoliation do not occur until the year following infection.

• *Control.* Except in wet seasons, needle cast diseases rarely cause sufficient damage to large trees to warrant preventive practices. See under leaf spots, Chapter 12.

NEEDLE BLIGHT. *Dothistroma septospora.* This fungus causes slightly swollen dark spots or bands (Fig. III-91) in late summer on 1-year-old needles of ponderosa and Austrian pines. The distal portion of the swollen needle turns light brown and dies. Severely affected trees show sparse foliage as a result of the premature drop-

Fig. III-90. Brown spot needle blight of pine.

Fig. III-91. *Dothistroma* fruiting bodies on pine needles.

Fig. III-92. Pine needle rust.

ping of needles. The disease is favored by cool, moist weather.

- *Control.* See under leaf spots, Chapter 12.

NEEDLE RUST. *Coleosporium asterum.* This rust is most common on pitch pine in the eastern states, but other species such as Austrian, loblolly, mugo, ponderosa, red, Scots, and Virginia are also susceptible. Infestations may be serious enough to cause defoliation. Blisters break out as $1/16$- to $1/8$-inch-high pustules (Fig. III-92), which open to discharge the bright orange-colored aeciospores of the rust. The spores from the pustules on pine infect either aster or goldenrod, on which the fungus forms golden or rust-colored pustules during summer. The rust is often able to overwinter on the crown leaves of goldenrod and asters, so it can perpetuate itself on these hosts. The rust is not perennial on pine, so pines must be reinfected with spores from goldenrod or aster.

- *Control.* The destruction of wild asters and

goldenrod near valuable pines and spraying or dusting the pines with sulfur early in the season will provide control.

SCRUB PINE NEEDLE RUST. *Coleosporium pinicola.* Virginia pine and occasionally other species may be seriously infected with this rust, which breaks out in spring with reddish pustules on the needles. The pustules, up to $1/2$ inch long, fade and become inconspicuous later. Defoliation may occur. This is a short-cycle rust, having no alternate host stage; the spores that develop in the pustules germinate there and shed secondary spores (basidiospores), which infect nearby pines.

This is not a serious rust on ornamental pines except for Virginia pine. No control measures have been suggested. Many other species of *Coleosporium* also cause pine needle rusts.

EASTERN GALL RUST. *Cronartium quercuum.* This is a gall-forming stem rust. Scots, pitch,

Fig. III-93. Eastern gall rust of pine.

jack, and Virginia pines are all very suscepti-ble. A characteristic distortion or kink of the trunk (Fig. III-93) is formed several feet from the ground or on main branches. More or less spherical galls 1 or 2 inches in diameter ap-pear on the branches, and eventually the parts above the galls are killed. These galls may be so numerous on jack pines that entire trees are killed. On the trunks of small scrub pines infection results in very striking spherical galls that entirely encircle the trunks. Galls 5 or 6 inches in diameter are not unusual. The rust galls should not be mistaken for galls caused by insects.

The fungus may cause the growth of a deep canker instead of galls. Since the canker can appear at the base of a tree near the ground, it

could be inconspicuous even though the infected area may be a foot across.

This rust has oak species as the alternate host. Numerous long horns of brownish or dark brown spores are formed on the infected oak leaves. The basidiospores which develop on these horns reinfect pines in spring.

WESTERN GALL RUST. *Endocronartium hark-nesii.* Stem galls caused by this fungus produce spores which infect other nearby pines.

FUSIFORM RUST. *Cronartium quercuum* f. sp. *fusiforme.* A related oak/pine rust, fusiform rust, damages landscape and forest pines in the South. Pruning infected cankers is advised, where practical. Fertilization should be avoided until trees are 10 years old, to reduce infection of young, succulent tissue.

SWEETFERN RUST. *Cronartium comptoniae.* This rust can attack stems of several pine species. Sweetfern is the alternate host.

• *Control.* Other than pruning, no practical controls for these rusts have been suggested. See under rust diseases, Chapter 12.

WHITE PINE BLISTER RUST. This disease is probably the most important disease of white pine in the northeastern United States. Other pines having five needles in the leaf cluster, such as Western white pine and sugar pine, are also susceptible. Blister rust also attacks both wild and cultivated forms of currants and gooseber-ries but produces less damage on these than on pines.

• *Symptoms.* On pines, reddish brown, drooping needles appear on one or more dead branches. Rough, swollen cankers are found at the bases of affected branches or on the trunk. In late spring, orange-yellow blisters filled with powdery spores break through the bark of the cankered areas (Fig. III-94). These blisters are the most positive sign of the disease and are most conspicuous from April to June; after that time they may be eaten by squirrels or covered with resin. The disease is fatal when trunk cankers or branch cankers completely girdle the affected parts.

The symptoms are less conspicuous on currants and gooseberries, where minute, orange-yellow pustules or brown, curved ten-

Fig. III-94. Blister rust of white pine caused by the fungus *Cronartium ribicola.*

drils appear on the lower surface. Most of the leaves are stunted when the pustules become numerous.

• *Cause.* White pine blister rust is caused by the fungus *Cronartium ribicola,* which overwinters in cankered bark of living pines. The orange-yellow spores produced in these cankers in late spring are blown to and infect susceptible currant and gooseberry leaves. A different type of spore, produced on these leaves in fall, is blown to pines and infects the needles, completing the fungus's life cycle. After the fungus enters the needle it penetrates downward into the twig and eventually the trunk. The whole process

takes place so slowly that several years may elapse from the needle infection stage to the appearance of the orange-yellow blisters on the branch or trunk.

• *Control.* Because currants and gooseberries are absolutely essential for the survival of the fungus, blister rust can be controlled by eradication of these undesirable alternate hosts. However, even where pines and these species are grown near each other, the environment may not be suitable for infection. Control measures are needed only in high-hazard areas, such as those having cool, moist summers. Local control of the disease can be accomplished by destroying all currants and gooseberries within 900 feet of the pines, but the European or cultivated black currant should not be grown within a mile of pines. This species, which is more susceptible to rust than are other currants and wild gooseberries, has been found to be the chief agent in the extensive spread and establishment of the disease in previously uninfested areas.

Many valuable pines showing moderate infection can be saved by prompt surgical treatments. Branches with cankers more than 4 inches away from the trunk should be removed. Cankers closer than 4 inches or on the trunk may kill the tree, so excessive surgery may not be justified. Forest geneticists have been breeding pines for resistance to this disease. Macedonian pine, *Pinus peuce*, one of the five-needle pines, is said to be resistant to blister rust. See under rusts, Chapter 12.

COMANDRA BLISTER RUST. *Cronartium comandrae*. This rust is widespread on at least 16 species of hard pines in the United States and can be severe on lodgepole pine. Spindle-shaped swellings are formed on branches and trunks of young trees, but the girth of older infected trunks may be constricted. Trunk cankers seldom exceed $3^1/_2$ feet in length.

• *Control.* Control measures have not been developed for this rust.

OTHER RUSTS. Many other rusts, too numerous to mention in this book, also occur on pines. These appear on all the aboveground parts of the tree, producing many and varied symptoms.

However, they rarely become important parasites on pines in ornamental plantings.

PINE WILT. A nematode, *Bursaphalenchus xylophilus*, causes wilt of pine. The nematode inhabits epithelial cells of resin ducts, causing infected pines to turn pale green, then brown, and then die. Longhorn beetles, which use dead and dying trees for breeding purposes, carry the nematodes to healthy trees. Emerging beetles may carry large numbers of nematodes in their breathing pores to healthy tree branch sites, where the nematodes enter through wounds made by beetle feeding. In some circumstances wilt-causing blue-stain fungi such as *Ceratocystis ips* and *C. pilifera* are associated with the nematode. In such cases it is not certain that nematodes are the sole cause of wilt. Scots and Austrian pines in the landscape are especially susceptible to wilt.

• *Control.* Remove and destroy nematode-infected trees.

WESTERN DWARF MISTLETOE. *Arceuthobium campylopodum*. Several species of pine in the western United States are subject to the debilitating effects of this pest.

• *Control.* Pruning infested branches is advised to reduce spread within the tree crown.

LATE DAMPING-OFF AND ROOT ROT. *Pythium debaryanum, Cylindrocladium scoparium, Fusarium discolor, Pellicularia filamentosa,* and *Phomopsis juniperovora*. In nurseries, coniferous seedlings too old to be susceptible to ordinary damping-off are subject to a root rot and a top damping-off by five fungi. Seedlings of jack and red pine are are especially susceptible.

• *Control.* Disinfest seeds and then plant them in steam-pasteurized soil.

LITTLE LEAF. *Phytophthora cinnamomi*. This highly destructive disease occurs on shortleaf and loblolly pines in forest plantings of the southeastern United States. The disease is most easily recognized in its advanced stages when the crown is sparse and ragged in appearance and the branches, lacking the mass of normal foliage, often assume an ascending habit. Leaves are only half their normal length. Trees die prematurely.

• *Control.* Ornamental pines in home plantings appear to escape this disease even in areas where the causal fungus is known to be present in the soil.

WOOD AND ROOT DECAY. Because pine wood contains a high percentage of resins, wood decays are not as frequent as in hardwoods and in some other evergreens. Some of the fungi associated with wood and root decay are *Armillaria mellea, Haematostereum sanguinolenta, Heterobasidion annosum, Phellinus pini, Fomes officinalis, F. roseus, Inonotus circinans,* and *Phaeolus schweinitzii*.

• *Control.* See under wood decay, Chapter 12.

ROOT DECLINE. *Verticicladiella procera*. White pine root decline is a problem in the Ohio River valley and mid-Atlantic states. It causes a

Fig. III-95. White pine root decline canker at the tree base .

dark brown canker under the bark at the base of the tree (Fig. III-95). Symptoms also include resin flow at the tree base and reduced shoot growth. The fungus is often vectored by bark beetles, but tree-to-tree spread, possibly through root grafts, has also been observed. *V. wageneri* causes a serious black stain root disease of pine and other conifers in the West.

• *Control.* Remove and destroy infected trees. Avoid planting white pines in poor growing sites.

Abiotic Diseases

WHITE PINE DECLINE. A needle blight of white pine (*P. strobus*) that appears as a browning of entire needles or tips of needles of the current season's growth has been prevalent throughout the northeastern United States. Needle browning is often accompanied by wrinkling of bark on branches (Fig. III-96) and thin, weak foliage. The primary cause of the trouble is unknown, although in some areas it has been associated with high soil pH, high soil clay content, and soil compaction.

Fig. III-96. Branch wrinkling associated with white pine decline.

Fig. III-97. Normal yellowing of previous season's white pine needles.

STUNT. Stunting and death of 5- to 40-year-old red pines (*P. resinosa*) may be due to poor soil drainage. To avoid injury, select a well-drained planting site at the start.

AIR POLLUTION. Most pines are sensitive to ozone, and when damage occurs it is sometimes called emergence tipburn (see Chapter 10). Emergence tip burn of white pine may actually be caused by a needle cast fungus, *Canavirgella banfieldii*. White and ponderosa pines are sensitive to sulfur dioxide.

SALT. White pines planted within 50 feet of heavily used highways may be severely injured or killed by the salt used to melt snow and ice. Trees growing close to salt water may also suffer. Spraying trees in such locations with an antidesiccant on a mild day in December and again in February will reduce injury.

GIRDLING ROOTS. Pines can develop this problem. See Chapter 10.

NEEDLE YELLOWING and DROP. In some areas pine needles are shed at the end of their second growing season. These second-year needles may all yellow at once (Fig. III-97), giving tree owners cause for alarm. This yellowing is normal and merely precedes shedding. After the discolored needles drop the tree is green again.

Insects and Mites

PINE BARK ADELGID. *Pineus strobi.* Though commonly called the white pine bark louse or aphid, this insect is actually an adelgid. The insects usually work on the undersides of the limbs and on the trunk from the ground up. They may be recognized by the white cottony material (Fig. III-98) that collects in patches wherever they are present. In the eastern states white pine may be seriously injured. Adult female adelgids overwinter and produce the white, cottony material in early spring, then lay their eggs into this protective matrix. The eggs hatch in spring, and soon the young insects may be seen crawling about on the trunk and branches. Several generations develop each summer.

• *Control.* Masses of adelgids can be dis-

Fig. III-98. Pine bark adelgid secretions on trunk and branches of white pine.

lodged from the trunk by a forceful stream of water. See under aphids, Chapter 12.

WHITE PINE APHID. *Cinara strobi.* These aphids feed on the smooth bark of the twigs and smaller branches of young trees and cause a winter injury due to drying out of the twigs. Sometimes several hundred aphids are clustered together. Sooty mold often develops on the honeydew.

Powdery pine needle and spotted pine aphids (*Eulachnus* spp.) also feed on pine.

• *Control.* The same as for pine bark adelgid.

PINE LEAF CHERMID. *Pineus pinifoliae.* This insect infests species of pines and spruces. It overwinters on pine and then moves in spring to spruces, on which it causes terminal galls. In summer the aphids move to white pine and give birth to nymphs, which suck out the sap from the new shoots, resulting in either the development of undersized leaves or the death of the new shoots.

• *Control.* This insect is vulnerable to treat-

Fig. III-99. Work of the pine false webworm (*Acantholyda erythrocephala*) in a branch of pitch pine *(Pinus rigida)*; infested branch is at the left.

ment just after the first galls open on the spruces in June. See under aphids, Chapter 11.

PINE WEBWORM. *Tetrolopha robustella.* Masses of brown frass at the ends of terminal twigs result from infestations of the webworm, a ⁴/₅-inch-long yellowish brown larva with a black stripe on each side of the body. Needles may become bound by webbing.

• *Control.* See under caterpillars, Chapter 11.

PINE FALSE WEBWORM. *Acantholyda erythrocephala.* The larvae of this pest are about ³/₄ inch long, greenish to yellowish brown. They feed on the leaves and tie masses of excreta and leaf pieces together into loose balls (Fig. III-99).

• *Control.* The same as for pine webworm.

EUROPEAN PINE SHOOT MOTH. *Rhyacionia buoliana.* This is a very serious pest of mugo pines and red pines in ornamental plantings. Austrian, Scots, and Japanese black pines also may be badly damaged. The dark brown to black caterpillars attack the tip ends of young shoots in May or the early part of June causing them to turn over and become deformed and die or causing the lateral buds to be blasted. Their presence is usually indicated by quantities of resin.

The moths that are the adult stage of this caterpillar emerge about June 15 and lay their eggs in August on the new buds.

• *Control.* For small plantings and low trees, hand-picking of the pest or pruning of infested shoots and buds is probably the most satisfactory method. For treatments, see under caterpillars, Chapter 11. Apply treatments in mid-April and again in late June.

NANTUCKET PINE MOTH. *Rhyacionia frustrana.* This species attacks two- and three-needled pines in the eastern United States. Mature larvae, tan with a dark head, feed at the base of needles and then bore into the shoots, which have become covered with webbing and pitch. The adult moth is reddish brown with gray markings. A relative, the subtropical pine tip moth, spreads pine pitch canker disease in the South.

• *Control.* See under caterpillars, Chapter 11. Treatments are applied in May and July.

ZIMMERMAN PINE MOTH. *Dioryctria zimmermani.* The cambium of twigs and branches on many species of pines is invaded by white to reddish yellow or green ³/₄-inch larvae. The

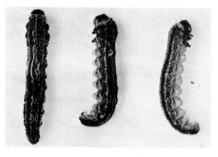

Fig. III-100. Larvae of the European pine sawfly *Neodiprion sertifer.*

adult moth is reddish gray in color and has a wingspread of 1 to 1¹/₂ inches. Branch tips turn brown, and the entire tops of trees may break off as a result of boring by the larvae.

• *Control.* See under caterpillars, Chapter 11.

SAWFLIES. *Diprion similis.* The caterpillars of the European pine sawfly are about 1 inch long and have black bodies with yellow dots; they feed in groups when young. They feed on the leaves of various species of pine during May and June and later again during September. In severe infestations pines may be entirely defoliated. There are over a dozen species of sawflies whose larvae feed on pines. Several important sawfly species include the southern, *Neodiprion excitans*; the red-headed, *N. lecontei*; the red pine, *N. nanulus nanulus*; the European pine, *N. sertifer* (Fig. III-100); and the jack-pine, *N. pratti banksianae.* Red pine sawflies feed on the previous year's growth, leaving tufted foliage. European pine sawfly larvae emerge annually, in early spring, and are most vulnerable while young.

• *Control.* See under sawflies, Chapter 11.

PINE NEEDLE MINER. *Exotelia pinifoliella.* Needles of ornamental pines turn yellow and dry up as a result of mining by a ¹/₅-inch-long larva.

• *Control.* See under leaf miners, Chapter 11.

PINE NEEDLE SCALE. *Chionaspis pinifoliae.* Pine needles may appear nearly white (Fig. III-101) when heavily infested with this scale, an elongated insect ¹/₁₀ inch long, white with a yel-

low spot at one end. The insects overwinter as eggs under female scales. The eggs hatch in May. The black pine needle scale, *Aspidiotus californicus,* is also widespread on pine and is often found in association with pine needle scale.

PINE TORTOISE SCALE. *Toumeyella parvicornus.* This cherry red or reddish brown scale, about ¹/₈ inch long, attacks jack pine in reforested areas, at times destroying 50 percent of the trees. It is closely related to *T. pini,* which occurs on Scots and mugo pines in some eastern states. *T. pinicola,* the irregular pine scale, attacks many pine species in California. Crawlers are vulnerable in early summer.

RED PINE SCALE. *Matsucoccus resinosae.* The current season's needles on infested trees are yellow and then turn brick red; later the tree dies. The young and adult stages are yellow to brown in color. They are difficult to detect because they are hidden in the bark or inside the needle clusters.

WOOLLY PINE SCALE. *Pseudophillipia quaintancii.* This scale, which resembles a mealybug, feeds on several pine species.

Fig. III-101. Pine needle scale on white pine.

• *Control.* See under scales, Chapter 11.

PINE SPITTLEBUGS. *Aphrophora parallela.*
These insects are perhaps most common on
Scots pine, but white pine also is seriously
attacked at times. They cause injury to smaller
twigs by drawing the sap from them. A foamy
substance is formed about the young insects,
giving the branches a whitish appearance (Fig.
III-102). The adult insects are about $^1/_2$ inch
long, grayish brown in color, and resemble small
frogs. Another species, the Saratoga spittlebug,
A. saratogensis, seriously damages jack and red
pines. The western pine spittlebug, *A. permutata*
feeds on several conifers in western states.

• *Control.* Most spittlebugs are relatively
harmless and can be washed from pines with a
strong stream of water. Spray with Asana,
Decathlon, Orthene, Sevin, Tempo, or Turcam
in mid-May and again in mid-July, directing the
spray forcefully to hit the little masses of spittle
that cover the insect.

WHITE PINE SHOOT BORER. *Eucosma glori-*
ola. The whitish caterpillars are about $^1/_2$ inch
long. They burrow down the centers of the lateral shoots, causing them to wilt and to die back
several inches. The insect spends the winter in
the soil and appears as a moth in spring.

• *Control.* Cut off and destroy infested
branches as soon as they are discovered, though
this is practical only for young trees.

PINE SHOOT BEETLE. *Tomicus piniperda.*
This introduced pest is affecting pines in Christmas tree plantations in the Midwest, in some
cases restricting tree shipment. Integrate sanitation, log trapping, and insecticide applications
to manage this pest.

WHITE PINE TUBE MOTH. *Argyrotaenia*
pinatubana. Greenish yellow larvae make tubes
by binding the needles together side by side and
eating the free end off squarely. Lodgepole and
whitebark pines in the Rocky Mountain region
are susceptible to a similar insect, *A. tabulana.*

• *Control.* Spray with Talstar in early May
and again in mid-July, if needed. This insect is
not normally a serious pest.

PALES WEEVIL. *Hylobius pales.* The bark of
young white, red, and Scots pines may be

Fig. III-102. Spittlebug on Austrian pine.

chewed by a night-feeding reddish brown to
black weevil $^1/_3$ inch long. Twigs and young trees
may be girdled completely and die.

WHITE PINE WEEVIL. *Pissodes strobi.* This
weevil is a very common pest of white pine
in estates and on lawns where trees are planted
as ornamentals. It is also damaging in forests
of white pine. The larvae feed on the inner bark
and the sapwood of the leading branches and
terminal shoots of the main trunks. The leader
is girdled and killed, requiring the tying up of
a lateral shoot to replace the leader. Some of
the branches that grow out to replace the leader
may be distorted. The beetles, $^1/_4$ inch long,
reddish brown and somewhat white-mottled,
begin to emerge in July, leaving characteristic holes in the bark. The larvae are pale yellowish grubs about $^1/_3$ inch long. This insect also
attacks spruce, as shown in Figure III-115
(p. 483).

DEODAR WEEVIL, *P. nemorensis,* causes
similar damage and also vectors the pitch
canker fungus in the South.

Fig. III-103. Top: Life stages of the black turpentine beetle. Actual size is approximately ¹/₄ inch. Bottom: Pitch tubes of the beetle at the base of a pine tree.

PINE ROOT COLLAR WEEVIL. *Hylobius radicis.* Austrian, mugo, red, and Scots pines may be severely damaged by this borer. The symptoms on infested trees are sickly and dead foliage and masses of pitch around the base of the trunk 3 or 4 inches below the soil surface.

• *Control.* See under weevils, Chapter 11.

BLACK TURPENTINE BEETLE. *Dendroctonus terebrans.* This insect (Fig. III-103) can cause the death of Austrian, Japanese black, and pitch pines. Red turpentine beetle, *D. valens,* attacks most pines and occasionally spruce and fir. Once considered secondary invaders that attacked pine weakened by some other agent, turpentine beetles are capable of infesting apparently healthy pines and killing them.

The beetles are $^1/_5$ to $^3/_8$ inch long, reddish brown to black, and robust. They bore into the lower 4 feet of the trunk, making galleries $^1/_2$ to 1 inch wide that extend downward for several inches to several feet. Resin flows out of the injured area, forming so-called pitch tubes easily visible on the bark surface (Fig. III-103). Destruction of the inner bark and wood by the larval stage eventually results in the death of the tree.

• *Control.* Maintain pines in good vigor by mulching, fertilizing and watering.

BARK BEETLES. *Scolytidae* (*Ips* spp. and *Dendroctonus* spp.). Trees in weakened or dying condition are subject to infestation by bark beetle larvae, small worms that mine the bark and engrave the sapwood. Tiny "shot-holes" varying from $^1/_{20}$ to $^1/_3$ inch in size (depending on the species) are present on the bark surface where beetles emerge.

• *Control.* No effective control measure is known. Trees should be kept in vigorous condition by mulching, fertilizing, and watering. Severely infested trees should be cut down and destroyed, or, if left standing, they should be debarked.

PITCH MOTH. *Synanthedon* spp. Weakened pines may be attacked by the larvae of these clear-winged moths. Pitch and frass accumulate on the surface of the tree.

PINE LOOPER. *Lambdina athasaria pelluci-daria.* This 1-inch-long moth larva is periodically a serious defoliator of red pine and pitch pine in the eastern United States.

PINE TUSSOCK MOTH. *Dasychira plagiata.* The colorful larvae are defoliators of jack, white, and red pine in the Midwest.

SPRUCE SPIDER MITE. See under Spruce.

ROSETTE MITE. *Trisetacus gemmavitians.* Infestation of pine buds by this eriophyid mite causes new shoots to fail to elongate, resulting in a rosette of stunted needles clustered at the shoot tip. The mite infests shoots as buds begin to swell.

• *Control.* Prune out infested shoots. See under eriophyid mites, Chapter 11.

PITTOSPORUM
(*Pittosporum*)

Some pittosporum species may develop into small trees. They are relatively pest-free.

PLANETREE, LONDON
See Sycamore.

PLUM, FLOWERING
See Cherry.

POINCIANA (*Delonix*)

Flame-tree or flamboyant-tree is a striking red-flowered tree grown in frost-free areas of the United States.

Diseases

CANKER. This tropical tree is subject to relatively few diseases. Perhaps the most destructive is canker, caused by the fungus *Botryosphaeria dothidea.*

• *Control.* Branches that show cankers should be pruned and destroyed.

OTHER DISEASES. Other diseases reported on this host are crown gall, caused by the bacterium *Agrobacterium tumefaciens*; anthracnose by a fungus belonging to the genus *Gloeosporium*; and root rots by two fungi, *Clitocybe tabescens* and *Phymatotrichum omnivorum.*

• *Control.* Control measures are rarely necessary.

Insects

LESSER SNOW SCALE. *Pinnaspis strachani.* The lesser snow scale frequently infests this host. The female is pear-shaped, white, semitransparent, sometimes speckled with brown. The male is narrow and white. Silk-tree, hackberry, palm, citrus, avocado, and other trees in the Deep South are also infested.

• *Control.* See under scales, Chapter 11.

POPLAR (*Populus*)

Poplars are fast-growing trees hardy in temperate regions. Some, such as cottonwood, are adapted to the relatively harsh growing conditions of the Great Plains. Some, such as Lombardy poplar, are short-lived because of their high susceptibility to disease. Poplars are generally not suitable for urban areas because their weak branches break easily and they require much maintenance to retain an attractive appearance. Some poplars are commonly known as aspens.

Diseases

Several destructive diseases affect the trunks and limbs of poplars. The genus to which the causal fungus belongs is used to differentiate the various types of cankers produced. A dieback, which is not clearly understood, and a number of leaf diseases, which rarely cause much damage, also occur on poplars.

CRYPTODIAPORTHE (DOTHICHIZA) CANKER. The dying and dead Lombardy poplars (*P. nigra italica*) in the eastern United States are ample evidence of this destructive disease. Black and eastern cottonwoods and balsam, black, and Norway poplars also may be affected, although these species are much less susceptible than Lombardy poplar (Fig. III-104).

• *Symptoms.* Elongated, dark, sunken cankers occur on the trunk, limbs, and twigs (Fig. III-105). The bark and cambium in the cankers are destroyed, and the sapwood is invaded and discolored. When the cankers completely girdle the trunks or branches, the distal portions die. Trees less than 4 years old are usually killed outright in a relatively short time. Older trees do not succumb as rapidly but are readily disfigured and become virtually worthless as ornamentals or windbreaks.

• *Cause.* The fungus *Cryptodiaporthe (Dothichiza) populea* causes this disease. The sexual stage of this fungus is *Discosporium populea.* On diseased bark it forms raised small pustules, which ooze cream-colored masses of spores in April and May. The spores are splashed onto leaves and into bark injuries, where they germinate and infect. Invasion often advances down the leaf petiole into the twig, causing death of the latter.

• *Control.* No effective control measures are known. Wounds of all sorts should be avoided. Pruning of diseased parts, as suggested for most canker diseases, does not appear to help control this disease, but in fact often spreads it.

Since leaf infections lead to twig infections, some investigators have recommended repeated applications of copper sprays to reduce leaf infections. As a rule, individual trees are not sufficiently valuable to justify four or more applications of a fungicide each year. Such a practice may be warranted, however, in nurseries where the disease threatens young stock.

The Japan poplar (*P. maximowiczii*) appears to show some resistance. Fremont and black cottonwoods, Carolina and gray poplars, and trembling aspen are considered resistant. Conflicting reports exist concerning the susceptibility of the Simon poplar (*P. simonii*); some assert that it is very resistant, while others consider it very susceptible.

CYTOSPORA CANKER. Another destructive canker disease of poplar, Cytospora canker, primarily affects trees poor in vigor. Among the factors that most commonly contribute to the weakened condition are drastic pruning and the unfavorable environment of city soils. Carolina and silver-leaf appear to be the most susceptible of the poplars. Maples, mountain-ash, and willows are also susceptible.

• *Symptoms.* Brown, sunken areas covered with numerous red pustules first appear on young twigs. The fungus moves down the stem

Fig. III-104. The Lombardy poplars on the left died as a result of infection by the fungus *Discosporium populea.*

Fig. III-105. Trunk of Lombardy poplar showing elongated sunken cankers caused by the fungus *Cryptodiaporthe.*

and invades larger branches or even the trunk. Circular cankers are formed, which continue to expand until the entire member is girdled and the parts above die.

• *Cause.* The fungus *Cytospora chrysosperma,* a stage of the fungus *Valsa sordida,* is the causal agent. It lives on dead branches but can attack live branches and trunks when the tree is weakened. Spores develop in the black pinpoint fruit-

ing structures on dead wood. They ooze from these bodies and are carried by rainsplash, birds, and insects to bark wounds or weakened and dead branches, where they germinate and infect.

• *Control.* Because Cytospora canker is primarily a disease of weak trees, the most effective preventive is the maintenance of the trees in high vigor by fertilization, watering, and the control of insect and fungus parasites of the leaves. In addition, dead and dying branches should be removed and all unnecessary injuries avoided. 'Noreaster' hybrid poplar and 'Platte', 'Mighty Mo', and 'Ohio Red' cottonwoods are resistant to this disease.

CRYPTOSPHAERIA CANKER. This disease is destructive to aspens and poplars throughout the Great Plains.

• *Symptoms.* A stem canker causes branch and tree mortality. Sapwood discoloration and decay are also associated with this disease.

• *Cause. Cryptosphaeria populina* (imperfect, *Libertella* sp.) causes canker and decay.

• *Control.* Prune infected branches.

HYPOXYLON CANKER. A highly destructive disease, Hypoxylon canker occurs less commonly than Dothichiza and Cytospora cankers. In the northeastern states it affects quaking and large-toothed aspens and balsam poplar. It is most harmful to young trees and is more common on forest trees than on ornamentals.

• *Symptoms.* Gray cankers of varying sizes appear along the trunk, but never on the branches. The color changes to black as the outer bark falls away from the surface of the canker. When the bark is peeled off, blackened sapwood is visible around the edge of the canker. Wefts of fungal tissue, resembling the chestnut blight fungus but different in color, are also visible beneath the peeled bark. Many cankers attain a length of several feet and cause considerable distortion of the trunk before complete girdling of the trunk and death occur.

• *Cause.* This disease is caused by the fungus *Hypoxylon mammatum.* Spores are believed to initiate infections through bark injuries.

• *Control.* This disease is so highly contagious and destructive that infected trees should be cut down and destroyed as soon as the diagnosis is confirmed. Injuries should be avoided as much as possible.

SEPTORIA CANKER AND LEAF SPOT. A serious stem canker disease of hybrid poplars, Septoria canker may also infect leaves of all species of native poplars.

• *Symptoms.* Leaves on young shoots or lowermost branches of native poplars show spots early in the season. Cankers appear later on the stems of hybrid poplars that have black, balsam, or cottonwood parentage. When cankers girdle the stem the distal portion dies.

• *Cause. Mycosphaerella populorum* causes leaf spot and stem canker. The asexual stage of this fungus is *Septoria musiva,* hence the common name for the disease. Fruiting bodies of the Septoria stage are frequently found in the cankered areas.

• *Control.* The most effective way of combating the canker stage of this disease is to use hybrid poplar clones that have proved to be naturally resistant. For leaf spot control, see Chapter 12.

FUSARIUM CANKER. *Fusarium solani.* This disease occurs on eastern cottonwood in the South, Midwest, and Canada.

• *Control.* Control measures have not been developed.

PHOMOPSIS CANKER. *Phomopsis macrospora.* Cottonwood suffers dieback and canker caused by this fungus.

SOOTY BARK CANKER. *Encoelia pruinosa.* This disease is especially damaging to aspen in the Rocky Mountain States.

BRANCH GALL. Small globose galls up to $1^1/_2$ inches in diameter occasionally occur at the bases of poplar twigs. Primarily the bark is hypertrophied, although some swelling of the woody tissues also occurs. Twigs and some branches may be killed, but the disease rarely becomes serious.

• *Cause.* Small black pinpoint fruiting bodies of the fungus *Macrophoma tumefaciens* are embedded in the bark of the gall, especially along the fissures. The fungus penetrates developing buds in early spring, and the swelling of

the tissue probably results from some stimulant secreted by the fungus.

• *Control.* Pruning the galled branches and dead twigs to sound wood and destroying the removed wood are usually sufficient to hold the fungus in check.

LEAF RUSTS. Several species of rust fungi attack poplars in the eastern United States. These produce yellowish orange pustules, usually on the lower leaf surface. The two species most common in the East are *Melampsora medusae* and *M. abietis-canadensis.* They require larch and hemlock, respectively, as alternate hosts to complete their life cycles. Two other leaf rusts are caused by *M. albertensis* and *M. occidentalis.*

• *Control.* Except for the rust fungus *M. medusae,* which causes heavy losses of young eastern cottonwoods in nurseries in the Midwest and South, leaf rusts rarely cause enough damage to necessitate special control measures. Several rust-resistant cottonwood clones are available.

LEAF BLISTER. Brilliant yellow to brown blisters of varying sizes occasionally appear on poplar leaves following extended periods of cool, wet weather. The blisters, which may be more than an inch in diameter, result from stimulation of the leaf cells by the fungus *Taphrina populina.* Another species, *T. johansoni,* causes a deformity of the catkins.

• *Control.* Spraying the trees with Ferbam before bud break in early spring will control leaf blister. This is suggested only where particularly valuable specimens are involved.

POPLAR INKSPOT. *Ciborinia whetzelii.* Thick round or elongated black spots form on leaves and then drop out, leaving holes. Aspen, Lombardy, and Carolina poplars are most subject to inkspot, whereas the large-toothed aspen (*P. grandidentata*) appears to be immune.

• *Control.* See under leaf spots, Chapter 12.

LEAF SPOTS. *Cercospora populina, C. populicola, Linospora tetraspora, Marssonina populi, M. castagnei, M. brunnea, Mycosphaerella populicola, Apioplagiostoma populi, Phyllosticta alcides, Venturia populina,* and *V. tremulae* (imperfect, *Pollaccia*). The latter two also cause a shoot blight. Many species of fungi, including *Mycosphaerella populorum* mentioned under the Septoria canker disease, cause leaf spots of poplars. Of these, *Marssonina,* which produces brown spots with a darker brown margin and premature defoliation, is by far the most common. It also invades the leaf petioles and twigs, causing tan spots with dark margins. A leaf blotch caused by *Septonina podophyllina* also affects poplars.

- *Control.* See under leaf spots, Chapter 12.

POWDERY MILDEW. *Uncinula adunca.* This is a common disease, appearing as a white mildew on both sides of the leaves; the damage is usually not serious.

- *Control.* See under powdery mildew, Chapter 12.

DIEBACK. Top dieback and death of Lombardy poplar, eastern cottonwood, and the goat willow (*Salix caprea*) have been reported. On Lombardy poplar, the wood first appears water-soaked, then red, and finally brown. When the entire cross section of the trunk is stained brown the tree dies

- *Cause.* The cause of this disease in not known. Bacteria have been isolated from the margins of the discolored area. There is a strong possibility that the water-soaked appearance and the subsequent staining actually result from extremely low winter temperatures and that the bacteria enter after the injury occurs.

- *Control.* No control measures are known.

HEART ROT. *Phellinus igniarius* and *Phellinus tremulae* are the most important of many poplar heart rot fungi. See Chapter 12.

WETWOOD. Poplars are especially susceptible to this condition. See Chapter 12.

CROWN GALL. *Agrobacterium tumefaciens.* See Chapter 12.

MOSAIC VIRUS. Poplars are susceptible to a virus that causes mosaic-type symptoms.

Abiotic Disease

AIR POLLUTION. Many poplars are sensitive to sulfur dioxide and are very sensitive to ozone. See Chapter 10.

Insects and Mites

APHIDS. *Pemphigus populitransverus* and *Mordwilkoja vagabunda.* The former produces galls on the leaf petioles of certain poplars; the latter is responsible for convoluted galls at the tips of twigs. Several other species of leaf- and bark-infesting aphids such as *Pterocomma* and giant bark occur on this host.

- *Control.* See under aphids, Chapter 11.

BRONZE POPLAR BORER. *Agrilus liragus.* This flatheaded borer attacks weakened trees.

- *Control.* Provide good growing conditions, mulching and watering as needed. Trunk sprays with persistent insecticides may help. See under borers, Chapter 11.

POPLAR BORER. *Saperda inornata* and *S. populea.* Swollen areas on small branches usually result from attack by the poplar borer, a white larva 1 inch long when fully grown. Hollowed out stem galls weaken and deform the young tree, and damaged branches may break off. The adult female is a bluish gray beetle with black spots and yellow patches and is $1/3$ to $1/2$ inch in length. Egg-laying scars made in late spring on young stems are brown and horseshoe shaped. This pest also attacks willow.

- *Control.* See under borers, Chapter 11.

WESTERN POPLAR CLEARWING BORER. *Paranthrene robiniae.* This borer, also called locust clearwing, is a pest of stressed birch, poplar, and willow.

- *Control.* See under borers, Chapter 11.

MOTTLED WILLOW BORER. See under Willow.

HORNET MOTH. *Aegeria apiformis.* This clear-winged moth larva is an important wood borer pest of poplar in some locations in the northern United States.

CARPENTERWORM. See under Ash.

AMERICAN PLUM BORER. See under Sycamore.

FLATHEADED and LEOPARD MOTH BORER. See under Maple.

COTTONWOOD TWIG BORER. *Gypsonoma haimbachiana.* This caterpillar feeds in growing cottonwood shoots.

TWIG GIRDLER. See under Hackberry.

COTTONWOOD LEAF BEETLE. *Chrysomela scripta.* This light green and black-spotted larva skeletonizes cottonwood leaves. Willows are also attacked by this pest. *C. interrupta* also feeds on poplar leaves.

IMPORTED WILLOW LEAF BEETLE. See under Willow.

RED-HUMPED CATERPILLAR. *Schizura concinna.* The larvae, which are yellow and black striped with red heads and a reddish body segment, gather in clusters and feed on the leaves. The adult, a grayish brown moth with a wingspread of 1¼ inches, deposits masses of eggs on the lower leaf surface in July. Winter is passed in a cocoon on the ground.

• *Control.* Populations of this pest are often reduced by naturally occurring parasitic wasps. See under caterpillars, Chapter 11.

SATIN MOTH. *Stilpnotia salicis.* The larval stage, black with conspicuous irregular white blotches, feeds on the leaves of poplar, willow, and sometimes oaks in late April or May. The adult, satin-white with a wing expanse up to 2 inches, emerges in July.

• *Control.* See under caterpillars, Chapter 11.

TENTMAKER. *Ichthyura inclusa.* Chewed and silken nests appear on twigs as a result of infestations of the tentmaker, a black larva with pale yellow stripes, which is 1¼ inches long at maturity. The adult female is a moth with white-striped gray wings. The pupae overwinter in fallen leaves.

• *Control.* Remove and destroy nests.

FOREST TENT CATERPILLAR. See under Maple.

FRUIT TREE LEAF ROLLER. See under Crabapple.

GYPSY MOTH, MOURNING-CLOAK BUTTERFLY. See under Elm.

WHITE-MARKED TUSSOCK MOTH. See under Elm.

IO MOTH. See under Sycamore.

FALL WEBWORM. See under Ash.

EASTERN TENT CATERPILLAR. See under Willow.

SCALES. Many species of scale insects infest poplars. Among these are black, brown soft, cottony maple, European fruit lecanium, greedy, lecanium, oystershell, San Jose, soft, terrapin, walnut, and willow.

• *Control.* See under scales, Chapter 11.

COMSTOCK MEALYBUG. See under Catalpa.

HONEY LOCUST PLANT BUG. See under Locust, Honey.

COTTONWOOD GALL MITE. *Eriophyes parapopuli.* This eriophyid mite causes warty, woody swellings on the twigs of cottonwoods and poplars.

• *Control.* See under eriophyid mites, Chapter 11.

PORT ORFORD CEDAR
See False Cypress.

QUINCE, FLOWERING
(*Chaenomeles*)

Bacterial and Fungal Diseases

CROWN GALL. *Agrobacterium tumefaciens.* This disease occasionally affects this host.

• *Control.* See Chapter 12.

FIRE BLIGHT. *Erwinia amylovora.* Flowering quince, like other rosaceous hosts, is subject to this disease.

• *Control.* See under fire blight, Chapter 12.

BROWN ROT. *Monilinia fructicola* and *M. laxa.* These fungi are usually more destructive on fruit trees than on the ornamental varieties. They cause a leaf blight and a blossom and twig blight of flowering quince.

• *Control.* Periodic applications of Captan or wettable sulfur sprays during the early growing season are effective.

RUST. *Gymnosporangium claivipes.* This bright orange-colored rust is more commonly found on fruiting trees than on the flowering varieties. It attacks the fruit as well as the leaves and young twigs. It also attacks apples and hawthorns. The alternate host is the eastern red cedar, *Juniperus virginiana.* Though very destructive to the common quince, it does little damage to the cedar. (See Juniper and also under Hawthorn.)

Another rust, *G. libocedri,* also attacks flowering quince leaves.

• *Control.* See under rusts, Chapter 12.

LEAF SPOTS. *Diplocarpon mespili* and *Cercospora cydoniae.* These leaf spots can become troublesome and cause premature defoliation in rainy seasons.

• *Control.* See under leaf spots, Chapter 12.

OTHER DISEASES. Among other diseases occasionally found on flowering quince are twig blight caused by *Botryosphaeria dothidea*; cankers by *Nectria cinnabarina, Phoma* sp., and *Botryosphaeria obtusa*; and root-knot nematode, *Meloidogyne* sp.

• *Control.* These diseases are rarely serious enough on flowering quince to warrant control measures.

Insects

COTTON APHID. *Aphis gossypii.* Young leaves and upper ends of tender branches may be heavily infested with these aphids. Apple aphid also feeds on quince.

• *Control.* See under aphids, Chapter 11.

YELLOW-NECKED CATERPILLAR. See under Oak.

FOREST TENT CATERPILLAR. See under Maple.

JAPANESE BEETLE. See under Linden.

SCALES. Cottony-cushion, pitmaking, pittosporum, Japanese wax, black, greedy, and California red scales feed on flowering quince.

REDBUD (*Cercis*)

Redbud is recognized by early spring flowers that appear on bare branches, just before flowering dogwood blooms. Redbud is tolerant of shade, sun, and soils that are alkaline, acid, or moist.

Diseases

CANKER. *Botryosphaeria dothidea.* This, the most destructive disease of redbud, also affects many other trees and shrubs. On redbud the cankers begin as small sunken areas and increase slowly in size. The bark in the center of the canker blackens and cracks along the edges. The wood beneath the cankered area becomes discolored. When the canker girdles the stem, the leaves above wilt and die. The causal fungus is easily recovered from discolored wood by standard tissue culture techniques in the laboratory. The causal fungus was formerly known as *Botryosphaeria ribis.*

• *Control.* Prune and destroy branches showing cankers. Surgical excision of cankered tissue on the main stem is occasionally successful if all infected bark and wood are removed. Periodic applications of a copper fungicide during the growing season may help to prevent new infections.

LEAF SPOTS. A conspicuous leaf spot disease occurs on redbud throughout the range of this tree in the eastern half of the United States. Scattered circular to angular brown spots, about $1/4$ inch in diameter, appear on the leaves early in summer. The spots increase in size until they attain a diameter of nearly $1/2$ inch by fall.

The fungus *Mycosphaerella cercidicola* produces circular to angular spots with dark brown borders on the leaves. Spores from overwintered fallen leaves initiate spring infections. The fungi *Cercosporella chionea, Plasmopara cercidis, Kabatiella*, and *Phyllosticta cercidicola* also cause leaf spots on redbud.

• *Control.* See under leaf spots, Chapter 11.

WILT. *Verticillium albo-atrum.* This fungus causes a serious wilt disease (Fig. III-106) in landscape plantings. See Chapter 12.

OTHER DISEASES. Among other fungal diseases of redbud are root rots caused by *Clitocybe tabescens, Ganoderma lucidum*, and *Phymatotrichum omnivorum.*

Abiotic Disease

AIR POLLUTION. Redbuds are sensitive to ozone. See Chapter 10.

Insects

CATERPILLARS. The larval stage of the California tent caterpillar, *Malacosoma californicum*, and the grape leaf folder, *Desmia funeralis*, occasionally infest redbuds.

• *Control.* See under caterpillars, Chapter 11.

COTTON APHID. *Aphis gossypii.* The young leaves and upper ends of tender branches may be heavily infested with these aphids.

• *Control.* See under aphids, Chapter 11.

Fig. III-106. Verticillium wilt of redbud showing vascular discoloration.

Fig. III-107. Redbud leafhopper injury.

LEAFHOPPER. An *Erythronema* species causes stippling of redbud leaves (Fig. III-107).

TWO-MARKED TREEHOPPER. See under Hop-Tree.

SCALES. Many species of scales infest the twigs and branches of redbud, including European fruit lecanium, greedy, ivy, and white peach.

• *Control.* See under scales, Chapter 11.

REDBUD LEAFROLLER. *Fascista cercerisella.* Also called redbud leaffolder, the larvae of this moth skeletonize leaves from within a nest. Leaves forming the nest appear to be stitched together.

• *Control.* This insect does relatively little damage, thus no controls are suggested.

OTHER INSECTS. The Rhabdopterus beetle, *Rhabdopterus deceptor*; the green fruitworm, *Lithophane antennata*; and the greenhouse whitefly (see whiteflies, under Citrus) all attack redbud.

• *Control.* Sevin sprays will control these pests.

ASIATIC OAK WEEVIL. See under Oak.

REDWOOD (*Sequoiadendron* and *Sequoia*)

These two West Coast natives, the legendary biggest and tallest trees, are grown in landscapes where they do not attain the immense size of those in native groves. Coast redwood requires more fog and moisture than giant redwood.

Diseases

CANKER. *Dermatea livida.* The cankers that develop on the bark contain the brown to black fruiting bodies of the fungus. These contain one-celled colorless spores. Another canker is caused by the fungus *Botryosphaeria dothidea.*

• *Control.* No satisfactory control measures have been developed.

NEEDLE BLIGHT. *Chloroscypha chloramela* and *Mycosphaerella sequoiae.* Redwood leaves may be blighted by these fungi. The giant sequoia (*Sequoiadendron giganteum*) is susceptible to two other leaf-blighting fungi, *Cercospora sequoiae* and *Pestalotiopsis funerea,* and to twig blight by *Phomopsis juniperovora.*

• *Control.* Small, importantly placed trees infected with these fungi should be sprayed with a copper or a dithiocarbamate fungicide to prevent new infections.

ARMILLARIA ROOT ROT. See Chapter 12.

Insects

CEDAR TREE BORER. *Semanotus ligneus.* The larvae of this beetle make winding burrows in the inner bark and sapwood of redwood, cedars, arborvitae, Douglas-fir, and Monterey pine, occasionally girdling and killing the trees. The adult beetle is $1/2$ inch long, black with orange and red markings on its wing covers.

• *Control.* See under borers, Chapter 11.

SEQUOIA PITCH MOTH. *Synanthedon sequoiae.* The cambial region of redwoods and many other conifers in the West is mined by an opaque, dirty-white larva. The adult moth is black with a bright yellow segment at its lower abdomen and resembles a yellow jacket wasp.

• *Control.* This pest is rarely abundant enough to warrant control measures.

MEALYBUGS. Three species of mealybugs may infest redwoods: citrus, *Planococcus citri*; cypress, *Pseudococcus ryani*; and yucca, *Puto yuccae.*

SCALES. Two scale insects have been reported on this host: greedy, *Aspidiotus camelliae*; and oleander, *A. hederae.*

• *Control.* See under scales, Chapter 11.

REDWOOD, DAWN (*Metasequoia*)

This tree is a native of China, and, although present in this hemisphere 15 million years ago, it was reintroduced as recently as 1945 by the late Dr. E. D. Merrill of the Arnold Arboretum. It resembles bald cypress in appearance.

Diseases

CANKER. *Botryosphaeria dothidea.* This is the only disease known to occur on dawn redwood (see Botryosphaeria canker, under Sycamore). An early symptom is the wilting of leaves on individual lateral branches. Cankers develop on the trunk at the base of the infected branches. Resin exudes from the infected area.

The canker disease also occurs on *Sequoiadendron giganteum* and on *Sequoia sempervirens* planted outside the native range.

• *Control.* Control measures have not been developed.

RUSSIAN-OLIVE (*Elaeagnus*)

Russian-olive is an excellent tree for very difficult situations and is relatively free of fungal parasites and insect pests. This low-growing tree tolerates cold temperatures.

Diseases

CANKERS. *Lasiodiplodia theobromae, Nectria cinnabarina (Tubercularia vulgaris), Fusicoccum elaeagni, Fusarium* sp., *Phomopsis arnoldiae,* and *Phytophthora cactorum.* Cankers on the branches and trunk of Russian-olive may be caused by any of these fungi. *L. theobromae* has killed large numbers of Russian-olives in shelter belts in the Great Plains states. Bark, cambium, and phloem tissues are killed in strips along the trunk and major branches. Infection is rapid up and down stems but slow around them, so complete girdling and death may take several years. *Phomopsis* appears to be more prevalent in the central and northeastern states.

• *Control.* No control measures have been developed; however, trees should be managed for optimum vigor.

LEAF SPOTS. Several fungi including *Cercospora carii, C. elaeagni, Phyllosticta argyrea; Septoria argyrea* and *S. elaeagni* cause spots on leaves of this host.

• *Control.* These fungi are rarely severe enough to warrant the use of protective fungicides.

RUST. *Puccinia caricis-shepherdiae* with the alternate host sedge (*Carex*), and *P. coronata* f. *elaeagni* with the alternate host reedgrass (*Calamagrostis*) occur on Russian-olive in the midwestern United States.

• *Control.* These diseases are not important enough to warrant control measures.

WILT. *Verticillium albo-atrum.* This disease was first found on Russian-olive in the West.

• *Control.* See Chapter 12.

OTHER DISEASES. Russian-olive is occasionally affected by crown gall caused by *Agrobacterium tumefaciens* and hairy root by *A. rhizogenes.*

• *Control.* Control measures are rarely necessary.

Insects

OLEASTER-THISTLE APHID. *Capitophorus braggii.* This pale yellow and green aphid lives on thistles during the summer and overwinters on Russian-olive.

• *Control.* Control measures are unnecessary.

SCALES. *Parlatoria oleae* and *Lepidosaphes beckii.* The olive (see under Kentucky Coffee Tree) and the purple (see under Oak) scales occasionally infest this host. Many other species of scales, including European peach, ivy, and dictyosperm, have also been recorded on Russian-olive.

• *Control.* See under scales, Chapter 11.

COMSTOCK MEALYBUG. See under Catalpa.

SASSAFRAS (*Sassafras*)

Sassafras, an excellent tree for landscape plantings, is extremely susceptible to Japanese beetle infestations. Like American holly, it is dioecious; the pistillate and staminate flowers grow on different trees. The slate-blue fruits are ornamental and provide food for birds. It is native to the eastern United States, where it grows in acid soils.

Diseases

CANKER. Two fungi, *Nectria galligena* and *Botryosphaeria obtusa*, are frequently associated with branch cankers (Fig. III-108).

• *Control.* Pruning affected branches to sound wood is suggested.

LEAF SPOTS. *Actinopelte dryina, Diplopeltis sassafrasicola, Glomerella cingulata, Metasphaeria sassafrasicola, Phyllosticta illinoisensis,* and *Stigmatophragmia sassafrasicola.* These species of fungi cause leaf spots on sassafras.

Fig. III-108. Cankers on sassafras stem caused by the fungus *Nectria galligena*.

Fig. III-109. Japanese beetles feeding on a sassafras leaf.

They are rarely serious enough to warrant control measures.

OTHER DISEASES. Powdery mildew caused by the fungus *Phyllactinia guttata*; wilt by a species of *Verticillium*; and shoestring root rot by *Armillaria mellea* have been recorded on sassafras. A yellows type disease, probably caused by a phytoplasma, causes bunching and fasciation of the branch tips, leaf rolling, and leaf dwarfing. This host is also subject to curly top caused by the beet curly top virus.

Insects

JAPANESE BEETLE. (Fig. III-109.) See under Linden.

PROMETHEA MOTH. *Callosamia promethea.* The leaves are chewed by the promethea moth larva, a bluish green caterpillar that grows to 2 inches in length. The adult female has reddish brown wings with eyelike spots near the tips; the wingspread is nearly 3 inches. The cocoons of this pest are suspended in the trees.

The caterpillars of the hickory horned devil, *Citheronia regalis*, io moth (see under Sycamore), gypsy moth (see under Elm) and polyphemus moth, *Antheraea polyphemus*, also chew sassafras leaves.

• *Control.* See under caterpillars, Chapter 11.

SASSAFRAS WEEVIL. *Odontopus calceatus.* This small snout weevil begins to feed as soon as the buds break and before the leaves have expanded, making numerous holes in the l eaves. The female adult then deposits eggs on the midribs of the leaves. The larvae that hatch from these mine into the leaves causing blotches. Tuliptree and magnolia are also susceptible.

• *Control.* See under weevils, Chapter 11.

SCALES. The oystershell and San Jose scales occasionally infest sassafras. In addition to these two scales, sassafras is occasionally infested by the European fruit lecanium, *Lecanium corni*; terrapin, *Lecanium nigrofasciatum*; Florida wax, *Ceroplastis floridensis*; and pyriform, *Protopulvinaria pyriformis* scales.

• *Control.* See under scales, Chapter 11.

SCREW-PINE
(*Pandanus*)

Diseases

LEAF BLOTCH. *Melanconium pandani.* Large leaf spots up to 2 inches wide and 3 or 4 inches long may develop from the leaf margin inward. Black fruiting pustules develop in a light gray zone along the inner margin of the spot. Similar spots may develop along the leaf blade at any point.

• *Control.* Prune off infected leaves and spray with a copper fungicide. Badly infected plants should be destroyed.

LEAF SPOTS. *Heterosporium iridis, Phomopsis* sp., and *Volutella mellea.* These fungi have been reported on screw-pine.

• *Control.* Leaf spots are rarely serious enough to warrant control measures.

Insects

The long-tailed mealybug, *Pseudococcus adonidum,* and thirteen species of scales attack screw-pine.

- *Control.* See under scales, Chapter 11.

SEQUOIA. See Redwood.

SERVICEBERRY (*Amelanchier*)

Serviceberry, also known as Juneberry, shadblow, and shadbush, blooms early, has good fall color, and is an excellent low-growing tree of the rose family. It is subject to many of the diseases and insects common to members of that family.

Diseases

FIRE BLIGHT. *Erwinia amylovora.* This bacterial disease, which is serious on certain hawthorns and other pomaceous trees, occasionally occurs on serviceberry.

- *Control.* See Chapter 12.

FRUIT ROTS. *Monilinia amelanchieris* and *M. fructicola.* The fruits of serviceberry are rotted in rainy seasons by these fungi.

- *Control.* Infections on particularly valuable specimens can be prevented by periodic applications of Captan or wettable sulfur sprays.

LEAF BLIGHT. *Diplocarpon mespili.* Serviceberry is occasionally attacked by this fungus.

- *Control.* See under Hawthorn.

POWDERY MILDEWS. *Erysiphe polygoni, Phyllactinia guttata,* and *Podosphaera clandestina.*

- *Control.* See under powdery mildew, Chapter 12.

RUST. More than 10 species of rust occur on serviceberry, with alternate hosts of juniper, white cedar, and incense cedar.

- *Control.* See under rusts, Chapter 12.

WITCHES' BROOM. Several species of serviceberry are subject to the witches' broom disease though it rarely causes much damage.

- *Symptoms.* The production of many lateral branches from a common center results in a bunching or broomlike effect. The lower surfaces of leaves attached to such twigs are covered with sooty fungus growth.

- *Cause.* The fungus *Apiosporina collinsii* causes witches' broom. It penetrates the twigs in spring and stimulates the production of the many lateral branches. The black growth of the fungus on the leaves contains small round fruiting bodies in late summer. These produce spores for the following season's infections.

- *Control.* Prune and destroy affected leaves and twigs.

Insects and Related Pests

BORERS. A number of borers, including the lesser peach tree, *Synanthedon pictipes*; apple bark, *Thamnosphecia pyri*; roundheaded apple tree, *Saperda candida*; and the shot-hole, *Scolytus rugulosus,* occasionally attack serviceberry.

- *Control.* See under borers, Chapter 11.

LEAF MINER. *Nepticula amelanchierella.* Broad irregular mines, especially in the lower half of the leaf, result from infestation by the larvae.

- *Control.* See birch leaf miner, under Birch.

PEAR LEAF BLISTER MITE. See under Mountain-Ash.

PEAR SLUG SAWFLY. See pear slug, under Cherry.

WILLOW SCURFY SCALE. See under Willow.

CAMBIUM MINER. See under Holly.

WOOLLY APPLE AND WOOLLY ELM APHIDS. Serviceberry roots are injured by these pests. See aphids, under Elm.

MITES. Spider and eriophyid mites attack serviceberry. See Chapter 11.

SHADBLOW, SHADBUSH See Serviceberry.

SILK-OAK (*Grevillea*)

Silk-oak, a native of Australia, has become naturalized in southern Florida. It is fast growing, but its wood is brittle. This tree prefers well-drained soil.

Diseases

Few diseases have been recorded for this tree. These include dieback, said to be caused by a species of *Diplodia*; leaf spot by the alga

Cephaleuros virescens; root-knot by the nematode *Meloidogyne* sp.; and root rot by *Phymatotrichum omnivorum*.

- *Control.* Measures are rarely taken to control the diseases on this host.

Insects

Leaf-eating insects on silk-oak include the tobacco budworm, *Heliothis virescens*; and the omnivorous looper, *Sabulodes caberata*. Sucking insects include the citrophilus and grape mealybugs, at least nineteen species of scales, and two species of mites—avocado red and Grevillea.

- *Control.* See Chapter 11.

SILK-TREE (*Albizzia*)

Silk-tree (*Albizzia julibrissin*), also known as mimosa, is used as a street and lawn tree in the South and as an ornamental in protected places along the Atlantic Coast as far north as Boston. It is a fast-growing, relatively short-lived tree. Its most destructive disease is wilt, and its most destructive insect the webworm.

Diseases

WILT. This highly destructive disease is prevalent throughout the range of this tree.

- *Symptoms.* Wilting of the leaves, which soon die, is the most striking symptom. A brown ring of discolored sapwood, usually in the current annual ring of the roots, stem, and branches, is another positive symptom. Bleeding along the stems occurs occasionally. Large numbers of spores are formed in the lenticels on the trunk and branches of affected trees, at times even before actual wilting of the leaves occurs. Such spores may account for widespread outbreaks of the disease. Trees usually die within a year or so of infection.

- *Cause.* The fungus *Fusarium oxysporum* f. *perniciosum* causes wilt. The fungus lives in soil and spreads in this medium. It also can be carried over in seed collected from diseased trees. There are two races of this fungus. One will infect *Albizzia julibrissin* and not *A. procera*, a silk-tree relative. The other, from Puerto Rico,

will infect *A. procera* and not *A. julibrissin.* Other species react to this disease as follows: *A. pudica* and *A. thorelii*, resistant to both races; *A. lebbek*, resistant to just one race; and *A. lophantha* and *A. kalkora*, susceptible to both races.

- *Control.* Dead and dying trees should be cut down and destroyed to avoid the spread of the disease. Until recently the only control known was to use wilt-resistant trees such as cultivar 'Union'. The cultivars 'Tryon' and 'Charlotte', at first resistant to Fusarium wilt, are now considered susceptible, probably because new fungal races have appeared or because of co-infection with root knot nematodes. Root-knot nematodes have been shown to negate host resistance to Fusarium wilt in tomatoes, so a similar mechanism may be working in mimosa.

OTHER FUNGI. Several other fungi are reported on silk-tree. *Nectria cinnabarina* (imperfect, *Tubercularia vulgaris*) causes dieback and canker; *Thyronectria austro-americana*, while destructive to honey locust, is weakly pathogenic to silk-tree twigs. *Ganoderma lucidum*, *Armillaria mellea*, and *Clitocybe tabescens* are associated with root rots. The rust fungus, *Sphaerophragmium* sp., has been reported on *A. lebbeck* in Florida.

Insects

WEBWORM. *Homadaula anisocentra.* The larvae of this pest, also called mimosa webworm, appear on the foliage in late spring or early summer. The larvae are $3/5$ inch long when fully grown and are gray-brown, sometimes pinkish, with five narrow, white, lengthwise stripes. At first they feed as a group in a web, but later they spread throughout the tree, tying the leaves in conspicuous masses and skeletonizing them. When mature, they drop to the ground on silken threads and spin cocoons in various cracks and crevices. The moths, gray with a wingspread of about $1/2$ inch, appear in May or June to lay eggs on the silk-tree flowers and leaves. A new generation of this pest may reappear once or even twice more during the growing season. These insects are also highly destructive to honey locust.

• *Control.* See under caterpillars, Chapter 11.

SCALES. Lesser snow (see under Poinciana), Asiatic red, and several other scales infest silk-tree.

JAPANESE WEEVIL. See under Holly.

BAGWORM. See under Juniper.

ALBIZZIA PSYLLID. See under Acacia.

OTHER INSECTS. Blister beetle, thornbug, and citrus mealybug also infest this host.

NEMATODES. *Meloidogyne arenaria* and *Trichodorus primitivus.* The former, known as the peanut root-knot nematode, causes small knots or galls on silk-trees in the South. The latter, known as the stubby root nematode, infests roots of A. *julibrissin* and azalea.

• *Control.* See under Boxwood.

SILVERBELL (*Halesia*)

Also called Carolina silverbell, snowdrop tree, and wild olive, this tree is an attractive native of the southeastern United States. Silverbell is subject to few diseases and has no significant insect problems.

Diseases

LEAF SPOT. *Cercospora halesiae.* In rainy seasons, this fungus causes brown circular spots to form on the leaves.

• *Control.* See under leaf spots, Chapter 12.

WOOD DECAY. *Polyporus halesiae.* A decay of the wood has been reported.

• *Control.* See under wood decay, Chapter 12.

SMOKE-TREE (*Cotinus*)

The smoke-tree, a close relative of ordinary sumac, is drought-tolerant and easy to grow. It is occasionally used in ornamental plantings. When grown in soils where nitrates tend to become deficient, its leaves show red spots or blotches. The application of a commercial fertilizer high in nitrogen usually helps overcome this trouble.

Diseases

WILT. *Verticillium albo-atrum.* This is the only serious disease of smoke-tree.

• *Control.* See Chapter 12.

RUST. Two rusts have been reported on this tree: *Puccinia andropogonis* f. *onobrychidis* and *Pileolaria cotini-coggyriae.* The latter produces conspicuous spots with hypertrophied centers surrounded by dead tissue.

• *Control.* These diseases are not serious enough to warrant control measures.

LEAF SPOTS. Four species of fungi are known to cause leaf spots on smoke-tree: *Cercospora rhodina, Pezizella oenotherae, Septoria rhodina,* and *Gloeosporium* sp.

• *Control.* See under leaf spots, Chapter 12.

POWDERY MILDEW. *Erisyphe cichoracearum.* This fungus also attacks smoke-tree.

Insects

OBLIQUE-BANDED LEAF ROLLER. *Choristoneura rosaceana.* The leaves may be mined and rolled in June by the pale yellow larvae. The adult moth is reddish brown with a wingspread of 1 inch. The front wings are crossed by three distinct bands of dark brown.

Other trees subject to this pest are apple, apricot, ash, birch, cherry, dogwood, hawthorn, horsechestnut, linden, maple, oak, peach, pear, plum, and poplar.

• *Control.* Spray early in June with Sevin.

SAN JOSE SCALE. *Aspidiotus perniciosus.* This scale occasionally infests smoke-tree.

• *Control.* See under scales, Chapter 11.

SORREL-TREE (*Oxydendrum*)

Also called sourwood, this tree is a native of the eastern United States. It grows to medium height and prefers acid soils. Its flowers are attractive to bees.

Diseases

TWIG BLIGHT. *Sphaerulina polyspora.* This fungus occasionally causes blighting of leaves at the tips of the branches. Trees injured by fire or in poor vigor appear to be most subject to this disease.

• *Control.* Pruning infected twigs is usually sufficient to keep this disease under control.

Fertilize and water when necessary to keep trees in good vigor.

LEAF SPOTS. *Cercospora oxydendri* and *Mycosphaerella caroliniana*. These two fungi occasionally cause spots on the leaves of sorreltree. The latter produces reddish or purple blotches with dry brown centers in midsummer.

• *Control.* See under leaf spots, Chapter 12.

Insects

No insects of any consequence attack this tree.

SOURGUM. See Tupelo.

SOURWOOD. See Sorrel-Tree.

SPRUCE (Picea)

Native spruces grow in the northern United States and in Canada. Besides being valuable forest trees, the spruces are highly prized as ornamentals. They thrive in moist, sandy loam soil and can tolerate more shade than most other conifers. The varieties most commonly used in ornamental plantings are Norway spruce and selected cultivars of Colorado blue spruce. They are subject to a number of destructive fungal diseases and insect pests.

Diseases

CYTOSPORA CANKER. The most prevalent disease of Norway spruce and Colorado blue spruce, especially those planted outside their natural range, is known as Cytospora canker. Koster's blue spruce and Douglas-fir are also susceptible, but to a lesser extent. Although the

Fig. III-110. Blue spruce affected by *Cytospora* canker disease. Note the dead leafless branches, which contrast with unaffected branches.

disease occurs on young trees, those over 15 years old appear to be most susceptible.

• *Symptoms.* The most striking symptom is the browning and death of the branches, usually starting with those nearest the ground and slowly progressing upward. Occasionally, branches high in the tree are attacked even though the lower ones are healthy.

The needles may drop immediately from infected branches or may persist for nearly a year, eventually leaving dry brittle twigs that contrast sharply with unaffected branches (Fig. III-110). White patches of pitch or resin may appear along the bark of dead or dying branches (Fig. III-111). Cankers occur in the vicinity of these exudations. The cankers are not readily discernible but can be found by cutting back the bark in the area that separates diseased from healthy tissues. In the cankered area, tiny black pinpoint fruiting bodies of the causal fungus are a positive sign of the disease. Trunk cankers may cause complete or partial girdling of some spruces.

• *Cause.* The fungus *Leucostoma kunzei* causes this disease. The fruiting structures protrude slightly during wet weather but are concealed beneath the bark scales during dry spells (Fig. III-112). In wet weather they release long,

Fig. III-111. Close-up of *Leucostoma*-infected branch showing white, resinous exudation.

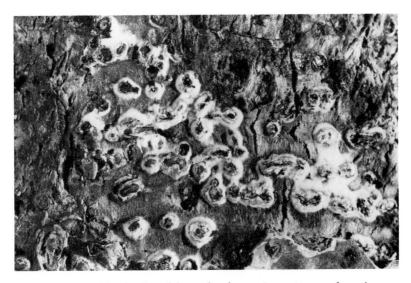

Fig. III-112. Fruiting bodies of the canker fungus *Leucocytospora kunzei.*

curled, yellow tendrils, which contain millions of spores. Wind and rain splash the spores to bark wounds on branches where infection takes place. When the infection girdles the branch the distal portions die. The fungi *Valsa* (imperfect, *Cytospora*) *abietis*, *V. friesii*, and *V. pini* also attack spruce, but are not as important as *Leucostoma kunzei.*

• *Control.* Infected branches cannot be saved. They should be cut off a few inches below the dead or infected parts, or at the point of attachment to the main stem. Because of the danger of spreading spores to uninfected branches, pruning should never be undertaken while the branches are wet. Available evidence indicates that the fungi enter the branches only through wounds, so bark injuries by lawn mowers and other tools should be avoided. As the disease appears to be most prevalent on trees in poor vigor, trees should be mulched regularly, fertilized at least every few years, and watered during dry spells, to increase or maintain their vigor.

SIROCOCCUS SHOOT BLIGHT. *Sirococcus conigenus.* This fungus causes cankers and tip dieback of the current season's growth. Foliage of affected twigs turns yellow, then reddish brown, and then drops. Small black fruiting bodies form on the leafless dead shoots.

• *Control.* Prune and destroy affected branches.

RHIZOSPHAERA NEEDLE CAST. *Rhizosphaera kalkhoffi.* Diseased needles begin yellowing in late summer or fall, turn brown the following spring, and drop the following summer. Trees under poor growing conditions or growing in a moist, humid environment are more prone to disease. Blue spruce is susceptible to this disease, Norway spruce is more resistant.

NEEDLE CASTS. *Lophodermium piceae, Isthmiella crepidiformis,* and *Lirula macrospora.* The lower branches of Colorado blue, Engelmann, red, black, and Sitka spruces may be defoliated by one of these fungi. The needles are spotted and turn yellow before they drop.

• *Control.* See under leaf spots, Chapter 12.

RUSTS. *Chrysomyxa ledi f. cassandrae, C. empetri, C. ledicola, C. chiogenis, C. arctostaphyli, C. piperiana, C. roanenis,* and *C. weirii.*

Several species of rust fungi occur on spruce needles. Each requires an alternate host to complete its life cycle. On the spruce, the fungi appear as whitish blisters on the lower leaf surface. Affected needles turn yellow and may drop prematurely. The names of the alternate hosts for each of the fungi listed can be obtained from state or federal plant pathologists.

• *Control.* Removal of the alternate hosts in the vicinity of valuable spruces is suggested. Protective sprays containing sulfur or Ferbam can be used but are rarely justified.

WITCHES' BROOM. Eastern dwarf mistletoe, *Arceuthobium pusillium,* causes compact masses of branches to form on black spruce.

WOOD DECAY. Spruces are subject to a number of wood decays caused by various fungi. Among those most commonly associated with various types of decay are *Haematostereum sanguinicola, Phaeolus schweinitzii, Laetiporus sulphureus, Inonotus tomentosus, Fomitopsis pinicola,* and *Phellinus pini.* Root decay is caused by *Armillaria mellea.*

• *Control.* See under wood decay, Chapter 12.

Insects and Mites

SPRUCE GALL ADELGID. *Adelges abietis.* Elongated, many-celled, cone-shaped galls, less than 1 inch long (Fig. III-113) result from feeding and irritation by this pest. Trees are weakened and distorted when large numbers of these galls are formed. Norway spruce is most seriously infested; white, black, and red spruces are less so. In spring the adult wingless females deposit eggs near where the galls later develop. Immature females hibernate in the bud scales. Spruce gall adelgid is sometimes called spruce gall aphid.

• *Control.* See under aphids, Chapter 11.

COOLEY SPRUCE GALL ADELGID. *Adelges cooleyi.* This insect is sometimes referred to as Cooley spruce gall aphid. Galls ranging from $1/2$ to $2^1/2$ inches in length on the terminal shoots of blue, Englemann, and Sitka spruce are produced by feeding and irritation by an adelgid closely related to one that infests Norway spruce. The adult female overwinters on the bark near the twig terminals and deposits eggs in that vicinity

Fig. III-113. Galls of spruce gall adelgid, *Adelges abietis,* on Norway spruce.

in spring. On Douglas-fir the adelgid insect secretes a white, waxy substance and causes needle distortion. .

• *Control.* Because Douglas-fir is an alternate host for this pest and is also injured by it, avoid planting that species near spruces. See under aphids, Chapter 11.

APHIDS. The green spruce aphid, *Elatobium abietinum,* and the pine leaf adelgid, *Pineus pinifoliae,* also infest spruce leaves. The balsam twig aphid, discussed under Fir, sometimes attacks this host.

• *Control.* See under aphids, Chapter 11.

SPRUCE BUD SCALE. *Physokermes piceae.* Globular red scales, about ¹/₈ inch in diameter, may occasionally infest the twigs of Norway spruce. The young crawl about in early July, and the winter is passed in the partly grown adult stage. Hemlock scales (see hemlock Fiorinia scale and hemlock scale, under Hemlock) and

pine needle scale (see under Pine) also attack spruce, especially trees in poor vigor.

• *Control.* See under scales, Chapter 11.

SPRUCE BUDWORM. *Choristoneura fumiferana.* One of the most destructive pests of forest and ornamental evergreens, this pest attacks spruce and balsam firs in forest plantations and also infests ornamental spruce, balsam fir, Douglas-fir, pine, larch, and hemlock. The opening buds and needles are chewed by the caterpillar, which is dark reddish brown with a yellow stripe along the side. The adult female moths are dull gray marked with brown bands and spots. They emerge in late June and early July.

WESTERN SPRUCE BUDWORM. *C. occidentalis.* This pest defoliates various conifers in western North America.

JACK PINE BUDWORM. *C. pinus.* This causes considerable damage to jack pine and to a lesser extent to red and white pine in the Great Lake states.

• *Control.* See under caterpillars, Chapter 11.

SPRUCE EPIZEUXIS. *Endothemia aemula.* Ornamental spruces in the Northeast may be attacked in late summer by the small brown larvae with black tubercles. They web needles together and fill them with excrement.

• *Control.* See under caterpillars, Chapter 11.

SPRUCE NEEDLE MINER. *Taniva albolineana.* The light greenish larvae web the leaves together and mine the inner tissues. They enter through small holes at the bases of the leaves, and as they feed inside, the leaves turn brown. The adult is a small grayish moth.

Other species of spruce needle miners, *Recurvaria piceaella* and *Epinotia nanana,* also mine the leaves of spruce, killing groups of needles along the stem.

• *Control.* See under leaf miners, Chapter 11.

SAWFLIES. *Pikonema alaskensis* and *Diprion hercyniae.* The former, known as the yellow-headed sawfly, sometimes defoliates trees completely. Ordinarily, however, infestation results merely in a ragged appearance with damaged needles that are thin, brown, and shriveled by the larval feeding. The latter, the European spruce sawfly, feeds on the old foliage, and consequently the trees are not killed, although the growth may be stunted. The larvae overwinter in cocoons on the ground, where they are subject to attack by mice.

• *Control.* See under sawflies, Chapter 11.

SPRUCE SPIDER MITE. *Oligonychus ununguis.* Yellow, sickly needles, many of which are covered with a fine silken webbing, indicate a severe infestation of the spider mite, a tiny pest only $1/64$ inch long (Fig. III-114). The young are pale green and the adult female is greenish black. Winter is passed in the egg stage on the twigs and needles. Although spider mites such as the two-spotted spider mite cause greatest damage in hot, dry seasons, spruce spider mites are most active in fall and spring, during the cool part of the season.

• *Control.* See under spider mites, Chapter 11. To determine if there is a need for treatment, spider mites must be monitored. Detect spruce spider mites by striking an affected branch sharply with a stick and allowing the mites and other debris to fall on a white piece of paper or cardboard. After a few seconds, tilt the paper so that the debris and needles fall to the ground. Spider mites left on the white surface should be visible, about the size of a period on this page, and will move slowly. Predatory mites and thrips may be present, but they move quickly. When a spruce spider mite is squashed with a finger, an olive-green streak results. Other mites leave a yellow or orange streak, or no streak. If one strike of the branch yields two dozen or more mites on a sheet of paper, treatments should be considered. If spruce spider mites are not present, treatments are not needed. If they are present in low numbers, treatments might be put off until further mite detection a few weeks later indicates a need for treatment. Blue spruce should not be sprayed with dormant oils.

SPRUCE RUST MITE. *Nalepella halourga.* This eriophyid mite, like its close relative on hemlock, is active during the cool part of the season and causes browning of interior needles. This mite, a pest only of spruce, overwinters as tiny red eggs deposited in groups at the base of needles of the underside of the tree branches.

• *Control.* See under eriophyid mites, Chapter 11.

WHITE PINE WEEVIL. See under Pine. The work of the weevil is easily distinguished from

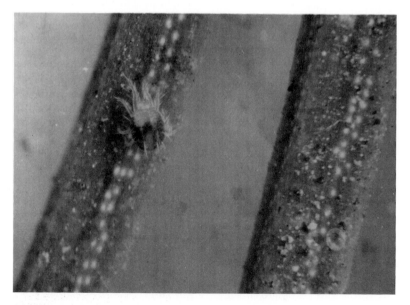

Fig. III-114. Spruce spider mite, *Oligonychus ununguis.*

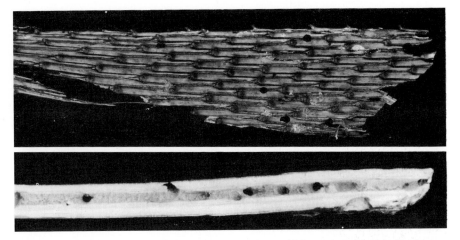

Fig. III-115. Work of the white pine weevil, *Pissodes strobi*, in a spruce branch. On the top may be seen the small holes through which the adults emerge; on the bottom is a twig split to show their work inside.

that of the pine shoot moth by the greater length of the shoots killed and by the holes in the bark from which the beetles have emerged (Fig. III-115).

- *Control.* See under Pines.

BAGWORM. See under Juniper.

PITCH MOTH. See under Pine.

BLACK VINE WEEVIL. See under Yew.

STEWARTIA (*Stewartia*)

These natives of Asia, which grow in the South, have attractive bark, flowers, and foliage. They are relatively problem free.

STRAWBERRY-TREE (*Arbutus*)

This low-growing evergreen requires frequent pruning. Another *Arbutus* species, madrone, is a slow-growing, medium-sized tree of the Pacific coast.

Diseases

LEAF SPOTS. *Septoria unedonis* and *Elsinoë mattirolianum*. The former fungus produces small brown spots on leaves of *Arbutus unedo* in the Pacific Northwest, and the latter a spot anthracnose in California. In addition, madrone is infected by *Ascochyta hanseni, Cryptostictis*

arbuti, Didymosporium arbuticola, Exobasidium vacinii, Mycosphaerella arbuticola, Phyllosticta fimbriata, Pucciniastrum sparsum, and *Rhytisma arbuti.*

- *Control.* See under leaf spots, Chapter 12.

CROWN GALL. *Agrobacterium tumefaciens.* This bacterial disease occurs occasionally on strawberry-tree in California and Connecticut.

- *Control.* See Chapter 12.

CANKER. Crown thinning is caused by *Phytophthora cactorum* trunk infections. See Chapter 12.

Insects

CALIFORNIA TENT CATERPILLAR. *Malacosoma californicum.* This species occasionally feeds on the leaves of strawberry-tree in California. It makes large tents in the trees like those made by the eastern tent caterpillar.

- *Control.* Clip out nests and destroy them. See under caterpillars, Chapter 11.

SCALES. *Saissetia oleae, Aspidiotus camelliae,* and *Coccus hesperidum.*

These three species of scales—black, greedy, and brown soft—are known to attack strawberry-tree in the West.

- *Control.* See under scales, Chapter 11.

MADRONE SHIELD BEARER. *Coptodisca arbutiella.* The larvae of this tiny moth mine and cut circular discs from infested leaves, leaving behind holes.

• *Control.* Pick off and discard infested leaves.

SUMAC (*Rhus*)

African sumac, which has a weeping growth form, is adapted to hot, dry climates with relatively mild winter temperatures. It is susceptible to few pests.

SWEETGUM (*Liquidambar*)

Sweetgum is native to the eastern and southern United States and to Mexico. It grows best in rich clay or loam soils but will tolerate a wide variety of sites. Winter hardiness has been a problem for some cultivars growing in the Ohio River valley region.

Diseases

BLEEDING NECROSIS. P. P. Pirone studied the bleeding necrosis disease of sweetgum in New Jersey and on Staten Island, New York, as far back as 1941.

• *Symptoms.* The most striking symptom, readily visible at some distance, is the profuse bleeding of the bark. This usually occurs at the soil line or a few feet above, but bleeding may occasionally be observed as high as 20 feet up on the main trunk and lower branches. The exudation looks like heavy motor oil poured over the bark.

The condition of the inner bark and sapwood beneath the oozing area presents a more positive symptom of the disease. The bark is dark reddish brown with an occasional pocket containing a white crystalline solid. The outer layer of sapwood may be brown or olive green in color. Infected trees may exhibit undernourished foliage and dying-back of terminal branches by the time profuse bleeding is visible on the trunk.

• *Cause.* The fungus *Botryosphaeria dothidea* causes bleeding necrosis, but when first reported the fungus was classified as the asexual stage, *Dothiorella.* There is some evidence that raising

Fig. III-116. Bacterial leaf scorch of sweetgum.

the grade around the tree makes it more subject to this fungus. Drought may also predispose trees to bleeding necrosis.

• *Control.* No control measures are known. Removal and destruction of diseased trees is suggested.

LEAF SPOTS. *Actinopelte dryina, Cercospora liquidambaris, C. tuberculans, Gloeosporium nervisequam, Exosporium liquidambaris, Leptothyriella liquidambaris,* and *Septoria liquidambaris.* These seven fungi may occasionally spot the leaves of this host.

• *Control.* Leaf spots are rarely serious enough to warrant control measures.

OTHER DISEASES. Sweetgum is susceptible to bacterial leaf scorch (Fig. III-116). See Chapter 12. Other diseases of sweetgum include leader dieback and sweetgum blight. The cause of leader dieback is still unknown, but it has been associated with drought and secondary fungi. Sweetgum blight has caused the death of many trees in the East; the cause is still not known.

• *Control.* Control measures are unavailable.

Abiotic Diseases

COLD INJURY. Extremely cold winter temperatures can kill phloem and cambium tissues of sweetgum, resulting in dieback and decline. See Chapter 10.

AIR POLLUTION. Sweetgum is sensitive to ozone. See Chapter 10.

Insects

SWEETGUM WEBWORM. *Salebria afflictella.* The leaves are tied and matted together by small larvae.

FALL WEBWORM. See under Ash.

FOREST TENT CATERPILLAR. See under Maple.

RED-HUMPED CATERPILLAR. See under Poplar.

OTHER CATERPILLARS. The leaves of sweetgum are occasionally chewed by larvae of several species of moths including the luna, *Actias luna*; polyphemus, *Antheraea polyphemus*; and promethea, *Callosamia promethea.*

• *Control.* See under caterpillars, Chapter 11.

BAGWORM. See under Juniper.

ASIATIC OAK WEEVIL. See under Oak.

AMERICAN PLUM BORER. See under Sycamore.

SWEETGUM SCALE. *Aspidiotus liquidambaris.* This scale is occasionally found on sweetgums in the eastern United States. Calico scale also attacks sweetgum (Fig. III-117).

Fig. III-117. Calico scale on sweetgum.

• *Control.* See under scales, Chapter 12.

COTTONY-CUSHION SCALE. See under Acacia.

WALNUT SCALE. See under Walnut.

SYCAMORE (*Platanus occidentalis*) and LONDON PLANETREE (*P. x acerifolia* or *P. occidentalis x P. orientalis*)

Also known as buttonwood and American planetree, sycamore is found from Maine to Minnesota, Florida to Texas. It is occasionally planted as a street tree but is not as desirable as the London planetree. Widely used in Europe, the London planetree is also commonly planted in cities in the northeastern United States because it is more tolerant to air pollutants and other unfavorable growing conditions than most trees. However, it does not do well where extremely low winter temperatures occur.

Diseases

BLIGHT, also known as anthracnose and scorch, is the most serious disease of sycamore. Two western species, *Platanus racemosa* and *P. wrighti*, are also susceptible. The disease is locally severe every year in some sections of the United States (Fig. III-118). The London planetree, though much more resistant, shows some injury during epidemic years. The destructiveness of this disease is often underestimated. While a single attack seldom results in serious harm, repeated annual outbreaks will eventually so weaken the tree that it becomes susceptible to borer attack and winter injury.

• *Symptoms.* The first symptom, often confused with late frost injury, is the sudden browning and death of single leaves or clusters of leaves as they are expanding in spring. Later, brown, dead areas along and between the veins appear in other leaves. The dead tissues assume a triangular form as a result of the death of the veinal tissues. The leaf falls prematurely when several lesions develop or when the petiole is infected. Many trees are completely defoliated and remain bare until late summer, when a new crop of

Fig. III-118. Left: This sycamore tree lost nearly all its leaves in June as a result of anthracnose. Right: The same tree with a new set of leaves by early October.

Fig. III-119. Twig canker of sycamore anthracnose. Note fruiting bodies protruding from the bark.

leaves is formed. Small twigs are also attacked and killed. Cankers appear on leaf spurs and twigs as sunken areas with slightly raised margins (Fig. III-119). When these completely girdle the infected parts, the distal portions, including the leaves, are killed. This is known as the shoot blight stage. Moderately infected trees show clusters of dead leaves scattered over the entire tree; these contrast sharply with the unaffected foliage.

• *Cause.* Blight is caused by the fungus *Apiognomonia veneta.* The fungus has several distinct types of spores (Fig. III-120). It may overwinter in the vegetative stage (mycelium) in fall-

en infected leaves and in twig cankers. In the South it passes the winter in the spore stage on dormant buds. Initial infections of the young leaves in spring may originate from spores that developed on fallen leaves on the ground, on infected twigs, and on the dormant buds. Additional crops of spores for late infections are formed on the lower surfaces of the newly infected leaves in early spring (Fig. III-121).

The prevalence and severity of attack are governed by weather conditions; frequent rains and cool temperatures favor rapid spread. The severity of the shoot blight stage is largely governed by the average temperature

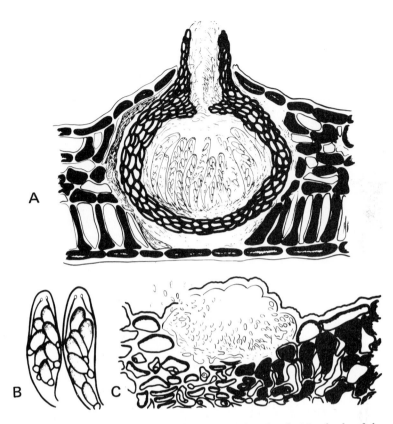

Fig. III-120. A, Cross section of sycamore leaf showing fruiting body of the fungus *Apiognomonia veneta*. B, Close-up of ascospores. C, The summer spore *Discula platani* stage of the fungus.

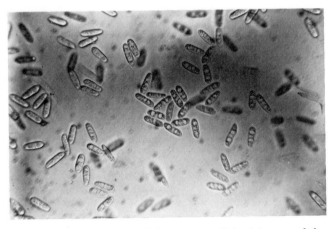

Fig. III-121. The spores of the summer *(Discula)* stage of the anthracnose fungus as they appear through a microscope.

Fig. III-122. London planetree affected by the cankerstain disease.

Fig. III-123. Trunk of London planetree infected with the cankerstain fungus *Ceratocystis fimbriata* f. *platani*.

during the 2-week period following the emergence of the first leaves. If the average temperature during this period is below 55° F (13° C), the injury will be severe. If it is between 55 and 60° F (16° C), it will be less severe, and if over 60° F, little to no injury will occur.

• *Control.* Control measures are rarely attempted on sycamores growing in open fields or in woodlands. They are justified, however, on particularly valuable specimens and on those used for shade and ornament. See under anthracnose, Chapter 12.

CANKERSTAIN. Thousands of London planetrees have died from this disease in the United States since the early 1930s.

• *Symptoms.* Conspicuous reduction in both amount and size of foliage is the most obvious symptom (Fig. III-122). This is followed by death of the tree within one to two years.

Sunken cankers appear on trunks, large limbs, and occasionally on small limbs (Fig. III-123). The cankers frequently have longitudinal cracks and roughened bark. Bluish black or brown discolorations appear on freshly exposed bark over the cankers. The callus tissue formed at the margin of a canker usually dies. Dark-colored streaks extend from the cankers inward through the wood to the central pith. Radiating web-shaped brown streaks frequently extend from the pith outward through the sound wood in other regions (Fig. III-124). The rays are mostly dark-colored. As the disease progresses, there is a gradual thinning of the leaves, which are smaller than normal. The disease is spread largely by pruning and is confined mainly to landscape trees. The sycamore, *P. occidentalis,* is less susceptible to this disease.

• *Cause.* Cankerstain is caused by the fungus *Ceratocystis fimbriata* f. sp. *platani.* Several closely related species cause blue stain in lumber. There is some evidence that as many as five species of nitidulid beetles are capable of disseminating the causal fungus.

The fungus enters the trunk or branches through bark injuries and grows toward the center. From this point it grows outward radially along various lines, forming cankers wherever it reaches the bark. Large numbers of spores are formed beneath recently killed bark or on freshly exposed wood surfaces.

Many cankers center on injuries made by saw cuts, pole pruner cuts, rope burns, accidental saw scratches, and injuries by climbers' boots. Observations indicate that cankers may also start from pruning cuts that have been painted with a wound dressing containing fungus-infested sawdust. The fungus can apparently survive in some of the commonly used asphalt wound dressings. The disease has been experimentally transmitted from infected to healthy trees by means of contaminated pruning saws. It is highly probable that the use of contaminated tools and tree paints constitutes the chief means of spread on streetside and shade trees. Injuries to the roots and trunk during curb and sidewalk

Fig. III-124. Cross section of planetree infected by *Ceratocystis fimbriata* f. *platani*. The discoloration shows the spread of the fungus through the wood.

installation and those made by vandalism are also important foci for fungal invasion. Natural bark fissures and frost cracks may also constitute entrance points.

• *Control.* Diseased trees should be removed and destroyed as soon as the diagnosis is confirmed. All injuries to sound trees should be avoided. Saws and other implements used in pruning planetrees should be thoroughly disinfested by washing in denatured alcohol, Lysol, or some other strong disinfectant after use on each tree in the more heavily infested zones. Because of the possibility of spreading the fungus by means of infested wound dressings, it is best not to apply a dressing on fresh wounds.

Plant pathologists once believed that the disease could not be spread by way of contaminated saws if pruning was done in winter, from December 1 to February 15. It is now known that this is not the case. Hence all pruning tools should be sterilized after use on each tree regardless of the season.

BOTRYOSPHAERIA CANKER. *Botryosphaeria dothidea.* This fungus, like *Verticillium alboatrum* and *Armillaria mellea*, is far more widespread than most professional arborists and nurserymen realize. It had long been known to cause cankers and dieback of redbud, but P. P. Pirone first showed that its asexual stage, *Dothiorella*, caused a highly destructive disease of London planetrees. Among other trees known to be susceptible to this fungus are apple, avocado, beech, flowering cherry, flowering dogwood, Japanese persimmon, hickory, horsechestnut, pecan, poplar, quince, sourgum, sweetgum, sycamore maple, and willow.

Leaf fires are no longer allowed in most municipalities, but when leaves are burned beneath streetside trees, they may become more susceptible to infection by the conidial stage, *Dothiorella gregaria*, and also to infections by the cankerstain fungus discussed earlier.

• *Control.* No control is possible once the

fungus has invaded the main trunk. When infections are limited to the branches, pruning well below the cankered area may remove all the infected material. Wounds and damage to the bark by fires should be avoided. The trees should be kept in good vigor by fertilizing, watering during dry spells, and spraying to control leaf-chewing and leaf-sucking insects.

POWDERY MILDEW. *Microsphaera platani* and *Phyllactinia guttata*. Powdery mildew normally occurs late in the growing season and does little damage to sycamores. London planetrees are especially susceptible to attacks of powdery mildew. Leaves and young twigs are covered with a whitish mold to such an extent that much of the foliage is destroyed. The seriousness of the attack appears to be weather dependent; during some seasons no mildew appears while in others the disease is serious, especially on young trees of a size for transplanting.

• *Control.* It is practical to spray trees in nursery plantings, but landscape specimens are rarely sprayed to control mildew. See under Powdery mildew, Chapter 12.

BACTERIAL LEAF SCORCH. *Xylella fastidiosa*. This disease causes leaf scorch, dieback, and decline of sycamores in the South and East. See Chapter 12.

OTHER DISEASES. Several diseases of minor importance, but potentially serious locally, are leaf spots caused by the fungi *Mycosphaerella platanifolia*, *M. stigmina-platani*, *Phloeospora multimaculans*, *Phyllosticta platani*, and *Septoria platanifolia*.

• *Control.* See under leaf spots, Chapter 12.

In the South, bark and twig cankers associated with the fungi *Botryosphaeria rhodina*, *Diaporthe scabra*, and *Hypoxylon tinctor* have been reported.

• *Control.* Control measures are rarely adopted.

Abiotic Disease

AIR POLLUTION. Sycamore is very sensitive to ozone. See Chapter 10.

DOG CANKER. Urine causes injury to trees planted along streets where dogs are commonly walked. The damage is confined to the lower 2 feet of the trunk. Trees up to 6 inches in trunk diameter may be killed in this way.

• *Control.* Placing a collar around trees visited by male dogs to deflect urine will help to eliminate cankers, but the dogs' urine could still seep into the soil and root area to cause severe damage to the roots and premature death of the tree.

Insects and Other Pests

APHIDS. *Longistigma caryae*. The giant bark aphid, up to $1/4$ inch in length, frequently attacks the twigs of sycamore and may gather in clusters on the undersides of the limbs. These insects are also called planetree aphids. They are the largest species of aphids known and exude great quantities of honeydew.

Another species, *Drepanosiphum platanoides*, infests maples in addition to sycamores, throughout the country.

• *Control.* See under aphids, Chapter 11.

SYCAMORE PLANT BUG. *Plagiognathus albatus*. This bug in its adult stage is $1/8$ inch long, tan or brown in color, with dark eyes and brown spots on the wings. The young bugs are yellow-green with conspicuous reddish brown eyes. Yellowish or reddish spots develop where these insects pierce the leaf. As the leaves grow, the injured areas drop out leaving holes. By midseason the leaves may appear tattered and yellowed.

• *Control.* See under plant bugs, Chapter 11.

LACE BUG. *Corythucha ciliata*. Parts of the leaves turn pale yellow and many drop prematurely as a result of the sucking by the lace bug. The adult, $1/8$ inch long, white and with lacelike wings (Fig. III-125), deposits eggs on the lower leaf surface. There are several generations per year. The adult passes the winter in bark crevices and other sheltered places in the vicinity of the trees. London planetree is much more susceptible to this pest than is sycamore.

• *Control.* See under lace bugs, Chapter 11.

SYCAMORE TUSSOCK MOTH. *Halisodota harrisii*. The caterpillar stage of this moth is yellow and has white to yellow hairs on its body. It occa-

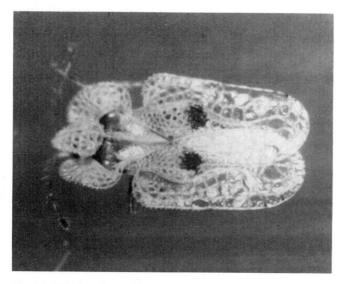

Fig. III-125. Lace bug adult.

sionally becomes abundant on sycamores in the northeastern United States. White-marked tussock moth (see under Elm) also attacks sycamore.

• *Control.* See under caterpillars, Chapter 11.

IO MOTH. *Automeris io.* The larva of this moth feeds on sycamore. The larva is covered with spines, which sting when touched.

ASIATIC OAK WEEVIL. See under Oak.

JAPANESE BEETLE. See under Linden.

SYCAMORE BORER. *Synanthadon resplendens.* This clearwinged borer also attacks oaks in southwestern states. Adults resemble yellowjacket wasps. Sawdustlike frass and roughened bark appear at the base of infested trees.

• *Control.* Apply parasitic nematodes to gallery openings, or use persistent insecticides on the trunks of valuable trees. See under borers, Chapter 11.

AMERICAN PLUM BORER. *Euzophera semifuneralis.* London planetrees, particularly streetside specimens with damaged bark, are quite susceptible to this pest. Damage to the inner bark and cambial regions by the larval stage, a dusky white, pinkish, or dull brownish green caterpillar, may be so extensive that the tree dies prematurely.

• *Control.* See under borers, Chapter 11.

FLATHEADED BORER. See under Maple.

SYCAMORE SCALE. *Stomacoccus platani.* This pest causes early spring leaf distortion and chlorotic spots where scales feed on leaf undersides.

Several other species of scale insects including black, cottony maple, grape, oystershell, and terrapin occasionally infest the sycamore and other species of *Platanus.*

• *Control.* See under scales, Chapter 11.

OTHER INSECTS. *Platanus* species are also infested by bagworms, borers, and whiteflies.

OTHER PESTS. Mites, root-knot nematode (*Meloidogyne sp.*), and several other species of parasitic nematodes may attack sycamore.

• *Control.* Control measures are rarely needed.

TANOAK (*Lithocarpus*)

Tanoak grows in the western United States, where it is used occasionally as a landscape tree. It is especially susceptible to *Armillaria mellea,* which causes basal breakage of the tree through root and butt rot.

TREE-OF-HEAVEN (*Ailanthus*)

Tree-of-heaven, also called ailanthus, is not used as a streetside tree but thrives where no other tree will grow. It frequently sprouts in parks and yards of large cities. Only the seed-bearing kinds should be planted, as the male or pollen-bearing form has an offensive odor when it blooms. It prefers neutral to alkaline soil and moderate winters and is tolerant to most diseases.

Diseases

WILT. *Verticillium albo-atrum, V. dahliae.* This is the most destructive fungus parasite of this host.
- *Control.* See Chapter 12.

ARMILLARIA ROOT ROT. *Armillaria mellea.* This fungus also has been destroying ailanthus in the northeastern United States.
- *Control.* See Chapter 12.

LEAF SPOT. *Cercospora glandulosa, Phyllosticta ailanthi,* and *Gloeosporium ailanthi.* These leaf-spotting fungi occasionally attack ailanthus.
- *Control.* Control measures are rarely applied for leaf spots.

TWIG BLIGHT. *Fusarium lateritium.* This fungus also causes branch and twig cankers on Japanese pagoda-tree. The sexual stage of this fungus is *Gibberella baccata.*
- *Control.* Control measures are rarely necessary.

OTHER DISEASES. Other fungal diseases causing dieback or cankers of ailanthus include *Botryosphaeria dothidea, Cytospora ailanthi, Coniothyrium insitivum, Nectria coccinea, Botryosphaeria obtusa,* and *B. rhodina.*
- *Control.* Control measures are rarely needed.

Insects

CYNTHIA MOTH. *Samia cynthia.* The larvae of this moth can completely defoliate a tree-of-heaven in a few days. Mature larvae are 3^1/$_2$ inches long, light green, and covered with a glaucous bloom. The adult moth is beautifully colored and has a wingspread of 6 to 8 inches (Fig. III-126). A crescent-shaped white marking is present in the center of each of the grayish brown wings.

AILANTHUS WEBWORM. *Atteva punctella.* Olive-brown caterpillars with five white lines feed in webs on the leaves in August and September. Adult moths have bright orange forewings, with four cross-bands of yellow spots on a dark blue ground. This webworm usually does little harm to the tree.

OTHER CATERPILLARS. Among other caterpillars that attack this host, particularly in cities, are the fall webworm, *Hyphantria cunea,* and the white-marked tussock moth, *Hemerocampa leucostigma.* Another species, the pale tussock moth, *Halisidota tessellaris,* attacks a wide variety of deciduous trees.

Fig. III-126. Cynthia moth, *Samia cynthia.* The larval stage of this insect feeds voraciously on ailanthus (tree-of-heaven) foliage.

• *Control.* See under caterpillars, Chapter 11.

OTHER INSECTS. Oystershell scale and citrus whitefly occasionally infest ailanthus.

• *Control.* See Chapter 11.

TULIP POPLAR. See Tuliptree.

TULIPTREE (*Liriodendron*)

Native to the eastern United States, this tree is also called yellow poplar or tulip poplar. It is a large tree that needs space. Tuliptree does not tolerate drought or construction grade changes.

Fungal Diseases

CANKERS. *Botryosphaeria dothidea, Cephalosporium* sp., *Fusarium solani, Myxosporium* sp., *Nectria magnoliae,* and *Nectria* sp. These species of fungi cause cankers in tuliptree. The first mentioned is perhaps the most destructive.

• *Control.* Prune and cart away infected branches. Valuable trees should be mulched, fertilized, and watered to increase their vigor.

LEAF SPOTS. *Cylindrosporium cercosporioides, Gloeosporium liriodendri* (Fig. III-127), *Mycosphaerella liriodendri, M. tulipferae, Phyllosticta liriodendrica,* and *Ramularia liriodendri.* These species of fungi cause leaf spots of tuliptree.

• *Control.* Leaf spots rarely become sufficiently destructive to warrant control measures other than the gathering and destroying of infected leaves.

POWDERY MILDEWS. *Phyllactinia guttata* and *Erysiphe polygoni.*

Two species of fungi produce a white coating over the leaves of this host (Fig. III-128).

• *Control.* See under powdery mildew, Chapter 12.

ROOT AND STEM ROT. *Cylindrocladium scoparium.* This disease has been associated with decline of large tuliptrees in Georgia and North Carolina. This may also be the cause of a root and lower trunk rot that has led to the decline of landscape tulip poplar in other areas.

• *Control.* Control measures have not been developed.

SAPSTREAK. *Ceratocystis coerulescens.* Sapstreak occasionally occurs on tuliptree.

• *Control.* No effective control has been developed.

WILT. *Verticillium albo-atrum.* This disease can be very damaging to landscape tuliptrees. See Chapter 12.

Abiotic Diseases

LEAF YELLOWING. In midsummer during dry, hot, periods many tuliptree leaves turn yellow and drop prematurely. The yellowing is due to climatic conditions and not to any destructive organism. The leaves of recently transplanted trees exhibit yellowing more frequently than those of well-established trees. Very often, small,

Fig. III-127. Anthracnose of tuliptree caused by the fungus *Gloeosporium liriodendri.*

Fig. III-128. Sooty mold (left) and powdery mildew (right) on tuliptree.

Fig. III-129. Adult stage of tuliptree scale.

Fig. III-130. Crawling stage of tuliptree scale, which appears in late summer.

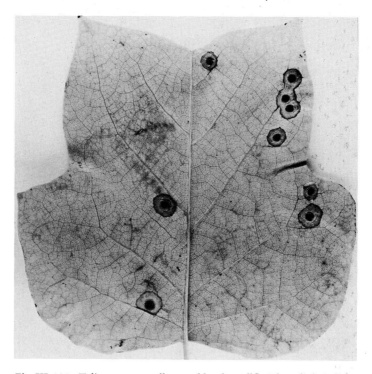

Fig. III-131. Tuliptree spot gall caused by the gallfly *Thecodiplosis lirio-dendri.*

angular, brownish specks appear between the leaf veins prior to yellowing and defoliation.

AIR POLLUTION. Tulip poplars are very sensitive to ozone. See Chapter 10.

Insects

TULIPTREE APHID. *Macrosiphum liriodendri.* This small green aphid secretes copious quantities of honeydew that can coat the leaves of the tree as well as those of other plants growing beneath the tree. The sooty mold fungus then covers the leaves, using the honeydew as a food source (Fig. III-128).

• *Control.* See under aphids, Chapter 11.

TULIPTREE SCALE. *Toumeyella liriodendri.* Trees may be killed by heavy infestations of oval turtle-shaped, often wrinkled brown scales, $^1/_3$–$^1/_4$ inch in diameter (Figs. III-129 and 130). The lower branches, which are usually the first to die, may be completely covered with scales.

Like the tuliptree aphid, this scale insect secretes much honeydew that drops onto leaves, which soon become covered with sooty black mold. Tuliptree scales overwinter as partly grown young, which grow rapidly until they mature in August. Females produce young scales in late August and then die and fall from the tree.

OTHER SCALE INSECTS. In addition to the very common and destructive tuliptree scale, this host is also subject to oystershell, walnut, and willow scale.

• *Control.* See under scales, Chapter 11.

TULIPTREE SPOT GALL. *Thecodiplosis liriodendri.* A gallfly produces purplish spots about $^1/_8$ inch in diameter on the leaves (Fig. III-131). These are frequently mistaken for fungal leaf spots.

• *Control.* Since gallfly damage is more unsightly than detrimental to the tree, sprays are rarely used for control.

SASSAFRAS WEEVIL. See under Sassafras.
ASIATIC OAK WEEVIL. See under Oak.
FLATHEADED BORER. See under Maple.

TUNG (*Aleurites*)

Native to China, tung is used as a source of oil and occasionally as a street tree. Tung normally does not tolerate freezing and requires deep, moist soils.

Fungal Diseases

FOLIAR DISEASES. Angular leaf spot (*Mycosphaerella aleuritidis*) and thread blight (*Ceratobasidium stevensii*) are the two most important. The latter kills foliage, and the leaves are left hanging by fungus threads.

ROOT ROT. *Clitocybe tabescens* causes root rot similar to *Armillaria* but does not produce rhizomorphs.

Mineral Deficiencies

Leaf chlorosis and malformation can be caused by mineral element deficiencies.

TUPELO (*Nyssa*)

Tupelo, also known as black gum, sourgum, and pepperidge tree, makes a fine lawn or park tree, but it is not commonly used because large specimens are difficult to transplant. Its fall coloration is outstanding. Tupelo is native to the eastern United States.

Diseases

CANKER. *Botryosphaeria dothidea, Fusarium solani, Nectria galligena, Strumella coryneoidea,* and *Septobasidium curtisii.* These five species of fungi produce stem cankers on tupelo.
- *Control.* Prune and destroy infected branches. Keep trees in good vigor by mulching, watering, and fertilizing when necessary.

LEAF SPOTS. *Mycosphaerella nyssaecola.* Irregular purplish blotches, which later enlarge to an inch or more in width, are commonly scattered over the upper surfaces of the leaves of young tupelo trees in the southeastern states. Minute black fruiting bodies of the fungus are visible in the infected areas.

Three other fungi, *Cercospora nyssae,*

Actinopelte dryina, and *Ceratobasidium stevensii,* also cause leaf spots and blights.
- *Control.* See under leaf spots, Chapter 12.

RUST. *Aplospora nyssae.* This rust fungus occasionally infects tupelos from Maine to Texas.
- *Control.* Control practices are not warranted.

WILT. Tupelo is susceptible to Verticillium wilt. See Chapter 12.

Insects

TUPELO LEAF MINER. *Antispila nyssaefoliella.* Leaves are first mined by the larval stage. When mature, the larvae cut oval sections out of the leaves and fall to the ground with the severed pieces. Adult moths emerge to lay eggs in May.
- *Control.* See under leaf miners, Chapter 11.

TUPELO SCALE. *Phenacaspis nyssae.* This scale is nearly triangular in shape, flat, and snow white. Cottony and San Jose scales may also be found on Tupelo.
- *Control.* See under scales, Chapter 11.

OTHER INSECTS. The azalea sphinx moth and San Jose scale have been reported on tupelo.
- *Control.* Malathion or Sevin sprays will control the larval stage of the moth and the crawling stage of the scale.

TURKISH FILBERT (*Corylus*)

This drought-tolerant tree grows well in almost any type of soil and has few pest problems.

WALNUT (*Juglans*)

Black walnut and butternut are rarely used as street trees as they require exacting soil conditions, are subject to wind and ice damage, and are untidy.

In ornamental plantings, the black walnut may produce detrimental effects on other types of plants growing nearby. The authors have observed extensive damage to native and hybrid rhododendrons, mountain laurel, and tomatoes grown in close proximity to black walnut roots. Similar damage has been reported on apple trees, alfalfa, potato, and a number of perennial ornamental plants grown near black walnut as well as butternut. In most instances, the removal of the nut trees resulted in the disappearance of the harmful effects once the roots had decomposed completely.

A number of important diseases and pests occur on the various species of walnut.

Diseases

CANKER. This disease is also common on apple, aspen, beech, birch, dogwood, butternut, and pignut hickory, as well as mountain, red, and sugar maples and red, white, and black oaks.

• *Symptoms.* The cankers are scattered, often numerous, rough, sunken, or flattened with a number of prominent ridges of callus wood arranged more or less concentrically on the trunks or branches. Canker size increases with age; on the trunk and larger branches a canker can be 4 feet long and $2^{1}/_{2}$ feet wide, with up to 24 concentric ridges. Complete girdling and death of the distal portions may occur within a few years on smaller branches, but may take 20 or more years to occur on large trunks. The concentric pattern of the canker is less evident on twigs and smaller branches.

In late fall and winter, the small, globose, red fruiting structures of the causal fungus on dead bark and wood are just visible to the unaided eye.

• *Cause.* Canker is caused by the fungus *Nectria galligena*. In late winter and in spring, the red fruiting bodies produce spores that are forcibly ejected into the air and are carried to the trees by wind and rain. Bark injuries made by insects, fungi, ice, low temperatures, wind, or rubbing plant parts, as well as dead branch stubs serve as entry points for the fungus. The spores germinate and produce mycelium, which penetrates the bark and wood. Penetration progresses slowly and kills about $^{1}/_{2}$ inch of tissue annually around the developing canker.

• *Control.* All badly diseased trees should be felled and the cankered tissues cut out and destroyed.

Valuable trees that are only mildly infected can be saved by surgical removal of the cankers. In addition, the trees should be fertilized, watered, and sprayed to control leaf pests and ensure rapid callusing.

Sites exposed to full sunshine should be used for new walnut plantings as the fungus does not appear to thrive under such conditions.

BRANCH WILT. *Hendersonula toruloidea (Exosporina fawceti).* This fungus kills the bark and wood of English walnut branches, causing their leaves to wither and die. It spreads down the branch, ultimately invading larger limbs. Infection occurs through wounds including those made by sunscald.

DIEBACK. Though black walnut occasionally dies back, this disease is most prevalent and destructive on butternut and the Japanese walnut (*J. ailantifolia*).

• *Symptoms.* On butternut, dieback occurs first on the smaller branches and progresses slowly to the main branches. The bark on affected branches changes from the normal greenish brown to reddish brown and finally to gray. Small black pustules soon cover the dead bark but disappear within a year or two, leaving small, irregular holes in the loose outer bark.

• *Cause.* The fungus *Sirococcus clavigignenti-juglandacearum* causes elliptical cankers with dark discoloration under the bark and eventual death of the tree. Spores are exuded in a tan, sticky mass from the black pycnidia on the bark to initiate new infections. The fungus *Melanconis juglandis* is often found as a secondary invader.

• *Control.* Remove and destroy severely affected trees. If the disease is still confined to the outer ends of the branches, prune back to sound wood to help check further spread. Fertilization, watering, sanitation, and insect and leaf-disease control measures should be adopted to help the tree regain its vigor.

BACTERIAL BLIGHT. English and Persian walnuts grown for ornament and for nut production in the eastern United States are susceptible to this destructive blight.

• *Symptoms.* Small, water-soaked spots that turn reddish brown appear on the young, unfolding leaves in spring. The spots are usually isolated, but they may coalesce, causing consid-

Fig. III-132. Walnut bunch in Japanese walnut tree, caused by a phytoplasma.

erable distortion of the leaves. Twigs may be completely girdled by black, sunken lesions, causing death of the parts above the girdle. Infection of husks usually results in premature dropping of fruit or in browning and decay of any fruit that remains attached.

• *Cause.* The bacterium *Xanthomonas campestris* pv. *juglandis* causes blight. The bacteria overwinter mainly in diseased buds and are spread by rain splash to various parts of the tree in spring. Pollen from blighted catkins may also spread disease.

• *Control.* Cut out and destroy badly infected shoots.

BARK CANKER. *Erwinia nigrifluens.* This bacterial disease was first found in California in 1955 on Persian walnut. Large, irregular, shallow, dark brown necrotic areas appear on the trunk and scaffold branches. The cankers enlarge in summer and are inactive in winter. A phloem canker of *Juglans regia* in California is attributed to the bacterium *Erwinia rubrifaciens.*

• *Control.* Control measures have not been developed for these diseases.

BROWN LEAF SPOT. *Gnomonia leptostyla.* The leaves of both butternut and walnut are affected by this walnut anthracnose fungus. It infects the leaflets early in summer, producing irregular dark brown or blackish spots and eventually causing premature defoliation. *Marssonina juglandis* is the conidial stage of brown leaf spot. *Ascochyta juglandis, Cercospora juglandis, Phleospora multimaculans, Marssonina californica,* and *Cylindrosporium juglandis* can also cause leaf spots. *Cristulariella pyramidalis* forms brown spots with white concentric rings (sometimes referred to as bull's-eye or zonate leaf spot). This fungus has been reported to cause complete defoliation of black walnut in the Midwest.

• *Control.* See under leaf spots, Chapter 12.

YELLOW LEAF BLOTCH. *Microstroma juglandis.* This disease is also called white mold or downy leaf spot. The fungus causes a yellow blotching on the upper sides of the leaves while coating their undersides with snow white fungus mycelium that contains enormous numbers of spores. *M. juglandis* also causes the witches' broom disease of shagbark hickory.

- *Control.* Same as for brown leaf spot.

TRUNK DECAY. The trunks of black walnut and butternut often have decayed heartwood. *Phellinus igniarius,* the so-called false tinder fungus, causes a white decay and forms hard, gray, hoof-shaped fruiting bodies. These may be up to 8 inches in width and are found along the trunk near the decayed area. A brown decay is caused by the fungus *Laetiporus sulphureus.* This fungus forms soft, fleshy, shelflike fruiting structures that are orange-red above and brilliant yellow below but become hard, brittle, and dirty-white as they age. Both fungi usually enter the trunk through bark injuries and dead branch stubs.

- *Control.* See under wood decay, Chapter 12.

WALNUT BUNCH DISEASE. Formerly called witches' broom, this disease is caused by a phytoplasma. Symptoms (Fig. III-132) include the appearance of brooms or sucker growths on main stems and branches, tufting of terminals, profuse development of branchlets from axillary buds, and leaf dwarfing. At times, death of the entire tree may occur. Symptoms can vary from mild to severe but are particularly pronounced on Japanese walnut; butternut, Persian walnut, and eastern black walnut are less susceptible.

- *Control.* Walnut bunch can be transmitted by grafting, hence propagating material should be pathogen-free.

OTHER DISEASES. A leaf spot and blasting of the nutlets of *Juglans* in California is attributed to the bacterium *Pseudomonas syringae.* Blackline, caused by the cherry leafroll virus, is a disorder characterized by formation of a narrow, dark brown, corky layer of nonconducting tissue at the graft union that results in girdling.

- *Control.* Control measures are rarely necessary.

Insects and Mites

WALNUT APHID. *Chromaphis juglandicola.* This pale yellow species is common on the undersides of leaves of English walnut on the West Coast.

The giant bark aphid (see aphids, under Sycamore) occasionally infests butternut and walnut.

- *Control.* See under aphids, Chapter 11.

SYCAMORE PLANT BUG. See under Sycamore.

WALNUT CATERPILLAR. *Datana integerrima.* Trees are defoliated by this caterpillar, which is 2 inches long and in its later stages is covered with long grayish white hairs. The wings of the adult female moth are dark buff with four brown lines and span 1$^{1}/_{2}$ inches. Eggs are deposited in masses on the lower sides of the leaves in July. Winter is passed in the pupal stage in the soil. Walnut caterpillar also feeds on hickory, oak, and honey locust.

OTHER CATERPILLARS. The following caterpillars also attack walnut: hickory horned devil, orange tortrix, omnivorous looper, red-humped, yellow-necked, hickory tussock moth, fall webworm, and fruit tree leaf roller.

- *Control.* See under caterpillars, Chapter 11.

BUTTERNUT WOOLLY WORM, *Eriocampa juglandis.* This larva feeds on hickory, walnut, and butternut. The white, cottony tufts covering the worm give it a distinctive appearance.

AMERICAN PLUM BORER. See under Sycamore.

LEOPARD MOTH BORER. See under Maple.

WALNUT LACE BUG. *Corythucha juglandis.* This pest, occasionally abundant on park trees, causes stippling and spotting of the foliage.

- *Control.* See under lacebugs, Chapter 11.

WALNUT SCALE. *Quadraspidiotus juglansregiae.* Masses of these round brown scales, each $^{1}/_{8}$ inch in diameter with a raised center, can severely weaken part or all of a tree. The adult female is frequently encircled by young scales. The pest overwinters on the bark.

OTHER SCALES. Nut trees are susceptible to many other scales, including black, California red, calico, citricola, cottony-cushion, greedy, oystershell, Putnam, scurfy, tuliptree, hickory lecanium, obscure, and white peach.

- *Control.* See under scales, Chapter 11.

MITES. The walnut blister mite, *Eriophyes erinea,* causes yellow or brown feltlike galls on the undersides of leaves. The following species of mites also infest *Juglans:* European red, black walnut pouch gall, platani, and southern red.

- *Control.* See under spider mites, Chapter 11.

WILLOW (*Salix*)

Willows are rarely used as street trees or in locations where the breaking off of their weak-wooded branches is likely to cause injury. Some species, however, are used extensively in landscape plantings, especially around ponds and streams.

Diseases

LEAF BLIGHT. Blight, or scab, is the most destructive disease of willows in the United States. The causal fungus is almost always associated with the fungus that causes black canker disease.

• *Symptoms.* Soon after growth starts in spring, a few small leaves turn black and die. Later, all the remaining leaves on the tree suddenly wilt and blacken, as if they had been burned by fire. Cankers on twigs may result from the leaf infections. Following rainy periods, dense olive-brown fruiting structures of the causal fungus appear on the undersides of blighted leaves, principally along the veins and midribs.

• *Cause.* Leaf blight is caused by the fungus *Venturia saliciperda,* which overwinters in infected twigs on which it produces spores during early spring. The spores are splashed by rain onto newly developing leaves where they initiate primary infections. Large numbers of spores of the asexual stage, *Pollaccia saliciperda,* are produced on these leaves. These spores are responsible for the severe blighting of the remaining leaves which occurs later in the season.

• *Control.* Pruning of dead twigs and branches will eliminate the most important source of inoculum for early-season infections. Spraying with mancozeb, maneb, or fixed copper three or four times at 10-day intervals, starting when the leaves begin to emerge in spring, will usually give control.

These practices are recommended only for valuable specimens.

A number of species, including weeping, bay-leaved, osier, purple, and pussy willows, appear resistant to blight. These should be planted in place of the more susceptible species, such as the crack and the heart-leaved willow. Golden willow has been reported by some investigators to be susceptible and by others to be resistant.

BLACK CANKER. Closely associated with leaf blight is the disease known as black canker.

• *Symptoms.* The symptoms of black canker resemble those of leaf blight but usually appear later in summer. Dark brown spots which often have concentric markings appear on the upper leaf surface. Whitish gray to gray elliptical lesions with black borders appear subsequently on the twigs and stems and clusters of minute black fruiting bodies develop in the stem lesions. Successive attacks over a 2- or 3-year period usually result in death of the entire tree.

• *Cause.* Black canker is caused by the fungus *Glomerella miyabeana.* The fungus overwinters on diseased twigs. In early spring spores are released from tiny black spherical bodies to cause the primary infections. Secondary infections occur in early summer, caused by spores exuding from fruiting bodies as small pink masses.

• *Control.* Prune and spray as recommended for control of leaf blight. Choose resistant cultivars such as bay-leaved, osier, and weeping willow, and avoid susceptible ones like crack, heart-leaved, white, purple, and almond willow. As with leaf blight, golden willow has been reported as both resistant and susceptible to black canker.

CYTOSPORA CANKER. Willows are subject to this canker disease caused by the fungus *Cytospora chrysosperma.* The perfect stage is *Valsa sordida,* already discussed under Poplar. The control measures listed under that host hold true also for willows. The use of resistant varieties offers a means of avoiding serious problems. One investigator has observed that the disease occurs rarely on black or peach-leaf willow, whereas crack and golden willows appear to be extremely susceptible.

OTHER CANKERS. Willows are also susceptible to canker diseases caused by *Botryosphaeria dothidea, Cryptodiaporthe salicella, Cryptomyces maximus, Discella carbonacea, Diplodina* sp., *Leucostoma niveum,* and *Macrophoma* sp.

• *Control.* Control measures are rarely needed when trees are provided with good growing conditions.

GRAY SCAB. *Sphaceloma murrayae.* This disease affects many species of willow including *Salix fragilis, S. lasiandra,* and *S. lasiolepis.* Round, irregular, somewhat raised, grayish white spots with narrow, dark brown margins appear on the leaves. Affected portions of the leaves frequently drop away.

• *Control.* See under leaf spots, Chapter 12.

LEAF SPOTS. *Ascochyta salicis, Asteroma capreae, Cercospora salicina, Cylindrosporium salicinum, Marssonina* sp., *Myriconium comitatum, Phyllosticta apicalis, Ramularia rosea, Septogloeum salicinum,* and *Septoria didyma.* At least ten species of fungi cause leaf spots on willow. Some may also cause premature defoliation.

• *Control.* See under leaf spots, Chapter 12.

POWDERY MILDEW. *Uncinula adunca* and *Phyllactinia guttata.* Leaves infected with mildew are covered with a whitish feltlike mold that develops chains of white spores that are shed as clouds. The small black fruiting bodies formed later in the season are characterized by microscopic appendages curled at the end like a shepherd's crook. This is not a serious disease of willows, but it may cause some loss of leaves.

• *Control.* See under powdery mildew, Chapter 12.

RUST. *Melampsora* sp. Three or four species of rust attack willow leaves and cause lemon yellow spots on the lower surfaces. Later in the season the spore-bearing pustules are dark colored. The disease may be severe enough to cause leaf drop. Rust-causing fungi and their alternate hosts include *M. paradoxa* (larch); *M. abieti-capraearum* (balsam fir); and *M. arctica* (saxifrage).

• *Control.* Although rust infections are not considered serious, they may result in heavy defoliation of young trees. Gather and destroy fallen leaves to help prevent serious outbreaks.

TAR SPOT. *Rhytisma salicinum.* This fungus causes well-defined, slightly raised, jet black spots about 1/4 inch in diameter. Tar spot is more common on maple.

• *Control.* Since this fungus overwinters on the old leaves, rake up and destroy them, especially where the disease is serious.

WITCHES' BROOM. Phytoplasma. This organism causes early breaking of axillary buds and subsequent growth of numerous spindly, erect branches with stunted leaves on *Salix rigida.* The witches' brooms die the winter after they are formed.

• *Control.* Ways of preventing this disease have yet to be developed.

WOOD DECAY. Species of *Daedalea, Phellinus,* and *Trametes* cause decay of living willow trees. Their characteristic fruiting bodies protrude through the bark. Unnecessary wounding should be avoided.

BACTERIAL TWIG BLIGHT. Pseudomonas *saliciperda.* Leaves turn brown and wilt, and blighted branches die back for some distance. Brown streaks can be seen in sections of the wood. The parasite winters in the cankers, and the young leaves are infected as soon as they unfold. It may cause the death of a large number of trees by serious defoliation. The damage has sometimes been attributed to frost injury.

• *Control.* Prune out infected twigs.

CROWN GALL. *Agrobacterium tumefaciens.* See Chapter 12.

Insects

APHIDS. Several species of aphids infest willows, the most common being the giant bark aphid, *Longistigma caryae.*

• *Control.* See under aphids, Chapter 11.

EASTERN TENT CATERPILLAR. *Malacosoma americanum.* Eggs of this insect hatch in April and in spring and early summer, the larvae chew the leaves and make small silken nests in branch crotches. This caterpillar is about 2 inches long, black with white lines down the back, and has yellow hairs. It prefers to feed on wild cherry but will feed on willow as well as apple, peach, and plum. When these trees are scarce, the insects will also defoliate ash, beech, birch, elm, maple, oak, poplar, and many shrubs. The adult female is a fawn-colored moth with a wingspan of 2 inches. Eggs are laid in cylindrical clusters on twigs in July.

Forest tent caterpillar (see under Maple) also attacks willow.

• *Control.* See under caterpillars, Chapter 11.

FOLIAR-FEEDING CATERPILLARS. Dagger moth (*Acronicta americana*). This hairy white caterpillar is common but causes little damage. Other hosts include maple and oak. Gypsy moth, mourning-cloak butterfly, and white-marked tussock moth (see under Elm), red-humped caterpillar (see under Poplar), io moth (see under Sycamore), and fall webworm (see under Ash) all feed on willow.

CALIFORNIA TENT CATERPILLAR. See under Strawberry-Tree.

CARPENTER WORM. See under Ash.

HEMLOCK LOOPER. See under Hemlock.

GIANT HORNET. See giant hornet wasp, under Franklin-Tree.

OMNIVOROUS LOOPER. See caterpillars, under Acacia.

WALNUT CATERPILLAR. See under Walnut.

MOTTLED WILLOW BORER. *Cryptorhynchus lapathi.* The white, legless, $^1/_2$ inch long larvae eat through the cambium, sapwood, and heart-wood, producing swollen and knotty limbs. The adult beetle is $^1/_3$ inch long with a long snout and grayish-black mottled wing covers. Eggs are laid in fall, and the larvae overwinter in tunnels beneath the bark. This pest is also known as the poplar and willow borer.

• *Control.* See under borers, Chapter 11.

BORERS. The following attack willow: dogwood borer (see under Dogwood) and the leopard moth and flatheaded borer (see under Maple).

POPLAR BORER. See under Poplar.

BASKET WILLOW GALL. *Rhabdophaga salicis.* Swollen, distorted twigs may be produced by the yellowish jumping maggots of the basket willow gall midge. The adults appear in early spring.

• *Control.* Prune and destroy infested twigs.

PINE CONE GALL. *Rhabdophaga strobiliodes.* Cone-shaped galls at the branch tips that hinder bud development are produced by small maggots. The adult, a small fly, deposits eggs in the opening buds. The larvae hibernate in cocoons inside the galls.

• *Control.* Remove and destroy galls in the fall or spray thoroughly with malathion when the buds are swelling in spring.

IMPORTED WILLOW LEAF BEETLE. *Plagiodera versicolora.* These beautiful metallic blue beetles, about $^1/_8$ inch long, live through the winter under the bark scales and in dead leaves around the tree. They emerge and lay lemon yellow eggs in early June. The ugly larvae or grubs feed on the undersides of the foliage, leaving only a network of veins. The adult beetle develops during July and produces a second brood in August. This beetle is so much smaller than the Japanese beetle (*Popillia japonica*) that it can scarcely be mistaken for it.

GRAY WILLOW LEAF BEETLE. *Pyrrhalta decora.* Larvae and adults feed on willow leaves in the northern states. Other leaf beetles such as *Chrysomela scripta* and *C. interrupta* skeletonize willow leaves.

• *Control.* See under foliage feeding beetles, Chapter 11.

WILLOW FLEA WEEVIL. *Rhynchaenus rufipes.* This beetle overwinters as an adult and emerges in the middle of April. During the latter part of May, it excavates a circular mine on the underside of the leaf and deposits its eggs. The adults feed on the foliage, which becomes brown and dry. The larvae begin to mine the leaves about the middle of June. By the end of July, where infestation is heavy, the trees appear as if scorched by fires.

California casebearer, *Coleophora sacramanta,* is also a leaf miner of willow.

• *Control.* See under leaf miners, Chapter 11.

ASIATIC OAK WEEVIL. See under Oak.

WILLOW LACE BUG. *Corythucha mollicula.* Willow leaves may be severely mottled and yellowed by this sucking insect.

• *Control.* See under lace bugs, Chapter 11.

WILLOW SHOOT SAWFLY. *Janus abbreviatus.* The female lays its eggs in the shoots in early spring and then girdles the stem, causing it to wilt and die. The young borers feed in the pith of the shoots, which eventually die as well.

Leaf gall sawflies (*Pontania* sp.) make spherical, reddish galls, $^1/_3$ inch in diameter, on willow, with little apparent tree damage.

• *Control.* See under sawflies, Chapter 11.

WILLOW SCURFY SCALE. *Chionaspis salicisn-igrae.* Branches and even small trees may be killed by heavy infestation of this pear-shaped, $^{1}/_{8}$-inch-long, white scale. The insect overwinters in the egg stage underneath the scale of the female.

Other scale insects that attack willows are black, brown soft, California red, cottony-cushion, cottony maple, European fruit, greedy, lecanium, hickory lecanium, azalea, latania, Glover, dictyospermum, obscure, oys-tershell, Putnam, soft, and terrapin.

• *Control.* See under scales, Chapter 11.

BAGWORM. See under Arborvitae.

WINGNUT, CAUCASIAN
(*Pterocarya*)

This Chinese native is related to walnut and is relatively free of pest problems.

YELLOW POPLAR.
See Tuliptree.

YELLOWWOOD
(*Cladrastis*)

This medium-sized tree is native to Kentucky and North Carolina. Narrow crotch angles make yellowwood subject to breakage.

Few diseases and only one insect pest have been recorded on this host. A powdery mildew caused by *Phyllactinia guttata,* canker by *Botryosphaeria dothidea,* wilt by *Verticillium albo-atrum,* and a decay of the butt and roots of living trees by *Polyporus spraguei* and *Xylaria mali* occur occasionally. An unidentified species of scale may occasionally infest yellowwood. Control measures are rarely necessary.

YEW (*Taxus*)

Yews are among the most useful evergreens for ornamental purposes. They withstand city conditions better than most other evergreens. The English yew, the Japanese yew, and a large number of other species and cultivars are planted extensively.

Diseases

NEEDLE BLIGHT. *Herpotrichia nigra.* Blighting of needles of Pacific yew in the western United States may be caused by this fungus.

• *Control.* This disease is rarely destructive enough to warrant control measures.

TWIG BLIGHT. Several fungi are associated with a twig blight of yew during rainy seasons. One of the most common is *Phyllostictina* (*Phoma*) *hysterella,* whose perfect stage is *Physalospora gregaria.* Others are *Pestalotiopsis funerea* and a species of *Sphaeropsis.*

• *Control.* Prune diseased twigs and destroy them. Spray with a copper fungicide several times at 2-week intervals during rainy springs.

ARMILLARIA ROOT ROT. *Armillaria mellea.* This disease may be a problem in soil formerly occupied by apple or oak trees. Affected plants wilt and die; white wefts of fungal mycelium are present beneath the bark at the base of the plant. See Chapter 12.

Phytophthora cinnamomi also causes root rot as well as dieback of yews, especially in wet locations.

OTHER FUNGAL DISEASES. Among other fungal diseases of yews are a root rot caused by *Pythium* sp. and premature leaf drop in which a species of *Alternaria* is implicated.

• *Control.* Control measures have not been developed.

Abiotic Diseases

The most prevalent problem on yews is dieback. The first symptom is yellowing of the growing tips; this is followed by general yellowing, wilting, and death. Several months may elapse from the time the first symptoms appear to the complete wilting and death of the plant. A below ground symptom is decay of the bark on the deeper roots, causing it to slough off readily.

This trouble is not caused by parasitic organisms. Studies by P. P. Pirone showed that it is associated with unfavorable soil conditions. In nearly every case investigated the soil pH was very acidic, from 4.7 to 5.4, and the soil was heavy and poorly drained.

Fig. III-133. Black vine weevil adults.

Fig. III-134. The larval stage of the black vine weevil causes severe injury to and even death of yews.

Fig. III-135. Rhododendron leaves chewed by adult black vine weevils.

Fig. III-136. Mealybug *Dysmicoccus wistariae* on yew.

• *Control.* Improve drainage by embedding tile in the soil, or move plants to a more favorable area. Add ground limestone, according to the recommendation of a soil specialist, to increase the pH to about 6.5.

TWIG BROWNING. This condition is caused by snow and winter damage and not by fungal parasites. The combined effect of heavy snow cover and low temperatures causes many small twigs at the ends of branches to turn brown in late winter and early spring. Ice falling from rooftops can also injure yew twigs, causing browning of individual twigs in spring.

• *Control.* Shake off heavy snow covers as soon as the snowfall stops. Prune browned branches in late spring.

Insects and Other Animal Pests

BLACK VINE WEEVIL. *Otiorhynchus sulcatus.* Also known as the taxus weevil, this is by far the most serious pest on yews. It is so named because of its color and because it attacks grapes in Europe. In the United States, it also attacks false cypress, hemlocks, and such broad-leaved plants as rhododendrons and azaleas (see Figs. III-133, 134, and 135). The leaves of yew turn yellow and whole branches, or the entire plant may even die when larvae chew the roots. The larvae are white-bodied, brown-headed, and $^1/8$ inch long. As few as eight larvae are capable of killing a large-sized yew. The adult is a $^3/8$ inch long snout beetle that feeds on the foliage of yews and other hosts at night, leaving the leaf edges scalloped. Eggs deposited in the soil during July and August hatch into larvae that feed on the roots of susceptible hosts. Root feeding and girdling can cause plant collapse.

• *Control.* See under weevils, Chapter 11.

STRAWBERRY ROOT WEEVIL. *Otiorhynchus ovatus.* This insect is related to the black vine weevil and has similar habits but is smaller, $^1/5$ to $^1/4$ inch long. It feeds on hemlock, spruce, and arborvitae, in addition to yews. It often wanders into homes in search of hibernating quarters and thus may become a nuisance.

• *Control.* See under weevils, Chapter 11.

TAXUS MEALYBUG. *Dysmicoccus wistariae.* First reported from a New Jersey nursery in 1915, this pest has become increasingly prevalent in the northern United States. The $^3/8$-inch-long bugs are covered with white wax and can completely cover the trunk and branches of yews (Fig. III-136). Young mealybugs overwinter in bark crevices and mature in June. Although all species of *Taxus* are susceptible, those with dense foliage such as *T. cuspidata nana* and *T. wardii*

Fig. III-137. The scale *Parthenolecanium fletcheri* on yew. Some of the scales have been opened to show eggs.

Fig. III-138. Termite injury of yew.

are preferred hosts. This pest has also been seen on apple, linden, and rhododendron, though it probably does not breed on these plants.

GRAPE MEALYBUG. *Pseudococcus maritimus.* This pest occasionally infests the Japanese yew. *P. longispinus,* the long-tailed mealybug, and *P. comstocki,* Comstock mealybug, also attack yews.

• *Control.* See under scales, Chapter 11.

SCALES. Seven species of scale insects infest yews: cottony taxus, California red, Asiatic red, dictyospermum, Fletcher (Fig. III-137), oleander, and purple. The cottony taxus scale, *Pulvinaria floccifera,* is ¹/₈ inch long, light brown, and hemispherical. In spring, it produces long, narrow, fluted, cottony egg masses.

• *Control.* See under scales, Chapter 11.

TAXUS BUD MITE. *Cecidophyopsis psilaspis.* Feeding by this mite can cause the growing tips of yews to enlarge and possibly be killed. New growth is distorted. As many as a thousand mites can be found in a single infested bud.

• *Control.* See under spider mites, Chapter 11.

ANTS AND TERMITES. The black carpenter ant, *Camponotus pennsylvanicus,* and the eastern subterranean termite, *Reticulitermes flavipes,* occasionally infest the trunks of older yews.

Both are capable of excavating the trunk and making their nests therein. We have observed earthen termite tunnels extending several feet up the stem. Limbs associated with the termite activity turn brown (Fig. III-138).

• *Control.* Apply Diazinon in liquid form around the trunk base and soil surface.

NEMATODES. *Criconemoides* and *Rotylenchus* spp. Several species of nematodes attack the roots of *Taxus.*

• *Control.* See under Boxwood.

ZELKOVA, JAPANESE (*Zelkova*)

This tree is touted as an elm substitute as their leaves are similar. Although it is susceptible to Dutch elm disease, it is able to survive infections.

Disease

NECTRIA CANKER. *Nectria cinnabarina.* This fungus has been reported on this host. See under Maple.

Insects

Japanese zelkova is host to the elm leaf beetle, *Pyrrhalta luteola* (see under Elm), and calico scale.

Selected Bibliography

Anonymous. 1979. A guide to common insects and diseases of forest trees in the northeastern United States. U.S.D.A. Forest Service, Northeast Area State and Private Forests Publication NA-FR-4, Washington, D.C. 127 pp.

Anonymous. 1985. Insects and diseases of trees in the south. U.S.D.A. Forest Services Southern Region General Report R8-GR5, Washington, D.C. 98 pp.

Bobbitt, V., A. Antonelli, C. Foss, R. Davidson, R. Byther, and R. Maleike. 1997. Pacific Northwest landscape IPM manual. Washington State University, Puyallup, Wash. 169 pp.

Byther, R., C. Foss, A. Antonelli, R. Maleike, and V. Bobbitt. 1996. Landscape plant problems, a pictorial diagnostic manual. Washington State University, Puyallup, Wash. 144 pp.

Cooper, J. I. 1993. Virus diseases of trees and shrubs. Chapman & Hall, London. 205 pp.

Dreistadt, S. H., J. D. Clark, and M. L. Flint. 1994. Pests of landscape trees and shrubs: An integrated pest management guide. Publication 3359. University of California Division of Agriculture and Natural Resources. 327 pp.

Drooz, A. T. 1985. Insects of eastern forests. U.S.D.A. Forest Service Misc. Publ. 1426, Washington, D.C. 608 pp.

Farr, D. F., G. F. Bills, G. P. Chamuris, and A. Y. Rossman. 1989. Fungi on plants and plant products in the United States. APS Press, St. Paul, Minn. 1252 pp.

Hanson, E. M., and K. J. Lewis (eds.). 1998. Compendium of conifer diseases. American Phytopathological Society, St. Paul, Minn. 124 pp.

Hartmann, G., F. Nienhaus, H. Butin, and K. Winter. 1988. Farbatlas Waldschaden: Diagnose von Baumkrankheiten. Eugen Ulmer GmbH and Co., Stuttgart. 256 pp.

Hepting, G. H. 1971. Diseases of forest and shade trees of the United States. U.S.D.A. Forest Service Agriculture Handbook 386, Washington, D.C. 658 pp.

Horst, R. K. 1982. Westcott's plant disease handbook, 4th ed. Van Nostrand Reinhold, New York. 803 pp.

Johnson, W. T., and H. H. Lyon. 1988. Insects that feed on trees and shrubs, 2nd ed. Cornell University Press, Ithaca, N.Y. 556 pp.

Jones, R. K., and R. C. Lambe (eds.). 1982. Diseases of woody ornamental plants and their control in nurseries. North Carolina Agricultural Extension Service AG-286, Raleigh. 130 pp.

Lloyd, J. 1997. Plant health care for woody ornamentals. International Society of Arboriculture. Champaign, Ill. 223 pp.

Partyka, R. E., J. W. Rimelspach, B. G. Joyner, and S. A. Carver. 1980. Woody ornamentals: Plants and problems. Chemlawn Corporation, Columbus, Ohio. 427 pp.

Peace, T. R. 1962. Pathology of trees and shrubs. Clarenden Press, Oxford, England. 722 pp.

Peterson, G. W. 1981. Pine and juniper diseases in the Great Plains. U.S.D.A. Forest Service General Technical Report RM-86, Rocky Mountain Forest and Range Experiment Station, Fort Collins, Colo. 47 pp.

Peterson, J. L. 1982. Diseases of holly in the United States. Holly Society of America Bull. 19, Baltimore, Md. 44 pp.

Pirone, P. P. 1978. Diseases and pests of ornamental plants, 5th ed. Wiley, New York. 566 pp.

Riffle, J. W., and G. W. Peterson (eds.). 1986. Diseases of trees in the Great Plains. U.S.D.A. Forest Service General Technical Report RM129, Washington, D.C. 149 pp.

Rose, A. H., and O. H. Lindquist. 1982. Insects of eastern hardwood trees. Can. For. Serv. For. Tech. Rep. 29. 304 pp.

Sinclair, W. A., H. H. Lyon, W. T. Johnson. 1987. Diseases of trees and shrubs. Cornell University Press, Ithaca, N.Y. 574 pp.

Solomon, J. D., F. I. McCracken, R. L. Anderson, R. Lewis, Jr., F. L. Oliveria, T. H. Filer, and P. J. Barry. 1980. Oak pests: A guide to major insects, diseases, air pollution and chemical injury. U.S.D.A. Forest Service General Report SA-GRI I, Washington, D.C. 69 pp.

Stipes, J. R., and R. J. Campana (eds.). 1981. A compendium of elm diseases. American Phytopathological Society, St. Paul, Minn. 96 pp.

Strouts, R. G., and T. G. Winter. 1994. Diagnosis of ill health in trees. Forestry Commission, Department of the Environment, HMSO, United Kingdom. 307 pp.

Végh, I. 1987. Champignons des arbres et arbustes d'ornement. INRA, Paris. 121 pp.

Index